UNIT PROCESSES
IN
DRINKING WATER
TREATMENT

Environmental Science and Pollution Control Series

Series Advisors

Dr. Suzanne Lesage
Environment Canada
Burlington, Ontario, Canada

Dr. Hans van Bergen
International Flavors and Fragrances
Hilversum, The Netherlands

Dr. Willy J. Masschelein
Consultant
Brussels, Belgium

Dr. Rene Wijffels
Wageningen Agricultural University
Wageningen, The Netherlands

Dr. Constantine Yapijakis
The Cooper Union
New York, New York, U.S.A.

Additional Volumes in Preparation

UNIT PROCESSES IN DRINKING WATER TREATMENT

Willy J. Masschelein

Consultant
Brussels, Belgium

Marcel Dekker, Inc.   **New York • Basel • Hong Kong**

Library of Congress Cataloging-in-Publication Data

Masschelein, W.
    Unit processes in drinking water treatment / Willy J. Masschelein.
        p.   cm. -- (Environmental science and pollution control ; 3)
    Includes bibliographical references and index.
    ISBN 0-8247-8678-5
    1. Drinking water--Purification.   I. Title.   II. Series.
    TD433.M34   1992
    628.1'66--dc20

92-20757
CIP

This material was produced privately by the author, who assumes the responsibility for the opinions or statement of fact expressed in the work. No endorsement by the Brussels Water Board is implied. This material is presented for informational purposes only.

This book is printed on acid-free paper

MARCEL DEKKER, INC.
270 Madison Avenue, New York, New York 10016

Current printing (last digit):
10 9 8 7 6 5 4 3 2 1

PRINTED IN THE UNITED STATES OF AMERICA

Preface

The aim of this book is to provide fundamental information on the theoretical basis involved in most drinking water treatment processes in current use, as well as practical experience and potential drawbacks of the use of these processes. Both aspects are intended to enable guidelines to emerge for the design of unit processes: theory, practice, and recommendations. We have aimed to meet the questions of a multiple audience in the professional field:

1. Scientists, officials, design engineers, and consultants
2. Lecturers in the field of water quality control and treatment
3. Senior and postgraduate students in sanitary engineering, including those from developing countries
4. Senior plant operators and all persons interested in drinking water, food chemistry, and technology

The book is exhaustive on some topics, less so on others; it is not intended to be an encyclopedia of water treatment. I am aware that there are risks associated with combining theory and practice, yet is this not a practice that we often face in reality?

Many colleagues in the United States and Europe suggested that I present an integrated summary of the experience acquired with the help of my co-workers when I was in charge of water quality and treatment at the Brussels Waterboard. Reference to their work is provided by the publications cited in the text. I would also like to thank Luc Reynaert for his daily assistance with documentation, as well as Edith Chad-Boyko and Roseline Mathu for their secretarial assistance. I assume complete responsibility for any defects that remain.

I also wish to recognize here the experience acquired through the conferences I gave at the International Institute for Hydraulic and Environmental Engineering in Delft (The Netherlands) and for the evaluation of material presented by students abroad.

Above all, I am indebted to my wife, Liliane, for her understanding during the trials that accompany endeavors such as this.

Willy J. Masschelein

Contents

UNIT PROCESSES
IN
DRINKING WATER
TREATMENT

1
Design Criteria in
Water Treatment Processes

1. WATER QUALITY CRITERIA, STANDARDS, AND GOALS

The basic role of water treatment facilities is to make available, in sufficient quantities, water having the appropriate quality for a variety of uses and at reasonable prices. The design of adequate facilities to attain this goal involves several phases:

1. Comparison of available sources with expected needs (evaluation of the master plan)
2. Flexible interconnection of unit processes into a global treatment scheme (definition of the flowsheet)
3. Accurate detailed design of each process to be operated (project formulations and design)
4. Provision made for future increased water needs or stricter quality standards

The fixing of standards and goals based on aesthetics is probably the oldest known method. Such standards are not based directly on health effects. Apparently, pure water may provide a false sense of security, as evidenced, in the mid-eighteenth century by outbreaks of cholera.

Drinking water is often expected to be wholesome or palatable. It is essential that drinking water not give rise to ill health, but wholesomeness is something more—it means that the water must be attractive to the drinker. Therefore, it must be free of color, taste, odor, and excessive amounts of mineral and organic matter, and consumer acceptance limits must be considered as a quality criterion and introduced into the standards. Few raw water sources meet the requirements of the consumer's concept of wholesome water. *The function of a water service is to serve the public*; thus treatment of the water supply is often necessary to produce adequate water on a continuous basis, thus protecting and promoting good health and well-being.

Quality criteria for raw water treated for use as drinking water are to be set in consideration of the quality requirements established for drinking water. For the definition of these standards, the constituents of water can be classified into two categories:

1. Maximum standards must be fixed for harmful compounds or organisms.
2. Minimum and maximum standards are required for necessary substances.

Obviously, in the first category, ideal standards would be zero, but this approach is unrealistic and can cause controversy. Some natural elements, such as iodine and fluoride, may have a favorable effect on health.

In addition to the standards, quality objectives or goals are sometimes formulated. They are a result of the wish to improve water quality and meet the requirements of wholesomeness in a more complete way. As Lee has stated (1): "Whatever the legal responsibilities, the moral responsibility for water quality remains within the competence of the utility." Consequently, each treatment plant must set quality objectives, as, according to Flentje (2): "Plant control can become an indefinite and uncertain procedure unless production is geared to definite specifications of quality. Unless the water flowing through a plant is scheduled to receive treatment to achieve a predetermined end, the plant cannot be considered to be under proper control."

Quality standards are normally indicated in the form of figures. These are necessary for legal definitions and serve to guide the interpretation of analytical determinations in the context of quality improvement during treatment. However, numbers do not usually have absolute validity, as they are generally obtained by considering 2 to 3 liters of water per day as part of the total daily intake of compounds. In that case, they can vary from one country to another, because the standard food pattern is not identical in all cases. Also, the existence of different standards can be grounded on a political decision about the subdivision of the allowable daily intake (ADI) among the various components of the food. Quality criteria are generally subdivided into several categories, the most common being microbiological and biological parameters, toxic substances (both inorganic and organic), physicochemical parameters, and radioactivity levels.

1.1   Standards for Bacteria and Other Biological Organisms in Drinking Water

Bacteria and other organisms, including viruses and parasites, in water were until recently the major health hazards in drinking water. Fortunately, by a combination of protective measures for water sources and appropriate disinfection, most of these risks have been overcome. However, one must not forget that if in the future waterborne diseases are likely to be less widespread, they will still remain endemic and give rise to limited, sporadic outbreaks. It is worth noting that in 1979 at least 45 epidemic outbreaks occurred in the United States, due principally to more resistant agents, such as cysts of *Giardia (Lamblia)*. Other cases are reported in Table 1, and various standards are given in Table 2.

These examples show that complacency is certainly out of the question and that continuous efforts remain absolutely necessary: *95% of the cases reported were due to deficiencies in water treatment*. Although the U.S. statistics are reliable, this is

Table 1   Waterborne Disease Outbreaks
in the United States (1961–1970) Due to
Water Distributed Through Public Systems

Illness	Outbreaks	Cases
Gastroenteritis	14	22,048
Hepatitis	8	239
Shigellosis	3	737
Salmonellosis	5	16,610
Giardiasis	2	157
Amebiasis	1	25

Table 2   Maximum Permissible Concentrations of Biological Organisms
Based on EEC Bacterial and Biological Standards for Drinking Water

Total coliforms	0/100 mL	Salmonella	(0/5 mL)
Fecal coliforms	0/100 mL	Pathogenic staphylococci	0/100 mL
Fecal stretococci	0/100 mL	Fecal bacteriophages	0/100 mL
Total count at 70°C	<10/mL	Enterogenic viruses	0/10 mL
Total count at 22°C	<100/mL	Pathogenic protozoans	Nil
Sulfite red. Clostridium	2/20 mL	Pathogenic animalcules	Nil

Source: Ref. 4.

not always the case for other countries, but it is well known that cholera, salmonellosis, and giardiasis are still widespread in many areas. For other standards, the reader may consult the literature (5–7).

1.2   Principal Waterborne Diseases

Several diseases are known to be transmitted by water, although not necessarily by water alone.

Typhoid fever.   Due to *Salmonella typhi*, discovered in 1880 by Eberth and often referred to as *Eberthella typhosa*. Under natural circumstances, the organism can remain alive for weeks or months in water, soil, and feces; in ice, it can survive for up to 5 months. The bacterium is introduced to the organism orally. Multiplication occurs in the liver, gallbladder, biliary tract, and duodenum. After a period of about 2 weeks, the microbe is discharged by both urine and feces. Typhoid carriers can spread the bacterium they carry even after their recovery. Not all people who have contracted the disease become carriers, but many do. Although typhoid fever has been almost totally eradicated by disinfection, continuous attention to the illness is necessary as long as the population does not develop further resistance to the organism.

Cholera.   Remains endemic in Asia and certain parts of Africa even though the danger has been greatly reduced through the efforts of the World Health Organization (WHO). The virulence of the causal organism, *Vibrio cholerae*, is reduced after 1 week's storage in water, and the organism is very sensitive to chlorine.

Gastroenteritis.   Breaks out sporadically as an intestinal disorder accompanied by such symptoms as vomiting, diarrhea, cramps, and even fever, nausea, and headache. The most significant organisms responsible for transmission of the disease are *Shigella* and *Salmonella* organisms. (*S. typhosa* and *S. paratyphi*). Other bacteria, such as the enteropathogenic *Escherichia coli* and even some microbes considered as "nonpathogenic," such as *Pseudomonas*, can sometimes cause similar infections.

Amoebae.   The causal organism of amebic dysentery is a cyst of *Endamoeba hystolytica*, which can be carried undetected until it causes chronic outbreaks. There have been several outbreaks of the illness since World War II (e.g., United States, 1955). The disease is a potential danger throughout the world, but is encountered more frequently in tropical and subtropical regions. Disinfection requires high doses of chemicals; however, sand filtration is also efficient. Ninety-two percent of cases of the illness transmitted by public distribution systems are the result of deficiencies in treatment but can also be caused by insufficient protection of the distribution mains (back-siphonage). *Giardia lamblia* is a flagellated protozoan causing giardiasis. Amoebae of the *Naegleria* type have been involved in amebic encephalitis. Both organisms are transmitted by water in recreational areas.

Schistosomiasis.   Parasitic disease transmitted by a blood fluke called *Schistosoma mansoni*, which lives in human abdominal veins and expels eggs in the urine and feces. The disease is widespread in tropical regions and appears to affect about 2 million people. The victim usually succumbs after years of gradual debilitation. Chlorination is effective, at least in the case of drinking water and swimming pool water.

Free-living *nematodes* have been reported to occur in water supplies. These worms are not known to be pathogenic but can ingest pathogenic bacteria such as *Salmonella, Shigella*, and *coxsackie* viruses. Although the problem of nematodes is basically aesthetic, when they multiply in a plant or distribution system, they can cause musty odors in the water. A free chlorine residual of even 2.5 to 3 ppm does not kill the organisms, but a permanent level of 0.5 ppm during treatment eliminates nematodes in the coagulation–flocculation settling process. There is growing concern regarding the gastrointestinal effects caused by the parasite *Cryptosporidium parvum*, particularly on immunodepressed persons. The cysts are resistant to chlorine but are killed by ozone and by chlorine dioxide (36).

Viruses.

Adenovirus: causes pharyngitis and swimming pool conjunctivitis. It is also the virus most sensitive to chlorine.

Enteroviruses: live in the alimentary tract. Several important groups can cause waterborne diseases.

Poliovirus: three types are known to cause a disease capable of invading the central nervous system. Transmitted by water, the virus is significant in drinking water practice despite the fact that vaccination has changed its incidence. Ozonization is particularly effective in this field.

Coxsackieviruses: these viruses, isolated at Coxsackie, New York, are the most resistant to chlorine of those actually known to be transmitted by water. They cause a variable number of infections characterized by fever, headache, and vesicular eruption in the throat.

Echoviruses: more than 20 different types have been isolated. They are known to cause acute respiratory symptoms.

Infectious hepatitis.   Although its causal virus is not yet completely known, the incidence of hepatitis is of growing importance. The illness is characterized by fever, nausea, and vomiting and eventually results in jaundice, due to inflammation of the liver. The best known outbreak of waterborne infectious hepatitis was one in New Delhi in 1955–1956, when more than 50,000 cases were reported. The drinking water was treated by conventional methods, but it was contaminated by domestic sewage. The prevention of infectious hepatitis is gaining importance in the treatment of drinking water.

1.3 Significance of Waterborne Diseases

When an infectious agent has reached an organism, the disease becomes apparent only after a lapse of time called the *incubation time*. The longest known incubation time for waterborne diseases is for hepatitis: up to 10 to 40 days. Dysenteria bacteria (e.g., *Shigella*, *Salmonella*, and some *E. coli*) can bring about the disease within 4 to 7 days after absorption of the bacterium. Cholera and waterborne meningitis can break out after 1 to 5 days' incubation time. The average residence time of water in the distribution system depends on the characteristics of the system, but delays of a week between the inlet of water at the treatment plant and outlet at the site of the consumer are not rare. This is one of the reasons why, in addition to the analysis of water at the site of production, a regular survey of water quality in the distribution system is essential. Furthermore, the minimum delay after which an indication of the bacteriological quality of the water is possible ranges from 16 to 20 h, which is equivalent to the culture time needed for coliforms. One thing is certain: *Preventive disinfection is essential*. Over the last 60 years there has been a marked decline in waterborne disease outbreaks, and this evolution corresponds to the increasing number of plants using chlorine disinfection. However, in the United States, no significant decrease has been reported since 1960.

1.4 Toxic Substances

There are at least two types of toxic substances to consider: acute toxicity factors and chronic toxicity factors. When toxicity is acute, ingestion produces a rapid reaction and a dose–effect relationship can be established. Acute toxicity is usually better known than chronic. Chronic toxic substances produce effects on long and repeated exposure to low levels of absorption. The effects are often difficult to detect because the substances can accumulate, making the effects cumulative and irreversible. In the latter case, an approach has been proposed based on the TLVs (threshold limit values) by inhalation (8). A worker inhales 10 m^3 of air during an 8-h day and is assumed to consume 2 L of water per 24 h. Some absorption factors by inhalation and ingestion are as follows:

Element	Inhalation	Ingestion
Cd	0.25	0.03
Cr	0.75	0.06
Pb	0.2	0.2

Where lead is concerned, noninjurious daily intake by air is (9)

$$10 \quad \times \quad 0.2 \quad \times \quad 0.2 \quad = 0.4 \text{ mg}$$
$$(\text{m}^3 \text{ air/day}) \quad (\text{TLV}) \quad (\text{absorption factor})$$

When the absorption factor by water is assumed to be 20%, for a 2-L daily consumption one would have

$$\frac{0.4 \times 100}{20 \times 2} = 1 \text{ ppm}$$

This is clearly too high a level, so the (air) TLV concept is not always applicable for generalization to water. This approach necessitates an accurate knowledge of the absorption factors through ingestion. Unfortunately, this is often lacking.

The quality standards for toxic substances are expressed in levels that are to be considered the maximum permissible concentrations. The quality standards for drinking water are often set at $\frac{1}{10}$ of the dosage permissible for the population in general so as to meet an acceptable value for the most sensitive individuals (e.g., babies, older persons, patients under special medication or subject to synergism) (see Fig. 1). On the other hand, a sequential increase in several effects can result from the ingestion of a given product [e.g., biochemical evidence (enzyme modifications), vomiting, gastric disease, diarrhea, dehydration, etc.]. The no-effect level (NEL) is always set on the basis of the less severe observable outcomes. If the NEL is not known for humans, $\frac{1}{10}$ of an animal NEL value is considered. If there is a risk of carcinogenicity, the reducing factor ranges up to 1/2000 of the NEL. The ADI

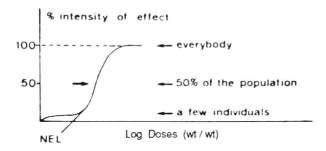

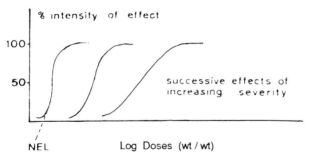

Figure 1   Dose–effect relations (schematic).

value in mg/kg body weight per day (ADI = average daily intake) is computed from the NEL values on the basis of the average body weight in a population (e.g., 60 to 65 kg). The permissible level (PL) for a given substance is obtained from the ADI based on an impact factor (IF) attributed to the particular product under consideration as part of the entire food intake or exposure to risks.

$$PL = \frac{ADI \times 60}{IF}$$

Present known ADI levels for some toxic substances (35) in foodstuffs and, more particularly, water are given in Table 3. A global approach as taken by the U.S. Environmental Protection Agency (EPA) is to consider water as having an impact of 20% in the global food package and to consider a security factor of 500. In the USSR the necessary security factor is considered to be 1000.

Generally, goals are not indicated; however, the basic aim is to keep the concentration as low as possible. Some elements are considered as exceptions: for example, fluoride, for which a goal of 1.0 to 1.7 has been fixed as a consequence of its possibly favorable effect on teeth. This is a typical example of the old saying that "there are no toxic substances, only toxic concentrations." Substances for which a minimum daily intake by food has been established include copper, manganese, molybdenum, selenium, zinc, and trivalent chromium.

Furthermore, in any discussion of minerals present in water, it is important to note that the majority of ionized minerals in the intestines cannot pass through the mucous lining in this form. They are absorbed only if they are not complexed or chelated by a variety of organic compounds. This fact should not be forgotten when studying drinking water because, by definition, virtually all mineral constituents of drinking water are present in the ionized state (9). However, it must be added that the water used in food processing readily makes inorganic ions available for absorption in the intestinal tract.

Table 3   ADI Levels for Toxic Substances

	ADI (μg/kg)	mg/person/day	mg/m^3 potable water
Arsenic	50	3.25	160
Cadmium	1	0.065	3.3
Copper	500	32.5	1,625
Mercury	0.7	0.045	2.2
Lead	7	0.45	22
Nitrate	5,000	325	16,250
Nitrite	200	13	650
Aldrin/Dieldrin	0.1	0.006	0.3
DDT	5	0.325	16
α-HCH	10	0.650	32
Heptachlor	0.5	0.03	1.5
HCB	0.6	0.04	2

Table 4   Toxic Factors: Standards for Maximum Permissible or Recommended Values for Concentrations in Drinking Water (mg/m^3)

Factor	ECC (3)	WHO European (4)	WHO Intern. (5)	USPHS–AWWA (6)[d] Standard	USPHS–AWWA (6)[d] Recommended limit[a]	USPHS–AWWA (6)[d] Mandatory limit[b]
Ag	10	—	—	50	—	50
As	50	50	50	50	10	50
Ba	100	—	—	1000	—	1000
Cd	5	10	10	10	—	10
CN$^-$	50	50–100	50–100	200	10	200
Cr	50	50	50	50	—	50
F	1500	1500	1500	—	800–1700	1400–2400
Hg	1	1	1	—	—	—
Ni	50	—	—	—	—	—
No$_3^-$ (mg/L)	50	45	45	45	45	—
NO$_2^-$	100	—	—	—	—	—
Pb	50	50	100	50	—	50
Sb	10	—	—	—	—	—
Se	10	10	10	10	—	10
PAH	0.2	0.2	—	—	—	—
Aldrine	⎫	—	—	17	—	—
Chlordane		—	—	3	—	—
DDT		—	—	42	—	—
Dieldrin		—	—	17	—	—
Endrin	To	—	—	1	—	—
Heptachlor	be	—	—	18	—	—
Heptachlorepoxide	defined	—	—	18	—	—
Lindane	in	—	—	56	—	—
Methoxychlor	the	—	—	35	—	—
Organic phosphates and carbamates	future[c]	—	—	100	—	—
Toxaphene		—	—	5	—	—
2,4-D; 2,4,5-T; 2,4,5-TP	⎭	—	—	100	—	—

[a]AWWA recommended limit: If the concentration of any of these constituents is exceeded, a more suitable supply or treatment should be sought.
[b]AWWA mandatory limit: If the concentration of any of these constituents is exceeded, further use of this water for drinking and culinary purposes should be evaluated by the appropriate health authority, because water of this quality represents a hazard to consumer health.
[c]For each pesticide individually a limit of 0.1 mg/m^3 is established, and the total of all pesticides may not exceed 0.5 mg/m^3. The parameter is under consideration.
[d]Completed by EPA National Drinking Water Regulations, May 29, 1989.

1.5   Effect of Most Toxic Factors in Water (10)

Silver: causes irreversible darkening of skin; argyrosis; lethal only at very high
        concentrations
Arsenic: recognized carcinogen (e.g., skin cancer); fatal by chronic toxicity at high
        doses [accumulates in some aquatic vegetables (e.g., watercress)]
Barium: toxic to heart, blood vessels, and nervous system, due to excessive muscle
        stimulation

Cadmium: causes nausea and vomiting; can accumulate in liver and kidneys

Cyanide: highly toxic, through irreversible interference with hemoglobin

Chromium: hexavalent form is toxic; causes ulcers after long-term exposure

Fluoride: above 4 g/m^3 causes mottled teeth, and more than 15 g/m^3 causes fluorosis

Mercury: causes gingivitis, stomatitis, and tumors; capable of invading the nervous system (Minamata illness)

Nickel: can accumulate in the kidneys and block them

Nitrate/nitrite: associated with methemoglobinemia and cancer of the stomach (can interfere with chlorine and coliform determination)

Lead: accumulates in bones; causes loss of appetite, anemia, and paralysis (saturnism)

Antimonium: similar to arsenic but active at lower levels

Selenium: poisoning symptoms similar to those of arsenic; also associated with dental caries

PAH (polyaromatic hydrocarbons): potentially carcinogenic

Chlorinated pesticides: neurotoxic; suspected carcinogens (e.g., Aldrin, Chlordane, DDT, Dieldrin, Endrin, heptachlor, heptachlorepoxide)

Methoxychlor: fatal at high doses

Organophosphates and carbamates: neurotropic, causing convulsions and death

Toxaphene: neurotoxic

Chlorinated herbicides (e.g., 2,4,-D): unpleasant rather than toxic

Remark. In both the WHO and American Public Health Association (APHA) approaches, standards for pesticide concentration are based on a specific toxicological evaluation. Maximum Contaminant Levels (MCL) have been defined by the EPA, May 1989.

1.6 Parameters of a Less Toxic Nature: Physicochemical Parameters

Substances dealt with in this category have physicochemical incidences, and often biological implications, without being specifically toxic. Important consequences may be organoleptic and aesthetic, including technical outcomes such as corrosion of a pipeline system and bacterial aftergrowth detrimental to the general quality of the water.

As standards are formulated, founded on a general technological basis, goals or objectives are proposed as aims to attaining wholesome drinking water. The parameters concerned are listed in Table 5. At excessive concentrations, some may be toxic, but at the usual levels these factors are considered simply undesirable. Although the justification for limits to toxic parameters is obvious, the standards for physicochemical parameters are often questionable and may need a more explicit explanation.

Organoleptic properties affect the acceptability of water to the consumer. This term covers such characteristics as color, taste, temperature, turbidity, foaming, and staining properties, all of which are generally agreed to be unpleasant. However, turbidity (as color and more particularly, suspended matter) not only promotes an unclear and dirty appearance but interferes with disinfection efficiency and may even be associated with the development of bacteria, larvae, crustacea, worms, mollusks, and snails. All these elements must, of course, be absent up to a significant reference level. Consequently, the aesthetic rejection limits for copper and zinc

Table 5 Physicochemical parameters: Standards and Goals for Drinking Water (μg/L)

Parameter	EEC (3) MAC	EEC (3) Goal	WHO European (4)	WHO International (5) MAC	WHO International (5) Goal	WHO International (5) Standard	USPHS-AWWA Recommended limit	USPHS-AWWA Mandatory limit	USPHS-AWWA Goal (17)
Cu	100	–	50	1,500	50	1,000	1,000	–	200
Fe	200	50	100	1,000	100	300	300	–	50
Mn	50	20	50	500	50	50	50	–	10
P	2,000	–	–	–	–	–	–	–	–
Zn	100	–	5,000	15,000	5,000	5,000	5,000	–	1,000
Phenols	0.5	–	1	2	1	1	–	–	–
Detergents	100	–	–	200	100	500	500	–	200
Color (Pt)	20	1	15 FCU	50	5	15	15	–	3
Turbidity (SiO$_2$)	10	1	1 NTU	–	–	(5 JTU)	(10)	–	1
Odor	2-3	0	–	–	–	3	–	–	None
Taste	–	–	–	–	–	–	–	–	None
Temperature (°C)	25	12	–	–	–	–	–	–	–
pH	9.5	6.5-8.5	<8	6.5-9.2	7.0-8.5	–	–	–	–
K (μS/cm)	–	400	–	–	–	–	–	–	–
Ca (g/m^3 as CaCO$_3$)	–	100	–	500	100	TH 2 mEq/L	–	–	32
Mg (g/m^3)	50	30	30	150	(30)	–	–	–	–
Na (g/m^3)	175	20	–	–	–	–	–	–	–
K (mg/m^3)	12	10	–	–	–	–	–	–	–
Al (mg/m^3)	200	50	–	–	–	–	–	–	50
CO$_3$H$^-$ (g/m^3)	–	120	–	–	–	–	–	–	–
SO$_4^{2-}$ (g/m^3)	250	100	250	400	200	–	250	–	–
Cl$^-$ (g/m^3)	200	50	600-200	600	200	–	250	–	–
NH$_4^+$ (mg/m^3)	500	50	–	–	–	–	–	–	–
O$_2$ (g/m^3)	–	>5	–	–	–	–	–	–	–
Dissolved matter (g/m^3)	–	–	–	–	–	500	–	–	200

are 1 and 5 g/m^3, respectively: the former for taste, the latter causing the appearance of a milky and greasy film. Iron and manganese have less direct toxicological significance but cause textile staining, are associated with pipe corrosion, and may support bacterial aftergrowth. Aluminum at concentrations higher than 0.05 g/m^3 can postflocculate in the mains. Ammonia, though nontoxic, must be limited to prevent bacterial aftergrowth and thus oxygen consumption. The aesthetic limit for chloride is detectable through taste at 250 mg CL$^-$/L. "Phenol" must be limited to avoid the noxious taste and odor of chlorophenols formed by the chlorination of water. In a general way, organic substances such as detergents, hydrocarbons, and extractable or oxidable "organic carbon" should not exceed levels that enhance objectionable taste and odor, not to mention their possible toxic effects.

Hardness, alcalinity, and pH are interrelated properties. Equilibrium among the carbonic acid species is to avoid aggressivity to calcium carbonate as well as corrosivity and excessive precipitation. It was with that in mind that the general AWWA goal for total hardness was set at 80 to 100 mg CaCO$_3$.

1.7   Standards for Radioactivity in Drinking Water

The basic philosophy in setting radiation standards is to assume that the population should receive the "as low as readily achievable" (ALARA) dosage. The most critical population (e.g., groups professionally exposed) may receive no more than 5 mrem per year. From this criterion a base standard is obtained based on each isotope existing separately in air, water, and so on, without any cumulative effect (18). For the population in general $\frac{1}{10}$ of the dose allowed for the professionally active population is accepted as the standard. From the base standard are obtained, in a more or less conventional way, maximal permissible concentrations of various isotopes (e.g., in water). A well-known approach to this problem is the Benelux formula for drinking water:

$$5\,\alpha(\text{total}) + \beta(\text{total}) + 500\,^{226}\text{Ra} + 30\,^{90}\text{Sr} + 2 \times 10^{-3}\,\text{H}^3 < 1000\,\text{pCi/L}$$

However, in any particular case (e.g., discharge of effluents of nuclear power plants, groundwater entrapment in radioactive soils, etc.), the problem must be considered specifically.

Since 1984, by Directive 80/836 Euratom, the permissible level of radioactivity in Europe is no longer defined in terms of volumetric concentrations (or activities) but in terms of dose of exposure (units of gray: 1 Gy = 1 J/kg). Limits for annual incorporation have been established for the various isotopes. When the exposure concerns different isotopes, the following condition must be satisfied for total incorporation:

$$\sum_j \frac{I_j}{I_{jL}} \leq 1$$

where I_{jL} is the yearly incorporation of isotope j and I_{jL} is the maximum permissible incorporation per year of isotope j. The position of European water treatment facilities is that the limit of incorporation through absorption of drinking water should be fixed at 10% of the total accepted risk of incorporation through ingestion. This position is not yet fixed by law, however.

1.8   Quality Criteria for Raw Water to Be Treated to Obtain Drinking Water

The most significant information on this point is to be found in an AWWA report (19), an IWSA survey (20), and principally, in a Directive of the European Communities (21). Three basis categories of raw water assigned to treatment for drinking water are defined, but we shall consider only raw water of quality A3 here. It corresponds to raw water that is to undergo advanced physical and chemical treatment, including disinfection and refining, chlorination (eventually up to breakpoint), coagulation, flocculation, sedimentation, filtration, refining (e.g., on active carbon), and disinfection (e.g., with ozone). Waters of categories 1 and 2 are submitted to more severe quality criteria so as to simplify their treatment.

In the United States the approach to raw water quality criteria is based on the definition of permissible criteria and desirable criteria. The type of treatment to which the definition of these criteria is related is called a *standard treatment*, involving coagulation with the eventual addition of coagulation aids and active carbon, sedimentation, rapid sand filtration, and disinfection with chlorine. It must be emphasized that the treatment is adapted to the particular water under consideration. The potential formation of organohalogenated compounds and more severe disinfection criteria (e.g., for giardiasis) determine any changes in the applicable criteria.

Permissible criteria are the characteristics and concentrations in the raw water that make it possible to obtain safe drinking water that meets the standards, by use of the procedures mentioned above. *Desirable criteria* are those characteristics of raw surface waters that represent a high-quality water than can be treated with a higher factor of safety or at less cost so as to meet the standards established for drinking water. The *EEC A3 and AWWA criteria* are grouped on a comparative basis in Table 6.

A disadvantage often associated with the formulation of raw water quality criteria is that they may be considered as permission to increase environmental pollution up to certain maximal limits. However, environmental parameters must be considered on a general basis. Consequently, according to the EEC directive, the criteria may not be misused to increase the current state of pollution of the surface waters.

Some parameters included in the raw water quality criteria are not corrected by the usual techniques for drinking water treatment [e.g., temperature, conductivity, total hardness (calcium + magnesium), alcalinity, sodium, sulfates, and chlorides]. Among the processes for treatment, no mention is made of aeration, softening, hardening, alcalinity correction, stabilization, and so on. Regarding radioactivity, WHO recommends a limit of 2.5 pCi/L (0.1 Bq/L) for gross α activity, and of 27 pCi/L (1 Bq/L) for gross β activity, as warning levels.

As outlined by the International Water Supply Association (IWSA) (20): "It is the responsibility of the waterworks to select treatment processes and to organize technical management in such a way that safe and wholesome drinking water can be delivered; the quality of the surface water to be treated for drinking water must be such that this responsibility can be accepted."

Advanced treatment systems use judicious combinations of unit processes to

Table 6   Quality Criteria for Raw Waters (g/m³)

Parameter	EEC A3 Recommended	EEC A3 Mandatory	AWWA[a] Permissible	AWWA[a] Desirable
Color	50	200	75	<10
Suspended material	—	—	500	<200
Temperature (°C)	22	25	Narrative	Narrative
Odor (dil.)	20	—	Narrative	Virtually absent
Turbidity	—	—	Narrative	Virtually absent
pH	5.5–9	—	6.0–8.5	Narrative
Conductivity (μS/cm)	1000	—	—	—
Nitrates	—	50	—	—
Fluorides	0.7–1.7	—	Narrative	Narrative
Fe	1	—	0.5	Virtually absent
Mn	1	—	1	Absent
Cu	1	—	1	Virtually absent
Zn	1	—	—	—
B	1	—	1	Absent
Be	—	—	—	—
Co	—	—	—	—
Ni	—	—	—	—
V	—	—	—	—
As	0.05	0.1	0.05	Absent
Cd	0.001	0.005	0.01	Absent
Cr	—	0.05	0.05	Absent
Pb	—	0.05	0.05	Absent
Se	—	0.01	0.01	Absent
Hg	0.0005	0.001	—	—
Ba	—	1	1	Absent
CN^-	—	0.05	0.2	Absent
SO_4^{2-}	150	250	250	<50
Cl^-	200	—	250	250
PO_4^{3-}	0.7	—	Narrative	Narrative
NH_4^+	2	4	0.64	0.01
Ag	—	—	0.05	Absent
NO_2^-	—	—	5	Absent
Zn	—	—	5	Virtually absent
Detergents (index)	0.5	—	0.5	Virtually absent
Phenols (index)	0.01	0.1	0.001	Absent
Oils and greases	0.5	1	Virtually absent	Absent
PAH	—	0.0001	—	—
CC extract	0.5	—	0.15	<0.004
Pesticides	?	0.005	(Individual limits)	Absent
Total β radioactivity	—	—	1000 pCi/L	<100 pc/L
Radium 226	—	—	3 pCi/L	<1 pc/L
Strontium 90	—	—	10 pCi/L	<2 pc/L
Total coliforms	500,000/L	—	100,000/L	<1000/L
Fecal coliforms	200,000/L	—	200,000/L	<200/L
Fecal streptococci	100,000/L	—	—	—

[a]Narrative, impossible to quantify a single numerical value for the criterion; absent, undetectable by the analytical methods used; virtually absent, present in very low, not objectionable concentrations.

reach the quality objectives of the treatment. Each unit process must be designed according to its own characteristics and technical specificity and be operated in the context of the treatment as a whole.

1.9   Environmental Water Quality

The achievement of safe drinking water cannot be considered as a social and/or technical action completely separate from the environment as a whole. Multiple use of available resources must be considered, and adequate protection of environmental resources merits considerable attention. Some of these aspects are discussed in this chapter, since quality standards for multiple uses of water can also contribute to the protection of raw water sources and be favorable for the production of safe drinking water.

   The quality of natural waters in which swimming or bathing is allowed by a public authority must be safe for the public health. Therefore, guide numbers (i.e., quality goals) and imperative values (maximum concentration limits) must be met. These involve quality objectives and requirements as indicated in Table 7. The values given in the table are derived from a directive of the EEC [22]. Other parameters not yet completely quantified concern floating materials normally absent (plastics, bottles, rubber materials, etc.), ammonia and total nitrogen, pesticides, cyanides, nitrates, phosphates, and when considering toxic properties, the most important heavy metals: arsenic, cadmium, chromium^{6+}, lead, and mercury. For these parameters specific limits must be set by the governmental health authorities to permit public use of the water surfaces under consideration.

   The water quality necessary for fish life can be defined for salmonides and cyprinides [23]. In general, the temperature should be between 10 and 20°C, the oxygen content between 50 and 100% of the saturation value, the pH between 6 and 9, and suspended matter less than 25 g/m^3. Other standards are listed in Table 8. (The strictest values are for salmonides; the more tolerant are for cyprinides.)

Table 7   Water Quality for Swimming and Bathing

Parameter	Unit	Guide value	Imperative value
Coliforms	number/100 mL	500	10,000
E. coli	number/100 mL	100	2,000
Fecal streptococci	number/100 m/L	100	—
Salmonella	number/L	—	0
Enteroviruses	PFU/10 L	—	0
pH	—	—	6–9
Color	—	—	Normal
Oil(s)	g/m^3	0.3	Not perceptible
MBAS	mg/L LaS	0.3	No foam
Phenol index	mg/m^3	5	50
Transparency (Secchi)	m	2	1
O$_2$	mg/L	10–14	—

Table 8 Essential Water Quality Parameters for Fish Life

BOD_5	<3	<6	(g O_2/m^3)
PO_4^{3-} (total)	<0.2	<0.4	(g PO_4^{-3}/m^3)
Nitrates	<4	<6	(g NO_3^-/m^3)
Nitrites	<0.05	<0.5	(g NO_2^-/m^3)
Ammonium (total)		<1	(g NH_4^+/m^3)
Phenol index		<0.005	(g/m^3)
Zinc	<0.03–0.5[a]	<0.3–2.0[a]	(g Zn/m^3)

[a]The higher values correspond to hard waters.

1.10 Wastewater Treatment

In wastewater treatment before discharge in natural waters the performance of treatment plants is to be fixed so as to meet specific needs for each industry. According to the literature, as a first approximation maximum discharge figures that can be accepted for processing in wastewater treatment plants are as listed in Table 9. The highest figures relate to advanced wastewater treatment. If these figures are exceeded, a preliminary specific purification step is necessary before the discharge can be accepted in the public sewage system. The average composition of domestic wastewater is given in Table 10.

Instant concentrations of the effluent of a given system also depend on the flow and settling conditions in the sewage collecting system. This factor must be considered when designing the intake system of a treatment plant. The loading of domestic sewage for an average 24-h period places demands on local conditions: for example, water consumption rates, consumers' customs, and the existence or absence of natural dewatering.

The figures in Table 11 relate to urban areas. In rural systems the general population often consumes only 100 L/person per day, but the needs for cattle can be high. On the other hand, part of the sewage can be discharged through a quay and not transmitted through the sewage system. In urban areas a sudden increase in suspended solids concentration can occur during rain and runoff periods, due to

Table 9 Treatability of Industrial Wastewater

pH	<6	
Conductivity	<250–500	mS/m
Alkali	<1–2	kg/m^3
Suspended matter	0.5–2	kg/m^3
Petroleum ether–extractable compounds	0.2–0.5	kg/m^3
Sulfides	25–50	g/m^3
Carbohydrates	0.5–1.5	g/m^3
Cyanides	10–20	g/m^3
Sulfates	1.8	kg/m^3
Metals		
Cr, Cu, Ni, Zn, Cd	20 g/m^3 individually	
	50 g/m^3 total	
As, B, Pb, Se, Hg	5 mg/m^3 individually	
	20 g/m^3 total	

Table 10   General Quality Parameters of Domestic Wastewater[a]

Parameter	Unit	Instant	After 1 h	Input/person/day (g)
pH	—	6.5–7.5	6.5–7.5	—
Dissolved solids	g/m^3	700–1000	—	100–160
Suspended solids	g/m^3	600–900	400–600	100
Sediments (settable solids)	vol %	5–10	<0.2	54–60
BOD	g/m^3	300–500	200–350	42–54
COD	g/m^3	600–1100	450–650	135
Organic nitrogen	g/m^3	30–50	10–20 ⎫	
Ammonia nitrogen	g/m^3	40–65	40–65 ⎬	12–15
Nitrate and nitrite nitrogen	g/m^3	0–2	0–2 ⎭	
Phosphorus compounds (P)	g/m^3	15–30	15–30	3–4
MBAS	g/m^3	10–20	—	—
Chlorides	g/m^3	200–400	200–400	20–100

[a]The equivalent inhabitant depends on local circumstances.

Table 11   Composition of Domestic Sewage

| Material | Domestic sewage (g/m^3) for two water consumption rates | | | | | | | |
| | 150 L/person/day | | | | 300 L/person/day | | | |
	Mineral	Organic	Total	BOD	Mineral	Organic	Total	BOD
Settable solids	130	270	400	130	50	130	180	65
Suspended solids	70	130	200	80	35	85	115	40
Dissolved matter	330	330	660	150	265	265	530	75
Total	530	730	1260	360	350	480	825	180

the transfer of deposits accumulating in the collectors during periods of low flow. In the Brussels area (e.g., Neerpedebeek), an increase in flow from 0.15 m^3/s to 0.9 m^3/s due to sudden rains can increase the suspended matter concentration temporarily from 250 g/m^3 to as much as 1000 g/m^3.

Ideally, the treatment of wastewater would maintain an environmental water quality level corresponding to the EEC standards for raw water to be treated to obtain drinking water (e.g., Table 6). A more generally accepted goal is to achieve the water quality necessary for fish life.

The true standard for quality is located in the mind.   (Bickerstaff)

2. QUANTITATIVE ASPECTS OF WATER TREATMENT

The basic water need, that of drinking water, amounts to 2.5 L/person per day. The quality standards as to toxicity are computed on this basis. However, the minimum needs for the use of water as a food and for general sanitary use is on the order of 20 L/person per day. Practical experience gained in countries with a high standard of living indicates that when for technical or local reasons, public water distribution is temporarily inadequate but water remains available locally (e.g., through free fire hydrants), the average consumption amounts to about 40 L/person per day. In native villages where water is carried by hand, the needs range from 15 to 35 L/person per day.

Under normal distribution conditions there is a slow evolution in the domestic use of water. Following are figures for the Brussels area:

1935	80 L/person/day
1950	123 L/person/day
1964	126 L/person/day
1970	153 L/person/day
1989	175 L/person/day

At present, the minimum consumption ranges from 80 L/person per day in older housing areas to 200 L/person per day in higher-standard residential locations.

For the United Kingdom as a whole (24) the consumption distribution curve as a function of population shows a distortion in the high-consumption fraction. The overall consumption is higher in Scotland (± 290 L/person per day) than in England (± 180 L/person per day) (Fig. 2). This is primarily the result of the flushing of toilets; in England, units are set at 9 L, and in Scotland, at 13.5 L.

2.1 Design for Future Needs

The prediction of future needs is the basis for process design. Evolution as a function of time of water needs is often expressed by a formula of the type

$$Q_n = Q_o(1 + x)^n$$

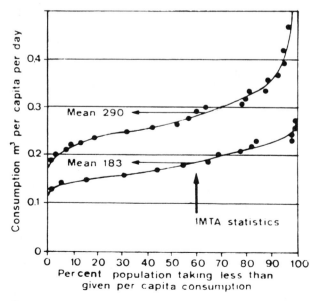

Figure 2 Distribution trends in water consumption in the United Kingdom.

where Q is the flow (average) in year n compared to the reference year, with an annual rate increase of x. During the period of the "golden sixties" the average x value in Europe was taken to be a 0.04; however, at present more realistic figures for domestic consumption are in the range 0.01 to 0.03. The maximum individual consumption reaches approximately 140 to 150 L/person per day (25). In the present analysis of trends it has become obvious, particularly in European countries, that the continuous growth in consumption will stop. A form to express the saturation tendency is

$$< Q > \ = \ \frac{< Q_{max} >}{1 \ + \ 10^{a-bt}}$$

where a and b are calculated or practically observed constants, and t is the time delay.

The global exponent $(a - bt)$ can be determined by linearizing the correlation:

$$\frac{< Q_{max} >}{< Q_t >} = 1 \ + \ 10^{a-bt}$$

$$a \ - \ bt \ = \ \log\frac{< Q_{max} > \ - \ < Q_t >}{< Q_t >}$$

Local parameters (e.g., lawn sprinkling or lack of water billing) can increase this figure to 250 L/person per day and more.

Another dominant design criterion is the forecast of population increase. In this difficult question the general trends are that in industrial countries, cities do not continue to develop after their population reaches 1 million. Towns and cities with a population of 20,000 to 100,000 often increase rapidly to 200,000. Local trends may be very different in the United States. On the other hand, many large cities in developing countries experience a fast, even explosive increase in population. Evaluation of this local parameter is important in design. Also, the living standard in rich residential areas is completely different from that in densely occupied areas with lower-income populations. In Asian countries where these situations exist, the average overall consumption is 25 m^3/ha per day, but figures as low as 10 m^3/ha per day may be encountered locally.

2.2   Water for Industrial Use

Most water for industrial use is taken from natural water reserves (e.g., rivers), and 80 to 90% of it involves cooling water that is restored at 90% of the amount taken in (26). Specific industrial needs are given in the literature (26,27): for example, 6 to 10 m^3/ton steel: 0.3 m^3/ton refined oil; 40 to 80 m^3/ton paper; 30 to 100 m^3/ton processed meat, including slaughter; and in the chemical industries, 450 m^3/ton acetic acid and 900 m^3/ton ammonium sulfate. Other sectors (e.g., electronical processing) often require high volumes (e.g., 200 m^3/worker per day).

As a global guideline, a first approximation for the needs of an industrial area (zoning) can be set on a level of 15,000 m^3/ha per year or 40 m^3/ha per day, inclusive of cooling water. This holds for industries in general which are not specifically high consumers of cooling water. In developing countries, the corresponding needs presently range around 50 m³/ha per day (28). In modern zoning the incidence of cooling water is somewhat less than in classical industrial areas and the average 750 m³/ha per year (approximately 2 m³/ha per day is equal to the *unrestored consumption*). Another approach to evaluating the water needs in industrial areas is to use a daily consumption figure of 0.1 to 0.15 m³ per worker employed.

In the past, evaluation of industrial water needs has been based on the continuous expansion trend of the national gross product by 2 to 4% yearly. The corresponding consumption increase was evaluated as

$$C_n = C_0(1.02)^n$$

giving the needs C_n after n years. According to present trends, these needs are overevaluated, and a more realistic approach is based on the maximum consumption per surface (e.g., hectares of zoning) and per worker.

The relative importance of industrial needs versus domestic consumption is highly dependent on local conditions. For developed agglomerations, industrial consumption can range from 10 to 50% of the total. Port areas often consume relatively more bulk industrial water (e.g., 40% of the total in Antwerp and up to 60% in Rotterdam compared to 14% in Brussels and 20% in Belgium as a whole). [By "bulk" industrial water we mean the needs of large industrial units for which the water must not necessarily be potable in all aspects (except cooling water for power plants).] The metering and billing conditions may also exert a dominant influence (e.g., for state or public-dependent sectors).

2.3   Water for Agricultural Use

Agricultural use most often relies on local sources rather than on public distribution systems. It can be assumed that the average daily needs of a pig are 0.4 m³/day and of a cow, 1.4 m³/day, the real consumption ranging from 50 to 100 L/day. For intensive horticulture in greenhouses 10 to 15 m³/ha per day may be required.

2.4   Water for Public Use

In urban distribution, general public water use ranges from 5 to 55% of total consumption for street cleaning, public fountains and park spraying, firefighting, and in some locations, unbilled water use in public and governmental offices. In Belgium, for example, the average public unmetered consumption is 17%, and in Brussels it is ±14% (including distribution losses). The distribution losses themselves can range from 3 to 40%, according to local circumstances, and must be added to the global evaluation of consumption rates and considered in the overall concept, which may also involve a reduction in rates through improvements in the distribution system.

2.5   Peak Factors

Peak factors are a significant parameter in the design of water treatment systems. The global approach can be given as a duration curve if appropriate reference data are available. Generally, the fluctuations in demands are in the range given in Table 12. In the Brussels area the peak factor (i.e., the maximum day/average day) is

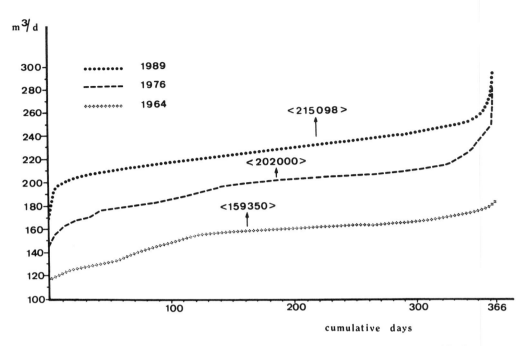

Figure 3   Duration curve of water consumption in the Brussels area (ordinate in $10^3 \, m^3$/day).

Table 12   Fluctuation in Demand

	Demand		
Period	Maximum	Average	Minimum
Annually	1.03	1	0.97
Monthly	1.15	1	0.85
Weekly	1.25	1	0.80
5 consecutive days	1.35	1	0.76
2 consecutive days	1.45	1	0.70
Daily	1.50	1	0.70
Hourly	2.50	1	0.15
Instantaneous	3.00	1	0.10

equal to 1.35 on a long-term basis. This peak factor can depend on local circumstances ($P_F \times <d>$ = maximum day): $P_F = 3$ for residential areas with a high living standard (lawn sprinkling) to $2 < P_F < 2.5$ for mixed zones, residential, and transforming industries, and P_F is always lower than 2 in industrial and rural areas. The overall peak factor is lowered by increased general, often unbilled use of water and by losses in the distribution system.

The fluctuations between summer and winter consumption for domestic use range to 1.6 (average summer to average winter). This parameter applied to industrial consumption is approximately 1.2 (29–31). Usually, there is a tendency for decreasing peak factors with increasing annual demand and increasing population. Tourism and vacationing lower the peak factor in normal residential areas and can increase it drastically in tourist locations (up to a factor of 4). In design the treatment capabilities must meet the maximum expected needs, including peak factors for 1 to 5 days, depending on the clear-water storage capabilities and global policy concerning the reliability of public water distribution.

2.6 Long-Term Planning for Water Needs

Long-term planning (32) of water needs relies on a systematic approach to various sectorial consumption needs [e.g., domestic (bathing, laundry, dishwashing)] but also on the more difficult to evaluate industrial consumption trends. These are related to several balance factors that are difficult to evaluate on a quantitative statistical basis: A limited number of industries consume the highest amounts, so that the basic data for prognosis cannot be assessed on a large number of potential consumers. Therefore, the evolution of industrial consumption in a given area is often a function of the economic activity in a specific sector. In general, the designs in countries with high living standards formulated during the "golden sixties" have been overestimated compared to the needs of the 1980s.

2.7 Water Cost Considerations

Water demand obeys the *law of demand*: The quantity of the commodity demanded decreases as its price increases, and if the real price decreases, the quantity demanded increases (33). According to the literature, the price-elasticity relationship in water demand is expressed as

$$Q = aP^b$$

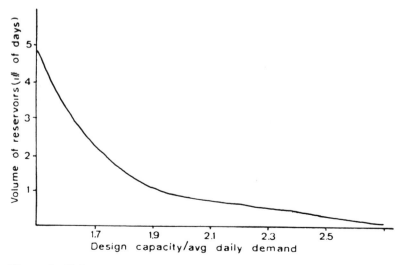

Figure 4 Volume of reservoirs as a function of treatment capacity.

where Q is the quantity of water demand, P is the price per unit volume (e.g., m³), a is a constant, and, b is the price elasticity coefficient: $(dQ/Q)/(dP/P)$. The price elasticity coefficient is *always* negative but changes considerably among different groups of users (see Table 13).

We note first that industrial demand is less price-elastic than domestic demand except in the United Kingdom, where most consumption is still unmetered. Large fluctuations can be observed in the industrial price-elasticity coefficient according to the technical feasibility of a reduction in water demand. At increasing prices for water distribution, the tendency to reuse industrial water through recycling in industrial plants is likely to increase and to diminish the b factor. Outdoor domestic use is more price-influenced than indoor use and the price impact is less pronounced in dry areas than in humid regions. This strictly economical approach does not consider minimum vital needs as a significant parameter in the price–demand relationship. It is likely that a more correct relationship for the price elasticity of demand is expressed as

$$Q = Q_0 + aP^b$$

where Q_0 is the minimum "unreducible" average demand (e.g., 20 to 40 L/person per day. Therefore, price-elasticity evaluations are best based on average consumption rates exceeding this minimum need.

Increasing water price influences average consumption but is hardly significant for maximum peak consumptions. Therefore, the peak factors can be increased at higher prices. Moreover, as the design capacities must take peak consumption into account, the effect of price on the maximum production capacity to be designed remains marginal. Therefore, there exists no unequivocal relationship between the price of water and the design capacity of treatment plants.

The cost–treatment capacity function is often described by a relationship of the type $FC = KQ^E$ (34), where FC is the first cost, K a constant depending on local conditions, Q the design capacity (in flow units), and E an economic scale factor. The E factor is usually less than unity and in given local circumstances:

Table 13   Price Dependency of Consumption

	b
Average	−0.86
Residential (high standard)	
Indoor use	−0.23
Outdoor use	
Humid areas	−1.6
Dry areas	−0.7
Industrial uses	
Average	−0.62
U.S.	−0.7
U.K.	−1.4
Food processing	−0.4

$$\frac{(FC)_1}{(FC)_2} = \frac{Q_1^E}{Q_2^E}$$

This means that the average first cost per unit capacity decreases with increasing total installed production capacity. Suggested values for the economic scale factors E are given in Table 14 for various structures.

2.8 Water System Design Report

The design is to be consigned in a report listing the following items:

1. Project definition
2. Population
3. Water demand $\}$ or wastewater
4. Available water resources $\}$ treatment objective
5. Basic design data and considerations
   a. Design period
   b. Capacity
   c. Water quality (raw and treated water)
6. Selection of water resource
7. Selection of site
8. Master plan
9. Flowsheet
10. Process calculations
    a. Change in water quality after each treatment step
    b. Required dosages of chemicals
    c. Costs of chemicals
11. General layout of the plant, including capacities and number of units in each treatment step
12. Dimensions of units
13. Hydraulic scheme with head loss and dimensions of pipes and channels

Table 14   Economic Scale Factors

Intake structures and pumping stations	
Small plants	0.2
Large plants	0.4
Storage and primary impounding	0.5–0.6
Static and primary sedimentation tanks	0.7–0.8
Rapid sand and activated carbon filters	0.6–0.8
Slow sand filters	0.9
Treatment plants (as a whole)	0.7–0.8
Pumping stations	0.3–0.5
Secondary wastewater treatment	0.7–0.8
Aeration systems	0.7
Conduits and channels	0.4–0.6
Offices and laboratories	0.4
Concrete reservoirs (clear wells)	0.6–0.7

14. Configuration of valves and gates
15. Design parameters, computations, and detailing of each phase
16. Relevant cross sections and lengths of various parts of the plant
17. Required total power
18. Costs (estimated)
19. Guidelines for plant operation

The process calculations (item 10) involve several sequences — particularly:

1. Nature of the chemicals used
2. Expected doses: minimum, maximum, average
3. Storage
   a. Form
   b. Period
   c. Quantity
   d. Equipment
   e. Delivery
4. Dosing form
   a. Makeup technique
   b. Dosing equipment for each product
   c. Number and type
   d. Reliability and standby equipment required
   e. Special requirements (e.g., corrosion)
5. Constructional aspects of work rooms
   a. Emptying and filling chemical storage areas
   b. Sluicing of work areas
   c. Ventilation
   d. Safety equipment
6. Reactional aspects
   a. Sequence of chemicals in the treatment
   b. Reaction time required for each chemical
   c. Mixing equipment
7. Safety aspects
   a. Stability and ignition/explosion levels
   b. Freezing and dosing problems
   c. Constructional protection and emergency safety equipment
   d. Safety regulations and laws
8. Costs per cubic meter
   a. Of the chemical (on the basis of average doses)
   b. Of the mixing energy (calculated on the G value, watthour basis)
   c. Of the chemical rooms, injection basins, storage, and dosing
9. Operational aspects
   a. Criteria for determining dosage
   b. Automatic control chain and/or total automation
   c. Dosing sequences in starting and stopping plant operation

3. CHECKLIST OF THE DESIGN PROCESS

In the fundamental phase of the design process a master plan and flowsheet must be developed. An example checklist is given below, showing the various parameters that are involved in the philosophy of the design process as a whole (34). Which factors are dominant depends largely on how local conditions affect the attainment of the universal scientific goal: to purify water to a good quality level for drinking water or to make available water appropriate for general environmental use at adequate pressure and at accessible costs.

3.1   Local Conditions

Sociological aspects
Standard of living
Population density
Political aspects
Local standards
Tradition
Availability of know-how and experience
Present and future socioeconomic development; proposed economic and physical
    planning and potential
Geography
Topography
Climate
Geological, hydrological, and hydrogeological aspects
Organization of the water supply and sanitation
Availability of regular workers and specialists, contractors, construction materials,
    chemicals, energy, and transport
Soil conditions and foundation aspects
Earthquakes; flooding
Existing water supply
Adequacy of existing wastewater disposal, sewage, and drainage systems
Legal aspects

3.2   Demand Quantity

Annual demand
Monthly demand
Weekly demand   minimum ⎫ past
Daily demand   average ⎬ present
12-h demand   maximum ⎭ future
Hourly demand
Instantaneous demand
Variation in demand and peak factors depending on several factors:
    Climate
    Size and type of distribution area
    Industrial demand
    Number of consumers supplied
    Population density
    Nature of distribution systems
    Pressure in distribution systems
    Leakage in distribution systems
    Demand for firefighting
    Demand for lawn sprinkling

Precipitation (raining) patterns
Urbanization
Type of sewage system
Accumulation zones in sewage collectors

3.3   Demand Parameters

Domestic use
Commercial and industrial use
Public services (e.g., firefighting, public works department)
Waterworks (e.g., flushing of distribution and transport systems, backwashing of
    filters)
Nondomestic purposes
Leakages and losses

The prediction for future demand is the basis for design. It is probably the
most hazardous part of the design process. The ultimate success of the project
depends on the goodness of forecasts.

3.4   Quality and Reliability

Standards and requirements from the point of view of public health, bacteriological
    safety of the water, toxic matter, protection of the consumer or user
Consumer's requirements: *constant* good quality—pressure, color, odor, hardness
Water company's requirements: physical, chemical, and bacteriological *stability* of
    the water to prevent difficulties during the transport of water in the distribution
    system (aggressivity, aftergrowth, deterioration of quality, etc.)
Requirements for efficient industrial and pleasant environmental use of the water
Reliability of the system as related to the permanent availability of water appro-
    priate to needs (this concept includes the fraction of time during which the
    system is "up-state")
Storage of chemicals
Installed or existing spare parts and equipment

3.5   Functional Efficiency

This is secured by:

Proper adaptation of the treatment process to the required quality and the available
    quality of the raw water
Proportioning the various component parts of the plant so as to ensure sufficient
    quantity, taking into account the demand pattern
Spare equipment installed and preventive maintenance schemes
Simplicity in execution and operation
Operator safety (e.g., chemical risks)
Automation and remote control
Possibility of manual operation
Available labor hours for operation and maintenance

3.6   Aesthetic Aspects

Architecture
Odor and noise problems
Ambient toxicity (TLVs)
Safety equipment
Sludge treatment and disposal facilities

3.7   Economic Aspects

Design period
Marginal costs
Investments and total annual costs
Minimum operation and maintenance costs
Benefit-to-cost ratio
Service life, economic amortization life
Automation and costs of employees

4.   PRESENTATION OF A DESIGN REPORT

1.   Introduction [short (e.g., historical)]
2.   Reviews of project area
     a.   Social background
     b.   Socioeconomical conditions
     c.   Political conditions
     d.   Environmental conditions
3.   Existing water supply system
     a.   Water demand and sources
     b.   Water supply facilities
     c.   Possibilities of operation and maintenance
     d.   Water tariff system
     e.   Organization
4.   Population, water demand, available sources
     a.   Population projection
          (1)   Service area
          (2)   Service connections
          (3)   Design period and target year
          (4)   Population served
     b.   Water demand
          (1)   Water consumption
          (2)   Water demand
     c.   Water sources
          (1)   Springs
          (2)   Groundwater
          (3)   Rivers
     d.   Water quality
          (1)   Analysis
          (2)   Adopted standards and recommended values

5. Comparative study of alternative plans
   a. General
   b. Water intake and treatment facilities
      (1) Design criteria (and flowsheet)
      (2) Layout and description
      (3) Preliminary cost estimation
   c. Water transport and distribution facilities
      (1) Design criteria
      (2) Layout and description
      (3) First-cost estimation
6. Proposed project
   a. Project formulation and design
      (1) Scope of the project
      (2) Water intake and treatment facilities
      (3) Water transport and distribution facilities
   b. Project implementation
      (1) Project implementation and schedule
      (2) More precise cost estimation
      (3) Recommended economic scale factor
   c. Organization and maintenance
      (1) Organization and staffing
      (2) Functions and responsibilities
      (3) Water analysis and monitoring schemes
7. Project justification
   a. General
   b. Financial analysis
   c. Socioeconomical analysis
      (1) Political impact of the project
      (2) Social impact of the project
      (3) Environmental impact
      (4) Cost–benefit evaluation
8. Conclusions and recommendations
   Appendixes: tables, graphs, and drawings

If the quality is no longer sufficient, the available quantity is zero.

REFERENCES

1. R. D. Lee, *J. AWWA, 64*, 216 (1972).
2. M. E. Flentje, Treatment plant control, p. 468 in *Water Quality and Treatment*, AWWA, Denver, Colo., 1971.
3. EEC, *Directive Relating to the Quality of Water for Human Consumption*, 80/778 CEE, July 15, 1980; (a version is in preparation, 1991).
4. WHO, *European Norms*, 1971, and working documents.
5. WHO, *International norms*, 1972, and working documents.
6. U.S. PHS and AWWA, *J. AWWA, 60*, 1317 (1968).
7. E. C. Lippy, *J. AWWA, 73*, 57 (1981).
8. H. E. Stokinger and R. L. Woodward, *J. AWWA, 50*, 515 (1958).

9. J. P. Buffle, *Human Intake of Minerals from Drinking Water in the European Communities*, Comm. E.C., EUR 5447, 1976, p. 207.
10. C. H. Tate and R. R. Trussel, *J. AWWA, 69*, 486 (1977).
11. J. F. Ferguson and J. Grovis, *Water Res., 6*, 1259 (1972).
12. Y. K. Chau et al., *Science, 192*, 1130 (1976).
13. P. P. St.-Amant and P. L. McCarty, *J. AWWA, 65*, 562 (1973).
14. J. D. Hem and W. H. Dierum, *J. AWWA, 65*, 562 (1973).
15. T. R. Camp, *Water and Its Impurities*, Reinhold, New York, 1963.
16. J. Gardiner, *Water Res., 8*, 23 (1974).
17. Anon. *J. AWWA, 60*, 1317 (1968).
18. A.R. 28.2.1963 (Moniteur Belge 16.05.1963); A.R. 23.05.1972; Directive EEC 15.07.1980, Journal L 246, 17.09.1980.
19. Anon., *J. AWWA, 61*, 137 (1969).
20. IWSA, Amsterdam Congress, 1976, p. L1.
21. EEC Directive 75/440, Journal L 194/26, 25.07.1975.
22. EEC Directive 76/160, Journal L 31/1, 5.02.1976.
23. EEC Directive 78/659, Journal L 222/1, 14.8.1978.
24. A. T. Twort, A. C. Hoather, and F. M. Law, *Water Supply*, Edward Arnold, London, 1974.
25. W. Breuer, *Gas- Wasserfach, 121*, 67 (1980).
26. D. Van Rijsbergen, *H_2O, 11*, 549 (1978).
27. J. Snel, *Tech. Eau, 325*, 15 (1974).
28. H. Aya and M. Nakao, *Industrial Water Supply and Water Quality*, IWSA Regional Conference, Singapore, 1979.
29. Anon., *J. AWWA, 53*, 459 (1961).
30. A. D. Henderson, *J. AWWA, 48*, 361 (1956).
31. L. S. Finch, *J. AWWA, 48*, 364 (1956).
32. P. L. Knoppert, *General Report 1*, IWSA, Amsterdam, 1976.
33. H. Hanke, *J. AWWA, 70*, 487 (1978).
34. C. Vaillant, *Lectures at the International Institute in Hydraulic and Sanitary Engineering*, Delft, The Netherlands.
35. R. L. Zielhuis, *H_2O, 14*, 290 (1981).
36. J. E. Peeters, E. Ares Mazas, W. J. Masschelein, I. Villacorta Martinez de Maturana, and E. Debacker, *Appli. Environ. Microbiol., 55*, 1519 (1989).

2

Use of Chlorine Dioxide

1. INTRODUCTION

Chlorine dioxide was discovered in 1811 by Sir Humphry Davy by reacting potassium chlorate with sulfuric acid. Chlorine dioxide is a relatively unstable gas (bp 11°C) which cannot be compressed and liquefied without danger of explosion. Therefore, it must be generated at the site of use.

Chlorine dioxide (ClO_2, MW 67.47) corresponds to the oxidation number IV of chlorine, has an angular structure, and contains an unpaired electron. Thus it must be considered as a free radical with a resonance structure.

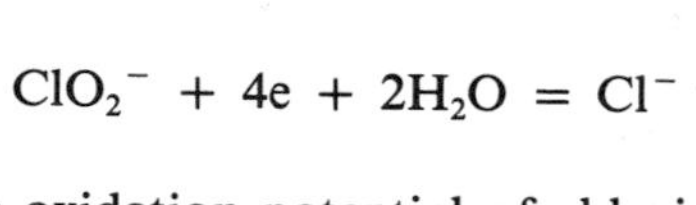

$$
\begin{array}{ccc}
Cl & & Cl \\
/\!/ \;\; \backslash\!\backslash \;\; \cdot & \leftrightarrow \; \cdot & /\!/ \;\; \backslash\!\backslash \\
O \qquad O & & O \qquad O
\end{array}
$$

Chlorine dioxide is an oxidant with two consecutive reactions:

$$ClO_2 + 1e = ClO_2^- \qquad E_0^{25°C} = 1.15 \text{ V as gas}$$
$$= 0.95 \text{ V aqueous solution}$$

$$ClO_2^- + 4e + 2H_2O = Cl^- + 4OH^- \qquad E_0^{25°C} = 0.78 \text{ V}$$

The oxidation potential of chlorine dioxide is aqueous solution decreases linearly by -0.062 V with each unit increase in pH.

An experimental equation that can be used to evaluate the temperature dependence of the first redox potential of chlorine dioxide is given by

$$E_0 \text{ (volts)} = -5.367 + 0.0613T - 19.4 \times 10^{-5}T^2 + 2 \times 10^{-7}T^3$$

The chlorite ion is a less significant oxidant, particularly where reactions with organic compounds are concerned. Reaction products are discussed later.

Chlorine dioxide can theoretically be considered as a mixed anhydride of chlorous acid and chloric acid:

$$2ClO_2 + H_2O = HClO_2 + HClO_3$$

However, in a neutral solution, chlorine dioxide hydrolyzes only poorly, although

$$K_H = \frac{|HClO_2| \cdot |HClO_3|}{|ClO_2|^2} = 1.2 \times 10^{-7} \text{ at } 20\,°C$$

The entire reaction is complex and involves disproportionations important in the design of chlorine dioxide generation systems.

Chlorine dioxide gas is potentially explosive when present in air at a partial pressure higher than 100 mbar or 10 vol percent at atmospheric pressure. The explosion may be caused by an electrical discharge, heat, or a source of ignition of any type. To be violent the decomposition requires a temperature above 45 to 50°C. The reaction is characterized by long induction periods.

The reaction velocity is proportional to the total pressure and to $(p_{ClO_2})^{1/2}$. The overall reaction is $ClO_2 = 0.5Cl_2 + O_2$. Because, after explosion, the off-gas contains chlorine and oxygen, there is an increased risk of fire.

Solid sodium chlorite is explosive on heating (or on contact with organic materials (limit of danger: $\pm 80\,°C$). The chemical is best not stored in the solid state but in solution (e.g., 300 to 400 g/L). When spilled, the area should be cleaned by abundant sluicing with water.

The maximum acceptable concentration (MAC) in working areas for gaseous chlorine dioxide in air established on the basis of experience is 1 ppm (vol) in air. This is a very conservative level of professional exposure.

For more detailed aspects of the general chemistry of chlorine dioxide, the reader is referred to the literature (1).

2.   DISINFECTION OF WATER WITH CHLORINE DIOXIDE

At present, the problems associated with the formation of THMs and other organochlorine compounds cast a new light on the potential advantages of the use of chlorine dioxide to disinfect drinking water. Since the beginning of the twentieth century, when it was used at the spa in Ostend, chlorine dioxide has been known as a powerful disinfectant of water. However, during the 1950s it was introduced more generally as a drinking water disinfectant, giving less organoleptic hindering than that of chlorine. Important European cities—Brussels, Zürich, Toulouse, Düsseldorf, Berlin, Monaco, and Vienna—and several cities in the United States—Evansville, Indiana, Hamilton, Ohio, and Galveston, Texas—make use of it. For residual concentrations the practical result, as far as disinfection is concerned, is even lower than 0.1 mg/L.

2.1 Mode of Action

A general observation is that the bactericidal and virucidal action of chlorine dioxide is not lowered by increasing the pH of the water (4–8) and is even increased in some instances by such changes as is true for bacteriofage f2 (4,5), amebic cysts (6), and polioviruses and other enteroviruses (7,8). In an alkaline medium, chlorine dioxide is known to disproportionate into chlorite and chlorate. However, the increase in bactericidal or virucidal action cannot be attributed to chlorite or chlorate which form on disproportionation of chlorine dioxide, although this mode of action had been suggested in the literature (9). The formation of $OH^{\bullet}$ radicals has been advanced as a possible mechanism but has not been proven (10). A more probable hypothesis is that in an alkaline medium the permeability of the lipid layer is increased and *chlorine dioxide free-radical gas* thus gets easier access to vital molecules in the cell (11).

Chlorine dioxide reaction with vital amino acids is supposed to be one of the dominant processes of bactericidal action. The amino acids that react significantly with chlorine dioxide are hydroxyproline, proline, histidine, cysteine, cystine, tyrosine, and tryptophane (5). In the case of viruses the capsid protein remains intact on disinfection with chlorine dioxide, and the action on tyrosine is considered to be an important pathway of virucidal action (8).

2.2 Decay Laws

Chick's law is applied to obtain pseudo-first-order constants from the linear part of the bacterial decay. Most often a lethal lag is observed, especially with the more resistant organisms. This is particularly the case with amebic cysts, which according to the multihit theory, respond to an estimated 80 to 100 lethal hits. Clumbing is another effect to be considered.

Bactericidal action is considered to be of first order, according to Chick's law:

$$\log_{10} \frac{N(t)}{N(0)} = -k_1 t$$

The pseudo-first-order kinetic constant k_1 is an implicit function of the concentration and mode of action of the disinfectant. According to Watson's law, $k_1 = kC^n$, where C is the concentration of the disinfectant. The decay factor D_{10} is the value of $kC^n t$ by which the original population of living organisms is reduced by 90%. Two D_{10} decay values correspond to 99% decay, and so on for higher D_{10} values. The pseudo-first-order decay constants are then correlated with the concentration of the disinfectant according to Watson's law:

$$C^n \times t = k$$

For *E. coli* the concentration factor n is of 1 for chlorine and about 2 for chlorine dioxide. For *Pseudomonas fluorescens (putida)* we observed a concentration factor of about 2.5 with chlorine dioxide. For *Naegleria gruberi* the n value has been established as 1, but the contact time would coincide with a power factor of 1.5 to 2 (6); in other words, by setting the contact time at power 1, n is on the order of 0.55.

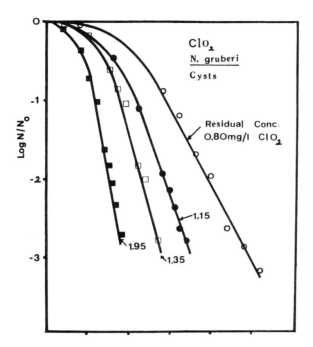

Figure 1   Inactivation curves of *Naegleria gruberi* cysts. (From Ref. 6.)

We have measured the same order of value for actinomycetaceae in water. The activity of a bactericidal agent in water can be expressed in terms of dosage in L^{-1} $\times$ mg $\times$ s to obtain a given decay value (e.g., 99% kill is expressed as $2D_{10}$).

Comments.   All other conditions being constant, the CT or D_{10} values are variable with temperature with sufficient linearity of D_{10} versus $1/T$ within 2 to 30°C (e.g., for *Giardia lamblia* and poliovirus the temperature coefficient is, respectively, 5.4 $\times$ 10^6 and 0.7×10^6 $L^{-1} \times$ mg $\times$ s $\times$ °C for $2D_{10}$ killing rates. The variability in data for poliovirus and the difference observed in very controlled conditions between purified and cell-associated rotavirus provide evidence of a significant difficulty in the fundamental evaluation of the resistance of the organisms to water disinfectants (i.e., the change in resistance depending on the initial condition of the germs). In this context it has been observed that cultured *E. coli* at a submaximal rate (i.e., at the exponential growth phase) are more resistant than those obtained at the stationary growth phase (14). In a general way, germs sampled or cultivated at a low population density are more resistant to disinfectants.

2.3   Effects of Residual Chlorite

Like chloride and chlorate, the stable secondary products of chlorine dioxide are of no direct value in disinfection. Chlorite has a weak bactericidal action on some germs possibly present in water (e.g., *Streptococcus faecalis*), but the action is too slow to be an operational method of disinfection (16,17). The action on *Pseudomonas putida* is more significant but still slow (see Fig. 2) and other species (e.g., *Ps.*

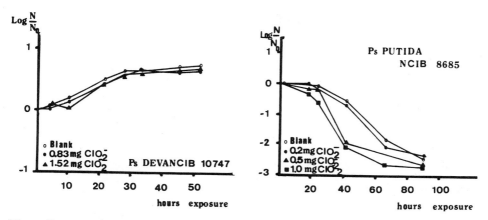

Figure 2   Action of chlorite ion on *Pseudomonas* species (indicative).

deva) are more resistant, but significant aftergrowth is eliminated by chlorite (17). The action on actinomyces and yeasts such as *Rhodontorula* is negligible. In summary, residual chlorite can contribute to bactericidal action but is not sufficient. However, it prevents aftergrowth in distribution systems significantly and in this way contributes to water quality.

Investigations are under way to check how far the "toxic properties" attributed to chlorite can be used to control the development of aquatic organisms that rely on respiration mechanisms in which substances similar to hemoglobin intervene. Figure 4 illustrate *tentative* data for adult *Asellus aquaticus*. The effect of chlorite is undoubtedly positive in preventing maintenance and development of such organisms in distribution systems. Further investigation is under way (18); age and stress conditions (e.g., light) may influence the quantitative results.

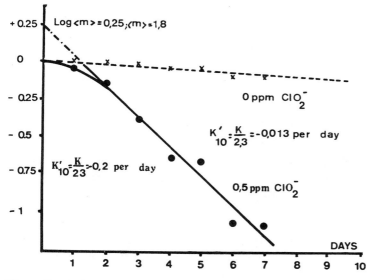

Figure 3   Action of chlorite ion on the survival of enterococci in water.

Table 1 Dosages for Various Decay Values

Typical $2D_{10}$ values ($L^{-1} \cdot$ mg $\cdot$ s)	References	Unit D_{10} or CT value[a] ($L^{-1} \cdot$ mg $\cdot$ min)
E. coli		
16 (pH 7, $t = 5°C$)	3	0.14
9 (pH 7, $t = 15°C$)	2, 7	0.075
8 (pH 7, $t = 25°C$)	3	0.07
Ps. fluorescens		
5–10 (pH 7–8, $t = 22°C$)		0.06[b]
Naegleria gruberi		
360–600 (pH 7, $t = 20°C$)	6	5
Poliovirus		
100 (pH 7, $t = 15°C$)	7	0.83
240 (pH 7, $t = 5°C$)	11, 12	2
750 (pH 10, $t = 20°C$)	9	6
1200 (pH 7, $t = 20°C$) (wild strain)	13	10
Coxsackievirus A9		
22 (pH 7, $t = 15°C$)		0.2
Rotavirus SA11		
30 (pH 6, $t = 5°C$, purified virus)	14	0.25
1000 (pH 9.5, $t = 5°C$, cell associated virus)	14	8.3
Bacteriophage f2		
16 (pH 7.1, $t = 3°C$)	5	0.13
Legionella pneumophila		
<5 (pH 7–8, $t = 22°C$)[b]		0.04
Giardia/Lamblia		
8400 (pH 7–8, $t = 15°C$)	3	70
Clostridium perfringens		
4000 (pH 7, $t = 20°C$)	13	33
(important lethal lag)		
Actinomyces		
(pH 7, $t = 20°C$)(wild strain)[b]		50

[a]CT values are computed on the basis of concentration factor n equal to unity.
[b]Tentative values under investigation.

3. DEVELOPMENT OF BY-PRODUCTS BY REACTIONS OF CHLORINE DIOXIDE IN WATER

The reactions of chlorine dioxide with organic substances cannot be considered separately from the global context of oxidation. In particular, the ammonium and bromide content of the water are of utmost importance to block secondary chlorination or bromination occurring as side reactions of chlorine dioxide treatment. Chlorine dioxide does not have a significant direct reaction with ammonium ions. In the dark bromide, ion is not oxidized with significant velocity by chlorine dioxide either. This is in contrast to reaction pathways with hypochlorous acid and ozone, which form hypobromous acid. Photosensitized oxidation of bromide is possible with chlorine dioxide.

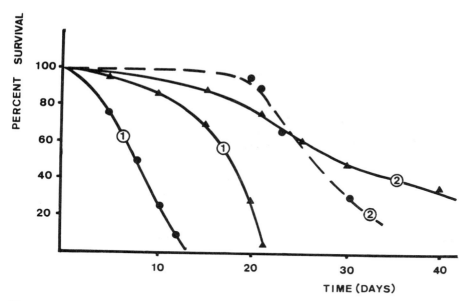

Figure 4 Sample set of data for the survival of *Asellus aquaticus* in the presence of chlorite ion. Food 1 mg/L Na acetate, 10^5 germs/mL; ●, daylight; ▲, dark; 1, 0.3 mg/L ClO_2^-; 2, no ClO_2^-. (From Ref. 18.)

3.1   Secondary Formation of Chlorine

Hypochlorous acid can be present in the medium either as a chemical present in the chlorine dioxide reagent or as the result of it being dosed into the water separately from chlorine dioxide. In some cases the secondary formation of hypochlorous acid has been suggested in the oxidation pathway of aromatic molecules by chlorine dioxide. This scheme corresponds to a concerted reaction of the following type:

Although speculative, this or similar reaction mechanisms could explain the formation of small amounts of chlorinated by-products, such as chloroquinones, on the oxidation of phenol even if pure chlorine dioxide is used.

Therefore, the presence of ammonium is very important to limit the secondary reactions of chlorine when chlorine dioxide is used. This point is often neglected in laboratory investigations, and it can be favorable to dose trace ammonium or monochloramine in water to be treated by chlorine dioxide. The simplified reaction schemes are indicated in Fig. 5. More detailed reaction pathways are summarized in

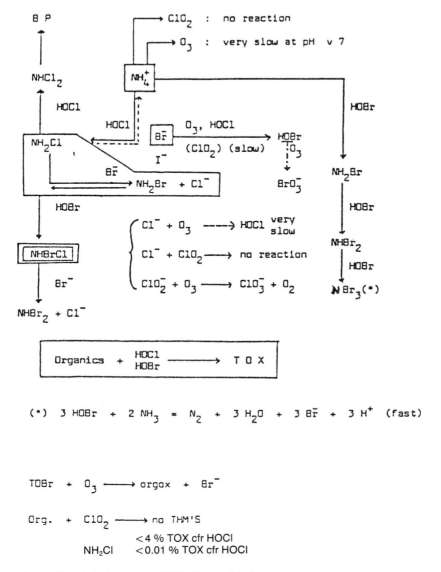

Figure 5 Global aspects of NH_3–Br^- oxidations.

Fig. 6 for chlorine–ammonia reactions and in Fig. 7 for chlorine–bromine–ammonia reactions. These "oxidation puzzles," reviewing the best experimental data available in the literature, indicate several orientations.

3.2 Observations

Monochloramine is the predominant product to be formed except when organic primary amino functions are present. The latter react faster with HOCl or HOBr than does NH_3. The oxidation of bromide to hypobromous acid is slower; however, it is faster than the secondary oxidation of monochloramine to dichloramine. The

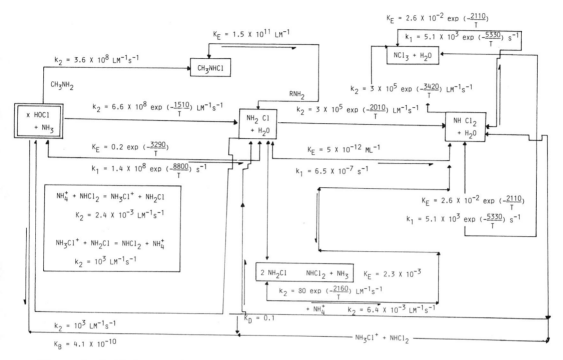

Figure 6 Oxidation puzzle of HOCl–NH$_3$ ($t = 25\,^\circ$C and pH 7 unless otherwise indicated).

exchange NH$_2$Cl + Br$^-$ = NH$_2$Br + Cl$^-$ is very fast and the equilibrium is in favor of monobromamine, which is likely to be an important part of the residual in bromide-containing waters.

The secondary formation of NHBrCl is a reasonable hypothesis, but further investigations are necessary to establish this molecule as a significant residual. Other oxidation reactions are indicated in the schemes, although they are less likely to occur in situations in which the initial concentrations of HOCl, as a secondary oxidant, are low.

Formation of organic bromamines occurs in the same order of velocity as the exchange giving monobromamine, resulting from monochloramine and bromide. Hence a limiting step for the formation of organic aminobromide compounds is the concentration of the bromide ion and to a greater or lesser extent, oxidation of the latter to hypobromous acid.

Little or nothing is known about the exchanges, such as

$$R-NHCl + NH_3 \rightleftharpoons NH_2Cl + RNH_2$$

In theory, as long as R is an electron donor, the equilibrium will be in favor of organic monochloro (or monobromo) amine. In the case of methylaminochloride an equilibrium constant of 1.5 × 10^{11} L/M in favor of the organic chloramine has

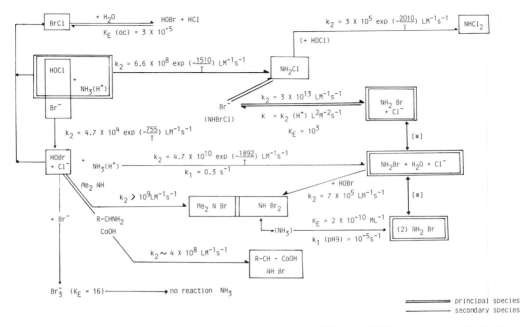

Figure 7 Oxidation puzzle of HOCl–Br⁻–NH₃ ($t = 25°C$ and pH 7 unless otherwise indicated).

been reported. Hence it is likely to be the dominant species in the presence of monochloramine. Further investigation is required to finalize conclusions concerning the organic aminochlorides.

However, the effects of pH must be considered further, as well as the relative proportions of ClO₂ to phenol. Moreover, the analytical techniques used are often iodometric without considering the side reactions or the reaction of quinones or similar oxidants in the analytical conditions applied. In all instances chlorite ion is one of the main secondary reaction products.

Secondary formation of chlorine has been considered in the reaction of indene and other aromatic hydrocarbons. In the case of indene in water at pH 7, 0.25 mol of chlorite is formed per mole of chlorine dioxide consumed. The experimental conditions also enable the disproportionation or intermediate formation of chloric acid according to the reaction (19)

$$2HClO_2 \rightarrow HClO + ClO_3^- + H^+$$

The reaction kinetics of chlorine dioxide with indene is of second order with a constant $k_2 = 2$ L M^{-1} s^{-1}. The reaction products are identified are diols and chlorhydrins.

The analysis of residual products by iodometric methods is often very questionable given that quinones can interfere significantly with iodometry. Specific analysis procedures for each residual reaction product, such as chlorine, chlorine dioxide, and chlorite, indicate, in the case of the oxidation of phenol, the formation of 65 to 90% chlorite versus the chlorine dioxide that has reacted. Hence the major part of the immediate inorganic reaction product of chlorine dioxide is the chlorite ion.

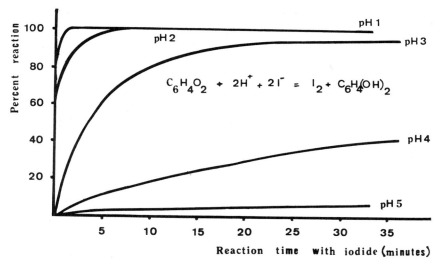

Figure 8   Iodometric titration curves of *para*-quinone (5×10^4 mol/L).

3.3   Reaction Products of Organic Compounds

The reaction products of chlorine dioxide with organic compounds in water have been reviewed extensively elsewhere (1,20,21).

Summary.

1. Practically no THMs are formed on the final treatment of drinking water with pure ClO_2. In the case of the raw water of the river Meuse at the Tailfer plant, treated exhaustively with either chlorine or chlorine dioxide, the level of TOXs formed with chlorine dioxide is less than 4% of that formed with chlorine (22). This formation can be minimized (e.g., if ammonium is present in the treated water source).

2. Phenols and phenolic compounds undergo ring-opening reactions. Di- and tricarboxylic acids and glyoxalic acid are stable end products. At higher concentrations of phenols, of which the 4-position is not substituted, transient formation of quinones is observed. If the ratio of chlorine dioxide is insufficient, secondary chlorination can eventually occur in the case where no ammonium is present. The reaction of the phenate ions is much faster than that of the corresponding phenols (about 10^5 to 10^6 times).

3. Chlorine dioxide reacts with humic acids. The reaction products identified are carboxylic acids, aldehydes, and glyoxal, and if the reaction is forced by over-dosing of ClO_2, there is also the formation of monochloracetic acid in addition to traces of monochlorosuccinic and di- and trichloracetic acid. The decarboxylation of fulvic acid represents about 30% of the original carbon content of the acid oxidized by ClO_2. Fulvic acid exhausts a ClO_2 demand at a ratio of 0.3 mol of ClO_2 per mole of carbon. Methyl esters are also secondary reaction products. Model hydroxylated phenylcarboxy acids are decarboxylated in the first phase of reaction, giving quinones in a following step if only a moderate excess of ClO_2 is used.

4. Aliphatic hydrocarbons do not react significantly with chlorine dioxide under practical conditions of water treatment. Aromatic hydrocarbons such as benzo[*a*]-

pyrene, anthracene, and benzo[a]anthracene react very fast with chlorine dioxide. Quinones are the primary first-formed reaction products. Pyrene, benzo[e]pyrene, naphtalene, and fluoranthene react only very slowly. Some olefins are sufficiently activated to react with chlorine dioxide (e.g., in styrenes, methyl styrenes, and stilbenes). The general reaction scheme then formulated is

$$2 >C=C< \,+\, 2ClO_2 \,+\, 2H_2O \xrightarrow{H^+} \overset{OH}{\underset{|}{>}}C \overset{OH}{\underset{|}{-}}C< \,+\, >\overset{OH}{C} \overset{Cl}{-} C< \,+\, H^+ClO_3^-$$

which, however, implicates an acid disproportionation of chloric acid.

$$2HClO_2 \rightarrow HClO + H^+ + ClO_3^-$$

5. Amino acids without specific reactive groups do not react with chlorine dioxide under normal conditions of water treatment. This is the case, for example, for glycine, alanine, phenylalanine, serine, and leucine. Reactive amino acids are tyrosine tryptophane, histidine, cystine, and methionine, by reaction of either the aromatic part of the molecule or of the sulfide groups. Sulfur-containing amino acids (e.g., cystine) react with chlorine dioxide, the parent sulfonic acid being the final product, with intermediate formation of bisulfoxide. Methionine is converted to the sulfone. The reactions of such amino acids with ClO_2 play an important part in the virucidal action of chlorine dioxide (23).

6. Primary amines do not react significantly with chlorine dioxide. Tertiary amines react on the α-carbon site next to the nitrogen, followed by cleavage of the $C-N$ link. For triethylamine the overall reaction is

$$H_2O + 2ClO_2 + (C_2H_5)_3N \rightarrow CH_3CHO + (C_2H_3)_2NH + 2H^+ + 2ClO_2^-$$

The second-order rate constant is approximately $2 \times 10^5 \, L \, M^{-1} \, s^{-1}$ (24).

Investigations on the potential mutagenicity of reaction products of chlorine dioxide with organic products occurring naturally in water (e.g., humic acids) have proved that this effect is always absent.

3.4 Inorganic Reaction Products

Chlorite is undoubtedly a principal reaction product. In the practice of treatment, between 40 and 60% of the chlorine dioxide reacted is found in the form of chlorite. In model studies this proportion ranges from 50 to 100% according to experimental conditions (e.g., excess, pH, presence of ammonia, etc.). Part of the ClO_2 is transformed into chlorate either through disproportionation or by photodecomposition through artificial irradiation or by sunlight.

The chlorite ion is a less significant oxidant, particularly when reaction with organic compounds or disinfection is concerned. However, some reactions of mineral products are of particular interest in the context of water treatment (e.g., oxidation of manganese salts into manganese dioxide with the formation of chloride). The oxidation does not proceed further than the Mn^{4+} stage, whence eventual overdosing of ClO_2 cannot hinder the demanganization, contrary to the use of other oxidants.

$$2ClO_2 + Mn^{2+} + 4OH^- \rightarrow MnO_2 + 2ClO_2^- + 2H_2O$$

Also,

$$2ClO_2 + 2S^{2-} \rightarrow 2Cl^- + SO_4^{2-} + S\downarrow$$

4. GENERATION OF CHLORINE DIOXIDE

There exist numerous methods of generating chlorine dioxide (1). The best suited for the scale of production capacities necessary for water treatment are based, at present, on sodium chlorite, $NaClO_2$, as a starting product. Commercial sodium chlorite usually contains 2 to 3% sodium chlorate on a weight basis versus chlorite.

4.1 Safety Criterion

A gas phase containing more than 10 vol % ClO_2 is spontaneously explosive and should be handled accordingly in chlorine dioxide generation. The solubility diagrams for chlorine dioxide in water, illustrated in Fig. 9, indicate that at a process water temperature of 20°C, the maximum "safe concentration" range of chlorine dioxide is 8 to 9 g/L. The gas density of chlorine dioxide is about 2.4 times that of air. If higher concentrations must be generated to obtain a global optimum yield, as, for example, 28 to 50 g/L in the acid generation process, the reactors must be operated under reduced gas pressure so as to lower the pressure of ClO_2 in the gas phase beyond the safe limits of explosion.

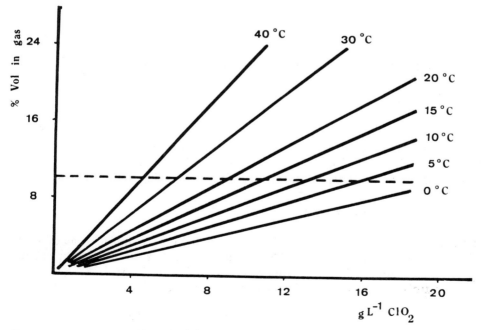

Figure 9 Solubility diagrams for chlorine dioxide.

A general approximate equation applying to the solubility of chlorine dioxide has also been formulated as follows:

$$L = \frac{ClO_2(aq)\ (g/L)}{ClO_2(g)\ (g/L)}$$

with values of L of 70 ± 0.7 at $0°C$, 45 ± 1 at $15°C$, and 26.5 ± 0.8 at $35°C$. The heat of dissolution evaluated by the log L function of T^{-1} is on the order of 26.8 kJ/mol.

4.2   Chlorite–Chlorine Reaction for Generation of Chlorine Dioxide

Hypochlorous acid and chlorine as Cl_2 both react with sodium chlorite solutions by fast reactions giving chlorine dioxide. Aieta has shown (25) that direct reaction of chlorine with dissolved chlorite is faster than chlorine hydrolysis.

$Cl_2 + 2ClO_2^{-1} = 2ClO_2 + 2Cl^-$
k_2 (zero ionic strength) $= 1.62 \times 10^4$ L M^{-1} s^{-1}; at 4 mol/L ionic strength:
$k_2 = 1.12 \times 10^4 M^{-1} s^{-1}$
Activation energy: 40 ± 5 kJ/mol
$$k_2 = 1.31 \times 10^4 \times \exp\left(-\frac{40}{RT}\right)$$

This second-order reaction is favored by chloride ions and assumes (25,26) an intermediate state such as

$$2\left|Cl - Cl \diagdown^{O}_{O}\right| = 2ClO_2 + 2Cl^-$$

Competition can occur through the secondary reaction

$$\left|Cl - Cl \diagdown^{O}_{O}\right| + H_2O = ClO_3^- + Cl^- + 2H^+$$

Under normal conditions the half-lifetime of the direct reaction of chlorine on chlorite has a value of 10^{-5} s.

When dissolved in water, chlorine is hydrolyzed according to the reaction

$$Cl_2 + H_2O = HOCl + HCl$$

The reaction is of first order, $k_1 \simeq 12.8$ s^{-1} (at $20°C$) (i.e., a half-lifetime of 10^{-2} s). The reaction kinetics is only slightly dependent on pH; $k_1 = 12.7$ s^{-1} at pH 10 and $k_1 = 12.9$ s^{-1} at pH 3. The activation energy is 60 ± 12 kJ/mol.

The hydrolysis constants of chlorine in water are reported as

$$Cl_2 + H_2O = HCl + HOCl \qquad k_H = \frac{|HOCl|\ |H^+|\ |Cl^-|}{|Cl_2(aq)|}$$

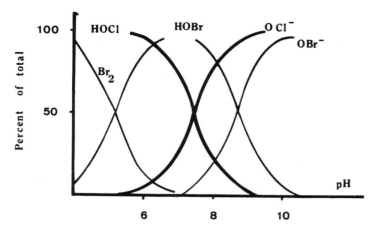

Figure 10   HOCl–ClO⁻ equilibrium in water (compared with dissolved bromine).

$t(°C)$	0	15	25	35	45
kH	1.46×10^4	2.81×10^4	3.94×10^4	5.1×10^4	6.05×10^4

Hence, in dilute solutions, hydrolysis is about quantitative.
    The acidity constants of hypochlorous acid are

$t°C$	5	10	20	30	35
pK_a	7.754	7.69	7.582	7.497	7.463

This constant determines the respective domains of existence of hypochlorous acid and hypochlorite as indicated in Fig. 10. Except at very low pH values and very high concentrations of total chlorine, "gaseous dissolved Cl_2" does not significantly exist in water, contrary to bromine, for example. No significant formation of chlorine dioxide is obtained on reaction of ClO^- with ClO_2^- (33). Marginal formation of chlorate is observed.
    Relevant redox potentials in the generation of chlorine dioxide are listed in Table 2.
    The reaction of hypochlorous acid with ClO_2^- can be summarized according to the literature (26) as

Table 2   Relevant Redox Potentials in Chlorine Dioxide Generation[a]

$HClO_2 + 3H^+ + 4e = Cl^- + 2H_2O$	1.57
$HOCl + H^+ + 2e = Cl^- + H_2O$	1.49
$Cl_2 + 2e = 2Cl^-$	1.36
$ClO_2(g) + e = ClO^-_2$	1.15
$ClO^-_3 + e + 2H^+ = ClO_2 + H_2O$	1.15
$ClO_2(aq) + e = ClO^-_2$	0.95
$ClO^- + 2H_2O + 3e = Cl^- + OH^-$	0.9
$ClO^-_2 + 2H_2O + 4e = Cl^- + 4OH^-$	0.78

[a]$E°$ in volts at 25°C.

$$2H^+ + 2ClO_2^- + HClO \rightarrow 2ClO_2 + H_2O + HCl$$

The authors also suggest $HClO_2$ as the reactive form of chlorite. At neutral pH, however, $HClO_2$ hardly exists:

$$HClO_2 \leftrightarrow ClO_2^- + H^+ \qquad K_a = 1.1 \times 10^{-2}$$
$$H_2CO_3 \leftrightarrow HCO_3^- + H^+ \qquad K_a = 4.07 \times 10^{-7}$$
$$HClO \leftrightarrow ClO^- + H^+ \qquad K_a = 3.3 \times 10^{-8}$$

The consequences of these dissociation constants are indicated in Fig. 11.

All experimental data converge to consider the formation of chlorine dioxide at neutral pH according to the global stoichiometry as probable:

$$HOCl + 2ClO_2^- = 2ClO_2 + Cl^- + OH^-$$

or

$$2NaClO_2 + HOCl + HCl = 2ClO_2 + 2NaCl + H_2O$$

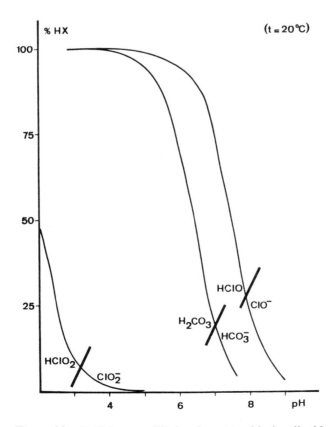

Figure 11   Acid–base equilibria relevant to chlorine dioxide generation. (From Ref. 28.)

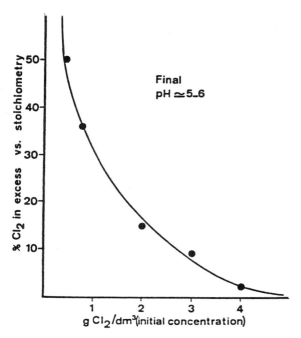

Figure 12 Excess chlorine necessary for > 95% conversion of chlorite to chlorine dioxide. (From Ref. 28.)

The question also relates to the final pH value and the excess chlorine necessary to obtain complete reaction of chlorite (see Fig. 12). Commercial sodium chlorite solutions contain excess alkalinity for stabilization. About 0.045 mol H^+ is currently necessary per mole $NaClO_2$ to neutralize the "excess alkalinity" of commercial chlorite. In tropical versions of $NaClO_2$ (i.e., at concentrations up to 400 g $NaClO_2$ per liter), 0.06 mol H^+ per mole $NaClO_2$ is sometimes necessary.

At too low chlorine concentration the stoichiometry is not obtained, and due to partial reaction, chlorite remains in the reactor effluent. A minimum concentration, or an excess of chlorine, is necessary (33) to obtain a yield higher than 95%. In most practical conditions the final pH is about 5 units and may depend slightly on local conditions, such as the alkalinity of the process water.

The gas-to-liquid chlorine concentrations at equilibrium are expressed by Henry's law as

$$P_{Cl_2} = H_0[Cl_2]$$

where P_{Cl_2} is the partial pressure in chlorine (atm), H_0 is Henry's constant at zero ionic strength of the solution (atm L mol), and $[Cl_2]$ is the dissolved chlorine concentration at equilibrium in mol/L. Direct chlorine solubilities (i.e., without hydrolysis) have been measured by Aieta (25). Henry's constant increases with temperature (Table 3) and is also influenced by the ionic strength of the solution. Dissolution of chlorine in water is exothermic, contrary to hydrolysis, which is endothermic.

Table 3   Henry's Constants for Chlorine[a]

$t(°C)$	H_0 (dist. H_2O)	h_{HCl}	h_{NaCl}	Heat of dissolution (kJ/mol)
5	(700)	(−0.039)	(0.084)	(32.5)
10	8.929	−0.03686	0.08052	30.75
15	10.929	(−0.0341)	(0.079)	(29.0)
20	13.210	−0.03395	0.07772	28.24
25	16.207	(−0.0338)	(0.076)	27.0
30	19.157	−0.03378	0.07555	25.73
40	(25.8)	(−0.033)	(0.0749)	23.0

[a]Values are measured except for those in parentheses, which are extrapolated.

In solutions of electrolytes of ionic strength I (mol/L) the Henry's constant can be evaluated by the equation

$$\log_{10} \frac{H}{H_0} = hI$$

in which h is the specific salting-out parameter (L/mol). In solutions of mixed electrolytes, log additivity of individual salting-out effects can be applied:

$$\log_{10} \frac{H}{H_0} = h_1 I_1 + h_2 I_2 + \cdots + h_n I_n$$

As for the ionic strength, the general formula applicable is $I = \Sigma_n C_n v_n^2$, where n is the valence number of the ions present; for example, $n = 1$ for HCl and NaCl, which are the dominant salts in the generation of chlorine dioxide starting from chlorite.

The concept of "dissolved chlorine" cannot be considered in practice without taking into account the chemical forms in which the dissolved chlorine exists. At low temperatures or at high partial pressures, chlorine hydrate may be dominant: $Cl_2 \cdot 6H_2O$. The most prevalent form is hypochlorous acid–hypochlorite, resulting from the hydrolysis. Practical workable solubilities as a function of temperature and "total partial pressure" (i.e., the sum of the partial pressures of chlorine and water vapor) are given in Table 4.

A method has been described in which the reaction medium is acidified by supplementary acid to obtain a final pH of about 2.5 (29). The absolute concentrations of reactants are, however, not specified in this investigation. The generation of chlorine dioxide with the ternary system $NaClO–NaClO_2 + H^+$ is possible in very controlled conditions. The maximum obtained yield is 95 to 96% in an optimal final pH range of 4.8 to 5.2 for effluent concentrations in ClO_2 between 3 and 5 g/L^{-1} (Fig. 13). With excess chlorine the yield over 95% is increased only marginally. The difference from 100% is due to chlorate (28). The question of indirect ($Cl_2 + H_2O$) and complementary acidification is of importance in considering numerous side reactions that can occur.

Table 4   Practical Solubilities of Chlorine

Partial pressure (mmHg)	Temperature (°C)					
	0	10	20	25	30	40
5	0.49	0.45	0.44	0.43	0.42	0.41
10	0.68	0.60	0.575	0.56	0.55	0.53
50	1.72	1.35	1.18	1.145	1.11	1.025
100	2.79	2.07	1.69	1.63	1.57	1.42
200	5.71	3.34	2.70	2.50	2.34	2.06
300	[a]	4.54	3.62	3.23	3.00	2.61
400		5.73	4.47	3.96	3.63	3.14
500		7.00	5.27	4.66	4.21	3.63
600		8.08	6.06	5.40	4.84	4.07
700		9.12	6.87	6.10	5.42	4.50
800		10.05	7.64	6.77	6.04	4.95
900		10.85	8.35	7.41	6.61	5.39
1000		11.62	9.06	8.12	7.18	5.83
1200		[a]	10.44	9.37	8.30	6.67
1500			12.43	11.19	9.95	7.96
2000			17.07	15.02	12.97	10.13
3000			[a]		18.70	14.45
4000					24.7	18.84
5000					30.8	22.3

[a]Formation of hydrate.

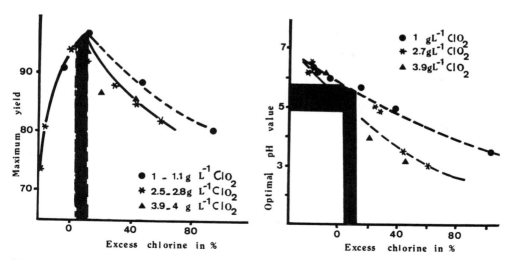

Figure 13   Excess chlorine and optimum pH in the ternary generation of chlorine dioxide. (From Ref. 27.)

Excess chlorine in the generation of chlorine dioxide may be detrimental due to oxidation into chlorate, particularly in an acid (pH < 5) medium.

$$H_2O + 2ClO_2 + HOCl = 2ClO_3^- + 2H^+ + HCl$$

$$\frac{d[ClO_3^-]}{dt} = 2k_2[HOCl][ClO_2] \qquad \text{with } k_2 = 0.021 \text{ L } M^{-1} s^{-1} \ (25\,°C)$$

The number of moles of chlorine dioxide that is reacted is, however, always slightly higher than the stoichiometric number of moles of hypochlorous acid. A proposed secondary reaction therefore is

$$2ClO_2 = Cl_2 + 2O_2$$

4.2.1 Design Rules for Cl₂ (HOCl) + Chlorite Reactors

The technologies of reactors for the generation of chlorine dioxide by oxidation of chlorite with chlorine are generally based on the use of prehydrolyzed chlorine (i.e., HOCl) (31). Figures 14 and 15 illustrate the components of a single flow-through reactor capable of producing 0.5 to 4.5 kg/h. Nominal dosing capacities are:

Chlorine: 0 to 3 kg/h
Dissolution water: 0 to 1m³/h } 1 to 4 g Cl₂/L
Sodium chlorite 25%: 0 to 20 L/h
Dilution water (optional): 0 to 5 m³/h, which can also be operated to rinse the reactor for standby

The chlorine is taken in from the general distribution lines of gaseous chlorine at 0.7 bar installed at the plant. The water for dissolution is available for ejector feed at an upstream pressure maintained in the range 2 to 4 bar. Chlorite is dosed with a Teflon-membrane metering pump into the ND 15-mm-diameter inlet pipe of the reactor. All necessary equipment to prevent backflow is provided for in the lines. The sodium chlorite is transferred from bulk storage tanks (2 × 25 m³) into an operational storage tank (PE or PVC) of 300 L pro reactor line installed. This subdivided storage is designed with a constant-level overflow for return of the excess chlorite into bulk storage. The flows of dissolution water and chlorite are measured with electromagnetic flowmeters (e.g., Fischer & Porter 10 D 1420 A and 10 D 1418 A) and the chlorine flow with a differential pressure flowmeter (e.g., Fischer & Porter 10 B 2490 S or Rota D 7867).

The flow of dilution water is measured with a conventional flowmeter. Chlorine is best dosed with positive-pressure chlorinators (e.g., Fischer & Porter 70 N 2919) or conventional reduced-pressure equipment provided with the necessary injector restrictions to reach the desired concentration range for chlorine concentration and water flow (e.g., Wallace & Tiernan U 91653). The reactor is a borosilicate glass column of total column volume 80 L filled with Raschig rings of diameter 12.5 mm or PE berl saddles 1 in. in size.

The flow-through pattern of the reactors is indicated in Fig. 16. The reactor effluent is dosed directly in the raw water, without intermediate storage. The flow of 0.75 m³/s to be treated is sufficiently constant to adopt this technique. The

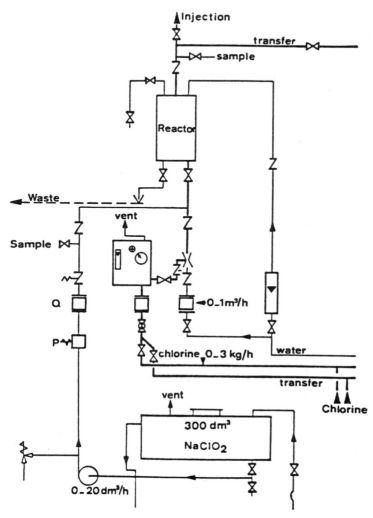

Figure 14 Schematic of the ClO$_2$ generator at the Tailfer plant of the Brussels Waterboard. (From Ref. 28.)

residual active oxidants are measured after reaction, the necessary dosing rate is adjusted manually, and the operation is controlled by recording the various flows of the chemicals involved in the generation. The analysis required to establish the ClO$_2^-$–ClO$_3^-$ balance is part of the daily routine work of plant operation control. Designed, installed, and operated continuously since 1973, the system has proven to be reliable over the long term. Flow-through patterns are best when close to plug-flow conditions, as illustrated in Fig. 16. A reaction time of 1 min is sufficient, but to account for mixing conditions a nominal retention time of 4 to 6 min is recommended.

4.2.2 On–Off Generating Systems

To adapt the chemical dosing rates as a function of variations in ClO$_2$ demand, the reactor is sometimes overdimensioned to obtain continuously correct mixing

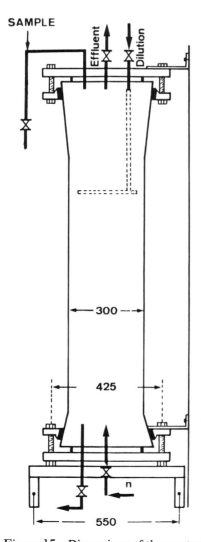

Figure 15   Dimensions of the reactor column at the Tailfer plant. (From Ref. 28.)

conditions. The reactor is operated on an on–off basis to fill an intermediate storage tank. High-performance, fast-acting security systems to control the start–stop sequences are necessary for the operation of intermittent generating reactors (a response time within 2% of the average nominal residence time in the reactor zone is necessary). The reactor operation time is kept at least at 25 to 30% (i.e., standby periods of less than 75% of the total cycle of operation). The dilution ratio of the reactor effluent must lower the concentration in chlorine dioxide to under 1 g/L.

The system installed at service reservoirs in the Brussels area is illustrated in Fig. 17. Under normal conditions, the reactor is operated at a maximum 0.75 m³/h flow of process water, which is equal to a theoretical transit time of 300 s. The

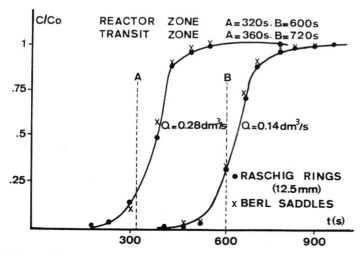

Figure 16 Flow-through curves for the Tailfer reactor. (From Ref. 28.)

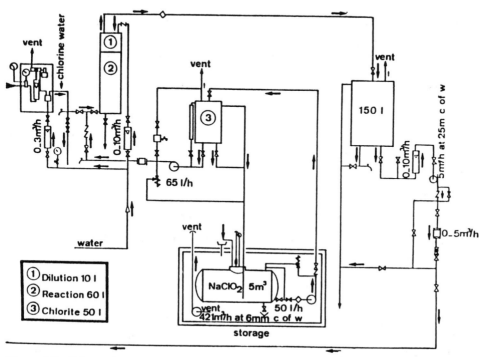

Figure 17 Schematic of discontinuous ClO$_2$ generators. (From Ref. 28.)

reactor is normally operated at 2 to 2.5 g Cl_2/dm^3 to produce a concentrated chlorine dioxide solution of 4 to 5 g ClO_2/dm^3, which is then diluted 10-fold for storage in the intermediate dosing vessel.

The general guidelines for start–stop procedures of $Cl_2/NaClO_2$ reactors are:

1. To start the process water for dissolving chlorine, no chlorine dosing may be possible without the availability of process water. A pressiostatic guard and/or minimum water flow detector ensures this function.
2. To dose chlorine and set the concentration at the necessary value. An analysis of the chlorine content before starting $NaClO_2$ dosing is mandatory.
3. To operate the dosing of $NaClO_2$ solution at an appropriate rate.
4. To equip all dosing lines with an accurate backflush valve system, avoiding any backflow from the reactor to the chemical dosing lines.
5. To ensure a stop procedure involving shutdown of the dosing of reagents in the order $NaClO_2$–chlorine–process water.
6. To rinse the dosing lines and reactor with water preceding standby periods.
7. To measure the flow of each reagent, including process water, and indicate data locally. Transmission and continuous recording into a remote operational laboratory center are optional.
8. In the most important treatment designs, all the start procedures are to be operated locally; the stop procedures are operated automatically either remotely or locally.

The published data for the direct reaction of molecular chlorine with sodium chlorite to form chlorine dioxide (25,32) always involves reactions with chlorine in excess to stoichiometry, although industrial claims have been made for production at stoichimetric proportions with a chlorine dioxide yield up to 95 to 98%. To investigate this alternative, one of the Tailfer reactors was modified according to the scheme shown in Fig. 18. In the experiment chlorite was injected through valve 2 and pump 2, and the chlorine dioxide concentration was analyzed at sampling points 1 (reaction time ± 5 s) and 2 (reaction time ± 400 s). No significant difference in yield was found between the two sampling points.

The yield curve is given in Fig. 19. With stoichiometric ratios at the usual concentration range of 4 g/L ClO_2 to be produced, the yield ranges about 95%. Nearly 100% can be obtained with a slight excess of chlorine (e.g., 6%). In summary: Direct reaction enables us to simplify the reaction columns if appropriate mixing conditions exist. Further developments of the process can open up interesting perspectives.

4.2.3 Acidification of Chlorite as a Process of Generation of ClO_2

An alternative process for chlorine dioxide generation is based on direct acidification according to the reaction

$$5NaClO_2 + 4HCl = 4ClO_2 + 5NaCl + 2H_2O$$

Chlorhydric acid introducing a common ion in the system is more favorable than for other acids. Competition by this reaction, giving four molecules of chlorine dioxide for five moles of chlorite engaged, is to be avoided in the chlorine–chlorite generation process.

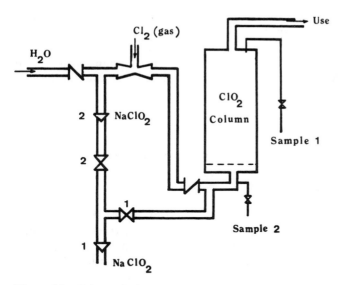

Figure 18 Schematic for injection of NaClO₂ prior to chlorine in the dosing line of a ClO₂ reactor.

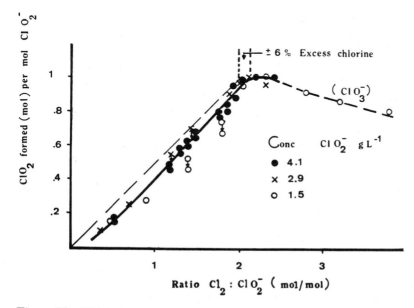

Figure 19 Yield of direct reaction of gaseous chlorine on a solution of sodium chlorite.

The reaction kinetics of the system $NaClO_2$ + HCl remains incompletely investigated, but the reaction is much slower than that of chlorine with chlorite and is temperature dependent (33). Moreover, to operate the process quantitatively, high concentrations are required. The reaction is operated under pH conditions in which $HClO_2$ is the dominant form of chlorite (28). Thus, in addition to the reaction of synthesis,

$$5HClO_2 = 4ClO_2 + Cl^- + H^+ + 2H_2O$$

a secondary reaction with less formation of chlorine dioxide has been reported:

$$4HClO_2 = 2ClO_2 + ClO_3^- + Cl^- + H^+ + H_2O$$

Increased concentrations in chloride ion lower the contribution of the latter reaction, hence decrease the concentration of chlorate as a secondary reaction product.

The most important decay reaction for neutral or slightly acid highly concentrated solutions of chlorine dioxide is summarized by the global equation

$$6ClO_2 + 3H_2O = 5HClO_3 + HCl$$

Therefore, the highly concentrated solutions are best diluted to concentrations below 1 g/L.

4.2.4 Design Rules for HCl + $NaClO_2$ Reactors

To approach a theoretical yield of $4ClO_2$ versus $5NaClO_2$, several conditions must be fulfilled:

1. An excess of HCl is required (e.g., by working at an equal weight of HCl and $NaClO_2$.
2. On dilution of $NaClO_2$ with process water, the precipitation of $CaCO_3$ hinders the transfers, whereas unless softened process water is used, a concentrated solution of $NaClO_2$ is best injected in prediluted HCl.
3. In laboratory investigations at atmospheric pressure and in open vessel reactors, the reaction is best carried out in a tubular vessel with a low ratio of liquid surface-to-volume proportion (e.g., a titration burette).
4. The best working conditions in full-scale continuous reactors are obtained when carrying out a reacting mixture containing ≥ 45 to 50 g of $NaClO_2$ and 50 g of HCl per liter (i.e., 300% of the stoichiometric quantity of the acid). Practical proportions can range from 250 to 300% of the stoichiometric amounts of HCl.
5. Instant mixing is essential to obtain a good yield. Therefore a built-in mixing baffle in the reacting zone favors the reaction.
6. The pH value of a reactor effluent containing up to 30 g ClO_2/L is in the range ≤ 0.5. If the pH is above 1, the reaction is too slow and only partial yields are obtained.
7. Owing to the high ClO_2 concentration, the entire process is to be operated under vacuum as indicated in Fig. 20. In the case of a lack of process water or vacuum, all security devices must stop the dosing.

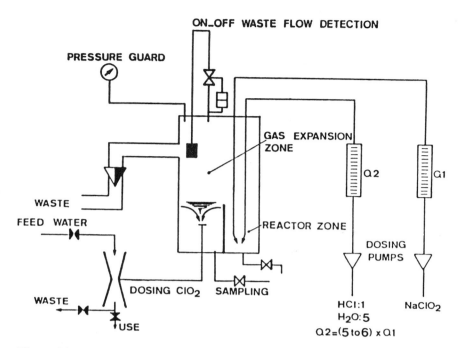

Figure 20 Schematic of ClO_2 generation by reaction of $NaClO_2$ with HCl.

A typical reactor is illustrated in Fig. 21. It has a generating capacity of 2.2 kg/h at nominal capacity and a residence time in the reactor zone of 300 s. The yield is higher than 95%. Part of the chlorine dioxide is sucked from the reactor as a gas, and another part as a solution. The nominal detention time in the reactor zone is a compromise between the generation of ClO_2 and its stability in highly concentrated acid solutions. The compromise ends in a reaction time between 200 and 600 s at a process water temperature of 15°C (Fig. 22).

A more recent study (33) of the reaction conditions of the synthesis of chlorine dioxide with sodium chlorite and a high excess of chlorhydric acid has produced evidence of the temperature effect on the reaction kinetics. Particularly at lower concentrations (e.g., 10 to 20 g/L ClO_2 in the process water), the reaction can be considerably slowed down (e.g., by a factor of 3) when the temperature is 10°C (respectively, 20°C). In all instances, at these concentrations, the reaction time was considerably longer at 10°C than the design zone we have recommended. At first sight the conclusions were advanced as possibly being in contradiction with our data (28).

However, a more careful examination of the data can give the representation in Fig. 23. The design zone—about 3 to 6 min reaction time for a concentration of 28 to 38 g/L ClO_2 in the reactor effluent—is in fact confirmed provided that the reaction is operated with 280 to 300% excess of chlorhydric acid versus the stoichiometric concentration. In these conditions the temperature of the process water remains of marginal importance at least as far as water temperatures in the range of drinking water are concerned. At lower concentrations of ClO_2 (respectively, of

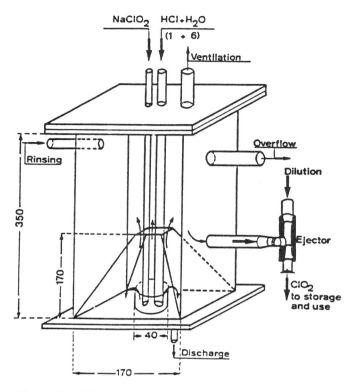

Figure 21 Diagram of truncated-pyramid reactor (Uccle reservoir, Brussels Waterboard). (From Ref. 28.)

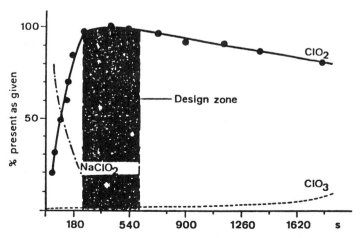

Figure 22 Reaction time for sodium chlorite (45 g/L) and hydrochloric acid (56 g/L) in a batch reactor ($t = 15°C$). (From Ref. 28.)

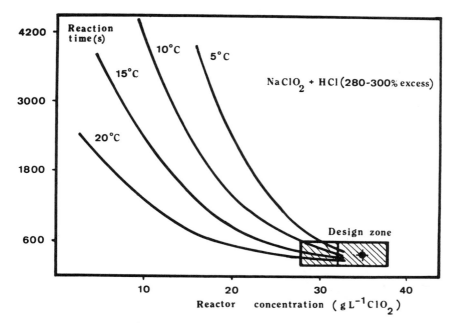

Figure 23 Temperature effect on the reaction $NaClO_2 + HCl$.

ClO_2^-) the time necessary to complete the reaction depends very significantly on the process water temperature. Moreover, the boiling point of chlorine dioxide is about 11°C.

To meet the needs of ClO_2 variable in time, discontinuous generation with storage after dilution is feasible, as in the case of the $NaClO_2$–Cl_2 process.

Constructions that make it possible to produce variable quantities of chlorine dioxide are, for example, the Ben-Ahin reactor of the Brussels Waterboard (Fig. 24) The reactor has two different regimes of operation, depending on the position (open–closed) of valve V_1. The dynamic range of the reactor is of 12 to 150 g ClO_2/h. The generation of high concentrations of ClO_2 in the reactor zone of acid-chlorite reaction needs high-level securities for starting procedures.

General Guidelines for Start–Stop Procedures.

1. Start the process water entering the reactor and set at the appropriate flow.
2. Start the ejector for dilution to obtain the necessary vacuum in the reactor system (to be obtained and controlled before step 3).
3. Start the dosing of chlorhydric acid, the reactor effluent being conducted to waste lines.
4. Start the dosing of chlorite, the effluent being conducted to waste lines.
5. After at least a transit of two reactor volumes, make the necessary analytical control and fine-setting of the dosings.
6. Direct the ClO_2 stream to the water to be treated.
7. Set all the automatic securities and relays in the appropriate positions, which means no dosing if vacuum or ejector flow is inadequate, stop all dosings if process dilution water is inadequate, continue ejector flow when dosings are inadequate, and switch over to a waste line in case of anomaly.

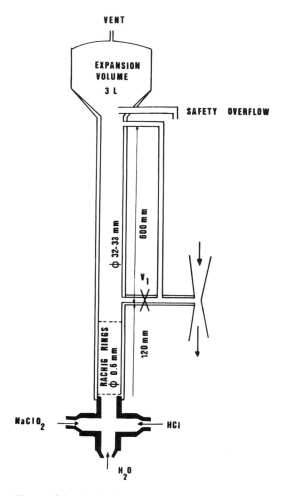

Figure 24   Variable-rate reactor for generation of chlorine dioxide at Ben-Ahin.

The stop procedure is as follows:

1. Direct the reactor effluent to the waste line.
2. Stop the dosing of chlorite and chlorhydric acid.
3. Continue to rinse the system with the process water, under eventual increased flow to speed up the cleaning by flowing-through two to three reactor volumes. (For prolonged standby, all the dosing lines should be purged and rinsed.)

4.2.5   Generation of Chlorine Dioxide by Reduction of Chlorate

Chlorate is widely available due to its use as a herbicide. There exist numerous methods to generate chlorine dioxide starting from chlorate (1). Most of these methods operate under strong acid conditions and the direct acidification methods deliver considerable amounts of chlorine due to secondary reactions:

$$2ClO_3^- + 4HCl = 2ClO_2 + 2H_2O + 2Cl^-$$
$$ClO_3^- + 6HCl = 3Cl_2 + 3H_2O + Cl^-$$
$$\text{overall:} \quad 8ClO_3^- + 24HCl = 6ClO_2 + 9Cl_2 + 12H_2O + 8Cl^-$$

Direct reduction of chlorate with sulfur dioxide has been described according to the stiochiometry (34):

$$2NaClO_3 + SO_2 = 2ClO_2 + Na_2SO_4$$

Gaseous sulfur dioxide is reacted with a concentrated solution of sodium chlorate (e.g., 40 wt%). The reaction time is in the range of 30 min. Water cooling of the reactor is necessary to prevent danger of explosion (e.g., to lower the temperature to 65 to 70°C). A reaction mixture of ClO_2 gas and ClO_2 liquid is sucked continuously from the reactor with an ejector, creating a vacuum of at least 0.5 to 0.7 bar negative pressure. This precaution is essential to avoid explosions in view of the fact that the concentration theoretically possible ranges up to 250 g/L (i.e., far above the limits of solubility and acceptable safe equilibrium concentration limits in the gas phase).

The method could be of some interest, given that chlorate is readily available and much cheaper than chlorite. Up to now there have been no significant small-scale applications such as those necessary for the disinfection of drinking water. If not injected into the water immediately, intermediate storage requires preliminary dilution to meet the stability conditions of stored ClO_2 stock as discussed for the HCl–chlorite method of generation.

5. TOXICITY OF CHLORINE DIOXIDE AND RELATED OXYCHLORINE COMPOUNDS

In view of the considerable interest in chlorine dioxide as an oxidant for water treatment, and eventually as an alternative to chlorination, extended investigations on the toxicity of residual chlorine dioxide, chlorite, and chlorate are considered. They can be classified broadly in two groups: the potential effects of the residual chlorine dioxide itself, and the impact of its inorganic reaction products.

5.1 General Context

The best evidence is that obtained by the use in Europe for over 20 consecutive years of chlorine dioxide to disinfect drinking water and also by experimental investigations carried out on human volunteers (35–38). Traditional toxicity tests on animals have been investigated as models, with some advantages claimed, such as the easier control of dietary intake and the daily activity of animal subjects. In most investigations on animals, the dosage level has been increased to accelerate and amplify the physiological responses. It does appear, however, that under such conditions sufficient care was not always taken to be in line with one of the fundamental principles of toxicology: There are no toxic products, only toxic doses. Some effects, produced under exaggerated or even forced-feeding conditions and diet, are unlikely to occur in normal food-intake situations.

5.2   Summary of Evidence

The major evidence obtained thus far is summarized in Table 5. It is surprising how very few toxicological evaluations concern the ingestion of chlorine as a residual disinfectant in drinking water (44), although the hypochlorite ion has been shown to promote sperm abnormalities in mice fed with drinking water. The same did not occur with chlorine dioxide, chlorite, or chlorate (52).

5.3   Existing Legislation for Water

The residual concentration of chlorine dioxide in Belgium is 0.25 mg/L; in Germany, 0.2 mg/L; and in Switzerland, 0.15 mg/L. Another approach is that of the *suggested no-adverse-response levels* (SNARLs), published in 1987 by the U.S. National Academy of Sciences in 1987. The values calculated are based on the assumption that 20% of the daily intake comes from ingestion of drinking water, and a safety factor of 10 was taken into consideration. On this basis figures proposed as the maximum admissible concentrations were 0.21 mg/L for ClO_2 and 0.024 mg/L for chlorite and chlorate ions. However, the experimental evidence was poor and not positively established. Indeed, known existing concentrations proven not to have any effect are accepted as such without considering that higher concentrations might have no effect either.

It is also remarkable that chlorite and chlorate ions are considered indistinctly, although the toxicity levels are probably very different. A lack of accurate knowledge of the real concentrations present is probably the reason for these assumptions. The EPA [*Fed. Regist.*, *52* (212), 42177–42222 (1987)] did not accept the approach of the U.S. National Academy of Sciences, given that none of the compounds considered is expected to determine the average daily intake other than through drinking water. On the other hand, a safety factor less than 10 could be adopted in the future. Assuming that nearly 100% of exposure comes from drinking water, the SNARL is estimated at 1 mg/L, including chlorine dioxide and its oxidation–reduction products, chlorite and chlorate.

The association of chlorate toxicity with that of chlorite is certainly questionable, and the analytical evidence for chlorate is often poor. Whatever the conclusions, it appears that at present much more is known on the toxicity of chlorine dioxide and related compounds in water than it is for chlorine in water.

6.   CONCLUSIONS

The disinfecting capacity of chlorine dioxide has been well recognized for three-quarters of a century and is currently used for postdisinfection or safety treatment of drinking water in several major European cities. All tests made thus far have indicated the absence of relevant toxicity of reaction or by-products of the treatment when performed properly. In addition to the reaction of inorganic compounds such as iron, manganese, and sulfide, organic phenolic compounds are modified to give no additional organoleptic hindering compounds. Disinfection capacity of chlorine dioxide is far higher than that of chlorine. When starting with commercial sodium chlorite, adequate generating methods exist to meet any needs existing in drinking water plants and distribution systems. Some precautions are necessary for safe handling.

Table 5 Summary of Toxicity Data on Chlorine Dioxide and Related Ions in Water Treatment (1987)[a]

Effect and Parameter Unit	ClO_2	ClO_2^-	ClO_3^-	Remarks
Lowest effect level (water mg/L)	12 (35–38)	1.2 (35)	1.2 (39)	Acute effect: 1000–15000 for ClO_3^- (40)
NOAEL (mg/kg/day)	1 (rats)	0.7–1 (rats)	—	
LD_{50} (mg/kg)	140 (rats)	140–200 (rats)	200 (humans)	
Chronic toxicity				
mg/day	None at 5 (37)	None at 5 (37)	None at 5 (37)	5 as sum of the three parameters (humans)
mg/L	None at 10 (41) positive at 30 (41)	—	—	All reversible up to 1000
Methemoglobinemia through water (mg/L)	100 (42,43)	50 (none at 10) (44)	10–100 (rats, chickens)	Distinction between methemoglobinemia and hemolysis is arguable
Hemolytic effects through water (mg/L)	None at 2, positive at (44) 15 pigeons 100 chickens (42) 30 monkeys	100 (rats) (42), none up to 70 (42), stress at 50 (44)	6–300 (monkey) (44)	
Glutathion loss (mg/L)	50 (42)	50 (42)	10–100 (42)	Tests in vitro; no effects observed on humans (38)
Effect on kidneys (mg/L)	—	None at 100 (mice)(45)	—	
Embryotoxicity (mg/kg/day)	—	None at 4 (46), positive at 14 (rats)	—	
Mortality pregnant rats (mg/kg/day)	—	None at 20 (46), positive at 100	—	
Teratogenic effects (mg/kg/day)	NOAEL: 1 (rats), 10 (rats) (42,46)	NOAEL: 1 (rats), 10 (rats) (46)	NOAEL: 1 (rats) 10 (rats) (46)	
Carcinogenic effects (mg/L)	—	—	None at 10 (44)	
Increased intestinal turnover (mg/L)	>10	>10	Not observed	
Body weight effect (mg/kg/day)	5 (rats) (47)	2 (rats) (46)	—	
Threshold waterbodies (mg/L)	0.25 (48)	—	10 (49,50)	
Organoleptic threshold water (mg/L)	0.4 (47)	0.3 (48)	5 (49,51)	For humans

[a]Doses and/or concentrations as indicated; literature references within parentheses.

NOTATION

h	salting-out parameter, L/mol
n	concentration factor, dimensionless
n	ionic number, dimensionless
v	valence, dimensionless

C	concentration, mol/L, g/L, or ppm
CT	concentration and time in bactericidal action
D_{10}	decade abatement factor, concentration $\times$ time
E°	redox potential in normal condition, volts
H_0	Henry constant, atm $\cdot$ L $\cdot$ mol^{-1}
I	ionic strength, mol/L
L	solubility ratio water to gas, dimensionless

REFERENCES

1. W. J. Masschelein, *Chlorine Dioxide*, Ann Arbor Science, Ann Arbor, Mich., 1979.
2. C. H. Rupp, *Gesund. Ing., 104*, 278 (1983).
3. R. C. Rice, *EPA Workshop*, March 1988.
4. C. I. Noss and V. P. Olivieri, *Appl. Environ. Microbiol., 50*, 1162 (1985).
5. C. I. Noss, W. H. Dennis, and V. P. Olivieri, Water Chlorination, in *Environmental Impact and Health Effects*, R. L. Jolley, Ed., Vol. 4, Book 2, 1983, p. 1077. Ann Arbor Science, Ann Arbor, Mich.
6. Y. Chen, O. Sproul, and A. Rubin, *Water Res., 19*, 783 (1985).
7. P. V. Scarpino, S. Cronier, M. L. Zink, and F. A. O. Brigano, *Proc. 5th Annual AWWA Conference on Water Quality in the Distribution System*, 2 B-3, 1977, p. 1.
8. V. P. Olivieri, F. S. Hauchmann, C. I. Noss, and R. V. Asi, Water Chlorination, in *Environmental Impact and Health Effects*, R. L. Jolley, Ed., Vol. 5, 1985, p. 619. Ann Arbor Science, Ann Arbor, Mich.
9. M. E. Alvarez and R. T. O'Brien, *Appl. Environ. Microbiol., 44*, 1064 (1982).
10. J. Ridgway, *WRC (U.K.), ER 481* (1977).
11. O. Sproul, Y. S. R. Chen, J. P. Engel, and A. J. Rubin, *Environ. Technol. Lett., 4*, 335 (1983).
12. G. R. Taylor and M. Butler, *J. Hyg., 89*, 321 (1983).
13. R. S. Fujioka, M. A. Dow, and B. S. Yoneyama, *Water Sci. Technol., 18*, 125 (1986).
14. M. S. Harakeh, J. D. Berg, J. C. Hoff, and A. Matin, *Appl. Environ. Microbiol., 49*, 69 (1985).
15. D. Berman and J. C. Hoff, *Appl. Environ. Microbiol., 48*, 317 (1984).
16. W. J. Masschelein, *Water S.A., 6*, 117 (1980).
17. W. J. Masschelein, G. Fransolet, and E. Debacker, *Eau Quebec, 14*, 41 (1981).
18. N. Hansen, G. Fransolet, and W. J. Masschelein, *Symposium organisé par l'Association Pharmaceutique Française pour l'Hydrologie*, Lille, Sept. 1989.
19. Ch. Rav. Acha and E. Choshen, *Environ. Sci. Technol., 21*, 1069 (1987).
20. Ch. Rav. Acha, *Water Res., 18*, 1329 (1984).
21. M. Doré, Univ. de Poitiers, Chimie des Oxydants et Traitement des Eaux, *Technique et Documentation*, Paris (1989).
22. R. Savoir, L. Romnee, and W. J. Masschelein, *Aqua, 2*, 114 (1987).
23. C. I. Noss, F. S. Hauchman, and V. P. Olivieri, *Water Res., 20*, 351 (1986).
24. D. H. Rosenblatt, A. J. Hayes, B. L. Harrison, R. A. Streaty, and K. A. Moore, *J. Org. Chem., 28*, 2790 (1963).

25. E. M. Aieta, Ph.D. thesis, Stanford University, 1984.
26. F. Emmenegger and G. Gordon, *Inorg. Chem.*, 6, 633 (1967).
27. R. Halleux, G. Fransolet, and W. J. Masschelein, *Trib. Cebedeau*, 484, 87 (1984).
28. W. J. Masschelein, *J. AWWA*, 76, 70 (1984).
29. R. W. Jordan, A. J. Kosinski, and R. J. Baker, *Water Sewage Works*, 44, (Oct. 1980).
30. R. G. Kieffer and G. Gordon, *Inorg. Chem.*, 7, 235 (1968).
31. W. J. Masschelein, *J. AWWA*, 77, 73 (1985).
32. E. M. Aieta and P. V. Roberts, *Environ. Sci. Technol.*, 20, 44, 50 (1986).
33. H. Overath, K. Th. Oberem, D. Wittich, and H. J. Ammann, *Vom Wasser*, 65, 236 (1985).
34. J. W. Sprauer, U.S. Patent 2,833,624 (1958). *Chem. Abstr.*, 52, 15857 (1958).
35. J. R. Lubbers and J. R. Bianchine, *J. Environ. Pathol. Toxicol.*, 5, 215 (1984).
36. J. R. Lubbers, S. Chauhan, J. K. Miller, and J. R. Bianchine, *J. Environ. Pathol. Toxicol.*, 5, 229 (1984).
37. J. R. Lubbers, J. R. Bianchine, and R. J. Bull, Chap. 95, in *Water Chlorination*, vol. 4, book 2, Ann Arbor Science, Ann Arbor, Mich., 1983, p. 1335.
38. J. R. Lubbers, S. Chauan, and J. R. Bianchine, *Environ. Health Perspect.*, 46, 57 (1982).
39. G. J. Tuschewitzki and J. U. Hohn, *38th Conférence Internationale du Cebedeau*, Brussels, 1985, p. 261.
40. J. O'Grady and E. Jarecsni, *Br. J. Clin. Pract.*, 25, 38 (1971).
41. J. Musil, Z. Knotek, J. Chalupa, and P. Schmidt, *Technol. Water*, 8, 327 (1964).
42. D. Couri, M. S. Abdel-Rahmen, and R. J. Bull, *Environ. Health Perspect.*, 46, 57 (1982).
43. G. S. Moore, E. J. Calabrese, S. R. Dinardi, and R. W. Tuthill, *Med. Hypotheses*, 4, 481 (1978).
44. L. W. Condie, *J. AWWA*, 78, 73 (1986).
45. G. S. Moore and E. J. Calabrese, *Environ. Health Perspect.*, 46, 31 (1982).
46. D. Couri, C. H. Miller, R. J. Bull, J. M. Delphia, and E. M. Ammar, *Environ. Health Perspect.*, 46, 25 (1982).
47. S. A. Fridlyand and G. Z. Kagan, *Hyg. Sanit.*, 36, 190 (1971).
48. Brussels Waterworks, practical experience.
49. I. I. Avezbakiev and N. M. Demidenko, *Gig. Sanit.*, 5, 11 (1979).
50. O. Pravda, *Hydrobiologia*, 42, 97 (1973).
51. D. Stofen, *Städtehygiene*, 24, 109 (1973).
52. J. R. Meier, R. J. Bull, J. A. Stober, and M. C. Cimino, *Environ. Mutagen.*, 7, 201 (1985).

3
Oxidation with Ozone

"Whatever new problems have arisen, ozone has always been part of the solution."

1. IMPROVING WATER QUALITY BY OZONE TREATMENT

1.1 Disinfection

The first full-scale treatment plant for drinking water, set up in Nice in 1906 by Marius Paul Otto, had as its objective disinfection. The bactericidal effects of ozone are only slightly influenced by changes in pH ranging between 7 and 10 (1). The effect is very fast, more or less instantaneous, and no resistant waterborne strains are known.

During experiments of historical significance in 1964, Coin et al. (2) brought to light viral inactivation by ozone. By their work it was proven that a residual ozone concentration level in water of 0.4 mg/L for 4 to 6 min presents a sufficient guarantee of deactivation of polioviruses (3). Moreover, this research implicitly introduced the CT concept—the complementarity of concentration and reaction time—as, for example, correlated by Watson's law:

$$\log \frac{N_t}{N_0} = -kC^n t$$

where N_0 and N_t are the volumetric concentration of germs at times 0 and t, respectively, C is the concentration of disinfectant, and k is the decay constant.

The *CT* concept is at present the underlying principle for the application of

67

ozone taken into consideration by the EPA. It should be pointed out, however, that in each of these experiments the chemical ozone demand of the water has been included in the treatment applied.

Other scientists (4) have sought to overstep this parameter by using a double-column arrangement (see Fig. 1). The first column is designed to satisfy the water's chemical ozone demand. The second is used to study the ozone concentration parameter. Some of the results obtained are shown in Fig. 1: With water of a high organic matter content, even where there is no residual ozone, a certain decrease in virus-level concentration is noted. It becomes even greater as soon as there is residual ozone, even in small quantities.

To summarize, the viral particles are quickly removed (30 s) with a very low level of residual ozone (0.05 mg/L); however, the WHO recommendations include a large coefficient of safety. Later it was demonstrated that other enteroviruses were sensitive as well as polioviruses, in the order: Polio II > ECHO I > Polio I > Coxsackie 85 > ECHO 5 > Coxsackie A9 (5).

The first published evidence of the efficiency of ozone in the killing of waterborne parasites appears to be that of Lagrange and Rayet (6), who experimented with the effects of ozonization on *Schistosoma mansoni* causing bilharziasis in central Africa. Applying a dosage of 0.9 mg/L for 6 min, provided for good disinfection of clear waters. Another term of reference is the killing of the cysts of *Entamoeba histolytica*. The inactivation of *Giardia lamblia* cysts (7) at pH 7 with 0.14 mg/L made it possible to obtain 90% inactivation, respectively, after 1 and 4 min at 25 and 5°C. This is equal to *CT* values of 0.17 mg/L · min at 25°C and 0.55 mg/L · min at 5°C. At present the inactivation of *Giardia* cysts is a criterion for appropriate disinfection proposed as a guideline by EPA (8): *CT* values corresponding to three decades (i.e., 99.9% removal of *Giardia* cysts) and 99.99% or four decades of removal of enteroviruses are required for water systems without filtration. If a filtration step is used, a credit of two decades of decay for *Giardia*

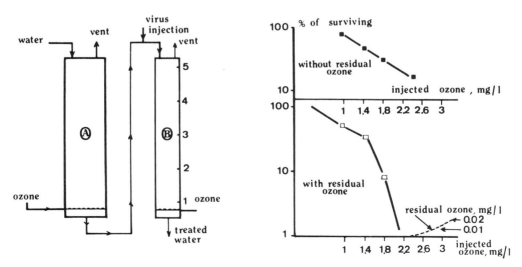

Figure 1   Inactivation of poliomyelitic virus. A, chemical ozone demand; B, inactivation of viruses B_1–B_3.

cysts and one decade for viruses is given for that operation. If a settling phase exists, an additional credit of 0.5 decade for *Giardia* can be given. In both cases (i.e., 1 and 0.5 decade of inactivation of *Giardia*) a four-decade inactivation of viruses is ensured. Accepted *CT* values in mg/L · min in the pH range 6 to 9 are at present set as follows:

t (°C)	0.5	5	10	15	20	25
0.5 log	0.53	0.33	0.27	0.27	0.2	0.13
1 log	1.13	0.67	0.53	0.47	0.33	0.20

This means that the rule used in Europe for the postozonization dosage — 0.4 mg/L residual concentration for at least 4 min (*CT* = 1.6) — also covers the criteria formulated by the EPA in the United States.

1.2 Taste and Odor Control and Color Improvement

Improvement in the organoleptic properties of the water is one of the most relevant effects of ozonization of clear waters through which ozone has gained consumer confidence. This is particularly the case with peaty waters (9); sample data are shown in Fig. 2. In Europe, wherever ozonization has been utilized, remarkable improvement in taste and odor has been observed. In the United States, a particularly well-documented case is the Monroe plant, treating Lake Michigan water (10). However, musty tastes associated with geosmin, including methylisoborneol, are not always improved by ozonization.

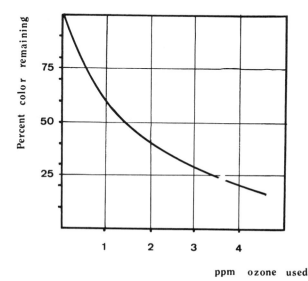

Figure 2   Color removal with ozone. (Data from Ref. 9.)

1.3   Removal of Iron and Manganese

The ozonization of water containing iron or manganese leads to insoluble oxides or hydroxides which are either precipitated or filtrable, as recognized in earlier experiments (11). The oxidization of ferrous iron takes place faster than that of organic substances and produces ferric hydroxide:

$$Fe^{2+} + O_3 + H_2O = Fe^{3+} + O_2 + 2OH^-$$
$$Fe^{3+} + 3H_2O = Fe(OH)_3\downarrow + 3H^+$$

The oxidation of manganese is even more complex (12–14). At stoichiometric ratios or slightly higher, in the pH range 5 to 7, formation of insoluble manganese (IV) oxide results:

$$Mn^{2+} + O_3 + H_2O = Mn^{4+} + O_2 + 2OH^-$$
$$Mn^{4+} + 4OH^- = MnO_2 \downarrow + 2H_2O$$

However, partial overoxidation is difficult to avoid, under which conditions soluble permanganate is formed:

$$2Mn^{2+} + 5O_3 + 3H_2O = 2MnO_4^- + 5O_2 + 6H^+$$

As ozone needs may be defined by other priorities, a treatment scheme based on deliberate oxidation of manganeous salts to permanganate followed by subsequent reduction on activated carbon filters can give the following results:

$$4MnO_4^- + 3C + H_2O = CO_3^{2-} + 2HCO_3^- + 4MnO_2 \cdot H_2O$$

(secondary reaction: $2Mn_2O_7 + 3C = 4MnO_4 + 3CO_2$)

1.4   Reactions with Organic and Inorganic Substances

This reactivity is discussed in a subsequent section. As far as humic and fulvic acids are concerned, the ozonization of natural waters containing humic or fulvic acids results in a decomplexation of metal ions, including calcium. Oxidation should normally be followed by a coagulation phase. Under practical conditions, oxidation usually requires up to 0.5 mol per mole of carbon in humic or fulvic acids. Usually, the ozone consumption decreases when the doses are higher than 0.7 mg of O_3 per milligram of carbon in the humic acid.

1.5   THM Control

Ozone does not produce THMs directly. If the water is rich in bromide, hypobromous acid can be formed transiently and thus potentially increase the formation of brominated THMs. However, the practical formation of brominated THMs depends on the properties of the organic compounds. If the reaction of ozone with the organic compound is fast and forms compounds that are not precursors for THM formation (e.g., phenols), no formation of brominated THMs is observed.

However, if the organic compound reacts with ozone more slowly than does the bromide ion (e.g., alkylaryl ketones) and acts as a precursor in THM formation, the formation of brominated THMs is increased.

1.6 Removal of Organic Matter

The reduction in ultraviolet (UV) absorbance of sample water is important as measured at both 254 and 270 nm. This can be explained by the reactions of ozone with unsaturated and aromatic structures. Figure 3 illustrates typical evolution of the absorbance at 254 nm as a function of the ozone dosage applied to a typical water sample from the river Meuse (Tailfer plant). The direct effect of conventional ozonization on the decrease in TOC values is usually very limited, on the order of 0.2 to 0.3 mg/L, due to the fact that only a very small part of the oxidation reactions lead to the formation of CO_2. This applies to raw waters in general. For model compounds (e.g., phenols), the decrease in TOC can eventually reach 30 to 50%.

1.7 Turbidity Removal and Microflocculation

When introducing ozone to raw waters, the structure and size of suspended particles can be modified. A typical example (river Meuse at Tailfer) is given in Fig. 4. One can observe that medium-sized (20 to 50 μm) and large (50 to 200 μm) particles first decrease in number while the smaller ones increase. For higher ozone dosings, the smaller particles decrease in number and the medium-sized particles increase, presumably due to microflocculation. In the case of the river Meuse water, this change concerns less than 5% of the total particle mass.

The state of technology is not yet well established (15–17). Generally well-recognized advantages of the method are: more rapid filtration rates with increased length of filter runs, savings in coagulation chemicals, and *better water quality*.

In preozonization, no significant residual is required. In general, very fast reactions consume 1 to 2 mol of ozone per mole of organic compound, and slower

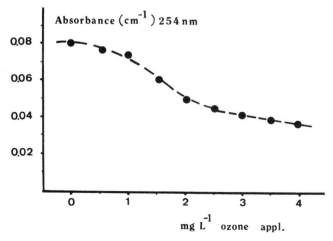

Figure 3   Decrease in UV absorbance on ozonization (typical example).

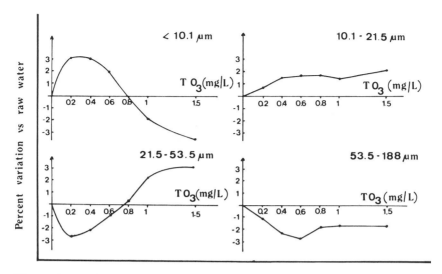

Figure 4 Particle-size distribution in ozonized raw water (river Meuse.)

reactions, about the same amount of ozone. This leads to the following rule of thumb: Inject 0.1 to 0.15 mg ozone per milligram TOC in the raw water to be treated. The bicarbonate content can play an additional role in the system: The higher the bicarbonate content, the lower the dosage required. The ozone dosing can be optimized on the basis of the coagulant dosage or through the particle-size pattern of the raw water (Fig. 4). In general, 0.8 to 1.2 mg/L ozone is necessary. For most raw river waters, a very comfortable design is to provide for 2 mg/L.

1.8 Promotion of Biodegradability

Very early ozonization has been assessed as a treatment process increasing the BOD of a raw water source. With humic acids, the maximum possible increase in BOD corresponds to a decrease of 80% of the original 254-nm UV absorbance. To achieve maximum biodegradability, high dosing rates of ozone may be necessary in the specific case of humic acids (e.g., 3 to 4 mg ozone/mg DOC) (18). In a general way, ozonization of organic compounds in water produces oxidized products which are suitable foodstuffs for bacteria and organisms, and thus enhances the feasibility of biodegradation. Figure 5 illustrates the effect of ozone on the biodegradability of a mixed wastewater.

The combination of ozone + biodegradation on activated carbon has been fully developed in Germany by the Engler-Bunte Institut and used in the Döhne plant at Mülheim (20) (see Fig. 6). Since then, numerous plants have been constructed that integrate both operations. One of the advantages of activated carbon versus other support material for biological treatment is that during the colder period of the year, when biological activity slows down, removal of organics can still be achieved by adsorption. Biological activity of adsorbed products can start again when temperature and global conditions are favorable (e.g., oxygen content, pH, etc.). The filters are operated at conventional surface loadings: for example, 8 m/h up to a high speed of 25 to 30 m/h at Mülheim. In some developments

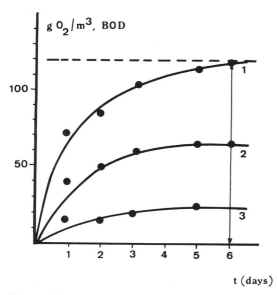

Figure 5 BOD of river Senne water (Brussels). 1, Senne downstream from Brussels; 2, after lime treatment at pH 11; 3, after preozonization at 30 mg/L and lime treatment at pH 11. (From Ref. 19.)

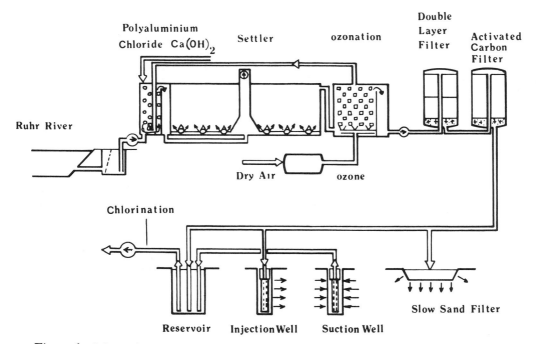

Figure 6 Schematic of the Döhne plant (Mülheim-Ruhr).

[Biocarbone filter (21)] the biological filtration layer is aerated to provide for sufficient oxygen consumed by the organic compounds as degraded.

Again, the treatment dosage requires local adjustment. No residual concentration is required. A first rule of thumb is to dose 0.15 to 0.2 mg ozone per mg/L^{-1} TOC. Generally, if precise data are not available, provision of 2 mg/L ozone dosing facilities are recommended.

2. GENERATION OF OZONE

Ozone is unstable and must be generated at the site of use. Ozonization requires that the following elements be available:

A gas source containing oxygen
Generating equipment
Ozone-to-water contact
Off-gas control

See Fig. 7.

2.1 Gas Source

The feed gas must be free of impurities such as dust and oil and have a moisture content below 25 ppm (wt) or a dew point lower than $-50°C$. The operational temperature of the process gas should be as low as possible. A typical flowsheet for the air preparation is shown in Fig. 8. It is recommended that a dust filter be mounted following the air-drying units. If lubricated compressors are used, an oil separator is necessary. Dry or water-ring compressors are a preferred technology. Appropriate drying materials are silica gel, activated alumina and molecular sieves (dew point from -55 to $-70°C$). Air drying is a determinant in the success of ozone generation.

It is of utmost importance to use a dry process gas for the corona discharge. Figure 9 illustrates the effect of moisture on the relative ozone production yield. It is also important to limit nitric acid formation to protect the generators and to

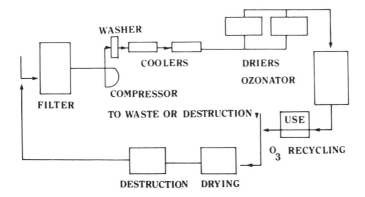

Figure 7   Typical flowsheet of an ozonization line.

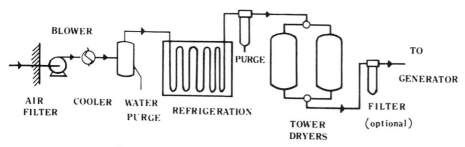

Figure 8 Flowsheet for air preparation.

increase the efficiency of the generation process. With air as the feed gas and under good operating conditions, about 1 mol of N_2O_5 may be produced per 100 mol of ozone.

The International Ozone Association (IOA) has published a checklist for the adequate design of feed gas preparation.

Checklist: Feed Gas Preparation for Ozone Generation.
General.
Concept of sizing and design basis
Standby capacity: installed or redundancy level
Degree of automation and control
Nature and composition of gas
Ambient air conditions (temperature, humidity, etc.)
Compressor system.
System (lubricated, oil-free, etc.)
Type (piston, rotating piston, screw, liquid ring, cooling system, etc.)
Size (dynamic range of use, pressure range, flow range)

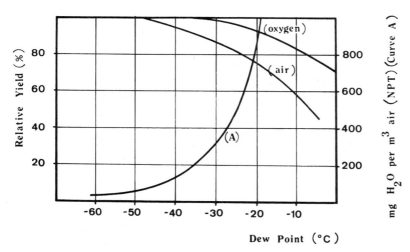

Figure 9 Incidence of humidity on ozone generation yield.

Control systems (flow, temperature, pressure, etc.)

Filter system for dust and oil removal (size and criteria handled)

Noise level [dB(A)] and control

Automatic controls for shut down

After-cooler systems (type, material of construction, performance criteria)

Drying system.

Size and degree of redundancy

Possibility of visual inspection

Acceptable gas conditions (inlet and outlet temperature, pressure, and flow)

Dew point at 1.013×10^5 Pa and continuous monitoring equipment, if any

Materials of construction

Requirement for refrigerated drying (type, size, capacity, gas outlet temperature, etc.)

Desiccant: type and specific adsorption capacity as a function of pressure

             danger of poisoning and flooding

             danger of dust formation (optional filtration)

             principle of regeneration procedure

             gas flow used in regeneration

             estimated cycle length of the various phases

Special requirements for oxygen use.

Storage or supply facilities

Built-in security

Materials of construction; gas-flow velocity in pipe system

Operating temperature and pressure

Specific control equipment and security

Cleaning procedures of components

Miscellaneous.

System for pressure control

Gas-flow control systems

Oxygen recycling

Costs.

Estimation of running costs

Remarks.

Standards and codes of practice being followed

Regeneration of the desiccant is generally obtained by external heating of the tower that is on standby, as two towers are being operated on an alternating basis. Pressure-swing regeneration is often too expensive for large units. The oxygen used in ozonization plants can be generated on site cryogenically or by pressure-swing absorption–desorption, or stored on site in a self-cooling storage tank filled automatically on a continuing basis by contract with a manufacturer. The liquid oxygen is evaporated and expanded to a pressure suitable to be fed into the ozone generator.

2.2 Ozone Generation by Electrical Discharge

In 1857, von Siemens developed the first industrial ozone generator based on corona discharge (22). Two concentric glass tubes were used, both covered on one side by a layer of tin, the outer externally, the inner internally. Air was circulated through

Figure 10 Storage of 15m³ liquid oxygen with evaporators of 300 m³/h (NTP) gas at a pressure of 3 bar (Tailfer plant).

Figure 11 Gas expansion and dosing units O₂ (3 bar → 0.7 bar).

the annular space. Later this technology was improved by circulating cooling fluids along the discharge air or oxygen gap, resulting in lower generation temperatures and less thermal destruction of the ozone.

The overall reaction of formation is described by the endothermic reaction

$$3O_2 \rightleftharpoons 2O_3 \qquad (\Delta H° \text{ at 1 atm, 284.5 kJ})$$

Also, the entropy of formation is large and unfavorable:

$$\Delta S° \text{ (1 atm)} = 69.9 \text{ J mol}^{-1} \text{ deg}^{-1}$$

Ozone cannot be generated by thermal activation of oxygen since the standard free energy of formation $\Delta G°$ (1 atm) = $+161.3$ kJ/mol. To summarize: Ozone can only be decomposed easily by heating, and adequate temperature control of the process gas is one of the determinant aspects in ozone generation efficiency.

The generation of ozone involves the intermediate formation of atomic oxygen radicals, which in turn can react with molecular oxygen. All processes that can dissociate molecular oxygen into oxygen radicals are potential ozone generation reactions. Energy sources that make this possible are electrons or photon quantum energy. Most commonly used are electrons obtained from high-voltage sources in the silent corona discharge.

Electron activation of oxygen leading to the formation of monoatomic oxygen ions, O^-, is, at present, not thought to contribute significantly to the formation of ozone. Moreover, the monoatomic oxygen ion can promote the destruction of ozone according to the reaction

$$O_3 + O^- \rightarrow O_2 + O_2^- \qquad (+2.98 \text{ eV or } 287.4 \text{ kJ/mol})$$

The intermediate oxygen radicals can recombine to re-form oxygen with the liberation of considerable heat:

$$O + O + M = O_2 + M \qquad (+5.1 \text{ eV or } 491.6 \text{ kJ/mol})$$

Consequently, if the concentration of oxygen radicals becomes too high, the relative yield of ozone generation will drop.

Moreover, electrons can decompose ozone, which also limits the acceptable discharge strength.

$$O_3 + e \rightarrow O_2 + O + e \qquad -1.084 \text{ eV or } -104.5 \text{ kJ/mol}$$

In summary: Ozone generation is an equilibrium process in which the conditions for generation also involve reaction schemes for destruction. An ozone generator is always a compromise wherein the designer has considered the relative importance of several factors in view of the objective to be reached, the relative local costs of the various components, and the technical skill available for operation of the systems.

The parasitic process leading to the formation of nitric acid is indicated by the following reactions:

$$O_2 + N_2 \xrightarrow{e} 2NO$$

$$2NO + O_3 \rightarrow N_2O_5$$

$$N_2O_5 \rightarrow 2NO_2 + \tfrac{1}{2}O_2 \quad \text{or} \quad N_2O_5 \xrightarrow{e} NO + NO_2 + O_2$$

$$N_2O_5 + H_2O \rightarrow 2HNO_3$$

In the normal operation of properly designed systems, a maximum of 3 to 5 g of nitric acid is obtained per kilogram of ozone produced with air. If increased amounts of water vapor are present, larger quantities of nitrogen oxides are formed when spark discharges occur. Also, hydroxyl radicals are formed, which combine with oxygen radicals and with ozone.

2.3   Corona Discharge

Ozone is currently produced by circulating dried air or oxygen on dielectric elements applied to a high-voltage alternating current. In practice, most voltages range from 5 to 20 kV. The most common frequency of the current is 50 to 60 Hz; however, medium (600 to 1000 Hz) = frequency current and even high frequencies (up to 3000 Hz) are occasionally used.

The formation of ozone through electrical discharge in a process gas is based on nonhomogeneous corona discharge in air or oxygen. There are a large number of distributed microdischarges wherein the ozone is effectively generated. It appears that each microdischarge lasts only several nanoseconds and is about 2.5 to 3 times longer in air than in oxygen (23). The current density ranges between 100 and 1000 A/cm^2. The capacitance of industrial corona cells is due to both the discharge gap and the dielectric material (Fig. 12).

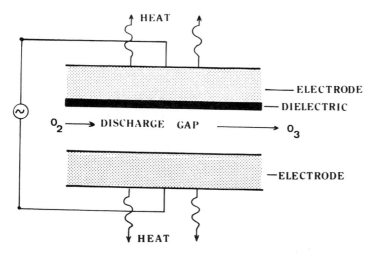

Figure 12   Schematic representation of ozone discharge gap.

Formulas proposed for the total cell capacitance are

$$C(f) = C(f) = \frac{2\pi\epsilon\epsilon_0 L}{\ln(r_e/r_i)}$$

and

$$C(f) = \epsilon\epsilon_0 \frac{A}{d}$$

for cylindrical geometry and parallel plates, respectively, where

r_e, r_i = outer and inner concentric electrode radius
d = distance between electrodes
L = electrode length
A = electrode area
ϵ_0 = absolute dielectric constant: 8.854×10^{-12} F/m
ϵ = relative dielectric constant (e.g., 6 for glass, 1 for air/oxygen)
C = capacitance (F)

The driving potential is a function of the current frequency:

$$V = V_0 \sin(2\pi\nu \times t)$$

where V_0 is the peak potential of the alternative current applied. At a value V_s the corona breaks down due to sparking. The following values for the sparking potential (in volts) have been reported in the literature (24):

For air: $V_s = 29.64 P d_g + 1350$
For oxygen: $V_s = 26.55 P d_g + 1480$

where P is the absolute gas pressure (kPa) and d_g is the thickness of the gas gap (mm). This indicates that for an equal value of d_g the sparking potential for oxygen is somewhat lower than for air. This can be corrected by using a slightly smaller gap in the design of oxygen-based generators.

The various parameters are correlated by an equation verified experimentally for the average power input (25):

$$P = 4C_d V_s \nu\left(V_0 - \frac{C_d + C_g}{C_d V_s}\right)$$

where C_d and C_g are the capacitance of the dielectric and of the gas layer, respectively. The power taken off by the corona discharge is increased with thinner dielectrics with high dielectric constants, by increasing the peak voltage, and by increasing the current frequency.

2.4 Types of Ozone Generators

2.4.1 The Otto Plate Ozonator

The Otto ozonator is composed of a certain number of "units," each containing two enanthiomorpous elements: a cast-aluminum water-cooled block, a glass plate

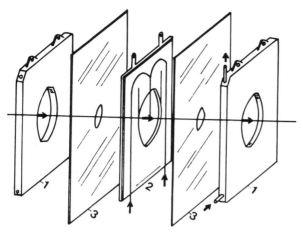

Figure 13 Otto plate generators. 1, Cast aluminum blocks, water-cooled on the low-potential side; 2, high-potential full-plate electrodes; 3, glass plate dielectrics.

dielectric, an air space, and a stainless steel electrode (see Fig. 13). The ozone is collected at the central outlet pipe. The dielectric glass plates are in contact with the low-potential electrode and distant by 2.5 mm from the high-potential electrode. The ozone containing air flows through circular exits located in the center of all glass-plate electrodes.

A typical standard package is composed of 40 vertically mounted pairs of electrodes for a nominal production capacity with air of 500 to 600 g/h. The high voltage ranges from 8.5 to 12.5 kW. The ozone concentration when using air is rarely higher than 15 g/m^3; unequal air distribution all along the electrode area limits the cooling system. At present, this technology is surpassed by the Welsbach tubular ozone generator. However, it is worth noting that in areas where the availability of spare parts may be a problem, the Otto plate system is still in operation.

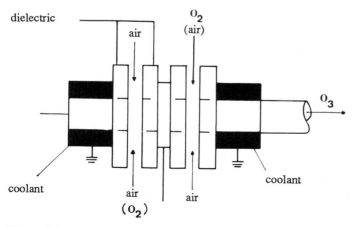

Figure 14 Plate ozonator discharge gap (modern version).

Discharge gap: 2 to 3 mm
Voltage: 7.5 to 20 kV
Frequency: 50 to 500 Hz
Dielectric: glass, 2.5 to 5 mm
Capacity: 2.5 kg/h at 50 Hz; 6.5 kg/h at 500 Hz
Energy requirements: (air, 50 Hz) 16 Wh/g O_3 at 15 g/m^3; 18 to 19 Wh/g O_3 at 20
    g/m^3
Pressure at outlet: 0 bar

The parallelism of the square or rectangular plates and the free dilatation of the discharge gap at variable temperature are both essential to ensure the success of the ozonator. The major problem in the operation of the Otto ozonator is plate breakage due to insufficient cooling or dilatation. To inject the ozonated air into the water, the outlet, being pressure-less, must be compressed or sucked into the water.

2.4.2 *The Welsbach Tube Ozonator*

The Welsbach tube ozonator, designed by Welsbach in the United States and licensed in Europe by Degrémont. Large systems are at present also manufactured by Emeri in USA, PCI in USA, Trailigaz in France and by ASEA-BBC in Switzerland.* These manufacturers all produce interesting adaptations of the device. In the tube ozonator the outer electrodes are stainless steel tubes surrounded by cooling water. The tubular glass dielectrics are coated inside with a conductor (generally aluminum) acting as a second electrode. The glass tubes are sealed at one end, forcing the gas to flow through the discharge gap (see Fig. 15).

Discharge gap: 2 to 3 mm
Voltage: 7.5 to 20 kV
Frequency: 50 to 600 Hz
Dielectric: glass, 2 mm (tends to be diminished)

---

*In 1991 ASEA-BBC and Degrémont (Ozone) combined to form OZONIA.

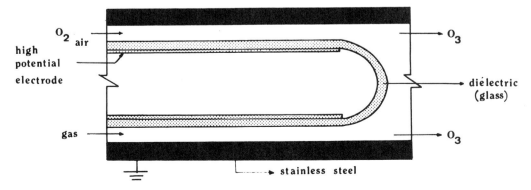

Figure 15 Tube dielectric for ozone generation (schematic).

Capacity: 25 kg O_3/h
Energy requirements: (air, 50 Hz) 16 Wh/g O_3 at 15 g/m^3; 18 Wh/g O_3 at 20 g/m^3
Pressure at outlet: approx. 1 bar

Horizontal tube ozonators are the most widespread type in use for water treatment. Cooling is essential for efficient operation. The increase in temperature of the cooling water should preferably not exceed 5°C, and the cooling-water temperature at the inlet is best maintained below 20°C. Increasing gas flow rates promote production by their effect on cooling. The exit pressure enables direct injection of the ozonized air into the water. Periodic cleaning of the contacts is necessary using trichlorethylene or ethyl alcohol.

2.4.3 *Megos and WEDECO Systems*

These systems are intended especially for the production of high concentrations of ozone in the process gas [e.g., 50 to 100 g/m^3 (NTP)]. It is composed of a metal central rod electrode and outer electrode. The dielectric is a glass tube plus gas gap (see Fig. 16).

Outer electrode: 5 to 12 mm in diameter
Inner electrode: 0.5 to 3 mm
Dielectrics: glass, 1 to 2 mm
Voltage: 7 kV (nominal)
Capacity: 1 to 300 kg/h (O_2)
Energy requirements: 16 to 20 kWh/kg (air); 8 kWh/kg (O_2)

2.5 Production Yield

The expressions for ozone yield can vary. Theoretically, 2960 J (i.e., 0.82 kW) is necessary to produce 1 g of ozone. The yield is sometimes expressed by G values (i.e., molecules of ozone produced by the consumption of 100 eV of energy). Reference values for ozone generation at 100% yield are thus 142.3 kJ/molecule O_3; 2965 J/g O_3; 38.5 Wh/molecule O_3 per hour; 0.803 Wh/g O_3; 1.47 eV/molecule O_3 or $G = 68$. Practical yields obtained with air are illustrated in Fig. 17. By using oxygen or enriching the process air in oxygen the generating capacity of a given ozonator

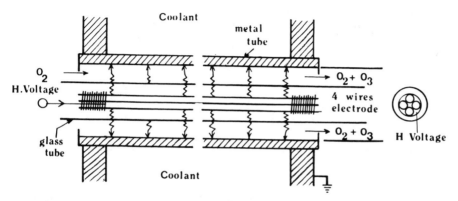

Figure 16 Schematic of the central electrode system.

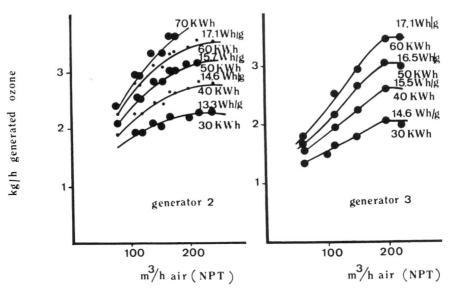

Figure 17 Production of a Welsbach ozonator (nominal 4 kg/h, 50 Hz with air; pressure, 0.54 bar positive; generator 2 is clean; generator 3 is dirty). (From Ref. 26.)

can be increased by a factor ranging from 1.7 to 2.5 versus air, depending on the design parameters, such as gas discharge gap and current frequency. Some examples are given in Fig. 18.

A distinction must be made between the uprating of ozone concentration and ozone quantity produced with oxygen/air at a variable process gas flow rate. This

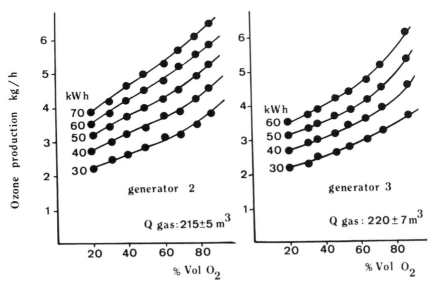

Figure 18 Increase in production using oxygen or oxygen-enriched air (generator 2 is clean; generator 3 is dirty). (From Ref. 27.)

means that the production capacity can be increased by a factor of 1.7 by running a classical air-based design with oxygen, while the production may be decreased by a factor of 2.5 when operating an oxygen-based design with air. All together, at least 20 to 30% of the nominal design capacity must be considered as the minimum that can be operated on a permanent basis in all cases. Similarly, an ozonator cannot be operated continuously at maximum nominal production capacity. Usual installed overcapacity ranges from 25 to 30% of nominal capacity. The yield obtained using an oxygen-enriched process gas is increased with a small gas space and a higher electrical current frequency.

Since all variations result in energy loss in the form of heat, cooling the process gas is very important. Figure 19 illustrates a typical effect of cooling-water temperature on the yield of ozone generation with air. The most efficient mode of cooling is the "both-sides" cooling system: that is, cooling on both the high-voltage and ground sides. However, in case of accidental breakage of the dielectric, the cooling liquid (e.g., water) enters the discharge gap and causes short-circuiting of the entire system. Therefore, cooling on the ground side only is the safer and more reasonable design.

IOA has also published a design checklist for ozone generators.
Checklist: Ozone Generation Based on Electrical Discharge in a Process Gas.
General sizing concepts.
Design capacity (kg/h)
Standby capacity
Degree of automation: gas flow, cooling water, and so on
Expected variation in ozone demand and dynamic range of production
Recommended stock and availability of spare parts
Ozone generator.
Type and number of dielectrics (plate, tube, central rod)
Maximum ozone production per unit (kg/h) at given temperature of the coolant
 and ozone concentration
Ozone concentration produced, for example in g/m^3 (NTP) at outlet and with
 correction for gas density

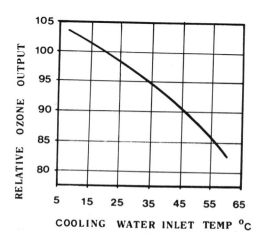

Figure 19   Effect of ozonator cooling on ozone generation efficiency. (From Ref. 28.)

Gas pressure at outlet of the ozone generator (bar · g)
Minimum and maximum operational voltage, breakthrough voltage (kV)
Operating frequency (Hz)
Maximum production capacity compared to average needs
Requirements for cooling of ozone generator at design capacity
Type and composition of the coolant (e.g., chloride content)
Inlet gas temperature: normal operating value and maximum admissible
Possibility of visual inspection of discharge
Gas flow measurements, equipment, and conditions of installation
Control of cooling water and feed gas failure
Possibility of variation of production
Electrical protection system
Requirements for maintenance
    Electrical power supply.
Power factor correction, if required
Low potential (mains) voltage and current frequency
Power consumed at point of mains supply (kW)
System and enclosure class of HT transformer and cabinet
Type of frequency converter (if any)
Cooling for transformer and frequency converter (if any)
Ambient temperature considered in design (°C)
Electrical balance of off-take phases
Method of measurement of energy consumption
Earth connection system
Stability of power supply (requirements)
    Materials.
Ozone generator
Piping system
Joints and gaskets
Valve system
    Miscellaneous.
Safety against risks of flooding
Safety for back-suction of water
Possibility of flushing the ozone generator with dry gas
Provisional planning for cleaning operations
Specific security measures with oxygen as a process gas
    Costs.
Estimation of running costs
    Remarks.
Standards and codes of practice being followed

2.6 Practical Aspects of Recycling Off-Gases into Ozone Generation Systems

The first aspect to consider when using oxygen as a process gas generally focuses on the economic conditions under which the oxygen can be obtained. Recycling the off-gases after contact of the ozone–oxygen with the water often appears as an attractive alternative. It is therefore necessary to consider carefully all aspects of the equilibria of dissolution of gases into the water, particularly oxygen and nitrogen, but even carbon dioxide; and this should be carried out as a function of temperature, pressure, relative concentration of the various gases, and finally, the salinity of the water.

The water submitted for ozonization should generally be assumed to be saturated and equilibrated versus air: that is, a gas mixture containing 21% oxygen and 79% nitrogen at the contact pressure, which is usually 1 atm in post-ozonization by bubbling into baffled chambers, for example. Contact with a process gas enriched in oxygen content compared to air achieves a higher dissolved oxygen concentration as well as partial stripping of the dissolved nitrogen. Hence recycling requires partial purge of the gas phase to allow evacuation of the excess of nitrogen and complementary enrichment with fresh oxygen to reinstate the initial composition of the process gas.

The technical and economical implications of these systems based on purge and complement depend on local conditions, such as pressure, temperature, gas-to-liquid ratio, and the method of contact (29–33). With contactors operating at atmospheric pressure or a similar pressure range, equilibrium is established according to Henry's law for the dissolution of gases. However, under dynamic conditions, nitrogen stripping is generally not quantitative during the residence time of the water in the ozone-contact system.

The saturation concentration of oxygen in water increases linearly with increasing pressure (Fig. 20). To exploit this characteristic one can operate at the exact pressure, which makes it possible to obtain the necessary dissolved oxygen for a given application: for example, raw water treatment or treatment of wastewater. At the same time, the nitrogen gas content is decreased in relation to the oxygen when the dissolved oxygen content is increased (Fig. 20). The function is of the reciprocal type.

Consequently, recycling of the process gas rarely, if ever, requires a nitrogen purge when the system that is providing contact with the water is under a large positive pressure. However, we must take into consideration the fact that at high pressures the dissolved oxygen content will usually be in excess of needs (e.g., 300 g/m^3 at 8 bar positive pressure). Therefore, the technique often used is to saturate a substream of the water under positive pressure and dilute the substream into the bulk of the water to be treated.

Criteria for the design and operation of the system are as follows:

1. The ozone needs determine the ratio flow and ozone concentration in the gas phase.
2. The necessary oxygen in the treatment determines the partial water flow and its pressure.
3. A compromise is obtained by integrating the costs of the nitrogen purge versus the addition of oxygen and the ozone concentration generated.

A typical example is the Wittlaer Waterworks in Duisburg, Germany (see Fig. 21).

Recycling the process gas in low-pressure systems does not require special materials of the type generally used in ozonization: for example, stainless steel, copper, brass, and cast aluminum. Alloys with a high magnesium content are to be avoided. Specifications for the transport of oxygen must be met, such as the absence of organic products (e.g., oils, lubricants, etc.).

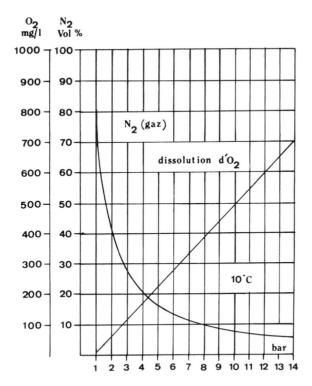

Figure 20   Dissolution equilibria at 10°C of O_2/N_2.

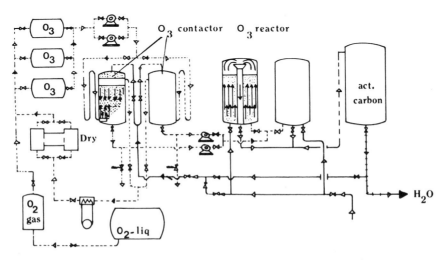

Figure 21   Schematic of the Wittlaer ozonization plant. (From Ref. 30.)

In high-pressure systems compression of the ozone-containing gas is necessary to allow diffusion into water under positive pressure. This operation necessitates particularly tight gas circuits. Water-ring compressors are practically the only units suitable and must be constructed of high-quality stainless steel such as AISI 316, 318, or similar. Such a system and its operation can be expensive.

The recycled gas should not contain organic impurities at levels that could reach the lowest explosion limits (LEL values). The monitoring of these limits is part of the system, particularly when it concerns the treatment of wastewater. The off-gas treatment must also be designed to cope with the possible presence of traces of organic contaminants in an oxygen-enriched gas.

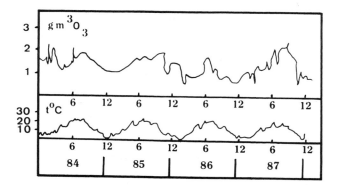

Figure 22 Seasonal variation of ozone demand for postdisinfection (Tailfer plant, river Meuse).

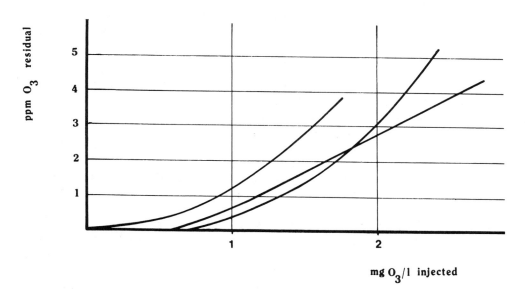

Figure 23 Minimum ozone demand (postdisinfection).

3. EVALUATION OF OZONE NEEDS

Fundamentally, the ozone needs for a given application are best evaluated by a pilot investigation, a semibatch reactor, or at least in a laboratory reactor without head space. The seasonal variation of the ozone demand can be high (e.g., variations by a factor of 2 in the Tailfer plant of the Brussels Intercommunal Waterboard) (Fig. 22). Small amounts of ozone injected in water react fast without giving significant residual concentrations. Typical examples are shown in Fig. 23. This indicates that even with very pure water, to obtain a significant residual concentration, at least 1 mg/L is necessary. The practical injection rates most often found in post-ozonization systems range from 2 to 4 mg/L.

3.1 Laboratory Tests

There are two classical methods but no standardized procedure at this time.

3.1.1 *The Flask Method*

In the flask method a closed flask is used, into which certain volumes of water and ozonized air are injected, providing the treatment rate. After stirring for a given time, the residual ozone and residual concentration of the pollutant in the water are measured. By varying the treatment rate, the ozone demand of the water can be determined under the conditions of the experiment. The characteristics of this test are:

1. A well-determined contact time (that can be varied experimentally)
2. A reduction in the concentration of ozone in the ozonized air during contact
3. A reduction in the concentration of pollutant if it is attacked by ozone

3.1.2 *The Column Method (Semibatch Reactor)*

In the column method a contact column is used with a porous disk at its base for the introduction of the ozonized air, and a vent at the top. The water can be injected at the top and flows out at the base (i.e., countercurrent to the ozonized air). The water can also be injected at the base and flow out at the top (i.e., concurrent to the ozonized air. Injection into the column can be done in two ways:

1. The column is filled with the water to be treated, or with a pollutant whose destruction by ozone is sought, and the liquid flow is then stopped. The ozonized air is injected continuously at the base of the column, and the evolution in the concentration of the pollutant in the water is measured as a function of time. This procedure is characterized by:
    a. A reduction in the pollution concentration as a function of time
    b. An increase in the concentration of dissolved ozone in the water
    c. At the end of the experiment, an equilibrium state that shows the dissolved ozone concentration that it is possible to achieve (temperature, ozone concentration in the ozonized air)

    It is therefore possible to examine the kinetics of oxidation of a pollutant by ozonized air whose concentration is relatively constant (because it is being renewed continuously). On the other hand, it is not possible to examine the efficiency of the actual dissolution of the ozone.
2. The water to be treated and the ozonized air are both injected continuously. This simulates actual operation. It is characterized by:

a. An apparently well-determined water contact time in the column
b. A reduction in the concentration of the pollutant after flow through the column
c. An increase in the ozone concentration in the water passing through the column

An equilibrium state is reached depending on the experimental conditions.

These methods are used either to study the kinetics of oxidation by dissolved ozone of a particular pollutant, or to study the transfer of the ozone from the ozone carrier air to the water to be treated. It is obvious that there is a certain degree of interference between these two objectives. In fact, as the ozone is transferred from the air to the water, the dissolved ozone is consumed, on the one hand, by the oxidation reaction of the pollutant, and on the other, by self-destruction of the ozone in the water.

4. SOLUBILITY OF OZONE IN WATER

4.1 Definitions

The dissolved concentration at saturation, C_s, can be expressed either in terms of the coefficient of solubility, also called the solubility ratio S, or as an absorption coefficient (i.e., β) (34). S is defined by the volume of gas dissolved per unit volume of liquid at the temperature and pressure under consideration and in the presence of the equilibrating gas at 1 atm pressure. β, often called the Bunsen absorption coefficient, is the volume of gas expressed at NTP which is dissolved at equilibrium by a unit volume of liquid at a given temperature when the partial pressure of the gas is the unit atmosphere; this is equal to the pressure of the gas itself minus the vapor tension of the liquid.

It is generally agreed in the literature that when ozone is dissolved in water, Henry's law is obeyed. This means that the C_s values are proportional to the partial pressure of ozone, P_γ, at a given temperature. The expressions for the saturation concentration of a dissolved gas considered to be thermodynamically ideal are related as follows:

$$C_s(\text{kg/m}^3) = \beta M \times P_\gamma$$

where M is the mass volume of the gas [i.e., 2.14 for ozone (kg gas/m^3) gas] and P_γ is the partial pressure of the gas in the given gas phase. Under normal conditions (i.e., 273 K and 1 atm pressure), the S value for ozone is 0.64. Data for the solubility of gases potentially associated with the use of ozone are given in Table 1.

Table 1   Solubilities of Gases Compared to Ozone

Gas solubility	Ozone	Oxygen	Nitrogen	Carbon dioxide	Chlorine	Chlorine dioxide
β (vol/vol)	0.64	0.049	0.0235	1.71	4.54	±60
β (O$_3$)/β gas	1	13.3	27.7	0.38	0.14	±0.01
C_s (kg/m^3) for $P_\gamma = 1$	1.4	0.07	0.03	3.36	14.4	180

4.2   Influence of the Water Temperature

The most important side effect, other than the partial pressure of ozone in the gas phase, which influences the solubility significantly is probably the water temperature. The literature on this subject has been reviewed extensively by Morris (35), and there exists some controversy among the various sources. Most of the former literature data are of questionable accuracy since less reliable analytical methods were used. Moreover, sampling is undoubtfully often an uncontrolled source of error.

A critical and selective examination together with an experimental investigation (36) leads to the data illustrated in Fig. 24. The data in the figure are measurements in distilled water and in water containing bicarbonate (e.g., up to 200 mg/L). The thermodynamic data that result are

$$\frac{d(\ln S)}{dT} = \frac{\Delta E}{RT^2} = -0.048$$

in which ΔE is the molar internal energy of dissolution. At 20°C, for example, this is equal to

$$\Delta E \,(293\ \text{K}) = (-0.048) \times (8.31) \times (293)^2 = -34.2\ \text{kJ/mol}$$

or an enthalpy change of exothermic dissolution (Van't Hoff's equation):

$$\Delta H^{\circ}_{293} = \Delta E + RT = -34.2 + \frac{293 \times 8.31}{1000} = -31.8\ \text{kJ/mol}$$

Henry–Dalton constants computed from the data are given in Table 2.

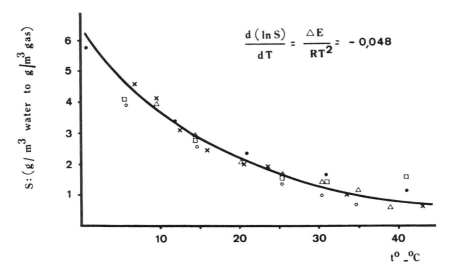

Figure 24   Solubility curve for ozone in water as a function of water temperature. ○, Caprio (39); ×, ●, Khadaroui (36); △, Rawson (37); □, Matsorov (38).

Table 2 Henry-Dalton Constants for Ozone Residuals in Water[a]

	t(°C)							
	0	5	10	15	20	25	30	35
S (water/gas)	0.64	0.50	0.39	0.31	0.24	0.19	0.15	0.12
H_a	1,945	2,490	3,190	4,200	5,190	6,555	8,302	10,375
k_H	34,990	45,600	59,450	76,180	100,000	128,600	164,800	210,400

[a]Concentration range 0.1 to 0.65 mg/L.

$$K_H = F(n_i;n_T) \times F(n_T \times R \times T;\ V_T \times C_L) \qquad H_a = F(P_i;x)$$

where

K_H = Henry's constant (atm · m^3/mol)
H_a = apparent Henry's constant in atm per molar fraction of ozone in the liquid
n_i, n_T = fractional and total number of moles in the gas phase
R = universal gas constant: 82 (atm · m^3/mol · K)
T = equilibrium temperature (K)
V_T = total volume of the gas phase (m^3)
C_L = equilibrium liquid-phase concentration (mol/m^3)
P_i = partial pressure of ozone in the gas phase (atm)
x = molar fraction of ozone in the liquid phase

An overall equation can be computed from the data:

$$\ln H_a = 22.3 - \frac{4030}{T}$$

4.3 Influence of the pH Value

Roth (40) has overviewed data on the effect of pH on Henry's constant. In the range 3.5 to 60°C and 0.65 to 10.2 pH units, the data converge to an equation:

$$H_a = 3.84 \times 10^{-7} \times [OH^-]^{0.035} \exp \frac{-2428}{T}$$

Specific correlations have been published (41) for two particular values of pH:

$$\text{pH} = 2: \quad \ln H_a = 20.7 - \frac{3547}{\text{T}}$$

$$\text{pH} = 7: \quad \ln H_a = 18.1 - \frac{2876}{\text{T}}$$

where H_a is in atm per molar fraction of ozone in the water. The dispersion of experimental data remains considerable.

4.4 Effect of Ionic Strength of the Solution

$$\ln K_H = 19.1 - \frac{2297}{T} + 2.659 \times 10^{-3} \times I - 0.688 \frac{I}{T}$$

(with K_H in atm $\cdot$ m^3/mol and I in mol/m^3) Consequently, at medium ionic strengths such as those found in drinking waters, the effect on ozone solubility remains marginal.

4.5 Effect of Thermal Decomposition of Ozone in the Gas Phase

In semibatch columns, ozone consumption can be observed in the gas phase even when the dissolved ozone concentration remains stable (38). The drop in ozone concentration in the gas phase at the exit of the contacting column increases with increasing temperature. The initiating step is assumed to be

$$O_3 + O_2 \rightarrow O_2 + O_2 + O \qquad k_2 = 2.02 \times 10^{-2} \exp \frac{2400}{T} \qquad M^{-1}s^{-1}$$

Once the formation of radical oxygen is obtained, the decomposition of ozone is very fast:

$$O_3 + O = 2O_2 \qquad k_2 = 3.37 \times 10^{10} \exp \frac{5700}{RT} \qquad M^{-1}s^{-1}$$

5. DECOMPOSITION OF OZONE IN WATER

5.1 Summary of Data Reported

See Table 3. With concentrations of dissolved ozone that are representative for those encountered in water treatment (e.g., lower than 1 mg/L dissolved ozone), the autodecomposition of ozone is an apparent first-order reaction to ozone and nearly first order to the hydroxyl ion concentration.

5.2 Reaction Mechanisms of Ozone Decomposition

Knowing the exact stability in water is important to guaranteeing the necessary reaction time. In water treatment, not only the rate of ozonation of dissolved substances but also the self-decomposition is influenced by several factors, such as temperature, pH, ionic strength, and type of ions present. In constant conditions the reaction rate observed is generally of first order to ozone:

$$-rO_3 = -\frac{d|O_3|}{dt} = k_D|O_3|$$

However, numerous discrepancies exist in the data summarized by several authors (58).

Basically, the reaction involves a five-step chain reaction:

Table 3 Reaction Characteristics of Ozone Decomposition in Water

Order	Concentration range (mg/L)	t (°C)	pH	Ions	E_{act} (kJ/mol)	Refs.
$|O_3|^1$	0.8–1.8	3.5–60	0.5	NaOH		42
	Batch reactor, 0.01–2.4	3.5–60	0.5–10.2			43,44
	Batch reactor		9.6–11.9			45
	Flow reactor, 0.5–5	25	10–13	NaOH		46
	Satur. vs. 0.5 g ozone/L oxygen	0.27	12–13		41	47
	0.5–0.6 (+O_2)	20–40	5–6		62	[a]
$|O_3|^2$	35–75	0	2–4			48
	5–40	20	2–10	Acids		49
$|O_3|^{1.5}$	14	25	0–7	$HClO_4$		50
			2.1–10.2			51
$|O_3|^0$						52
$|O_3|^1|OH|^1$	0.7–6	20	11–13	CO_3^{2-}/HCO_3^-		53
	1.2 (column)	18–27	8.5–13.5	CO_3^{2-}/HCO_3^-		54
$|O_3|^1|OH|^{0.75}$	0.06–1.5	1–20	7.6–10	CO_3^{2-}	96	55
$|O_3|^1 + |O_3|^{1.5}|OH|^{0.5}$	Saturated vs. 86 g/L (O_2)	10–40	2.5–9	(PO_4)	112	56
$|O_3|^1|OH|^1 + |O_3|^2|OH|^1$	Stop-flow		12		41.3	57

[a]Data of Brussels' Waterboard.

$$O_3 + H_2O \rightarrow 2OH^\circ + O_2 \quad k_2 = 1.1 \times 10^{-4} \, (L \, M^{-1} s^{-1}) \, (20°C)$$
$$O_3 + OH^- \rightarrow O_2^{-\circ} + HO_2^\circ \quad k_2 = 70 \, (2\times) \, (L \, M^{-1} s^{-1}) \, (20°C)$$
$$O_3 + OH^\circ \rightarrow O_2 + HO_2^\circ \rightleftharpoons O_2^{-\circ} + H^+$$
$$O_3 + (HO_2^\circ) \rightarrow 2O_2 + OH^\circ \quad k_2 = 1.6 \times 10^9 \, (L \, M^{-1} s^{-1})$$
$$2HO_2^\circ \rightarrow O_2 + H_2O_2$$

The production of hydrogen peroxide has been determined by Schultze and Schultze-Frohlinde (59).

All investigations converge to the evidence that in an alkaline medium the reaction is faster than in acid conditions. The global kinetical equation proposed now is

$$-rO_3 = k_D |O_3| = k_A |OH^-|^{1/2} \cdot |O_3|^{3/2}$$

and the k_D and k_A values determined in phosphate-buffered water at different pH values with a constant ionic strength of 0.15 are (58)

$$k_D = 5.43 \times 10^3 \exp\left(\frac{-4964}{T}\right) \quad s^{-1}$$
$$k_A = 9.5 \times 10^{16} \exp\left(\frac{-10,130}{T}\right) \quad L \, M^{-1} s^{-1}$$

This means that at pH 3 and lower, the OH^- ion has no practical influence. At pH values between 7 and 10 both steps contribute about equally to the self-decomposition of ozone. A typical half-life time of ozone in drinking water ranges from 15 to 25 min.

Initiation through the OH° radicals can be considerably decreased by OH° scavengers. Particularly significant are:

$$OH^\circ + O_3 \rightarrow O_2 + HO_2^\circ \quad k_2 = 3 \times 10^9 \, (L \, M^{-1} s^{-1})$$
$$OH^\circ + HCO_3^- \rightarrow OH^- + HCO_3^\circ \quad k_2 = 1.5 \times 10^7 \, (L \, M^{-1} s^{-1})$$
$$OH^\circ + CO_3^{2-} \rightarrow OH^- + CO_3^{-\circ} \quad k_2 = 4.2 \times 10^8 \, (L \, M^{-1} s^{-1})$$
$$OH^\circ + H_2PO_4^- \rightarrow OH^- + H_2PO_4^\circ \quad k_2 < 10^5 \, (L \, M^{-1} s^{-1})$$
$$OH^\circ + HPO_4^{2-} \rightarrow OH^- + HPO_4^- \quad k_2 < 10^7 \, (L \, M^{-1} s^{-1})$$

At present, the general scheme is best summarized by the cycle given by Staehelin and Hoigne (60) (Fig. 25). The oxygen superradical ($O_2^{-\circ}$) has a high specificity for reaction with ozone to degenerate into oxygen.

Hydrogen peroxide in its dissociated form can act as a chain carrier or even as an initiator by producing $O_2^{-\circ}$ or $O_3^{-\circ}$ radicals:

$$H_2O_2 \rightleftharpoons H^+ + HO_2^- \quad (pk_A = 11.6)$$
$$HO_2^- + O_3 \rightarrow HO_2^\circ + (O_3^\circ)$$

In addition to their reaction with ozone (direct reaction) or with the hydroxyl radical

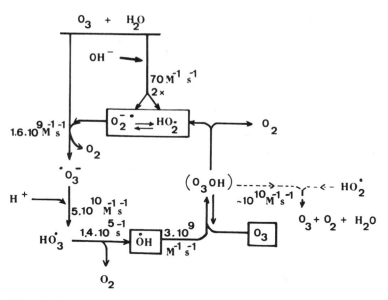

Figure 25 Global cycle of ozone decomposition in water. (From Ref. 60.)

(indirect reaction), dissolved substances may, by converting the radicals, act as promotors or inhibitors of the ozone decomposition chain. For example, a scheme for promotion by the formate ion, which forms the oxygen superradical, is the following:

$$HCOO^- + OH° \rightarrow H_2O + °COO^- \qquad k_2 \simeq 3 \times 10^9 \ (M^{-1}s^{-1})$$
$$°COO^- + O_2 \rightarrow O_2^-° + CO_2 \qquad (fast)$$

Other products, such as carbonate–bicarbonate scavengers of the OH radicals, do not restore an active radical of the ozone decomposition cycle, and therefore act as inhibitors of ozone decomposition.

A wide variety of compounds potentially present in natural waters can thus interfere with the ozone self-decomposition cycle. In practice, it has been assumed that about half of the ozone contacted with water is transformed into OH radicals (61), although, in theory, this percentage could be higher.

6. OZONE TRANSFER TO WATER

6.1 General Principles

Diffusion of ozone in water obeys Fick's law for molecular diffusion. At 20°C the diffusion constant is $D_{O_3} = 1.74 \times 10^{-9} \ m^2/s$ and can be corrected for different physical conditions according to the Nernst–Einstein relationship $(D \times \mu)/T =$ constant, where μ is the dynamic viscosity of water and T is the absolute temperature. Consequently, ozone transport is much faster in the gas phase than in the liquid phase. The latter determines the overall transport rate.

6.2   The Double-Film Model

The transfer of ozone to water without a reaction is currently accepted as occurring according to the double-film model (see Fig. 26). The driving force is $C_s - C$).

$$\frac{dC}{dt} = +k_L S(C_s - C) \qquad \log\frac{C_s - C}{C_s} = k_L S \times t$$

where S is the specific exchange surface in the liquid film and depends on practical conditions (e.g., mixing, pressure, and total gas and liquid volumes).

On the gas side, the concentration of ozone is expressed in terms of partial pressure:

$$y = \frac{P(O_3)}{P_{\text{total}}}$$

Part of y is transferred to the liquid phase to build up C. If y^* is the part transferred,

$$y^* = \frac{P^*(O_3)}{P_{\text{total}}}$$

and $y - y^*$ is the driving force in the gas phase for ozone transfer into the liquid. The practical transfer rate is given by

$$\frac{-dM}{dt} = k_{(g)} S_{(g)} (y - y^*)$$

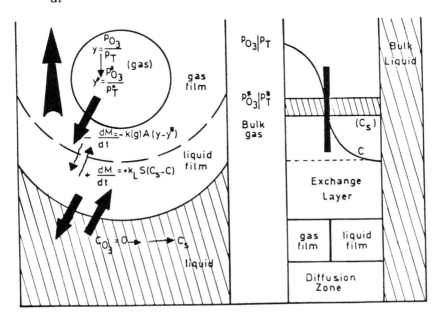

Figure 26   Schematic for double-film transfer model.

The equilibrium condition, (i.e., what the gas releases and what the liquid receives is given by

$$\frac{+dM}{dt} = +k_L S(C_s - C) = N_t$$

(i.e., the amount transferred per unit time).

In a gas–liquid system it is often difficult to determine the specific exchange area S; as a result, estimations for k_L are variable. By considering the total volume of a reactor (V), one has

$$d(N_t) = k_L S(C_s - C)\, dV$$

for a small-volume portion of the reactor. For a total reactor several approximations are necessary, depending on the type of reactor (62).

Numerous empirical, sometimes very complex equations have been formulated to evaluate the gas transfer constant as a function of various operational parameters. Only a simplified approach can be given here because few of them concern the transfer of ozone. The most reliable value of k_L for ozone is on the order of 2 to 3 $\times 10^{-4}$ m/s, which is about 2.5 times higher than for oxygen.

Many systems and devices have been applied to transfer ozone to water; these systems and devices were discussed earlier (63). The most widespread methods at present are the use of porous diffusors and gas–liquid dispersing turbines. Other methods involve injectors and liquid-packed towers.

6.3 Technical Devices for Ozone to Water Contacting

6.3.1 Porous Diffusors

Diffusors should produce bubbles with average effective radii ranging up to 2 mm. In practice, diffusors with pore sizes of 50 to 100 μm are installed at the base of a water column 4 to 6 m in height. The head loss of the immersed porous diffusors must be maintained at 300 to 500 mm water column. Average gas flow in each contacting column usually remains below 10% of the water flow. Based on bubbles of 4×10^{-3} m diameter and when using ozonized air, the average exchange surface ranges to 150 m^2/m^3 water, and in countercurrent contacters the average downflow velocity of the water is to be kept at 4 to 6 $\times 10^{-2}$ m/s.

An empirical relation for k_L can be taken into consideration for bubble columns with a relative gas to water velocity between 0.1 and 0.2 m/s and bubble diameters of 3 to 5 $\times 10^{-3}$ m:

$$k_L = 1.13 \left(\frac{D_{O_3} v_R}{d_B}\right)^{1/2} \simeq 3 \times 10^{-4}\, \text{m/s}$$

where D_{O_3} is the diffusion constant for ozone, 1.74×10^{-9} m^2/s, v_R the relative velocity gas versus water (m/s), and d_B the bubble diameter (m). The free bubble rise velocity depends on the bubble size (Fig. 27).

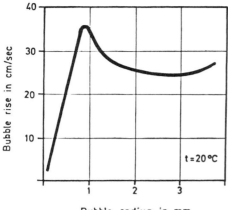

Figure 27 Bubble rise velocity as a function of dimensions (countercurrent water flow, 0.08 m/s).

By defining U_{sg} as the superficial gas velocity (i.e., gas flow divided by the cross section of the reactor), one can define experimental equations for $k_L S$ by establishing relations of the type $k_L S = \alpha U_{sg}{}^\beta$. Typical experimental relations are

$$k_L S\,(\text{s}^{-1}) = 2.32 \times 10^{-4} U_{sg}^{0.82} \qquad [\text{at } 5\,°\text{C (64)}]$$
$$k_L S\,(\text{s}^{-1}) = 7.91 \times 10^{-4} U_{sg}^{0.54} \qquad [\text{at } 20\,°\text{C (65)}]$$

with a value of U_{sg} of 4×10^{-2} m/s: for example, the equations give $k_L S$ values of 1.7×10^{-5} and 1.4×10^{-4}, respectively, at 5 and 20°C, thus indicating large temperature effects in gas-to-liquid transfer of ozone. However, no significant "dissolved concentration" should be reached versus the saturation value ($C \ll C_s$), so that the driving force for transfer remains practically independent of the residual dissolved concentration.

Different approximations have been formulated to evaluate the values U_{sg} to be used. The method of Hughmark can be applied to free-moving, noncoalescent bubbles:

$$\epsilon_g = \frac{U_{sg}}{0.3 + 2U_{sg}}$$

where ϵ_g is the volume fraction of gas in the reactor. If ϵ_g is 10%, one has

$$0.1 = \frac{U_{sg}}{0.3 + 2U_{sg}} \rightarrow U_{sg} \simeq 40 \; mm/s$$

Another approximation is given by the relation

$$U_{sg} = \frac{T_{O_3} \times h}{C_g \times T_R}$$

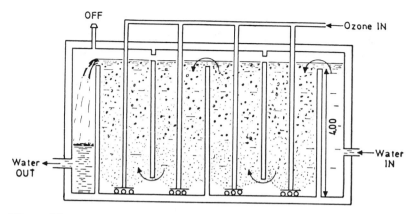

Figure 28 Baffled chamber system for repeated ozone contact.

in which T_{O_3} is the treatment dose (g/m³ water), h the water column height (m), C_g the ozone concentration in the gas at the interface between gas and water (eventually in the off-gas) (g/m³) and T_R the detention time of the water in the contactor (s). (A typical example in postozonization is $T_{O_3} = 1$ g/m³; $h = 4$ m; $C_g = 1$ g/m³, which gives $U_{sg} \simeq 33$ mm/s).

In postozonization ϵ_g usually ranges from 8 to 15% and U_{sg} from 20 to 60 mm/s. Diffusion is preferably applied in a countercurrent direction (i.e., with an upward gas flow and a downward water flow). Four to six baffled chambers are installed in series, as illustrated in Fig. 28. The downflow of the water versus the gas somewhat promotes "ballasting" of the bubbles, so increased contact time results. To keep the ballasting from coming to high, the average downflow velocity of the water (water flow divided by the cross section of the contact basin) must be kept below 0.1 m/s. Similar criteria hold for transit ports between baffled chambers (Fig. 29).

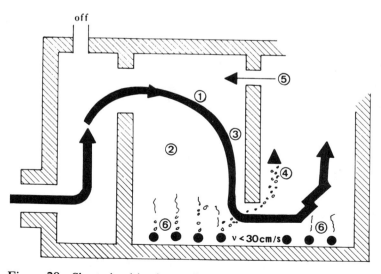

Figure 29 Short-circuiting in gas–liquid contact chambers. 1, 3, Potentially underozonized zones; 6, 2, potentially overozonized zone; 4, parasite substream. (From Ref. 63.)

The average velocity gradient throughout the injection basins ranges from 100 to 200 s^{-1}, but in bubble-formation zones, 8000 to 10,000 s^{-1} can occur. The "horizontal water velocity" within the baffle gates must be kept below 0.3 m/s. For disinfection a certain contact time is required, (i.e., the CT criterion to be met). When the dissolved concentration is stationary (i.e., there is neither decomposition nor consumption of ozone in the water, the following equation can be used:

$$\frac{C}{C_0} = \frac{n}{(n-1)!} \, n^{n-1}\left(\frac{T}{t}\right) e^{-n(t/T)}$$

For $n = 6$ and $C/C_0 = 0.1$, $t/T = 0.525$; this means that the efficient contact time is about half that of the nominal water residence time.

Contact by means of porous diffusors can be a problem with water containing iron and manganese or with raw waters containing suspended matter. In such cases other contacting systems should be used. One of these is the "ozoflotation unit" for the removal of algae and other suspended particles (66) (Fig. 30). Contact is maintained by a clear-water sweep of the porous diffusor disks by about 10% of the total water flow. Bubbles are to be produced as for flotation (i.e., of less than 1 mm diameter, particularly in this case in the size range 200 to 500 μm). The upflow velocity in the flotation compartment ranges from 16 to 22 m/h. The system is very useful for the control and removal of algae. The necessary ozone doses are variable as a function of the density of the algal population and range between 0.3 and 2 g/m^3. Air or oxygen can be used in the process, together with various classical coagulants. Up to 70 to 80% of the total average algal population can be expected to be removed, along with the other benefits of preozonization.

6.3.2 Ballasted Bubble Contact

When the downflow velocity of the water is increased, it is possible to obtain cocurrent contact, in which the dispersed ozone gas is drawn with the water flow.

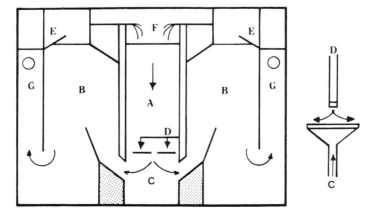

Figure 30 Principle of ozoflotation. A, ozonization compartment; B, flotation zone; C, porous diffusor; D, clear-water sweep; E, skimming gatters; F, raw water inlet; G, clear-water outlet. (From Ref. 66.)

The water velocity must be higher than 40 cm/s, preferably in the range 60 cm/s, to avoid hydraulic intermediate regimes with bumping. The system depends on a limited range of gas to water flow (67) (e.g., 5 to 10 volumes of water per unit volume of gas).

The depth of the tubular contactors is virtually unlimited. A schematic is given in Fig. 31. The system approaches the "deep-shaft" technology for aeration. An interesting aspect is that when moving downward, the total pressure on the gas phase (i.e., on the bubbles) increases, which also means that the pressure of ozone in the gas phase is increased, and of course also the driving force for ozone transfer. The system is most suitable to build into water towers for disinfection purposes (68).

6.3.3 Ejectors

Downflow ballasted bubble contacting can, in fact, be assimilated by an ejector with low counterpressure, except that in the latter the speed of the water is much higher (i.e., in the range of 1 m/s and more). The ejector injection system is often used in demanganization, in which the porous systems can easily become clogged. A typical example is the Düsseldorf process, which treats water of the river Rhine (Fig. 32). A reaction and buffer vessel with a hydraulic residence time of 30 min is incorporated in the process. The injector (A) is operated with clear water that is about 1.5% of the total flow.

High-counterpressure ejectors are more expensive to operate and can also deter-

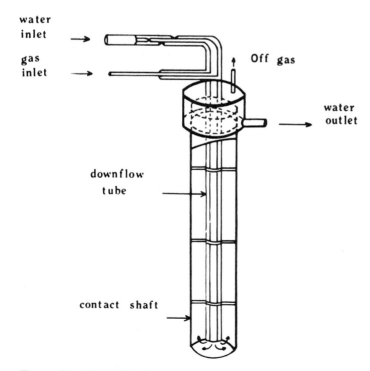

Figure 31 Hydrokinetic ozone contact.

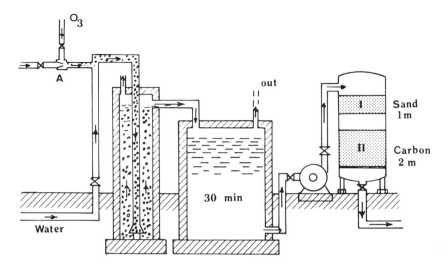

Figure 32 Principle of the Düsseldorf process for the removal of manganese. (From Ref. 83.)

mine "oversaturation" of the water in oxygen and nitrogen. These ejectors promote degassing on expansion to atmospheric pressure, with secondary phenomena such as foaming or airlifting of suspended particles.

6.3.4 Turbines

Turbines can sometimes be assimilated to multiple low-counter pressure ejectors, as is true of the turbines installed at the Brussels Waterboard Tailfer plant (Fig. 33). At nominal power the turbine circulates equal volumes of gas and water. When the gas flow is dropped, more water is circulated, up to the power tolerance limits of the motor (i.e., up to about 50% of the nominal design gas flow). The entire system is safeguarded by a dispersion tube installed in parallel with the main turbine. Standard systems are available up to 900 m³ NTP gas flow, and special designs can operate with higher flow rates.

6.3.5 Mechanically Stirred Reactors

Few equations relate to ozone transfer in mechanically stirred reactors. The transfer efficiency is usually rather low (e.g., 50 to 60%) (69). An empirical correlation has been formulated as follows (70):

$$K_L S = 2.57 \times 10^{-2} \times (Q_G \times N)^{0.67}$$

where Q_G is the gas flow in L/s and N is the number of rotations per second. Thus the $k_L S$ values for the stirred vessels are in the range of four times that of the free bubble contacting columns.

Another approach is based on power input per unit water volume (71):

$$k_L S = F \left(\frac{W}{V}\right)^{1/2} \left(\frac{Q_G}{V}\right)^{1/2}$$

Figure 33 Multiple dispersing turbine.

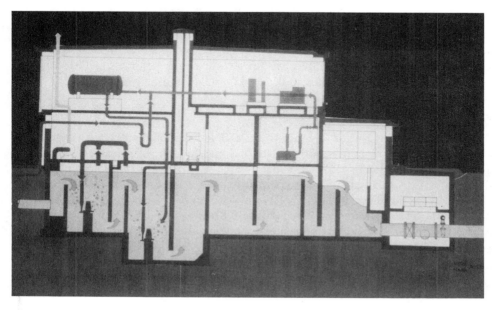

Figure 34 Cross section of the post-ozonization scheme at the Tailfer plant (Brussels Waterboard).

where F is an experimental factor (on the order of 0.02 under usual conditions), W is the power input per volume water (i.e., a function of the velocity gradient), and Q_G is the gas flow. Typical examples give $k_L S$ values in the range 2 to 10×10^{-4} (i.e., up to four times higher than for conventional bubble contact columns).

6.3.6 Packed Towers

The wet exchange surface of packed columns is at least three times higher than that in bubble columns. However, the technique is much more expensive. Equations have been formulated to be applicable to ozone transfer (72,73). They take the following form:

$$\frac{k_L}{a_p D_{O_3}} = 5.1 \times 10^{-3} (a_p d_p)^{0.4} \left(\frac{U_{SL}\rho_L}{a_p \mu_L}\right)^{4.3} \left(\frac{U_{SL} a_p}{g}\right)^{-1/3} \left(\frac{\mu_L}{\rho_L U_{SG}}\right)^{1/2}$$

where

$$
\begin{aligned}
a_p &= \text{theoretical specific surface of the packing elements} \\
d_p &= \text{nominal diameter of packing material} \\
U_{SL} &= \text{superficial water velocity} \\
\rho_L &= \text{density of the liquid} \\
\mu_L &= \text{dynamic viscosity of the liquid} \\
g &= \text{acceleration of gravity} \\
D_{O_3} &= \text{diffusion constant of ozone in water}
\end{aligned}
$$

In relation to the total surface of the packing elements, the specific surface is an implicit function of the packing material and of the operational conditions:

$$a = a_p \left\{ 1 - \exp\left[-1.45 \left(\frac{\sigma_c}{\sigma}\right)^{0.75} \left(\frac{U_{SL} S_L}{\sigma a_p}\right)^{0.2} \left(\frac{U_{SL}^2 a_p}{g}\right)^{-0.05} \left(\frac{U_{SL}\rho_L}{a_p \mu_l}\right)^{0.1} \right] \right\}$$

where σ_c is the critical surface tension related to the packing material versus the value σ for water. Typical values are (in kg/s^2): rough ceramics, 0.061; carbon, 0.056; rough or oxidized metals, 0.075; PVC, 0.04; polished metals, 0.035; glass, 0.03; polyethylene and similar plastics, 0.033.

6.3.7 Spray Towers

Spray tower ozone-to-water contact is less efficient and generally not used widely except in the case of the removal of iron and manganese. A typical example is the Jonchay plant (74) (Fig. 35). Of concern here is groundwater, which also has a deficit of dissolved oxygen. The water is sprayed into the off-gas of baffled contact chambers. The total contact time is 10 min, with an ozone treatment dose, varying according to the raw water quality, between 0.8 and 1.4 mg/L. The average manganese content of 0.45 mg/L is diminished to 0.02 mg/L and even less after filtration.

6.3.8 Ozone Contact and Chemical Reactions

When ozone is transferred to a liquid containing substances that react with the ozone, the specific transfer coefficient is no longer determined exclusively by diffusion but also by continuous exhausting of the liquid phase by reaction. This gives rise to an increased driving force for transfer and an increased transfer coefficient, defined by

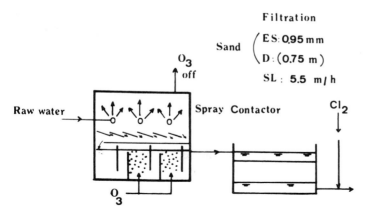

Figure 35 Manganese removal at the Jonchay plant. (From Ref. 74.)

$$\frac{k_L(R)}{k_L} = B$$

B being an empirical coefficient that is larger than unity. An approximation for the acceleration coefficient B is

$$B = \left(1 + D_{O_3} \times \frac{k_1}{k_L}\right)^{1/2}$$

in which k_1 is the first-order reaction constant of the oxidized substance. If the reaction is fast [e.g., oxidation of iodide ($k_1 \sim 10^8$ s^{-1})], B is on the order of 24. If k_1 is on the order of 10^2 to 10^4 s^{-1}, as for organic compounds for example, B, is about unity and the reaction is likely to take place in the liquid or eventually in the boundary layer. In semibatch contactors the reactions are often of observed zero order; typical examples have been published for nitrophenols (75).

Another approach is the one based on the Hatta number (H_a), which is suitable for second-order reactions, first order to ozone and first order to the reacting product:

$$H_a^2 = \frac{D_{O_3} k_2 |M|_0}{k_L^2}$$

If $H_a > 3$, the reaction will occur rapidly and essentially in the liquid film (76). For phenol, for example, $k_2 \sim 10^8$ M^{-1} s^{-1} at pH 7; hence

$$H_a = \left[\frac{1.74 \times 10^{-9} \times 10^8 \times M_0}{(2 \times 10^{-4})^2}\right]^{1/2} \gg 3$$

To realize the limiting condition, M_0 must be on the order of 0.2 mg/L or (2×10^{-6} m). This means that under the practical conditions of postozonization, most second-order reactions will take place in both the liquid and in the gas-to-liquid exchange film. This can be otherwise in model studies or for more polluted waters and effluents. As long as $H_a < 0.3$, the reaction will take place in the liquid.

Compounds that do not react directly with ozone in similar conditions can interfere by competitive inhibition mechanisms. Typical examples are acetic acid, ethanol, and ammonium chloride, which all slow down the oxidation of picric acid (77).

6.3.9. *Comparison of Ozone Contactors*

The data for comparison of the relative performance, advantages, and drawbacks of various ozone contactors are summarized in Table 4.

6.3.10. *Costs of Ozone Contact*

Costs for ozone-to-water contact can depend strongly on local conditions and the hydraulic line of the plant as a whole. Numerous systems have been used. General indications are summarized below.

1. *Dispersion with porous plates or pipes:* static system but can be subject to clogging. Plug-flow patterns should be approached. Average cost: 2 to 3 Wh/g O₃. This cost is often included in the ozone generation cost.
2. *Static mixing:* advantage of static systems, but ozone losses occur. Average cost: 4 to 5 Wh/g O₃.
3. *Direct injection in a pipe:* low cost but subject to bumping and eventually corrosion. Average cost: 0 to 5 Wh/g O₃. If operated at higher counterpressure, oversaturation of the water with oxygen and nitrogen may occur, causing subsequent "air lifting" of the impurities.
4. *Slow downflow injection shaft:* has no moving parts, can be subject to channeling, and needs a given water-to-gas flow ratio. Average cost: 4 to 5 Wh/g O₃.
5. *Eductor with total water flow:* high turbulence and good mixing but with risk of clogging. Average cost: 4 to 5 Wh/g O₃, but can range from 10 to 45 Wh/g O₃ in high-counterpressure systems.
6. *Recirculating propeller or turbine:* good mixing and high bubble contact rate which approaches the theoretical contact time but is based on moving construction parts. Average cost: 5 to 7 Wh/g O₃.
7. *Packed or plate columns:* have a good exchange rate and high ozonization yields but are subject to clogging and are water-pressure dependent. Average cost: 15 to 40 Wh/g O₃.

7. REACTION OF DISSOLVED INORGANIC IONS IN WATER WITH AQUEOUS OZONE

7.1 General Kinetic Aspects

Ozone dissolved in water is considered to react either through direct reaction of O₃ with the reacting partner R or through indirect mechanisms involving free radicals of the ozone decomposition cycle. Among them, until now only OH° has been

Table 4 Comparison of Ozone-to-Water Contracting Systems

System: example	Major advantage	Major disadvantage	Average estimated operating cost (Wh/g O_3)
Dispersion with porous elements:	Static	Clogging and channeling	2–3
Repeated static injection: Sauter	Static	Critically flow dependent	2–3
Static mixing: VAR mixer; Kenics; Ross ISB	Static	Losses	4–5
Sonic mixing: Yonkers	Static	Costs	Unknown
Total injectors (rapid): Otto; PPI	High contact rate	High losses	15–20
Pipe injection (hydrokinetic)	Low cost	Bumping	0–5
Partial injection	High turbulence, dissolution	Partial over- and under-ozonation	10–45
High counterpressure: chlorator			4
Low counterpressure: CEO France; Düsseldorf			
"Slow downflow" injection: Submers; Waagner-Biro	No moving parts	Channeling	4–5
Spray towers	Low investments	High losses	Unknown
Packed or plate column	High yields	Clogging and Pressure Dependence	15–40
Gas impeller	Suppleness of operation and low investments	Less ozone–bubble contact (losses)	2–3
Surface turbine: Kerag	Accessibility	Losses	7–10
Vortex system	Simplicity	Experimental; instability	4–6
Recirculating propeller: Tailfer, Frings	High bubble contacting	Moving turbines	5–7

recognized as a potential reagent. The direct reaction is considered to be of second order (i.e., first order in O_3 and first order in R). The ozone consumption rate is then

$$-r_{O_3} = -\frac{d|O_3|}{dt} = k_2 |O_3| \cdot |R| + k_D |O_3| + k_A |OH^-|^{1/2} |O_3|^{3/2}$$

The two second terms represent the self-decomposition (considered uncatalyzed and not inhibited by R). For example, at pH 7 and 20°C, these terms represent an r_{O_3} value on the order of 10^{-6} at an ozone concentration of 10^{-5} M/L.

While using a large excess of R and measuring the ozone consumption, we obtain the k_2 value of the pseudo-first-order consumption of ozone:

$$k_2 = \frac{2.303}{R_0 \times t} \times \frac{\Delta R}{\Delta O_3} \times \log \frac{|O_3|_0}{|O_3|_t}$$

In this case, R/O_3 is the stoichiometric factor of the reaction of ozone with R to be considered constant during the reaction time considered. Hence "initial rate constants" are measured and do not apply strictly in the case of consecutive reactions with variable stoichiometry. The k_2 values (see Fig. 36) nevertheless remain an adequate tool to evaluate competition between different compounds in reacting

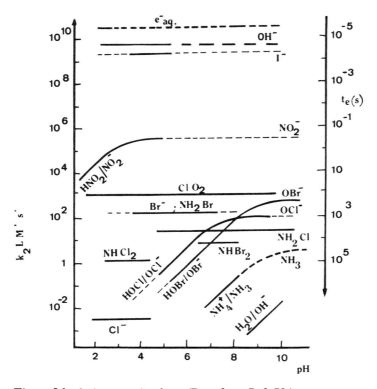

Figure 36 k_2 (apparent) values. (Data from Ref. 78.)

with ozone. In the direct reaction, the attack is electrophylic; hence anionic forms of a compound are more reactive than are neutral or cationic forms.

7.2 Stoichiometry of Inorganic Reactions

7.2.1 Oxidation of Chlorite with Ozone

The global oxidation of chlorite ion with ozone is expressed by a stoichiometric equation such as

$$ClO_2^- + O_3 = ClO_3^- + O_2$$

However, the process involves two consecutive stages:

$$ClO_2^- + O_3 \rightarrow ClO_2^\circ + (O_3^\circ) \qquad k_2 = 10^5 \ L \ M^{-1} \ s^{-1}$$
$$ClO_2 + (O_3^\circ) \rightarrow \quad ClO_3^- + O_2 \qquad k_2 = 10^3 \ L \ M^{-1} \ s^{-1}$$

This means that the second step is slower than the first, thus implicating the possible intermediate formation of chlorine dioxide. The generation of chlorine dioxide on a technical scale by ozonization of sodium chlorite at pH 4 has been described (79). Its intermediate formation in the conditions of water treatment has been demonstrated (80) (see Fig. 37). The final reaction product formed is the chlorate ion, which is not further oxidized significantly (e.g., up to perchlorate).

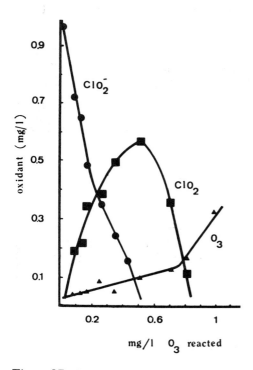

Figure 37 Intermediate formation of ClO_2 on the ozonation of ClO^-_2. (From Ref. 80.)

7.2.2 Oxidization of Nitrite to Nitrate

$$NO_2^- + O_3 = NO_3^- + O_2 \qquad k_2 = 4 \times 10^5 \text{ L } M^{-1} \text{ s}^{-1}$$

NX

$$+ \text{ org.}$$

$$\longrightarrow \text{ organonitro compounds}$$

However, in the presence of organic matter, organonitro compounds are formed. Nitrobenzenes, nitrophenols (81), and chloropicrine have also been identified. Therefore, at least in part, an intermediate NX occurs which can induce the formation of organonitro compounds (82).

7.2.3 Oxidation of Halides

No precise data are available on the potential oxidation of fluoride ion with ozone. Chloride is very slowly oxidized ($k_2 \sim 3 \times 10^{-3}$ L M^{-1} s^{-1}) into hypochlorous acid. Hypochlorous acid is further oxidized to chlorate but also partly disproportionated into chloride. Chloride is practically stable under the conditions of conventional postozonization. Bromide is oxidized first to hypobromite ($k_2 = 160$ L M^{-1} s^{-1}). The oxidation to form bromate proceeds ($k_2 = 8 \times 10^7$ L M^{-1} s^{-1}) in competition with the reaction of hypobromite with dissolved organic products or ammonium, if present. Iodide reacts very rapidly (estimated k_2 constant of 10^9 to 10^{10} L M^{-1} s^{-1}) to form hypoiodide iodine and eventually, iodate.

7.2.4 Oxidation of Chlorates, Bromates, and Iodates

Chlorates, bromates, and iodates are stable terms of ozonization. Further oxidation evolutes with reactions of a k_2 value of 10^{-3} to 10^{-4} L M^{-1} s^{-1}.

7.2.5 Oxidation of Ammonia

Ammonium is oxidized by ozone only in the deprotonated ammonia form. Nitrate is the common end product. The presumed stoichiometry is

$$NH_3 + 3O_3 = NO_3^- + H^+ + 2O_2 + H_2O \qquad k_2 \sim 20 \text{ L } M^{-1} \text{ s}^{-1}$$

The reaction of ammonium is much slower ($k_2 \sim 1$ L M^{-1} s^{-1}). The process is considered to occur through the hydroxyl radical. For particular ammonium species, the second-order reaction constants at pH ~ 8, calculated by composition of the kinetical constants of the protonated and deprotonated forms, are estimated as follows (all values in M^{-1} s^{-1}):

NH_3–NH_4^+	1	ClO_4^-	$< 2 \times 10^{-5}$
NH_2Cl	26 ± 4	ClO_3^-	$< 10^{-4}$;
$NHCl_2$	1.3 ± 0.5	BrO_3^-	$< 10^{-3}$
NH_2Br	160 ± 40	ClO_2	1000 ± 100
$NHBr_2$	40 ± 20	ClO_2^-	$> 10^4$

7.2.6 Oxidation of Chloramine and Bromamine

Chloramine and bromamine are oxidized into nitrate and there occurs liberation of the corresponding halide ions. The kinetic constants are ~ 1 and 160 L $M^{-1}s^{-1}$ for NH_2Cl and NH_2Br, respectively.

7.2.7 Oxidation of Sulfites and Sulfides

Other reductors readily oxidized with ozone are sulfites and sulfides, which form sulfates ($k_2 \sim 10^5$ to 10^9, depending on the pH).

7.2.8 Oxidation of Metals

Iron and manganese are oxidized rapidly into ferric hydroxide and manganese dioxide. However, on further treatment (83), permanganate can be formed:

$$2Mn^{2+} + 5O_3 + 4H_2O = 2MnO_4^- + 3O_2 + 6H^+$$

enters in competition with the more suitable formation for manganese removal:

$$2Mn^{2+} + 2O_3 + 4H_2O = 2MnO(OH)_2 + 2O_2 + 4H^+$$

Oxidation of the ions of manganese, cobalt, and nickel oxides is produced by reactions of which the velocity constants are on the order of 1 L $M^{-1}s^{-1}$. Precipitation of these metals is promoted on flocculation–filtration.

When present, chromium can be oxidized into the hexavalent state and eventually requires special attention. Lead is oxidized through a reaction with k_2 on the order of 10^5 to 10^6 L $M^{-1} s^{-1}$ into poorly soluble β-PbO_2. Metals complexed with EDTA, such as Pb, Ni, Cd, and Mn, are decomplexed and further oxidized. This reaction is a simulation of metals complexed with natural humic acids or of situations in which metals are discharged in a complexed form (e.g., in the photographic industries) (84). Borate buffers inhibit oxidation through ozone consumption.

7.2.9 Oxidation of Hydrazines

Hydrazines, which are of increasing importance in the aquatic environment due to spills of wash water from the cleanup of tank cars, are oxidized with ozone according to the following reaction:

$$N_2H_2 + O_3 \rightarrow 2H_2O + N_2$$

The formation of nitrates is marginal (85). Methyl-substituted hydrazines also give methanol, formol, and formic acid as secondary reaction products.

7.2.10 Oxidation of Cyanide

Cyanide (contrary to cyanhydric acid) is very rapidly oxidized into cyanate with ozone (mole per mole), which is in turn further oxidized, but very slowly. The reaction is known to be catalyzed by copper but hindered by iron. The course of the reaction is not detected in this case (86).

7.2.11   Oxidation of Ureum

Ureum is oxidized slowly (i.e., within 1 to 2 h under the conditions of swimming pool water). Nitrate, carbon dioxide, and water are the reaction products according to the stoichiometry (87)

$$(NH_2)_2CO + \tfrac{8}{3}O_3 = 2HNO_3 + CO_2 + H_2O$$

The reaction is catalyzed by metal ions such as those of copper.

7.3   Reactions with Organic Compounds

The electronic structure of ozone is dipolar with both a positive and a negative center.

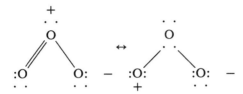

As a consequence, ozone can be both an electrophilic and a nucleophilic reagent. Usually, in water, the electrophilic reaction mechanisms predominate in direct ozone reactions. On the other hand, the decomposition of ozone into OH° radicals determines indirect ozone reaction schemes with organic compounds.

Standard oxidation reduction potentials are:

$$O_3 + 2H^+ + 2e = O_2 + H_2O \qquad E_0 = 2.07\,V$$
$$O_3 + H_2O + 2e = O_2 + 2OH^- \qquad E_0 = 1.24\,V$$
$$O° + 2H^+ + 2e = H_2O \qquad E_0 = 2.42\,V$$
$$OH° + 1e = OH^- \qquad E_0 = 2.80\,V$$

Hoigne has extensively studied the reaction kinetics of dissolved ozone consumption with different organic compounds. A partial summary is given in Fig. 38.

As a general rule, for the direct reaction of O_3, the following concepts can be assumed:

1. Saturated alkyl groups react very little. Most chlorinated hydrocarbons, even unsaturated hydrocarbons, do not react directly with ozone and require indirect ozonization through OH° radicals. Benzene reacts only very slowly; polycyclic hydrocarbons react more rapidly.
2. Phenolic compounds react within seconds and the possible pathways are shown in Fig. 39. Phenate ion reacts faster than protonated phenol.
3. Carboxylic acids, keto acids, and similar compounds are stable end products. Typical examples are given in Fig. 40.
4. Amines react very slowly with ozone except at high values of pH, where again the free-radical reaction (OH°) can predominate. Tertiary amines are oxidized more readily, as are aromatic amines (Fig. 41).

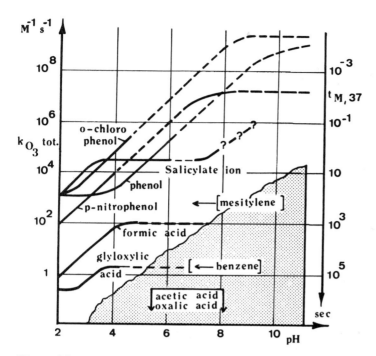

Figure 38 Typical kinetical data for ozone reactions with organic compounds (0.5 mg/L ozone in a batch reactor). (From Ref. 88).

5. Alcohols can react with ozone with intermediate formation of hydroperoxides. Carboxylic acids are the end products with primary alcohols, while ketones result from the reactions of secondary alcohols. Carbohydrates react hardly at all with ozone.

6. Mercaptans are transformed into sulfonic acids. Disulfides and sulfones are intermediates. Sulfur-containing amino acids (cysteine, cystine, and methionine) react similarly and fast.

7. Amino acids (constituents of proteins) react through the electrophilic mechanism illustrated in Fig. 42 for phenylanaline.

8. Among the phosphorus ester pesticides, parathion is most typical: Moderate ozonization leads to paraoxon, which is more toxic than parathion, but on further ozonization fewer toxic products are generated (e.g., nitrophenols), which are oxidized further to give nitrate and CO_2 as end products.

7.4 Direct Ozonization Kinetics

In drinking water treatments the direct mechanism, or reaction with ozone, considered to be O_3, is by far the most important pathway. Reaction rate constants of relevant chemical functions have been developed by Hoigne (88).

8. SAFETY

Ozone is toxic when inhaled. The threshold limit value (TLV) laid down in most countries is 0.1 ppm (vol). Temporary exposures of 0.3 ppm (vol) are acceptable

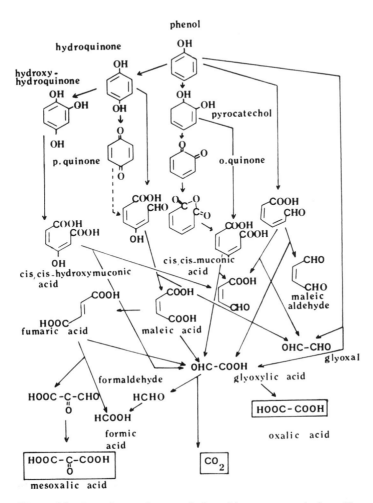

Figure 39   Reaction pathways of phenol in aqueous solution. (From Ref. 89.)

and no adverse outcomes have ever been recorded up to 1 ppm (vol). A sufficiently reliable measurement of ozone in the atmosphere often causes a problem under the normal conditions of water treatment plants. Therefore, for monitors of very low ozone concentrations in the gas phase, rigorous performance specifications have been formulated (91). A tentative list is given below in outline form.

Lower detection limit, mg/L
Time rate of response (s)
Span drift, mg/L day
Zero drift, mg/L day
Total interference equivalent response, mg/L
Control alarm set point concentration, mg/L
Full-scale response, mg/L
Precision at control point (or full scale), mg/L

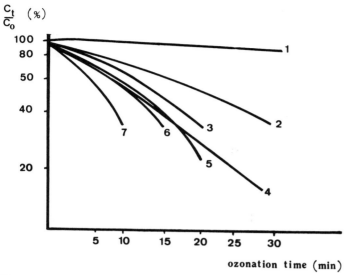

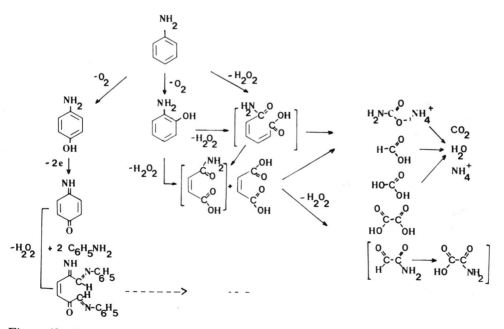

Figure 40 Ozonization of carboxylic acids (10^{-5} M; pH = 5; t = 20°C; ozone contacted 100 mg in 10 min). 1, Oxalic acid; 2, glyoxylic acid; 3, glyoxal; 4, phenol; 5, muconic acid; 6, formic acid; 7, maleic and fumaric acid.

Figure 41 Reaction scheme for the ozonization of analine.

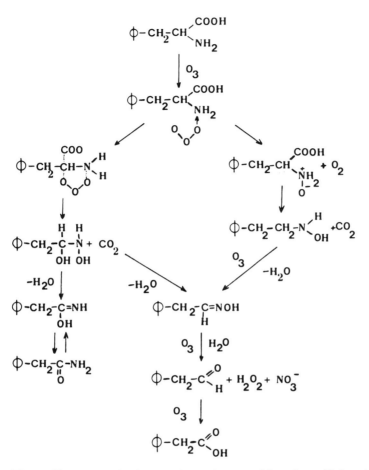

Figure 42   Schematics for reactions of ozone with amino acid functions. (From Ref. 90.)

Response linearity, %
Maximum span/zero temperature coefficient, mg/L · °C
Unserviced operation, months
Maintenance interval, months
Operating temperature range °C
Operating humidity range, °C
Water stream conditions
Calibration requirements
    Online
    Gas phase
    Dissolved phase
    Procedures
Input power
    Power, W
    Voltage, V
    Frequency, Hz

Output signal
Configuration
    Bench, rack
    Weight
    Dimensions
    Accessibility
Interface requirements
    Carrier air source and disposal
    Output recording
    Auto calibrate
    Remote operation
    Local and remote operational status

NOTATION

d_B	bubble diameter, m
d_g	gas discharge gap, mm
k_L	gas transfer constant, m/s
v_R	relative velocity, m/s
x	molar fraction of ozone in the liquid phase, dimensionless
C	capacitance, F
C_s	saturation concentration
CT	concentration (of disinfectant) multiplied by action time (usually mg · min/L)
D	diffusion constant, m^2/s
E	internal energy, J/mol
G	molecules ozone produced by 100 eV of energy
H	enthalpy (internal), J/mol
H_a	Henry's constant, atm per molar fraction of ozone in the liquid
I	ionic strength, mol/m^3
K_H	Henry's constant, atm · m^3/mol
M	mass volume of ozone, 2.4 kg/m^3
N	number of rotations per second, s^{-1}
P	power, W
P	absolute gas pressure, kPa
P_γ	partial pressure
Q_G	gas flow, L/s or m^3/s
S	entropy, J mol^{-1} deg^{-1}
S	specific exchange surface in the liquid films, m^2
U_{sg}	superficial gas velocity, gas flow/cross section
V, V_0, V_s	electrical potentials, V
β	Bunsen absorption coefficient, gas volume per liquid volume
ϵ	dielectric constants, F/m
ϵ_g	volume fraction of gas in the reactor, dimensionless
ν	current frequency, s^{-1}

REFERENCES

1. W. W. Smith and R. E. Bodkin, *J. Soc. Am. Bacteriol.*, *47*, 445 (1944).
2. L. Coin, C. Gomella, C. Hannoun, and J. C. Trimoreau, *Presse Med. (Paris)*, *72*, 2153 (1964); *75*, 1883 (1967).
3. P. Gevaudan, G. Bossy, C. Gulian, and Y. Sanchez, *Terres Eaux*, *67*, 25 (1967).
4. J. C. Block, Y. Richard, Y. Joret, Ph. Hartemann, and J. M. Folliguet, *Eau Ind.*, p. 69 (Oct. 1981).
5. D. Roy, R. S. Englebrecht, and E. S. K. Chian, *J. AWWA*, *74*, 660 (1982).
6. E. Lagrange and R. Rayet, *Rev. SEE (Belg.)* (1952).
7. G. B. Wickramanayake, A. J. Rubin, and O. J. Sproul, *Appl. Environ. Microbiol.*, *48*, 671 (1984).
8. U.S. EPA, *Fed. Regis.*, *52*(212), 42177 (1987).
9. A. T. Palin, *Water & Water Eng.*, *57*, 271 (1953).
10. W. L. Le Page, *J. AWWA*, *77*, 44 (1985).
11. V. A. Hamm, *J. AWWA*, *48*, 1316 (1956).
12. W. Hopf, *Förd. Gas. Wasser Elektrizitätsfaches*, *10*, 3 (1966).
13. P. Schenk, *Gas- Wasserfach*, *103*, 791 (1962).
14. W. Hopf, *Gas- Wasserfach*, *111*, 83 (1970).
15. Y. R. Richard, *Symposium IOA Wasser*, Berlin, 1981, p. 735.
16. M. R. Jekel, *Ozone Sci. Eng.*, *5*, 21 (1983).
17. D. A. Reckhow, P. C. Singer, and R. R. Trussel, Ozone as a coagulant aid, in *Ozonization: Recent Advances and Research Needs*, AWWA, Denver, Colo., 1986, p. 17.
18. E. Gilbert, *Water Res.*, *22*, 123 (1988).
19. B. Regnier and W. J. Masschelein, *Trib. Cebedeau*, *463–464*, 259 (1982).
20. H. Sontheimer, E. Heilker, M. R. Jekel, H. Nolte, and F. H. Vollmer, *J. AWWA*, *70*, 393 (1978).
21. A. Tiret, in *Ozonization Manual for Water and Wastewater Treatment*, W. J. Masschelein, Ed., Wiley, Chichester, 1982, p. 286.
22. W. von Siemens, *Poggendorff's Ann.*, *102* (1857).
23. U. Kogelschatz et al., Ozone generation from oxygen and air: discharge physics and reaction mechanisms in *Proc. 8th Congress IOA*, Zürich, 1987, p. A1.
24. J. D. Cobine, *Gaseous Conductors*, Dover, New York.
25. S. I. Reynolds, *Symposium on Corona*, Special Technical Publication 198, ASTM, Philadelphia, 1955.
26. W. J. Masschelein, *T.S.M. Eau*, *72*, 177 (1977).
27. W. J. Masschelein, *Ozonization Manual for Water and Wastewater Treatment*, Wiley, New York, 1982.
28. J. J. Carlins, *Environ. Prog.*, *1*, 113 (1982).
29. W. J. Masschelein, G. Fransolet, J. Genot, and R. Goossens, *T.S.M.-Eau*, *76*, 385 (1976).
30. M. Simon and H. Scheidtmann, *Gas- Wasserfach*, *109*, 877 (1968).
31. H. M. Rosen, *Water Sewage Works*, *119*, 114 (1972).
32. E. Albrecht, *Dechema Monogr.*, *75*, 343 (1974).
33. O. Leitzke and G. Grener, *Symposium IOA*, Nice, 1979, p. 76.
34. W. J. Masschelein, in *Handbook on Ozone*, Chap. 6, Vol. 1, R. C. Rice and A. Netzer, Eds., Ann Arbor Science, Ann Arbor, Mich., 1982.
35. J. C. Morris, *Ozonews*, *16*, 14 (1988).
36. A. B. Khadraoui, *Mémoire présenté à l'ISIB*, Bruxelles, 1988.
37. A. E. Rawson, *Water Waste Eng.*, *57*, 102 (1953).
38. V. I. Matsorov, S. A. Kashtanov, A. M. Stepanov, and B. A. Trugubov, *Zh. Prikl. Khim.*, *48*, 1838 (1975).

39. V. Caprio, A. Insola, P. G. Lignola, and G. Volpicelli, *Chem. Eng. Sci.*, *37*, 122 (1982).
40. J. A. Roth and D. E. Sullivan, *Ind. Eng. Chem. Fundam.*, *20*, 137 (1983).
41. A. Ouederni, *Ozone Sci. Eng.*, *9*, 1 (1987).
42. G. Gordon, *Proc. 8th Congress IOA*, Vol. 2, 1987, p. E27.
43. J. A. Roth and D. E. Sullivan, *Ozone Sci. Eng.*, *5*, 37 (1983).
44. J. Staehelin and J. Hoigne, *Environ. Sci. Technol.*, *16*, 676 (1982).
45. G. I. Rogozhkin, *Chem. Abstr.*, *73*, 81108y (1970).
46. G. Czapski, A. Samuni, and R. Vellin, *Isr. J. Chem.*, *6*, 969 (1968).
47. M. G. Alder and G. R. Hill, *J. Am. Chem. Soc.*, *72*, 1884 (1950).
48. V. Rothmund and A. Burgstaller, *Monatsh. Chem.*, *34*, 47 (1913).
49. M. O. Gurol and P. C. Singer, *Environ. Sci. Technol.*, *16*, 377 (1982).
50. M. L. Kilpatrick, Ch. C. Herrick, and M. Kilpatrick, *J. Am. Chem. Soc.*, *78*, 1784 (1956).
51. K. Y. Li, Ph.D. dissertation, Mississippi State University, 1977.
52. Sh. Sheffer and G. L. Esterson, *Water Res.*, *16*, 383 (1988).
53. L. Forni, D. Bahnemann, and E. J. Hart, *J. Phys. Chem.*, *86*, 255 (1982).
54. L. Rizzutti, K. Augugliaro, and G. Marrucci, *Chem. Eng. Sci.*, *31*, 877 (1976).
55. W. Stumm, *Helv. Chim. Acta*, *37*, 773 (1954).
56. W. J. Masschelein, *Manuel Pratique d'Ozonation*, 2nd ed., Lavoisier, Technique et Documentation, Paris, 1991.
57. H. Tomiyasu, H. Fukutomi, and G. Gordon, *Inorg. Chem.*, *24*, 2962 (1985).
58. J. L. Sotelo, F. J. Beltran, F. J. Benitez, and J. Beltran-Heredia, *Ind. Eng. Chem. Res.*, *26*, 39 (1987).
59. H. Schultze and D. Schultze-Frohlinde, *J. Chem. Soc. Faraday Transt.*, *71*, 1099 (1975).
60. J. Staehelin and J. Hoigne, *Environ. Sci. Technol.*, *19*, 1206 (1985).
61. J. Hoigne, p. 121 in *Process Technologies for Water Treatment*, S. Stücki, Ed., Plenum Press, New York, 1988.
62. M. Roustan and J. Mallevialle, Theoretical aspects of ozone transfer into water, in *Ozonization Manual for Water and Wastewater Treatment*, W. J. Masschelein, Ed., Wiley, Chichester, 1982.
63. W. J. Masschelein, p. 143 in R. G. Rice and A. Netzer, *Handbook of Ozone Technology and Applications*, Ann Arbor Science, Ann Arbor, Mich., 1982.
64. M. Roustan, J. P. Duguet, B. Brette, E. Brodard, and J. Mallevialle, *Ozone Sci. Eng.*, *9*, 289 (1987).
65. A. Laplanche, N. Le Sauze, and B. Langlais, *Proc. 9th Congress IOA*, New York, Vol. 2, 1989, p. 513.
66. N. Martin and M. M. Bourbigot, in W. J. Masschelein, *Manuel d'Ozonization des Eaux*, 2nd ed., Lavoisier, Paris (1991).
67. W. J. Masschelein, G. Fransolet, and J. Genot, *Water Sewage Works*, *122*, 57 (dec) (1975).
68. E. Brodard, M. Roustan, and J. Mallevialle, *Proc. 6th World Congress IOA*, Washington, D.C., 1983.
69. Ozone contacting, chapter in *AWWARF-CGE Ozonization Manual* (1991).
70. A. Ouederni, *Ozone Sci. Eng.*, *9*, 1 (1987).
71. J. Mallevialle, M. Roustan, and H. Roques, *Trib. Cebedeau*, *28*, 175, 1975.
72. K. Onda, *J. Chem. Eng. (Jpn.)*, *1*, 56 (1968).
73. S. S. Puranik, *Chem. Eng. Sci.*, *29*, 501 (1974).
74. Y. Richard, *IOA-NIWR Conference*, 1984, p. S-327.
75. W. J. Masschelein and R. Goossens, *Ozone Sci. Eng.*, *6*, 143 (1984).
76. E. Brodard, P. Kassen, M. Roustan, J. P. Duguet, and J. Mallevialle, *Proc. 8th Congress IOA*, Zürich, 1987.

77. W. J. Masschelein, in *Role of Ozone in Water and Wastewater Treatment*, D. W. Smith, Ed., University of Edmonton, Edmonton, Alberta, Canada, 1987.
78. J. Hoigne, *Gaz-Eaux Eaux Usees*, *65*, 773 (1985).
79. W. J. Masschelein, *Chlorine Dioxide*, Ann Arbor Science, Ann Arbor, Mich., 1979, p. 49.
80. Y. Richard and L. Brener, *T.S.M. Eau*, *76*, 627 (1981).
81. M. Dore and B. Legube, *J. Fr. Hydrol.*, *14*, 11 (1983).
82. Ch. Becke and D. Maier, *Vom Wasser*,. *59*, 269 (1982).
83. W. Hopf, *Gas- Wasserfach*, *111*, 156 (1970).
84. R. L. Shambaugh and P. B. Melnyk, *J. WPCF*, 113 (1978).
85. W. F. Cowen, R. A. Sierka, and J. A. Zirrolli, Chap. 5, p. 101, in *Chemistry in Water Reuse*, W. J. Cooper, Ed., Vol. 2, Ann Arbor Science, Ann Arbor, Mich., 1981.
86. M. D. Gurol, W. M. Bremen, and T. E. Holden, *Env. Prog.*, *4*, 46 (1985).
87. D. Eischelsdörfer and Th. V. Harpe, *Vom Wasser*, *37*, 173 (1970).
88. J. Hoigne, *Process Technologies for Water Treatment*, in S. Stücki, Ed., Plenum Press, New York, 1988, p. 121.
89. M. Jarret, Thèse, Institut National (FR) d'Agronomie, 1985.
90. M. Dore, *Chimie des Oxydants et Traitement des Eaux*, Ed., Lavoisier, Technique et Documentation, Paris, 1989.
91. D. P. Lucero, *Communication at the European IOA Seminar*, Madrid, 1990.

4

Ultraviolet Disinfection of Water

1. BASIC PRINCIPLES OF ACTION

1.1 Basic Principles

Disinfection of water with UV light is considered to be a photochemical process. Only photons absorbed through a quantified transition in the molecules of a system are active in producing photochemical changes in atoms and molecules. In other words, if the light is not absorbed, it is inefficient. Also, the energy of the photons must be sufficiently high to effect an energy-level change in the absorbing molecules. Bacteria exhibit a maximum absorption of light at 260 to 265 nm (Fig. 1). Nematode and worm eggs and tobacco mosaic viruses are more sensitive to 220-nm light [1]. Finally, whatever their origin (direct emitted, reflected, scattered, etc.), all photons absorbed are potentially active in disinfection. If more than one component of the solution absorbs the active quantum, the effect can be lowered due to competitive absorption.

1.2 General Effects

The bactericidal effect of radiant energy from sunlight was first reported in 1877 [2]; however, the UV part of the sunlight that reaches the earth's surface is merely confined to wavelengths higher than 290 nm. The UV radiation is subdivided into several regions (see Fig. 2). The range of practical importance in the germicidal effect is that of UV-C, from 280 to 220 nm wavelength.

The technical use of UV was advanced by Hewitt's discovery of the mercury vapor lamp in 1901. In 1910, Cernovedeau and Henri experimented with UV disinfection of water in Marseille. The disinfecting action of UV radiation in water is not based on the formation of ozone from oxygen, which requires energy corre-

123

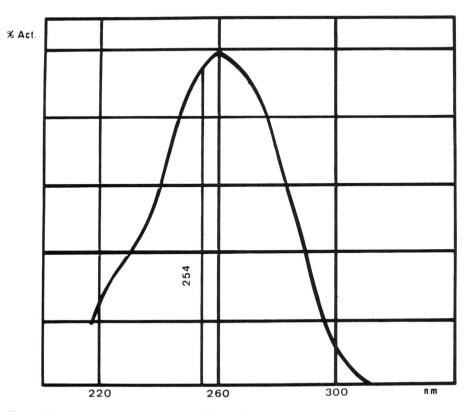

Figure 1   UV absorption spectrum of bacteria.

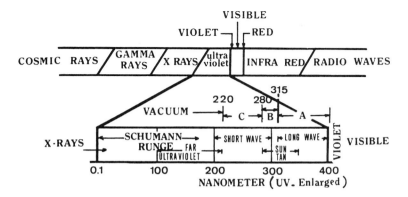

ELECTROMAGNETIC   SPECTRUM (Enlargement of ultraviolet region)

Figure 2   Subdivisions of UV range of electromagnetic waves.

sponding to wavelengths lower than 220 nm (e.g., the 185-nm band of a mercury vapor lamp) and also dry oxygen or air (3). It has also been suggested that when immersed into water, the light from UV sources produces hydroxyl or peroxy radicals, which in some instances have been considered to be the disinfecting agents. Moreover, photolysis of hydrogen peroxide at 254 nm can release hydroxyl radicals, which are strong oxidants:

$$H_2O_2 \xrightarrow[h\nu]{254\ nm} 2OH\cdot$$

(The quantum yield of 0.5 to 1 is still the subject of controversy.)

The absorption coefficient of H_2O_2 at 254 nm is in the range 17 to 18 L mol^{-1} cm^{-1}. At concentrations in the range up to 0.6 M/L, the effect of H_2O_2 and UV is usually synergetic. An enhancement factor of 4000 times the disinfection can sometimes occur. At higher concentrations of hydrogen peroxide, the effect can be lowered due to competitive adsorption of the 254-nm light. Bicarbonate interferes as a scavenger in oxyhydryl radical chemistry, as known in the field of ozone.

1.3 Actual Theory of Germicidal Action

The most advanced theory of the bactericidal action of UV supposes that photochemical alteration of DNA hinders the bacteria from normal reproduction. The DNA absorption spectrum is similar to the bactericidal efficiency distribution as a function of wavelength. Photochemical change in the pyrimidine bases of DNA have been reported, particularly formation of thymine dimers (Fig. 3). Secondary

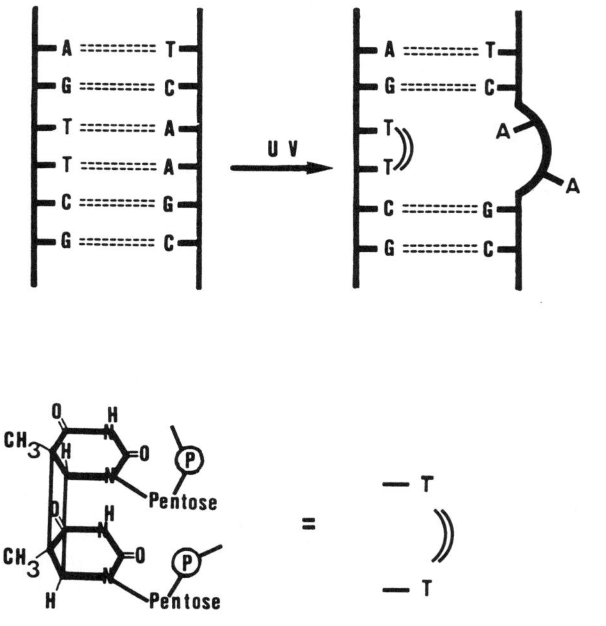

Figure 3 Formation of thymine dimer (schematic).

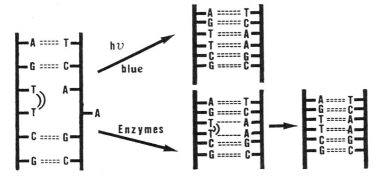

Figure 4   Repair mechanisms of irradiated DNA.

formation of thymine–cytosine and cytosine–cytosine dimers can also block cell division. If the irradiation has been insufficient, repair action can be undertaken, such as photorepair and dark or enzymatic repair (Fig. 4).

Sometimes bacterial aftergrowth has been reported (4). However, in practice, at sufficient disinfection rates this effect is marginal. Particular care must be taken to avoid warming up sampled water (5). If too-low dosing is to be avoided, considerable overdosing can also be detrimental, due to photochemical reactions. Yet the normal irradiation in disinfection is only in the range of 10% of that for photochemical oxidations. It is also worth noting that a hypertreatment with UV makes it possible to prepare apyretic injection water (6,7). [To complete the information, if dissolved ozone is present in the water, UV light induces photochemical decomposition, forming radicals that can conduct to further oxidations [e.g., $O_3 + h\nu = O_2 + O(^1D)$ and $O(^1D) + H_2O = 2\,OH\cdot$]. Finally, another possibility is the formation of Kolbe radicals from alkylcarboxy acids by UV irradiation in the presence of hydrogen peroxide.]

1.4   Equipment Design Considerations

In the design of UV equipment we must consider:

The choice of a UV source of known intensity
The quality level to be attained from the water available
Geometric criteria of reactor design
Mounting systems and safety conditions

2.   CHARACTERISTICS OF AVAILABLE UV SOURCES

2.1   UV Intensity

The intensity or irradiation density, *I*, is expressed in $J\,m^{-2}\,s^{-1}$ or in $\mu W\,cm^{-2}\,s^{-1}$. The useful intensity is influenced by various factors: the technology of the emission lamps themselves, emission yield influenced by lamp temperature, optical transmission of materials, electrical voltage, aging of the lamps, and formation of deposits on the surface. Furthermore, the conditions of emitter installation influence the useful intensity: the adsorption of light by the liquid in contact with the lamps, the reflectance of materials used to construct the reactors, the geometric factors of lamps to reactor, and superposed action of different lamps.

2.2 Emission Yield of Lamps

Table 1 indicates typical data for different technologies applicable at present for disinfection and oxidative treatment of water. The type in most common use up to now has been the hot cathode type. Manufacturers indicate a total UV efficiency (η_1) of 25 to 45% of the electrical power input (3,9,10), with an extreme lower limit of 15 to 30%. Cold cathode lamps of the preheated type are similar but have a lower power output (e.g., 15 to 20% UV efficiency), just as more modern lamps with low efficiency (e.g., 1.5 to 2%) which are used less in water treatment but their emission yield is generally accurately known. They are often of the single bulb type and most suited for research work, especially when low irradiation doses are studied. Medium- and high-pressure mercury lamps are polykinematic sources in which the 254-nm line is less dominant (see Fig. 5) and the total UV efficiency barely attains 5 to 10%. Like high-yield low-pressure lamps, the potential interest of this technology is in the high linear energy, which makes it possible to construct smaller reactors for a given irradiation dosage.

2.3 Effect of Temperature

For all Hg/A lamps the optimum emission temperature is about 40°C. Ambient (e.g., water) temperature in contact with the lamp, if lower than 40°C, can make the emission yield drop significantly (8) (see Fig. 6). If not indicated otherwise, the emission yield indicated by the manufacturers is at 40°C and must be corrected for temperature as necessary. A typical curve is given in Fig. 6.

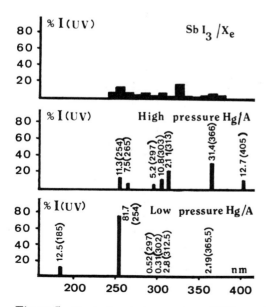

Figure 5   Typical emission spectra of UV germicidal lamps.

Table 1 Average Data of Typical UV-Source Technologies[a]

Type of source	Low-pressure hot cathode	Hg/A low-pressure cold cathode[b]		Hg low-pressures high yield	Hg/A medium pressure/high yield[c]	Sb/Xe High pressure
		Preheated type	Glow discharge			
Gas pressure	10^{-3}–10^{-2} mbar	10^{-3}–10^{-2} mbar	10^{-3}–10^{-2} mbar	(Carrier?)	1–3 bar	1 bar
Starting time	3–5 min	Immediate	Immediate	5 min	5–8 min (no reignition)	0 if (<50°C) (no reignition)
Specific electrical loading (W/cm glow zone)	0.45	0.4–0.6	0.85	30	>40	30–40 (60)
Linear output (W/cm discharge length)	0.2–0.3	0.06–0.1	0.013	6.15	3–7	3.5–4.5
UV efficiency (η_1) W(UV)/W(e)	0.25–0.45	0.15–0.2	0.015	0.2	0.07–0.12	0.12
UV-C% of total UV	80–90	85	85	85	≤10	≤50
Average lifetime (h)	7000–10,000	15,000	2500	2000	4000	7000
Optimal temperature (°C)	40	40	40	40	40	−20 to +70

[a]Lamps based on gas-continuum discharge are not yet applied in this specific technique, as they are more suited for the ozone-active range below the 200-nm wavelength zone. The subject is in full development. The low-pressure mercury–argon lamp is a monokinematic UV source from which about 80% of the UV intensity is due to the resonance line at 254 nm.

[b]The cold cathode type delivers more vacuum-UV light of 184.9 nm than the hot cathode.

[c]The maximum pressure mercury-arc lamps (30 to 100 bar) with $\eta_1 = 0.2$ to 0.3 are less applied in water disinfection.

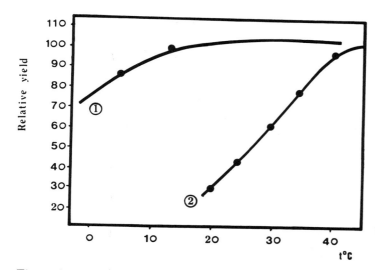

Figure 6 Drop in emission yield as a function of lamp temperature (low-pressure Hg/A — hot cathode lamp). 1, Contact with air; 2, contact with water.

Antimonium iodide/xenon lamps have been developed with a UV efficiency of up to 12%. The yield is not dependent on temperature in the range -20 to $+70°C$, and the source is polykinematic. Less than 20% of the total UV power is in the range directly applicable for disinfection (see Fig. 5). The linear electrical loading of the lamp ranges around 40 W/cm. High-pressure Hg/A and Sb/Xe lamps require high voltage during starting. They cannot be reignited immediately on an on–off basis.

2.4 Effect of Current Voltage

The emission yield of mercury lamps also depends on the electrical voltage available. This effect is more important for cold cathode than for hot cathode lamps (see Fig. 7).

2.5 Effect of Aging

The aging effect of electrodes and lamp materials causes a drop of about 10% during the first 100 to 200 h of service, which then gradually tends to stabilize at 80% of the nominal emission after ± 1000 h of service (see Fig. 8). The normal service time of 1 year for low-pressure Hg lamps can lower the emission intensity by 40% of nominal. It is advisable to check regularly using a photocell. One single start–stop operation causes aging equivalent to that resulting from 1 h of nominal operation.

2.6 Transmission Yield

Even though the transmission yield of optical material is generally taken into account in emission yields and indicated by the manufacturers, this point may require occasional verification. The UV transmittance of materials is important for the construction of isolating enclosures between the lamp and the water. Typical data are indicated in Fig. 9. The following comments can be made on the point of transmittance of optical material:

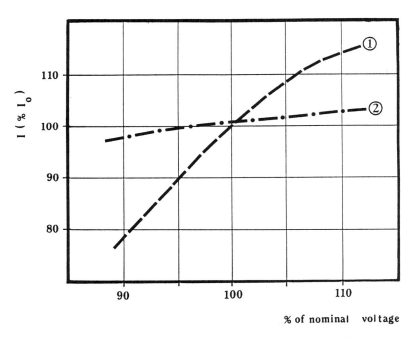

Figure 7   Relative UV emission intensity as a function of voltage changes (low-pressure Hg/A lamps). 1, Cold cathode type; 2, hot cathode type.

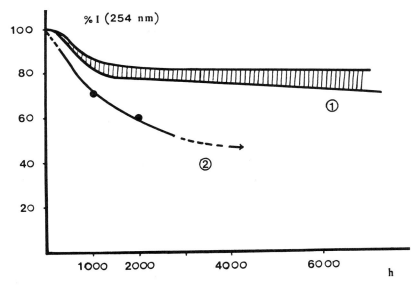

Figure 8   Drop in UV intensity on aging of a typical Hg-A source. 1, Low pressure; 2, high pressure.

Pyrex 9741: considered here as the material with unit cost index; transmittance 40 to 70% at 254 nm, no transmittance at 185 nm. Fast change in transmittance on aging is observed (e.g., 30% lowering after 100 h).

Vycor 791: relative cost factor 2.5; 85% transmission at 254 nm, no transmittance at 210 nm. Loss of 2 to 2.5% after 100 h and of 13% after 200 h of irradiation.

Crystalline quartz: relative cost factor 5; up to 87% transmission at 185 nm.

Fused quartz: relative cost factor 3.3; 87% transmission at 185 nm, 93% at 194 nm and 98% at 254 nm.

Kel-F: transmission of 254-nm light is good, but the material quickly deteriorates on irradiation.

PTFE: transmittance is of 45 to 65% at 265 nm, but the material is still of unknown aging on irradiation. The material is subject to biological fouling.

The transmission curves in Fig. 9 should be considered as indicative only. The results can also depend on the purity of the quartz. Moreover, if the application concerns disinfection only, it is suitable to avoid the 185-nm wavelength in order to limit the photochemical reactions of dissolved substances. However, if ozone formation in the gas phase is the goal, the material must demonstrate good transmittance of the 185-nm wavelength.

2.7 Reflection of 254-nm Light

The reflection of 254-nm light plays an important part in the indirect irradiation technique, in which lamps are placed outside the water stream and part of the light is directed toward the water with reflectors. In the submerged lamp technique, if deposits are formed on the lamps or on the enclosure tubes, some of the photons can be lost by reflection. When the absorbance of the water to be treated and the thickness of the water layer are of magnitudes that do not make it possible for them

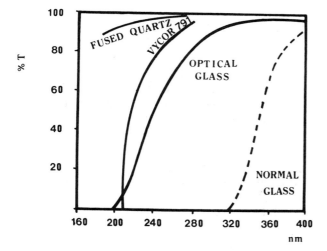

Figure 9   Transmission yield of optical material. (The optical transmittance of the liquids (e.g., waters of different quality) is to be considered most accurately. This is treated in Sections 3 and 4.)

to absorb some of the photons directly, it is important to take into consideration reflection on the inner walls of a tubular reactor or on the surface of the rods used to assemble such a system mechanically. Typical values are (in % reflection of the 254-nm wavelength):

Etched aluminum	88
Precipitated magnesium oxide with calcium carbonate	70 to 80
Aluminum paints	75 to 80
Aluminum foil	73
Plaster	40 to 60
Metallic chronium	45
Metallic nickel	38
White paper	25
Stainless steel, AISI 304 or 316	<25
Water paints	10 to 30
Porcelain, white	5
Normal window glass	4
Open water surface	4
PVC and miscellaneous organic precipitates	1
"Black" lamp coating	1

Precipitates and deposits on the outer side of the lamps or thermal isolating tubes can determine an important loss in yield. If the water is in stagnant flow conditions, this effect can be enhanced; therefore, adequate design rules must be observed.

2.8   Control of UV Intensity

The UV intensity can be estimated by calibrated photocells either of the cylindrical type (shaded area 2 in Fig. 10) or with cosinusoidal photocells which have a directional sensibility that corresponds approximately to the complement of a cosine curve (curve 1 in Fig. 10). The intensity values indicated by cylindrical cells depend most heavily on the direction of measurement; the cosinusoidal cells have a broader aperture of measurement. Cylindrical cells are ideal for operational control or for laboratory experiments in which small, even surfaces are irradiated.

The values of UV intensities or power as indicated in the manufacturers' catalogs are usually measured with cosinusoidal cells in dry and stagnant air at 20°C. A built-in photocell in disinfection units enables the units to control the lamp performance level over time. However, it does not make it possible to measure the irradiation dosage with sufficient accuracy.

2.9   Nominal Emission Intensity or Power

The nominal emission intensity is seldom given by manufacturers in terms of W(UV) but is often indicated as irradiance or power per unit surface at 1 m distance (W/m^2 or mW/cm^2). Expressed correctly, these tabulated values correspond to the power

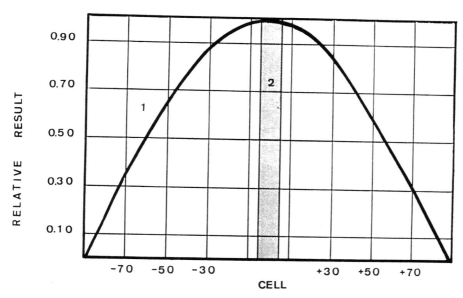

Figure 10 Sensitivity profile of cylindrical (curve 2) cells and cosinusoidal photocells (curve 1) (equivalent aperture 81° in example).

emitted per steradian, that is, the solid angle corresponding to 1 m² located on a sphere of 1 m radius around the source considered to be located in a point. Consequently,

$$\Omega = 4\pi \frac{R^2}{R^2} = 4\pi \qquad (\text{for } R = 1 \text{ m})$$

By multiplying the power per steradian by 4π, one should have the total energy emitted by the lamp considered as a point source. However, in practice, lamps are not point sources and have an internal angular distribution of photons. Therefore, the power measured per steradian must be multiplied only by a factor of about 8 (see Fig. 11), and the real value must be given by the manufacturer of the lamp. The total output can also be obtained by direct measurement with a rotating photometer as indicated in the experimental setup in Fig. 12.

2.10 Simplified Method of Photometric Measurement of Nominal Light Intensity

The power is given by

$$P = 2 \int_{90°}^{0} \int_{0}^{2\pi} I(\theta) d\theta \, dA(\theta)$$

$$\int_{0}^{2\pi} I(\theta) \, d\theta = 2\pi a(\theta) I(\theta)$$

$$dA(\theta) = r \, d\theta$$

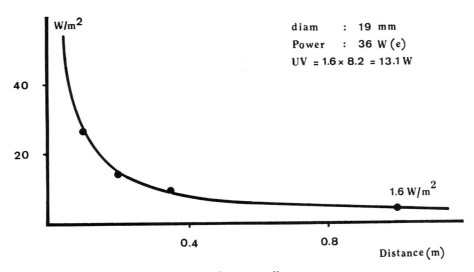

Figure 11   Nominal intensity expressed per steradian.

Hence

$$P = 2 \int_{90°}^{0°} 2\pi a(\theta) I(\theta) r \, d\theta$$

and

$$\frac{1}{2} \frac{dP}{d\theta} = 4.022 I \sin \theta \qquad (\text{for } r = 0.8 \, m)$$

This technique is also adequate for relative measurements of the power loss of a given lamp through aging. Graphical integration of the area under an experimental curve obtained by plotting the cumulative $dP/d\theta$ value as a function of θ between 0

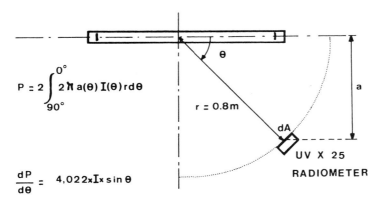

Figure 12   Intensity measurement by a rotating photometric detector.

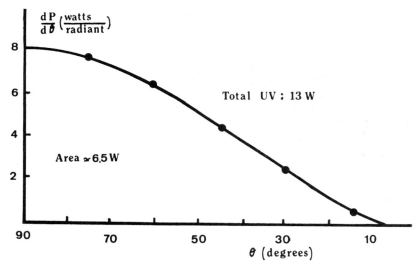

Figure 13   Graphical integration of nominal power measured by photocells. (From Ref. 9.)

and 90° delimits a surface of half the power emitted by the source. A typical example is shown in Fig. 13.

Zonal distribution of radiation power is less relevant: As long as the water flows coaxially to the lamp and with complete axial mixing, the exposure can be considered as homogeneous. The power emitted at the ends of tubular lamps is lower than that in the middle. The greater the distance from the lamp or the higher the absorption, the higher this difference is. If the flow is more or less perpendicular to the lamp axis, precautions (mixing, mounting the lamp ends outside the water zone, etc.) must be followed to avoid submitting to low exposure the part of the water circulating at the ends of the lamps. In water the intensity in these zones can be decreased below 50% of that in the central part of the lamp (Fig. 14).

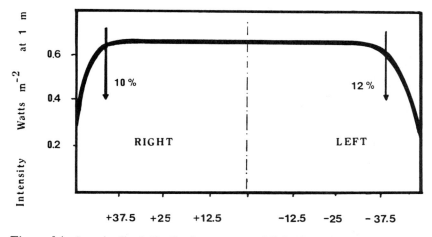

Figure 14   Longitudinal distribution pattern of light intensity of a tubular lamp (typical example).

2.11 Actinometric Measurement of Nominal UV Intensity

Actinometry supposes a quantitative and reproducible relation between the number of photons of a given frequency irradiating a reagent and a known single photon-induced chemical transformation. In the determination it is preferable that all photons be absorbed. For the analysis of 254-nm light, U^{IV}–sulfate 10^{-2} M coupled with oxalic acid 5×10^{-2} M developed by Leighton and Forbes (10) gives reproducible quantum yields of 0.6 mol/Es for low-pressure mercury lamps (9). The reaction is composed of a global pathway of oxidation of U^{IV} to U^{VI}, which in turn oxidizes oxalic acid. The U^{IV} is recovered. The decrease in oxalic acid concentration is of zero order in the range of applicability (Fig. 15).

The energy (Es = Einstein per second) is calculated from the decrease in concentration in oxalic acid according to the formula

$$I(\text{Es/s}) = \frac{\Delta(\text{oxalic acid}) \, (\text{mol/L}) \times \text{volume} \, (\text{mL})}{\Phi(=0.6)(\text{mol/Es}) \times t(\text{s}) \times 1000 \, (\text{mL/L})}$$

and $I \times 4.7 \times 10^5 = P(\text{W(UV)}$ at 254 nm).

2.12 Checklist Related to Lamps and Conditions of Installation

1. The intensity for design is best considered as that obtained after 1000 to 1200 h of emission, which means about 75% of the original intensity.
2. Low-pressure mercury lamps drop in intensity after 7500 to 10,000 h of operation. Replacement once a year is necessary.
3. Medium- or high-pressure mercury lamps or antimonium iodide lamps have a lifetime only about half that of low-pressure mercury lamps.

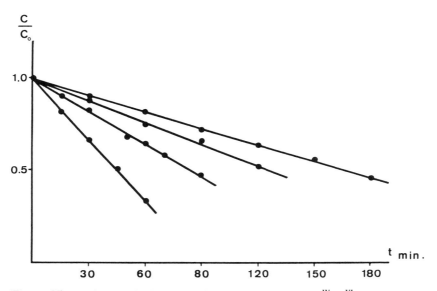

Figure 15 Actinometric decomposition of oxalic acid by U^{IV}–U^{VI}–sulfate.

4. One start–stop procedure is equivalent to aging by 1 h of service.
5. The lowest value of water temperature must be considered for emission intensity.
6. Full nominal emission yield is obtained only ±5 min after starting (hot cathode lamps); thus a time constant is necessary in automated start–stop procedures.
7. Under all circumstances, the lowest voltage of electrical current locally available is to be used when computing the emitted intensity.
8. The lamps are best installed at the site where water flow is the most constant (e.g., at the inlet of a water tower rather than at the outlet, although the water tower can pose a risk of recontamination).
9. Easy access to lamps and/or enclosures is necessary for cleaning, for which use of an optical fabric moistened with alcohol or dilute ammonia is recommended. If calcium carbonate deposits have formed, the use of a 6% citric or acetic acid is recommended. Abrasive detergents, abrasives, polishing oils or fats, and the like, are prohibited.
10. Overheating of ballasts must be avoided by installing them on a separate, vented panel.
11. If use of a bypass system is applied, absolute tightness of the valves is necessary to avoid the passage of untreated water into the treated stream. [Some official regulations do not accept the bypass system (11).]

2.13   Inventory of Existing UV Lamps
(Based on Belgian market, 1988.)

American Ultraviolet.

A great number of different types of cold and hot cathode lamps are available, with some models of fast-starting lamps.

Cold cathode lamps: 9 to 34 W(e) with UV yield from 20 to 30%; diameter 16 mm, length 22 to 73 cm, 20 to 75 μW/cm^2

Fast-start lamps: 16 to 65 W(e), UV yield of 23 to 35%, diameter 16 to 19 mm, length between 27.6 and 147 cm, 55 to 190 μW/cm^2 (at 1 m)

Hot cathode lamps: 4 to 30 W(e), yield 17.5 to 28% UV, length between 6 and 80 cm, diameter 13 to 25 mm, 5 to 85 μW/cm^2 (1 m)

Address: American Ultraviolet, 21 Commerce Street, Chatham, NJ 07928

ASEA-BBC.

Low pressure mercury vapor lamps with a high linear yield [1.35 W(UV)/cm]
U-shaped lamps
710 W(e); 22.8% UV, length 2 × 500 mm, diameter 14 mm, 1.9 mW/cm^2 (1 m)
895 W(e); 24% UV, length 2 × 700 mm, diameter 14 mm, 1.9 mW/cm^2 (1 m)
Further information: BBC-Druckschrift Nr. CH-E-S.0315.1-II-D
Address: BBC-AG, Geschäftsbereich E, CH-5401 Baden, Switzerland

General Electric.

Manufacturer of UV-incandescent lamps and UV-fluorescent lamps. General characteristics, power, and yields are as follows:
4 W(e); 9.5% UV; length 6 in.; diameter ½ in.; 7 μW/cm^2 (1 m)
8 W(e); 12.5% UV; length 8½ in.; diameter ⅝ in.; 14 μW/cm^2 (1 m)
15 W(e); 18.7% UV; length 14 in.; diameter 1 in.; 38 μW/cm^2 (1 m)

25 W(e).; 16% UV; length 14 in.; diameter 1 in.; 54 μW/cm^2 (1 m)

30 W(e).; 22% UV; length 32 in.; diameter 1 in.; 85 μW/cm^2 (1 m)

36 W(e).; 28.9% UV; length 29¼ in.; diameter ¾ in.; 135 μW/cm^2 (1 m) (single-pole lamp without ballast, immediate start)

65 W(e).; 22.2% UV; length 58 in.; diameter ¾ in.; 185 μW/cm^2 (1 m)

Further information: GE Black Light Bulletin

Address: E. Dirckx, c/o General Electrics Belgium, Chaussée de Haecht 1878, B-1130 Brussels, Belgium

Hanovia.

Low-pressure mercury vapor lamps of 30 W(e) and also medium-pressure lamps of 1.2 to 2.5 kW are available. Linear emission intensity at the lamp surface is 67 and 1720 W/m^2, respectively. Further details by request to the manufacturer.

Documentation: Hanovia fact file 3

Address: Hanovia Ltd., 145 Farnham Road, Slough, Berkshire SL1-4XB, England

Heraeus (Original Hanau).

Mercury vapor lamps of different types are available, often designed for air treatment and laboratory experiments

Sb I$_3$-Xe lamps of the polykinematic type with high intensity, of 14% yield between 300 and 280 nm, are marketed for water treatment

1000 W(e); 49% UV-C; length 25 cm; diameter 16 mm; 1 mW/cm^2 (1 m)

2000 W(e); 49% UV-C; length 60 cm; diameter 16 mm; 1 mW/cm^2 (1 m)

Address: Original Hanau Heraeus GMBH, Höhensonnestrasze, D-6450 Hanau, Germany

Katadyn (Multus).

Low-pressure mercury vapor lamps

45 W(e); 30% UV, length 765 mm; diameter 19 mm (the lamps are constructed with a quartz enclosure, diameter 40 mm, with 7% loss of UV intensity)

Address: Katadyn Produkte AG, Industriestrasze 27, CH-8304 Wallisellen, Switzerland

Osram.

Low-pressure mercury vapor lamps; relative intensity is well known at different wavelengths

10 W(e); 45% UV; length 142 mm, diameter 10 mm; 55 μW/cm^2 (1 m)

15 W(e); 23% UV; length 378 mm; diameter 26 mm, 40 μW/cm^2 (1 m)

30 W(e); 26.7% UV; length 835 mm; diameter 26 mm; 90 μW/cm^2 (1 m)

Further information: Entkeimungs-und Ozonstrahler, HNS: Osram-Produktinformation

Address: Osram-GMBH, Postfach 900620, D-8000 Münich 90, Germany

Philips.

Mercury vapor lamps with cold cathode (6 W) or hot cathode (15 to 40 W)

Lamps constructed of optical glass

6 W(e); 1.4% UV; length 75 mm; diameter 26 mm (cold cathode)

15 W(e); 23.3% UV; length 438 mm; diameter 26 mm; 37 μW/cm^2 (1 m)

25 W(e); 20% UV; length 438 mm; diameter 26 mm; 70 μW/cm^2 (1 m)

30 W(e); 30% UV; length 895 mm; diameter 26 mm; 83 μW/cm^2 (1 m)

40 W(e); 31.5% UV; length 1200 mm; diameter 38 mm; 94 μW/cm^2 (1 m)

Further information: Philips, TUV-Bulletin 611.2

Address: Philips Gloeilampenfabrieken, Eindhoven, The Netherlands

Ultra-Violet Products Inc.
Various technologies applicable in laboratory investigations
Address: Ultra-Violet Products, San Gabriel, CA 91778

Voltarc.
Mercury vapor lamps with hot or cold cathode; constructed of Vycor (also available
   in ozone generating alternatives)
39 W(e); 35.4% UV; length 76.2 cm; diameter 15 mm; 120 μW/cm^2 (1 m)
65 W(e); 41% UV; length 147.3 cm; diameter 15 mm; 190 μW/cm^2 (1 m)
24 W(e); 22.9% UV; length 47.6 cm; diameter 15 mm; 52 μW/cm^2 (1 m)
Further information: Voltarc G-UV-1173 Bulletin
Address: Voltarc Tubes Inc., 102 Linevood Avenue, Fairfield, CT 06430

Wedeco.
Fused quartz low-pressure mercury vapor lamps; flat lamps, 40 mm, large single-
   pole submersible probes
160 W(e); 46.9% UV; length 800 mm (irradiance not given)
130 W(e); 45.4% UV; length 800 mm
80 W(e); 46.2% UV; length 800 mm
65 W(e); 44.6% UV; length 800 mm
40 W(e); 45% UV; length 360 mm
32 W(e); 43.8% UV; length 360 mm
Flat high-intensity, low-pressure mercury vapor lamps simpsonized with trace metals are a recent development by Wedeco. These lamps, which have a constant
   emission intensity between +5 and +75(85)°C, cover a broad range of applications.
Further information: Wedeco Bulletin
Address: Wedeco, Daimlerstrasze 5, D-4900 Herford, Germany

3.   DESIGN OF UV REACTORS

3.1   Basic Principles

The basic principles in the design of UV reactors rely on the relation between
reactivity and irradiation dosage. The Bunsen–Roscoe law indicates that in static
systems the disinfection level is of first order as a function of irradiation dosage:

$$N_t = N_0 \exp[-k'(It)] \quad \text{or} \quad \ln \frac{N_t}{N_0} = -k'(It)$$

where N_t and N_0 are the average volumetric concentrations in germs or compounds,
respectively, after an irradiation time t and at zero time, I is the light intensity or
irradiation density, t is the irradiation time, and (It) is the irradiation dosage. In the
literature a D_{10} dosage is often mentioned. This is equal to that necessary to lower
the volumetric concentration by a factor of 10. If a 99.99% killing rate of germs is
required, a $4D_{10}$ dosage is necessary. Table 2 lists a series of $4D_{10}$ values as reported
in or calculated from the literature.

The Bunsen–Roscoe law holds strictly for a bacterial monolayer on an even
surface that is irradiated homogeneously, for instance, in a small cup having the
same dimensions as a photocell, capable of being irradiated alternatively in order to

Table 2 Killing Rates 254-nm UV ($4D_{10}$ Values)[a]

Organism	Optimal conditions (J/m²)	Practical conditions (J/m²)
E. coli	66	130–400
		<240>
Enterobacteria (general)	34–76	240
Legionella pneumophila	70	90
S. thyphimurium	150	250
Ps. aeruginosa	105	?
Clostridia	?	500
Ubiquitous (total plate count)	?	(500)
Resistant bacteria (Micrococcus radiodurans)	2,000	?
Brewer's yeast	66	250
Fungi	130–175	1,200
Mold spores	?	2,500
Coliphagi	40	?
Viruses (entero)	?	150–1500
Animals (micro)	?	(5,000)
Microalgae: phytoplankton	?	1,000
Algae (green-blue)	?	(10,000–25,000)

[a]Optimal conditions are as measured accurately, and the practical conditions are as in practice, which means not often precise in indicating dosage. These values also account for the flow-through patterns of the reactors.

measure the irradiation intensity. These conditions are significantly different from those encountered in the practice of water treatment, and the design must be adapted accordingly. Moreover, the first-order concept is valid only for a pure culture of a given microorganism without such protective mechanisms as encapsulation, spore forming, or shielding by suspended solids or bacterial clusters. In the monomolecular and first-order concept, the $x D_{10}$ value should vary linearly as a function of the irradiation dosage. Values observed in a typical cylindrical flow-through reactor are shown in Fig. 16. The general trend of these data is that for higher decay rates (e.g., at $4D_{10}$ a comparatively higher irradiation dosage is required versus D_{10} than would be expected in the strictly first-order dose-to-decay ratio concept. This is also illustrated in Fig. 17.

According to the logarithmic correlation between the remaining number of germs and the irradiation dosage, the residual number of germs can never be zero. However, at high decay rates, significant discrepancies often occur in the log–lin. relation between volumetric concentration in germs and irradiation dosage. This effect can be described by assuming that for a given bacterial population and strain, a number of organisms potentially resistant to disinfection exist (i.e., protected organisms, N_p). So the Bunsen–Roscoe law can be reformulated as follows (12):

$$N_t = N_0 \exp[-k'(It)] + N_p$$

assuming that $N_0 = N_0' + N_p$, in which $N_p \ll N_0$ and $N_0 \sim N_0'$ in the first term of the decay rate equation above.

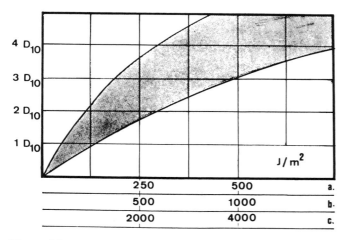

Figure 16   Distribution of the relation dosage to mortality for typical germs. a, Enterobacteria; b, sporulating bacteria; c, fungi.

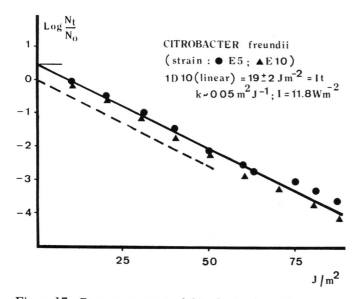

Figure 17   Decay rate curves of *Citrobacter freundii.*

3.2   Multihit/Multisite Killing Concept

Under experimental conditions more similar to practice, such as in plug-flow reactors or in completely mixed static reactors, a lethal lag phase is often observed at low irradiation dosages. Under these conditions the concepts of multihit and multisite killing should be used. A certain number of vital centers are to be hit, each by at least one photon, so as to obtain final deactivation of the organism. By accepting a pseudo-first-order kinetics for each elementary reaction of each individual vital center, of total number n for a given organism, the fraction of the total centers hit after an irradiation time of t equals to $1 - \exp(-kt)$. The probability of having hit n centers is then

$$P_t = [1 - \exp(-kt)]^n$$

and the fraction of the surviving organisms is

$$1 - P_t = \frac{N_t}{N_0} = 1 - [1 - \exp(-kt)]^n$$

Binomial expansion of this killing probability gives

$$P_t = [1 - \exp(-kt)]^n = 1 - n\exp(-kt) + \frac{n(n-1)}{2}\exp(-kt)^2 \cdots$$
$$= 1 - n\exp(-kt)$$

The decay rate is then expressed by

$$\frac{N_t}{N_0} = ne^{-kt} \qquad \text{or} \qquad \log\frac{N_t}{N_0} = \log n - \frac{kt}{2.3}$$

The n value can be deduced from the extrapolation to time $t = 0$ of the linear part of a semilogarithmic dosage-to-decay correlation (e.g., $n = 3$ for *Citrobacter feundii*) (see Fig. 17).

Following the concept where different vital centers in a single organism are each to be hit once in order to kill or deactivate the organism, this means multisite killing, the n value does not depend on the original volumetric concentration of organisms, and the linear parts of the decay curves (log N_t versus I_t) are parallel lines. In the multihit concept, in which a given vital center must be hit several times, the linear decay curves for different N_0 values are not parallel. It often remains difficult to clearly distinguish the mechanisms from one another. There also exists a normalized representation in which the time scale is given in t/t_r, t_r being a reference time in the linear part of the decay curves, after which the number of organisms is decreased to a fraction of $1/e$ or to 43.4% of the starting number. Preestablished curves then enable us to locate the experimental points and to interpolate the most probable n values. See Fig. 18 for some examples. Typical data obtained in a completely mixed laboratory reactor, designed to study the effect of low irradiation doses are also indicated in Fig. 19.

3.3   Step-by-Step Hit Theory

In the step-by-step killing scheme it is supposed that by continuous irradiation of the bacteria, partially hit bacteria can eventually repair after the irradiation (13). To avoid this effect, it is supposed that at least a number of consecutive reactions must occur (e.g., n).

$$B_0 \xrightarrow{\text{KI}} B_1 \xrightarrow{\text{KI}} B_2 \xrightarrow{\text{KI}} B_i \xrightarrow{\text{KI}} B_{n-1} \xrightarrow{\text{KI}} B_n$$

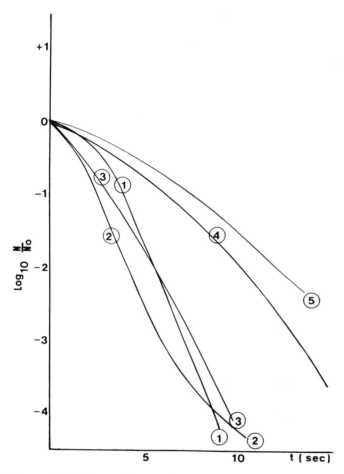

Figure 18 Direct and normalized representation of multihit killing of test organisms in a low-dosage reactor (reactor 1). 1, *Proteus mirabilis* ($n = 20$); 2, *Citrobacter freudii* ($n = 3$); 3, *E. coli* (sp.) ($n = 2$); 4, *E. coli* C ($n = 2$); 5, same as 4, no mixing.

When accepting that all elementary k values are identical, an intermediate state (e.g., B_x), which expresses the relation between the change in volumetric concentration of germs and irradiation time, according to (13), is given by

$$\frac{N_x}{N_0} = \frac{k(It)^x}{x!} \exp[-k(It)]$$

The fraction of organisms then surviving is given by

$$\frac{N_x}{N_0} = \exp[-k(It)] \sum_{x=0}^{n-1} \frac{k(It)^x}{x!}$$

(see curve 1 in Fig. 19).

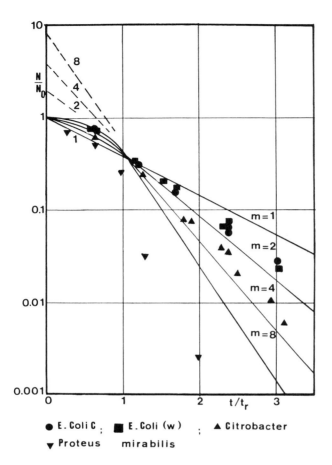

● E.Coli C ;   ■ E.Coli (w)   ;   ▲ Citrobacter

▼ Proteus   mirabilis

In all the concepts developed thus far the pseudo-first-order constants of all the elementary steps are supposed to be equal. Computation by successive approximation makes it possible to compute the most probable values of n and k. Applied to a completely mixed annular reactor, the same approach gives a decay rate equation as follows:

$$\frac{N_x}{N_0} = \sum_0^{n-1} \frac{k(It)^x}{1 + k(It)^{x+1}} = 1 - \left[1 + \frac{1}{k(It)}\right]^{-n}$$

In both of the proceding equations, authors (13) have assumed that I is equal to the average intensity in the reactor zone, in that the axial mixing is considered as complete in both approaches. It appears that for short reaction times, there is a zone in which the decay rate is slightly higher in a completely mixed reactor; however, overall performance is higher in a plug-flow reactor with axial mixing (see curve 2 in Fig. 19).

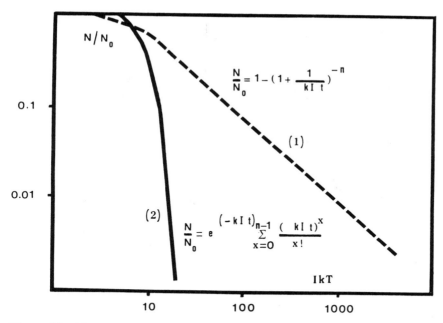

$$\frac{N}{N_0} = 1 - (1 + \frac{1}{kI\,t})^{-n}$$

$$\frac{N}{N_0} = e^{(-kI\,t)} \sum_{x=0}^{n-1} \frac{(kI\,t)^x}{x!}$$

Figure 19 Theoretical decay curves for step-by-step killing in plug-flow reactor (1) and completely mixed reactor (2). τ = reaction time for a single step. (Data from Ref. 13.)

3.4 Mixing Conditions in UV Disinfection

Mixing conditions are of utmost importance for the efficiency of reactors. For laboratory investigations in batch reactors, preference is to be given to complete mixing. In flow-through reactors, which are those mostly used in practice, the preferred technique is longitudinal plug flow with complete axial mixing. Different, all more-or-less empirical mixing techniques have been incorporated in commercial reactors. This makes it possible to place UV lamps orthogonal with regard to the water stream (14), as well as incorporating static mixers in a pipe reactor (8), mechanical wipers (15), conical dispersion elements promoting water passage close to the lamps (8), and finally, placement of the UV source in the turbulent area of an ejector (16).

The technical design concepts for mixing have been approached in a more rational way by Scheible (12) for the case of wastewater disinfection; however, these concepts can be generalized. Turbulent flow conditions (i.e., a Reynolds number [Re = $D/(v \times v)$] higher than 2000) are preferred to laminar flow conditions. However, in our opinion, this should not lead to constructions with a very low annular space between the lamp and the reactor wall. Under these conditions, a high number of photons could remain unabsorbed. This explains that turbulency conditions are justified in more heavily loaded wastewater than they are in drinking water treatment. Longitudinal mixing is further introduced into the decay rate constant in the form of a dispersion coefficient D, flow and dispersion both being supposed unidirectional along the lamp.

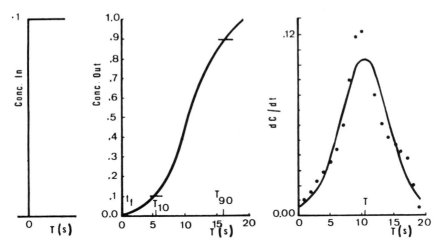

Figure 20   Typical flow-through pattern of an annular reactor.

Definitions.

1. v = water velocity in the reactor $= \dfrac{\text{length} \times \text{flow}}{\text{void volume}} = \dfrac{L \times Q}{V_v}$.

2.   Void Volume V_v = volume of the reactor − volume of the lamps.

3.   Variance in residence time, σ_t^2, of a tracer injected into the water stream.

4.   Dispersion coefficient D given by

$$\sigma_t^2 = \frac{2D}{vL} = \frac{\sigma_t^2}{T^2} \quad \text{and} \quad D = \frac{1}{2} \frac{v \times L \times \sigma_t^2}{T^2}$$

   A typical example is illustrated in Fig. 20.

5.   The Morrill dispersion index is equal to the ratio of the times of 90% and 10% of the tracer to pass. In perfect plug flow, $T_{90}/T_{10} = 1$. In UV systems the value must always be below 2.

6.   If the ratio of the theoretical residence time to the mean residence time (t/T) is lower than 0.8, there is probably an error in the void volume.

7.   The ratio of the time after which the first throughput occurs to the theoretical residence time (t_f/t) is a measure of most short-circuiting (1 in perfect plug flow, zero in complete mixed systems).

   By incorporating the residence-time distribution into the global decay constant one obtains, R being the removal rate in s^{-1},

$$N_t = N_0 \exp \left\{ \frac{v \times L}{2D} \left[1 - \left(1 + \frac{4 \times R \times D}{v^2} \right)^{1/2} \right] \right\} + N_p$$

$N_t' = N_t - N_p$ if N_p is significantly different from zero, but small compared to N_0; then

$$N'_t = N'_0 \exp\left\{\frac{v \times L}{2D}\left[1 - \left(1 + \frac{4 \times R \times D}{v^2}\right)^{1/2}\right]\right\}$$

or

$$\ln\frac{N'_t}{N_0} = \frac{v \times L}{2D} - \frac{v \times L}{2D}\left[1 + \frac{4 \times R \times D}{v^2}\right]^{1/2}$$

$$R = \frac{v^2}{4D}\left[\left(1 - \frac{2D\ln(N'_t/N_0)}{vD}\right)^2 - 1\right]$$

The light intensity, I, is an implicit function of R and can be evaluated by correlating log R with log I for different reactor configurations or on a relative basis, with log t for a particular reactor.

Further, in secondary wastewater treatment, the concentration of protected organisms can be correlated with the total suspended solids, which can eventually be substituted for N_p in the decay rate formula.

Example. Experiments were carried out to evaluate the killing rate of *E. coli* and *Clostridia* in a tubular reactor of void volume 2714 mL with an irradiation length of 32 cm. The measured dispersion coefficient was 2.53×10^4 m^2/s. The data obtained are given in Fig. 21.

Apparently, this approach makes it possible to estimate the relative resistance of different organisms in identical conditions. However, the approach is relatively imprecise where the evaluation of the dispersion coefficient is concerned and empiri-

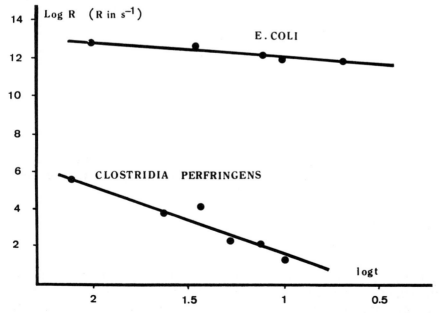

Figure 21 Indicative data for removal rates in an annular reactor (laboratory scale).

cal as far as the light intensity is concerned. It must be remembered here that the developments in this paragraph are based on the average intensity in the annular space of a tubular flow-through reactor. The UV-emitting tube is located along the axis of the reactor and is considered as a filamentous source (i.e., a series of point sources). These assumptions hold only for single-tube flow-through reactors with complete axial mixing, but they can hardly be applied to more complex multilamp reactors.

3.5   Diminishing of Light Intensity in Flow-Through Reactors

The intensity of the UV light at a given point of the water layer is determined by several factors:

The absorbance of light according to the Beer–Lambert law
The increase in solid angle and surface irradiated when the area of irradiation is not
    located near the source
The effects of the lamps as nonpunctual sources
The reflectance and recovery of UV intensity by the diffusion of light by turbidity
    or by the walls of the reactors
The geometrical distribution of the emission

A general equation that accounts for the light distribution efficiency in an annular space is formulated (17) as follows:

$$(N_0 - N)Q = \int_0^l \int_0^{2\pi} \int_{r_i}^{r_e} \frac{kNI_0 r_i}{r} \exp[-E(r - r_i)]r\, dr\, d\theta\, dl$$

and

$$(N_0 - N)Q = \frac{2\pi kNI_0 r_i l}{E} \{1 - \exp[-E(r_e - r_i)]\}$$

or

$$\frac{N}{N_0} = \frac{1}{1 + mkI_0 t} \qquad \text{where} \quad t = \frac{V_v}{Q}$$

and $V_v = \pi(r_e^2 - r_i^2) \times l$ for annular reactors. m is a dimensionless diminishing factor for annular space. It accounts for absorbance, reactor geometry, and lamp radius:

$$m = \frac{2r_i\{1 - \exp[-E(r_e - r_i)]\}}{E(r_e^2 - r_i^2)} = \frac{I_{rel}}{I_0}$$

The value of m gives the ratio of the relative average light intensity in the reactor (I_{rel}), to the nominal (corrected) emission intensity (I_0). The value of m may be considered as the geometrical absorbance factor, which in the case of single directed irradiation is always lower than the unit.

Typical values of the extinction coefficients (E) are:

Distilled water	0.015 to 0.02 cm^{-1}
Good-quality drinking water	0.05 to 0.25 cm^{-1}
Secondary clarified effluent	0.4 cm^{-1}

(The absorbance, A, equals to $0.43 \times E$; or $E \times 2.3 \times A$.)

3.6 Annular Space Reactors

The geometrical absorbance factors of some typical annular space reactors are given in Fig. 22. The first obvious remarks we have to make is that the larger the diameter of the lamp or of its quartz enclosure, the better the m value and the steeper the

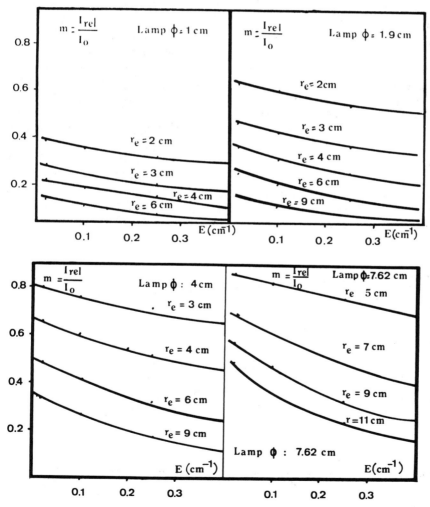

Figure 22   Geometric absorbance factors for annular space reactors.

drop as a function of increasing extinction. Possible variations in water quality must be taken into consideration when developing the design. As for the ratio of absorption itself, there is the transmittance factor:

$$E = \ln \frac{I_0}{I} = -\ln \tau$$

$E = 0.02$, $\tau = 0.98$; $E = 0.1$, $\tau\ 0.9$; $E = 0.25$, $\tau = 0.78$; $E = 0.4$, $\tau = 0.67$. From these values it appears that in most annular reactors only a limited proportion of photons are used effectively.

Dissolved ions such as Ca, Mg, Na, and Al have no observable effect on efficiency unless precipitates are formed on the lamp. Dissolved iron in concentrations up to 0.4 mg/L exerts no significant influence, but at 1 mg the extinction can range to 1 cm^{-1}. Organic compounds [e.g., amino acids (18) in concentrations of 3 to 4 mg/L, organ and tea extracts with a color index of 10°Pt (platinate degrees)] can lower the intensity by 50%.

The turbidity of values below 2.5 mg/L SiO$_2$ does not hinder the disinfecting action as long as no deposits are formed on the lamps (19). Even at high turbidity values reaching up to 40 to 100 mg/L SiO$_2$, disinfection are still operative, but less reliable (20). Protected organisms do increase under these conditions. The absorbance of the light is best measured with a spectrometer in which has been incorporated an integrating sphere, which accounts for the scattered light and remains available for germicidal action. In an experiment using 10 mg/L kaolin (Merck, *Art. 1906*) an increased dose effect of 15 ± 5% has been observed on coliphage f2 (21).

Fulvic acid can be added to the water tested as a natural product that absorbs the 254-nm light (17). Parahydroxybenzoic acid in the concentration range 5 to 15 mg/L has no direct effect on the living bacteria but is an optical competitor with extinction values in the range $E = 0.1$ to 5 cm^{-1} (see Fig. 23, for competition on f2). A BOD$_5$ value of 10 mg/L can decrease the available UV intensity by a factor of 4.

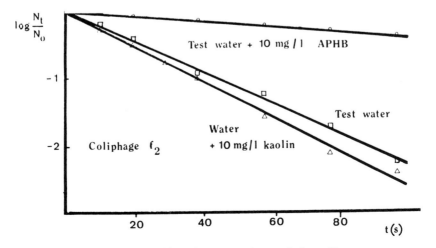

Figure 23   Effect of optical interferents on decay of phage f2.

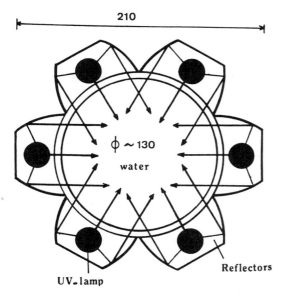

Figure 24 External arrangement of lamps with reflectors.

3.7 Indirect Irradiation Technologies

In view of the poor geometric absorbance factor of annular reactors, some authors have described them as negative configurations (22), in contrast to the arrangement in which several lamps are mounted around a tubular quartz reactor. In the latter case (see Fig. 24), an important part of the light intensity, up to 40 to 50%, can be lost due to the reflectance limitations of the material of which the reflectors are constructed. External irradiations exist in variable lengths (e.g., from 1.07 to 1.75 m with a volume range between 30 and 300 m³). Part of the cost is due to the expensive quartz tube through which the water is to be circulated.

Actinometric determinations of the indirect irradiation technique indicate that there is addition of the intensity of all of the tubes regardless of their relative position. However, the actinometer (uranyl sulfate–oxalic acid; section 2.11) is a strong absorbing liquid, which means that only a very small layer absorbs the photons without in-depth penetration. The overall intensity use in such configurations hardly attains 20%, as evidenced by actinometry.

In the indirect irradiation technique in which the lamps are placed outside the water circulating in an open channel, the absorption in the air can be neglected. The lamps are located at distance of 2 to 10 cm from the water layer, which has a thickness of 2 to 10 cm.

3.8 Multiple Submersed Lamp Technologies

The submersed lamp technologies installed remain by far the most economical setup, even if the geometry factor is unfavorable (see Fig. 22). It is possible to improve this parameter considerably by mounting several lamps in parallel. For a

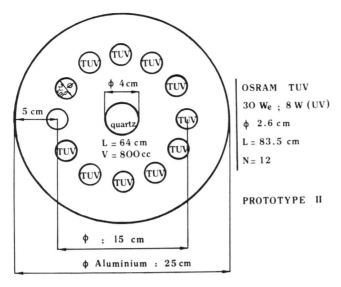

Figure 25   External-lamp reactor for experimental study.

square arrangement of four lamps ($r_i = 2$ cm) at a distance of 8 cm between each lamp, one obtains a distribution pattern of the *m* factor as shown in Fig. 27 for water with $E = 0.2$ cm^{-1}.

Within the square at the point defined by the junction of the lamp centers, the *m* factor is constant at a value of 0.95 (i.e., to nearly 100% of the intensity emitted by four quarters of the nominal intensity of each lamp). Inside the square, a cumula-

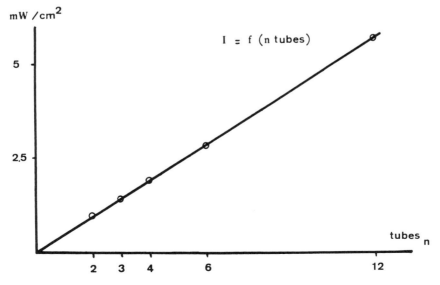

Figure 26   Additivity of irradiation in an external-lamp reactor.

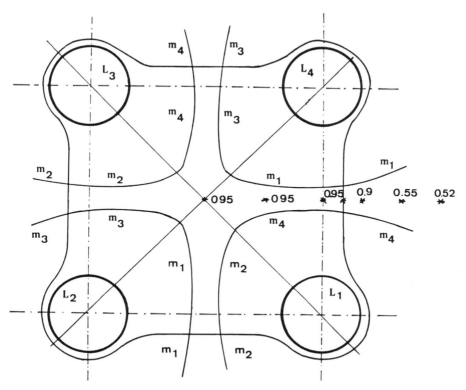

Figure 27 Distribution of *m* factors in a square pattern of lamps. *, *m* values at given location.

tive effect of the irradiation of the lamp is obtained. Outside the central square, there is a fast drop in geometry factor and, consequently, in utilizable light intensity. This determines the limits of the reactor area when only a small number of lamps are used. The distribution pattern also illustrates that axial mixing (which means perpendicular to the lamps) is less important in such an arrangement. Triangular ordering of lamps is shown in Fig. 28. For lamps with $r_i = 2$ cm located at 8 cm distance and considering an extinction value of 0.2, the central triangular area has a constant geometry factor of 0.8, while the direct exposure is that of $3 \times \frac{1}{6}$ of a lamp. Hence in this area there is a positive addition effect that is homogeneously distributed. The most correct and perfect light distribution of this type is obtained with seven lamps, six of them located in an hexagonal distribution with one lamp in the center. The distance between the lamps is to be chosen as a function of the extinction value of the water and the desired irradiation dosis.

Example. If one disposes of lamps with diameter 4 cm and emission length of 1.07 m, the nominal emission yield per lamp after 1000 operation is of 8 W at 254-nm light. The extinction value of the water is 0.125 cm^{-1}. Design a seven-lamp reactor for an active dosage of 240 J/m^2. $I_0 = 8$ W/2 $\times$ 3.14 $\times$ 0.02 $\times$ 1.07 = 60 W/m^2 per lamp and per second. For $r_e = 7$ cm and $r_i = 2$ cm, $m = 0.33$. For a three lamp combination system, $m = 3 \times 0.33 \simeq 1$. Consequently, $I_{rel} = I_0$ in the

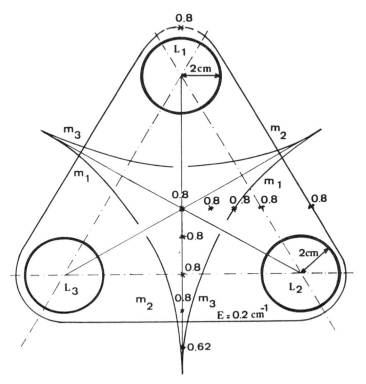

Figure 28   Distribution of *m* factors in triangular patterns of lamps. *, *m* values at given locations.

interlamp zone. By accepting an external layer of 0.4 cm outside the lamps, the relative intensity in this zone, computed for $r_e = 2.4$ and $r_i = 2$ in this zone, is $0.9 \times I_0$. The cross section is than equal to 0.106 m² and the useful volume to 0.114 m³.

The total intensity used (with average $m = 1$) is 420 (W/m²) · s. When dosing 240 J/m² a minimum exposure time of 0.57 s is required; this is equal to a theoretical flow capacity of 200 L/s through the reactor. This implicates that the necessary guaranty of suitable exposure required in practical mixing conditions and aging is to be safeguarded by a factor of 3.

The cross section of the reactor is illustrated by the hexagonal walls shown in Fig. 29. However, a more convenient cylindrical design where mechanical realization is concerned can be more appropriate. The lowest value of *m* at the longest apex is equal to 0.8 even without considering reflection. By increasing the interlamp distance to 14 cm ($r_e = 9$ cm), the *m* factor is equal to 0.242 and the relative intensity within the lamp area is $0.73 \times I_0$ instead of I_0. The cross section is increased to 0.157 m² and the volume to 0.168 m³. In this configuration, the necessary residence time that is to be obtained is 0.78 s. When computing on the basis of the active volume, the safety factor is therefore $168/(60 \times 0.78) = 3.6$.

Conclusions.   If the lamp location corresponds to an optimized pattern, variations in dimensions are possible, enabling us to obtain appropriate hydraulic conditions of pressure, head loss, and turbulency.

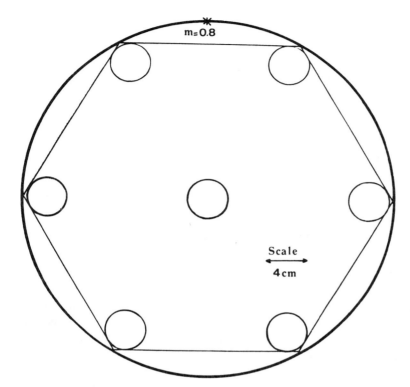

Figure 29 Hexagonal arrangement of the triangular reactor module.

A combination of square triangular locations can make an optimum design for high water flows as indicated in Fig. 30. Another, slightly different pattern that allows the reinforcement of the available intensity in the outside zone of the reactor is indicated in Fig. 31. The reactor wall can be located farther away from the outer lamps, in which case it approaches a circular distribution.

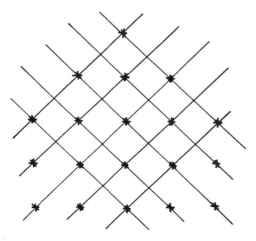

Figure 30 Lamp distribution patterns for high-flow-capacity UV reactors.

IRRIADIANCE AT P

$m_1 = 0.57 \times 2$
$m_2 = 0.28 \times 2$
$m_3 = 0.2 \times 1$
$m_4 = 0.17 \times 2$
$m_5 = 0.14 \times 2$

$l_r = l_0 \times 2.52$

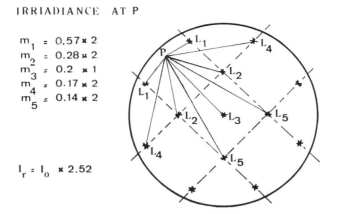

Figure 31   Increased symmetric lamp distribution scheme.

The entire methodology enables us to develop computer programs optimizing the flexibility in design. In the arrangement of Fig. 31 we used lamps of diameter 1.9 mm ($r_i = 0.95$ mm) disposed in the square module of 6 cm side ($r_e = 4$ cm) and water with an extinction value $E = 0.2$ cm^{-1}. The total irradiance at a point P is about $2.5 \times I_0$ because at least nine lamps contribute to the irradiation of this point. In this approach we must remind that a photon emitted by a given lamp, when irradiating a neighboring lamp, which is lighted, is absorbed and must therefore be considered lost. Mercury lamps that are not lighted may be assumed transparent to the 254-mm wavelength at a rate of 85 to 90%. If for mechanical reasons tie rods are necessary to assemble the reactor system, they are best constructed in UV-reflecting material or coated accordingly (see Section 2.7) and placed in the center between three or four lamps. Their best shape is square, to improve the global specular reflection.

4. AVAILABLE EQUIPMENT

Table 3 reports indicative data from the commercial documents of manufacturers of equipment available in Belgium. Further information available from the manufacturers' documents is as follows.

Actini-Stoutz.
Annular space reactor with orthogonal water flow
Materials: stainless steel
Standard water flow: 0 to 2 m^3/h (larger types up to 20 m^3/h)
Design dosage for 2 m^3/h: 250 J/m^2
$r - r_i$: not indicated (evaluated as 2 cm)
Lamp: low-pressure mercury vapor lamp
Address: Actini-France, B.P. 80, F-74202 Thonon, France

Advance-Ihle.
Antimony lamps, standard 1 to 12 kW
Standard reactor: up to 30 m^3/h, annular space reactor with transversal flow and
    equipped with a static mixer

Table 3 Nominal energy [W(e)] vs. Nominal Flows of Water Treated

Manufacturer[a]	W(e)	Q (m³/s)	W(e)/m³[b]	Remarks
Actini (Stoutz)	44–120	0.6–5.5 × 10⁻³	21.8	
Advance-Ihle	1000	8 × 10⁻³	34.7	Antimony lamp
Allveco	60	1.4 × 10⁻³	15.9	
Belgo-Nucleaire	39–1560	1.4 × 10⁻³	52	Modular: 5 m³/h
		28 × 10⁻³		
Berson	50	0.6 × 10⁻³	23.1	
	(2000)	7 × 10⁻³	80	High-pressure mercury lamp
	10,000	20 × 10⁻³	139	Indirect irradiation
BBC	800	30 × 10⁻³	7.4	High-intensity lamps
Duker	100	2.8 × 10⁻³	9.9	
Enerco	40–600	1.4–20 × 10⁻³	24	Teflon tubes
Grantzel	600–1200			Designed on size
Hanovia	1200–2500	10–55 × 10⁻³	12–34	Medium pressure mercury lamps
Jabay	30–960	0.8–13 × 10⁻³	22.2	
Multus-Katadyn	80	2.8 × 10⁻³	7.9	
Portacel	40–650	0.4–7.6 × 10⁻³	28	High-intensity lamps
Uvaudes (Holzli)	70	1.9 × 10⁻³	10.2	
Wallace-Knight	50–2600	0.13–7.6 × 10⁻³	132	High-intensity lamps
Wedeco	210	2.8 × 10⁻³	20.8	

[a]Based on products available in Belgium (1988).
[b]Indicative costs based on the average figures of the prospects.

Execution: stainless steel with inside ceramic coating
Optional: manual mechanical wiper for cleaning
Design dosis: 350 to 400 J/m²
Irradiation time: less than 1 s
$r - r_i$: not indicated
Address: Capital Controls, Route Axiale 4, B-5330 Assesse, Belgium

Allveco.
Annular space modular reactors of nominal flow of 2.5 m³/h
Low-pressure mercury vapor lamps of 30 W(e) each, standard length: 85 cm
Design dosage: 250 J/m²
$r - r_i$ 4 cm
Address: Alveco-Allfilters, Rue de l'Intendant 69, B-1020 Brussels, Belgium

Belgo Nucleaire.
Single- or multiple-lamp designs with optional mechanical cleaning; standard lamps
   42.9 or 91.4 cm with diameter 2.4 cm
Optional quartz enclosure or PTFE lining
Design dosage: 250 J/m² at $r = 0.8$
$r - r_i$: 70 mm for the single tube reactor (longitudinal or transversal flow as options)
Address: Belgo Nucleaire, Marsveldstraat 25, B-1050 Brussels, Belgium

Berson.
High- and medium-pressure mercury lamps up to 5 kW(e)
Channel reactor version or closed reactor version
Design dosage: 190 to 300 J/m^2, depending on r
$r = r_i$: 60 mm; length of lamps 0.7 m up to 1 m
Option: programmable automatic change in irradiation intensity
Address: Berson-Milieutechniek, P.O. Bus 90, NL-5670-AB Neunen, The Netherlands

Brown-Boveri (ASEA).
Built-in reactor into pipes equipped with a static mixer, U-shaped mercury vapor
   lamps with high electrical potential
Preheating necessary as well as isolation with quartz
Contact time 1 s; design dosage not indicated
Address: ASEA-Brown Boveri, Geschäftsbereich E, CH-5401 Baden, Switzerland

Düker.
Cylindrical reactors equipped with 1, 3, 6, or 12 lamps of standard length of 96 cm
   or 190 cm
Low-pressure mercury vapor lamps with quartz enclosure
Execution in steel or in stainless steel
Transversal flow; designs available for flows from 5 to 20 m^3/h; visual observation
   of lamps is possible
Design dosage and dimensions: not indicated
Address: Eisenwerke Düker-GMBH, D-8752 Laufach, Germany

Enerco.
Modular plates with four mercury vapor low-pressure lamps of 40 W(e) each
Standard: four plates per set
Traversal flow with automatical cleaning
Geometrical factors and dosage: not indicated
Address: Enerco, 2323 South Lipan, Denver, CO 80223

Gräntzel
UV systems designed on size using low-pressure mercury vapor lamps (specialist of
   UV photochemical reactors)
Address: A. Gräntzel, Durnerheimerstrasze 78, D-7500 Karlsruhe 21, Germany

Hanovia.
Cylindrical annular reactors in stainless steel AISI 316, equipped with a central
   medium-pressure mercury vapor lamp
Nominal power of the lamps: 1.2 or 2.5 kW(e)
Water flow: 305 to 200 m^3/h
Design dosage: variable between 160 and 700 J/m^2 (standard 320 J/m^2 after 3000 h
   of operation)
Address: Hanovia Ltd., 145 Farham Road, Slough, Berkshire SL1-4XB, England

Jabay.
Annular reactors with transversal flow-through
Execution: stainless steel equipped with mercury vapor lamps with high intensity
Design dosage: not given but based on transparency $\tau = 0.82$
$r - r_i$: 15 mm (small reactors) up to 14 cm (big reactors)
Address: Jabay Ltd., Rectory Lane, Kingston Bagpuize, Abington, Oxfordshire,
   OX13-5AS, England

Multus Katadyn.

Low-pressure mercury vapor lamps with quartz enclosure

Standard execution: galvanized steel with built-in static mixers

Basis model: annular reactor for 5 m³/h

As variant option: vessel reactors equipped for higher water flow (i.e., up to 100 m³/h) with several immersed lamps

$r - r_i$: between 3.3 and 9.5 cm (lower values possible for on-size designs in stainless steel)

Admissible pressure: 15 bar

Address: Multus Katadyn A.G., Industriestrasze 27, CH-8304 Wallisellen, Switzerland

Portacel.

Reactors in stainless steel, equipped with mechanical wipers

High-intensity lamps (unspecified)

Maximum capacity 185 m³/h with long residence time 11 to 33 s

(Particularity in description: electrical current of 110 V, 60 Hz)

Address: Portacel Ltd., Cannon Lane, Tonbridge, Kent TN-91-PP, England

Uvaudes-Holzi.

Standard unit tubular reactor 7 m³/h with 70 W(e) tube

Design in PVC

Dosage: 250 J/m²

$r - r_i$: 8 cm

Address: Holzli KG, Ginzkeyweg 12, A-4863 Seewalchen, Austria

Wallace-Knight.

Cylindrical reactors in stainless steel, each cylinder being equipped with one high-intensity mercury vapor lamp

Total power UV-C: 1.8 to 96 W for flow of 0.45 to 27 m³/h and reactor lengths of 21 to 65 cm

Design dosage: 357 J/m² (variable between 32 and 542 J/m²)

Particular aspects: resistance to temperature up to 100°C

Address: Wallace-Knight Ltd., 515 Ipswich Road, Trading Estate, Slough, Berkshire SL1-4EP, England

Wedeco.

Indirect irradiation system (standard).

Central quartz tube irradiated by six external low-pressure mercury lamps of 30 W(e) surrounded by reflectors

Options are possible

Design and geometrical factors: not indicated

Annular reactors.

Design dosage: 350 J/m² at $\tau = 0.9$

Flat low-pressure mercury lamp

$r - r_i$: 80 to 160 mm

Execution: standard in stainless steel

Address: Wedeco, Achenbachstrasze 55, D-4000 Dusseldorf 1, Germany

Lamps simpsonized with trace metals, emitting a high and constant intensity over a broad range of temperatures, are a recent development of Wedeco.

5. OPERATING, MAINTENANCE, AND COSTS

5.1 General Monitoring

Design of most equipment is based on an operational pressure of 10 or 16 bar. The reliability of the method requires special design considerations if:

The turbidity is higher than 40 g SiO_2/m^3
The color exceeds 10° Hazen
The iron concentration is higher than 4 g/m^3
The BOD exceeds 10 g/m^3
The suspended solids are higher than 15 g/m^3
The amino acids (+ proteins) exceed 3 g/m^3
Calcium carbonate deposits form on the lamps

    A built-in UV intensity meter (relative values) reaching the outside zones of the reactor can complete the equipment as well as a totalizing-recording system of the burning time of the lamps. Prefiltration is sometimes foreseeable (e.g., in a first stage at 25 to 50 μm porosity, followed by a phase at 10 μm). However, this pretreatment often generates trouble. Alternative cleaning of the UV sources or lamp enclosures is a valuable alternative. Cleaning is best carried out with a cloth damped with alcohol or ammonia and water, never with abrasive products.

5.2 Cost Evaluation Method

For annular reactors or cylindrical reactor zones, the UV energy necessary to obtain a given disinfection level is expressed by

$$U = tW_0 \frac{r_0}{r_e^2 E} (1 - e^{-Er}) \ln \frac{r_e + r_0}{r_0}$$

where

U = energy required by unit volume (J/m^3)
W_0 = UV power dissipated per unit volume (W/m^3)
r_0 = radius of the lamp (m)
r_e = external radius of the tubular reactor (m)
E = extinction (m^{-1})
t = irradiation time

A typical example is given in Fig. 32.

    The economical annular space that results from this evaluation is 2 to 4 cm. If a 10-s irradiation time is considered, the operation cost will range to 250 W/m^3. The equipment demands very little attention; maintenance is simple and does not require specially skilled staff. The lamps *and/or the quartz jacket surrounding the UV lamp* must be made accessible for easy cleaning without having to disassemble the entire unit. The quartz jacket should be wiped once a month.

    Renewal of the lamps determines an important part of the operational cost of the process and must be done strictly in the sequence recommended by the manufacturer (e.g., when the measured intensity has dropped at 70% of the value obtained with a new lamp). Although less relevant, professional risks associated with the use

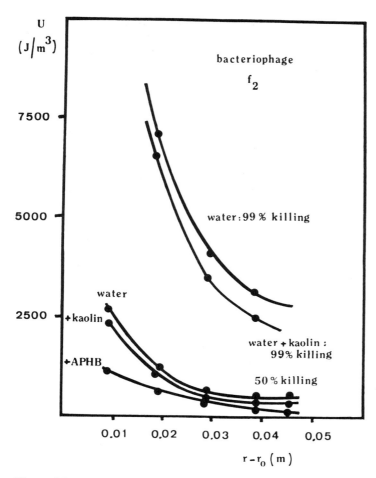

Figure 32   UV energy for disinfection (bacteriophage f2). (Energy data with APHB are to be multiplied by four).

of UV irradiation are those of erythema and conjunctivitis. Goggles with normal glass are adequate for protection of the eyes and the sources are best not directly visible. Replacement of lamps is a determining factor of operational cost: high-pressure or antimony lamps cost twice as much as low-pressure mercury lamps and have a lifetime of about half of the latter. However, the surrounding reactor construction is considerably simpler.

6. CONCLUSION

Disinfection through UV irradiation can be a simple and reliable method. It does not require specially skilled staff; however, the design must be very well adapted to the local circumstances. There is no need to supply, stock, or handle chemicals; consequently, the potential hazards of these operations are eliminated. For lamp stability, the operation of the system is best when continuous and permanent and there is no long-term residual germicidal effect; control of the efficiency of the process cannot be based on residual action and must be immediate when in current operation. Basis parameters for design as reported here as water temperature, available current voltage, global properties of the water, and mixing conditions in the contactors. The choice of lamps must also correspond to criteria that enable correct interpretation of the dose–effect relationship.

Representative test organisms are *E. coli*, coliforms, total bacterial count, and in case of severe requirements, *Clostridium perfringens* and phage f2. A safety factor of 1.6 to 1.7 in designing the dosage is recommended. Besides a weekly inspection and a monthly cleaning, at least a yearly replacement of the burners is to be anticipated. Mounting conditions must avoid stagnation and irregular flow conditions. Tentative design rules for multilamp, isointensity reactors are formulated here. They account for light-intensity amortization when remote from the lamp and for the additive effect of multilamp sources.

SYMBOLS AND UNITS RELATED TO UV DISINFECTION

$\dfrac{dp}{d\theta}$	power per radiant, watts per radiant
h	Planck's constant, 6.63×10^{-34} J $\cdot$ s
k, k'	first-order decay constant, m^2/J
l	reactor length, m
m	reactor geometrical factor, dimensionless
n	number of vital centers, dimensionless
r_e	radius of tubular reactor zone, m
r_o or r_i	radius of UV lamp, m
t	theoretical residence time, s
t_r	reference time, s
v	velocity, m/s
A	absorbance of light, cm^{-1} or m^{-1}
D	dispersion coefficient, m^2/s
D_{10}	dosage for one decade decay, J/m^2
E	extinction of light, cm^{-1} or m^{-1}
Es	Einstein per second
I	intensity of UV light, J/m^2 s
I_{rel}	relativated light intensity, J/m^2 s
It	dosage, J/m^2
L	length, m
N_p	protected germ concentration, number/L
N'_t, N_t	volumetric concentration of germs after time t, number/L
P	power, watts [W(e) and W(UV)]
P_t	probability of hit after time t, dimensionless
Q	water flow, m^3/s^1
R	removal rate, s^{-1}
T	residence time, s
U	energy per unit volume, J/m^3
V_v	void volume, L
W_o	UV power dissipated per unit volume, W/m^3

η_1	UV efficiency [W(UV)/W(e)], dimensionless
ν	frequency, s^{-1}
σ_t^2	variance of residence time, s^2
τ	transparency factor, dimensionless
ϕ	quantum field, dimensionless
Ω	steradian

REFERENCES

1. R. W. Legan, *Water Sewage Works, R 56* (1980).
2. A. Downes and T. P. Blunt, *Proc. R. Soc.*, 26, 488 (1877).
3. J. M. Dohan and W. J. Masschelein, *Ozone Sci. Eng.*, 9, 315 (1987).
4. J. D. Jepson, *Proc. Water Treat. Eng.*, p. 175 (1973).
5. W. J. Masschelein, *H₂O, 19*, 350 (1986).
6. F. A. J. Armstrong, P. M. Williams, and J. D. M. Strickland, *Nature (London)*, 211, 481 (1966).
7. F. A. J. Armstrong and S. Tibbitts, *J. Marine Biol. Assoc. U.K.*, 48, 1943 (1968).
8. J. R. Cortelyou, M. A. McWhinnie, M. S. Riddiford, and J. E. Semrad, *Appl. Microbiol.*, 2, 227 (1954).
9. M. Denis, G. Minon, and W. J. Masschelein, *Symposium IOA, Europe Wasser*, O.S.E., Berlin, Apr. 1989.
10. W. G. Leighton and G. Sh. Forbes, *J. Am. Chem. Soc.*, 55, 3139 (1930).
11. *Jugoslavenski Registrar Brodova, 40-003*, 666 (1979).
12. O. K. Scheible, M. C. Casey, and A. Fondran, *N.I.T.S.*, PB 86-145182 (1985).
13. B. F. Severin, M. T. Suidan, and R. S. Engelbrecht, *J. WPCF, 56*, 881 (1984).
14. D. Kuse, *Wasser Abwassertech.*, 6, 16 (1979).
15. G. Baer, *Public Works, 110*, 59 (1979).
16. F. Akkad, E. Pape, and F. Weigand, *Photodissociation of Halogens and Oxygen*, IHLE, Bericht, p. 24.
17. B. F. Severin, M. T. Suidan, B. E. Rittmann, and A. S. Engelbrecht, *J. WPCF, 56*, 164 (1984).
18. A. Dodin, J. Wiart, G. Escallier, and P. Vialat, *Bull. Acad. Natl. Med., 155*, 44 (1971).
19. J. Maurin and G. Escallier, *Bull. Acad. Natl. Med., 152*, 377 (1968).
20. G. Escallier and J. Maurin, *Bull. Acad. Natl. Med., 154*, 517 (1970).
21. W. J. Masschelein, E. Debacker, and S. Chebak, *Rev. Sci. Eau, 2*, 29 (1989).
22. G. O. Schenck, *Technologie der Wasseraufbereitung für pharmazeutische Zwecke*, Vol. 1, Concept, Heidelberg, 1979, p. 5.

5
Coagulation

1. INTRODUCTION

Coagulation is a derivation from the Latin word *coagulare*, meaning to "drive together." Carried out as a unit process in water purification, it consists of adding chemicals to aqueous dispersions in order to combine fine dispersed particles with larger agglomerates that may be removed after flocculation by subsequent processes, such as settling or filtration. Flocculation is presently considered as a separate process taking place at the end of the coagulation. It consists of promoting an increase in macroscopic flocs with or without the use of supplementary chemicals such as flocculation aids. Although flocculation is the natural consequence of coagulation, both processes correspond to specific technologies that must be carried out according to their specific design rules, which means that flocculation is to be considered as a separate process. This is discussed further in Chapter 7.

The coagulation process is generally set directly on the raw water. Therefore, with oxidation, it is one of the most important processes in surface-water treatment schemes. These waters contain substances originating both from nature and as a result of human activities: dissolved inorganic and organic substances, living organisms, and suspended material. Contaminations of natural origin result from land erosion and dissolution of minerals; they may also consist of living organisms, such as bacteria and algae, as well as decaying vegetation.

1.1 General Classification of Impurities

A preliminary classification of impurities based on their approximate unitary dimensions and most frequent composition is given in Table 1. The approximate sedimentation velocities indicated in Table 1 are based on the assumption of spheri-

Table 1 Classification of Impurities and Main Unit Processes for their Elimination

Approximate dimensions avg. diameter	Type	Material or composition	Main unit operation
10 mm	Gravel	Mainly quartz	Sedimentation, v ~ 1 m/s[a]
1 nm	Coarse sand	Mainly quartz	Straining & sedimentation, v ~ 0.1 m/s[a]
0.1 mm	Fine sand, vegetation	Mainly quartz, cells	Microstraining and/or coagulation, v ~ 0.1 m/s[a]
0.01 mm	Silt	Quartz Clay minerals Metal oxides Organisms	(Sedimentation v ~ 0.0075 m/s) coagulation (oxidation)
1000 nm to 5 nm	Bacteria	Living organisms	Adsorptive coagulation
200 nm	Colloids	Principally clay minerals Colloids coupled with organic substances (color)	Coagulation
Complexed compounds	Protected and hydrophilic colloids	Colloids coated by hydrophilic substances	Oxidation + coagulation
	Humic Compounds	Complex organometallic compounds	Oxidation + coagulation
Dissolved compounds $< 10^{-10}$ m	Dissolved organic compounds Dissolved inorganic compounds		Adsorptive coagulation and oxidation + adsorption softening

[a]Practical limits for direct sedimentation; all other entries in this column indicate principal uses of coagulation.

cal particles with a specific gravity of 2.65. The practical limit for direct sedimentation (e.g., in storage reservoirs) is a lower dimension of 0.01 mm particle size.

1.2 Basic Principles of Coagulation

Fundamentally, coagulation implies the removal of colloidal particles, that is, suspended particles, which on the basis of their average dimensions of 5 to 200 nm are considered as colloidal. The colloidal particles impart undesirable properties to the water, among which one can distinguish the following:

1. Turbidity, most often caused by inorganic clay minerals on which organic substances can eventually be adsorbed. Most turbidity particles are hydrophobic, or water repelling. The turbidity particles are of the high colloidal dimensions ranging from 0.2 to 10 μm and can settle by gravity after a sufficient delay.

Their elimination may be considered as easier than that of the total coagulation of the colloids as a whole.

2. Except for colloidal hydroxides of metal (e.g., iron), color in natural waters is generally caused by organic substances (i.e., humic and fulvic acids with molecular weight varying between 800 and 50,000 MW units). Consequently, color may be due to polar molecules in true solution with dimensions ranging between 3 and 10 nm. These are usually somewhat lower than for turbidity particles, and most of the particles responsible for color are hydrophilic (i.e., water attractive).

3. Complex organic compounds introduced to water as a consequence of the discharge of industrial wastewater can be considered as colloidal in their behavior, which is also the case for color.

4. Bacteria may be considered as colloids just like viruses and microalgae. Composed of polar organic molecules, they are hydrated and assumed to be hydrophilic. Although possibly removed by coagulation, it is more advantageous to kill the living organisms during the first stage of the coagulation in order to make them behave as "suspended matter." Also, when removed by coagulation-flocculation only, without disinfection, the sludge remains infectious by the presence of still-living pathogenic agents.

5. Owing to their emulsifying properties, detergents attach themselves to hydrophobic particles, rendering them hydrophilic, which means more difficult to eliminate by treatment. This type of colloid is designed as a "protected colloid," as illustrated schematically in Fig. 1.

Protected colloids are of growing importance, as domestic sewage contains essentially hydrophilic pollutants such as detergents, proteins, polyphosphates, and phosphate builders, which promote the formation of protected colloids. To facilitate their removal, coagulation should be preceded by or associated with oxidation treatment.

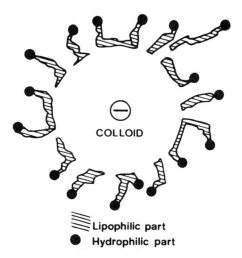

Figure 1   Protected colloidal system.

1.3   Hydrophilic and Hydrophobic Colloids

The difference between hydrophilic and hydrophobic colloids is important in the context of the treatment. Hydrophilic colloidal matter reacts spontaneously with water to form a suspension that can be dehydrated and rehydrated repeatedly. Owing to this characteristic, such colloids are sometimes named *reversible colloids*. When dehydrated, hydrophobic substances generally do not redisperse spontaneously in water. Therefore, they are sometimes called *irreversible colloids*. Pure hydrophobic systems do not react with the water phase, but real particles can combine with water on microsites, forming hydrogen bonds. Hydrophobic or irreversible colloids are sometimes named *caducous colloids*. The general differences in behavior of hydrophilic and hydrophobic colloids dispersed in water are indicated in Table 2.

2.   STABILITY OF COLLOIDAL SYSTEMS IN WATER

2.1   Mobility of Colloids

Owing to their small particle size, the specific surface area of colloids is very large, and because of this high ratio of surface to mass there is little or no tendency toward sedimentation. Consequently, properties such as solvatation and superficial ionization, more strictly related to the surface, become more important (2). In the case of hydrophilic particles, a relatively large number of solvent molecules are fixed to the surface of the particle. Interparticle contact is hindered by the "solvent sandwich." Stable hydrophilic sols can contain up to 10^{15} to 10^{18} particles/cm^3. Usually, river water contains less than 10^{12} particles/L.

In 1809, Reuss discovered that clay particles, when dispersed in water and placed in an electrical field, move toward the positive electrode. Electrically charged clay surfaces contain hydroxyl groups oriented on a matrix containing Si, Ca, and

Table 2   General Chemical Properties of Colloids

Effect	Type of colloid	
	Hydrophilic	Hydrophobic
Concentration	High concentration of the dispersed phase can be stable up to general gelling	Only low concentrations of the dispersed phase are stable
Dehydration	The residue after dessication takes up water	The residue after dessication is coagulated and rehydrated very slowly
Salt effect	Unaffected by small amounts of electrolytes: removed by large amounts of salts	Generally precipitated by electrolytes
Dispersion of light	Only weak Tyndall effect is observed	Marked light scattering and Tyndall effect
Hydration	Intervenes predominantly in the stability; gelling effect	Is not determinant for the stability; no gelling effect

Source: Ref. 1.

Al (2). The origin of the electrical charge carried by a colloidal particle can vary. The most frequent origin is the fixation of negative ions, especially OH^-, in the case of aqueous dispersions. When these ions are removed, the dispersion can become unstable. Other possible origins of the negative charge are the ionization of atoms at the particle surface (e.g., departure of H^+ ions to the medium) or replacement of elements in the crystal lattice by elements having a different charge.

Thermal agitation tends to distribute the suspended particles as well as the molecules and ions uniformly throughout the solution by Brownian motion, which is the observable movement of an individual particle resulting from the innumerable collisions between all particles, including molecules, of the system. The available energy to cause the movement is $3RT/2$ per particle, insufficient to move most of the turbidity particles (R is the gas constant and T the absolute temperature).

2.2 Zeta Potential

For colloidal suspensions involved in natural waters, the majority of the particles are negatively charged. The stability of such colloidal solutions is principally the result of the electrical characteristics of the particles composing the solutions. Although the colloids are charged, the entire solution remains electrically neutral. Oppositely charged ions (e.g., cations) are drawn to the colloid by long-range electrostatic attraction. As a consequence of this attraction of the counterions, a concentration gradient is established between the particle surface and the bulk of the solution. Hence two competing forces, diffusion and electrostatic attraction, spread the ionic charges contained in the water over a diffuse layer and promote a higher concentration of counterions adjacent to the surface of the particle. This concentration decreases gradually with increasing distance from the solid–water interface. This is illustrated in Fig. 2.

The thickness of the layer of the counterions is related to the ionic radius of the cations in aqueous solution (e.g., approximately 133×10^{-12} m for potassium, 95 to 98×10^{-12} m for sodium, 99×10^{-12} m for calcium, while for aluminum, magnesium, and iron ions, the corresponding values are, respectively, 45 to 50, 65, and 67×10^{-12} m). Hence the charge-to-surface ratio is higher for multivalent ions than for monovalent species. Owing to mutual repulsion of the counterions, species with high ionic charge density fix themselves preferably in the counterion layer.

The primary adsorbed ions are strongly attracted by the particle, and their electrical charge establishes an electromotive force between the particle and the water. A potential function results, as indicated in Fig. 3. The attracted counterions are disposed in a rigid layer called the *Stern layer* and move with the colloid in the liquid. Beyond the rigid layer lies a diffuse layer called the *Gouy–Chapman layer*. In this diffuse layer, starting from the plane of shear of the counterions, the concentration of the latter gradually decreases, while that of the iso-ions increases. The effective diameter of the particle includes the counterions in the double layer. As a result of the presence of these hydrated ions, the effective density of the particle is reduced and the settling velocity undergoes an important decrease.

The bulk of the solution with random distribution of the ions is located beyond the diffuse layer. In the Stern layer the electrical potential drops from ψ_0 to ψ_d. The thickness of the Stern layer is more or less equal to the dimensions of the ions. Just beyond this distance is the plane of shear separating the rigid Stern layer from the

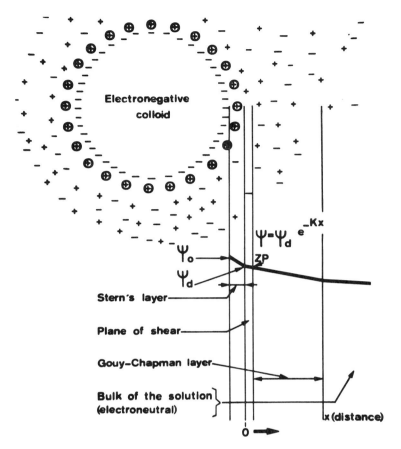

Figure 2   Schematic ionic charge distribution around a colloidal particle.

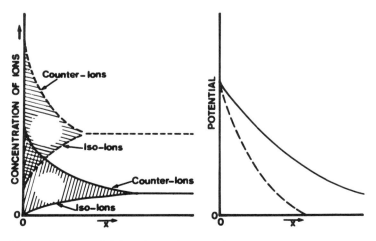

Figure 3   Potential functions of counterions to the bulk of the solution.

diffuse Gouy-Chapman layer. The difference in potential at this plane with respect to the water is called the *zeta potential* (ZP) or electrokinetic potential.

The decrease in potential in the diffuse part of the double layer as a function of the distance x from the wall of the particle is approximated by the following equation:

$$\psi = \psi_d \, e^{-Kx}$$

where (in the simplified assumption of ZP < 25 mV) ψ is the potential energy at a distance x from the wall of the particle including the counterions (i.e., immediately before the plane of shear). K is equal to the reciprocal of the double-layer thickness and for symmetrical electrolytes is expressed as follows:

$$K = \sqrt{\frac{8\pi e^2 \, \Sigma \, nz^2}{\epsilon k_B T}}$$

where

e = charge of the electron (1/1850 C)
n = number of ions per unit volume (m^3)
z = valence of the counterions
ϵ = dielectric constant (C/V · m)
k_B = Boltzmann's constant: 13.8×10^{-24} J/K
C = electrical charge Coulombs
T = absolute temperature

The value of the potential at the plane of shear is designed as the zeta potential of the colloid (ZP). Much discussion has taken place on the significance of the ZP in coagulation as applied in water treatment. In practice, the value of the ZP of the colloids is always negative in natural waters and can be measured according to the theory and methods described in Chapter 11. The standard deviation of the measurements applied must be considered as ±3 mV.

From the pH dependence of the ZP in the presence of aluminum salts as a coagulant, the suggested relationship of the ZP to the surface potential is ZP $\simeq$ $0.5\psi_0$ (3). Consequently, when ZP = 0 the surface potential is eliminated.

2.3 Mechanisms of Coagulation

In practice, there are several mechanisms of coagulation that intervene (e.g., electrostatic coagulation, chemical reaction with colloidal functional groups, adsorption-aggregation, and sweep coagulation). Only the first one mentioned is closely related to the ZP, which is also involved significantly in the adsorption–aggregation model for coagulation.

The zeta potential depends on the ionic strength of the solution and the electrostatic repulsion of the colloidal particles. The latter decreases when the double layer is compressed. Therefore, the potential function depends on the ionic strength of the solution, as indicated schematically in Fig. 4. This effect is the basis of *electrostatic coagulation*. The coagulation values for different electrolytes are

For monovalent electrolytes: 10 to 50 mol/m^3
For divalent electrolytes: ≤ 1 mol/m^3
For trivalent electrolytes: ≤ 0.1 mol/m^3

The number of ions necessary for electrostatic coagulation is generally considered as related to the reciprocal charge of the ions at the sixth power (4); $n_i \simeq (1/q_i)^6$. This relationship is known as the *Schulze–Hardy rule*.

In electrostatic coagulation, the charged ions are known to act as single species in the double-layer model. There exists no stoichiometric relation, and no restatilization of the system should occur when the coagulant dosage is in excess.

In most natural waters, and in particular for divalent electrolytes, concentrations of calcium exceed 1 mol/m^3, higher than the coagulation values. Nevertheless, colloids exist in such waters. Consequently, electrostatic coagulation is a secondary method in the coagulation process applied to water purification.

In coagulation through chemical reaction, insoluble or poorly soluble reaction products are to be formed on reaction of coagulants with colloids. This can be involved in color removal. This effect can be related to the Schulte–Hardy rule, as, according to the latter, the introduction of opposite charges can result in colloidal charge neutralization and consequent decrease of the ZP (e.g., eventually to zero), when, under these conditions, the colloidal system coagulates.

The coagulation reaction depends on the pH; the formation and precipitation of Fe or Al humates are optimal in the pH range 3.7 to 4.2 for Fe and 5 to 5.5 for Al. In this case there is a stoichiometric relation between the coagulant dosage and the removal rate (4).

Owing to the charge density for different ions, the surface potential drop depends on the valence of the ions introduced in the diffuse layer, which influences the compression efficiency of the rigid layer. Ratios observed for mono-, di-, and trivalent ions are approximately 1:8:600. The trend is illustrated schematically in Fig. 5.

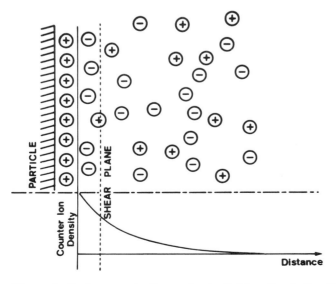

Figure 4   Ionic concentration and potential function in the diffuse layer.

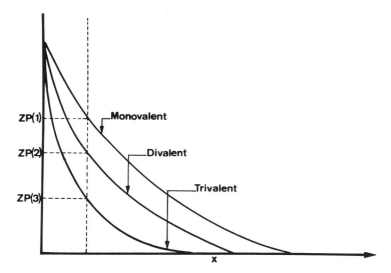

ZP(1) Monovalent

ZP(2) Divalent

ZP(3) Trivalent

x

Figure 5   Effects of charge of ions on double-layer thickness.

Turbidity in river waters is generally associated with suspended clay particles that have an ionic exchange capacity. Attempts to associate the aluminum dosage with the exchange capacity of experimental clay suspensions have indicated that the necessary dosages for coagulation are far in excess of the stoichiometrically expected quantities.

In *adsorption–aggregation coagulation*, coagulants (i.e., positively charged entities) are adsorbed primarily on the negative colloidal surface. Neutralization of colloidal charge can result, promoting precipitation of the colloids involved. Because the adsorption is nonspecific, it is possible that more charge may be adsorbed than necessary to neutralize the surface charge. It has been established (5) that the relatively narrow zone of good coagulation and low residual color is accompanied by reversal of the zeta potential of the floc particles from negative to positive (see Fig. 6).

In practice, efficient coagulation can be obtained even if the ZP is not exactly reduced to the zero value. This is explained by the bridging model developed by Stumm and O'Melia. In its simplest form, the chemical bridging theory proposes to fix a polymer molecule to the surface of a colloidal particle at one or several adsorption sites with the remaining part of the molecule extending into the solution. Polar functional groups in the colloid, as in the coagulant (e.g., hydroxyl, phosphatyl, or carboxyl groups) can be particularly effective in causing adsorption. The primary adsorption destabilizes the colloids according to the following mechanisms:

Adsorption coagulation:

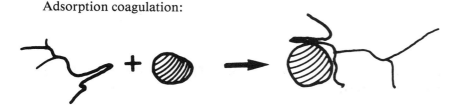

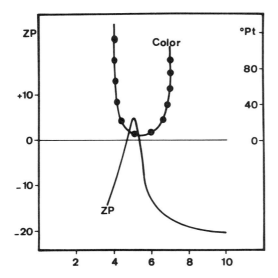

Figure 6   Color removal with aluminum sulfate in relation to ZP. (From Ref. 5.)

Optimum destabilization through adsorption coagulation occurs when a portion of the available adsorption sites of the colloidal particle is covered, leaving other positions available for a second adsorption with bridging:

Bridging coagulation:

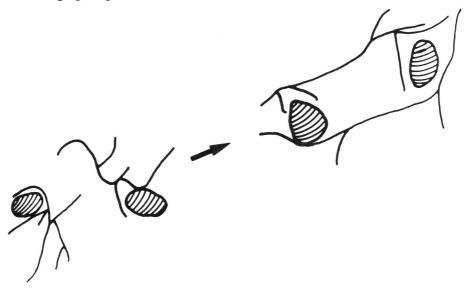

Coagulant doses that saturate the available surfaces of the disperse phase (colloids) produce a restabilized colloid:

Over-dosing:

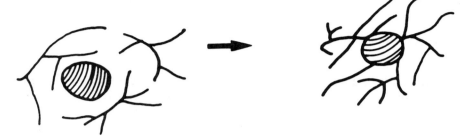

This is because there are no sites available for the formation of polymer bridges. This situation is obtained either by local deficiency of colloids or by overdosing of coagulants.

In some circumstances, mutual coagulation can occur by the collapse of negative colloids and the positive entities capable of being formed on local overdosing of positively charged coagulants. This phenomenon cannot be clearly distinguished from bridging coagulation; however, some essential characteristics are enumerated in Table 3. In homogeneous systems (e.g., SiO_2 solutions, the necessary concentration of coagulant can be related to the surface of the colloid, as given in Fig. 7 for SiO_2–Fe according to Stumm and O'Melia (6). If coagulation occurs out of the shaded zone in Fig. 7, the mechanisms are not those of adsorption-bridging.

The primary adsorption of the coagulant is considered to build up a monolayer (7) and the process then obeys the Langmuir equation (see Chap. 12). Furthermore, the adsorption is inversely proportional to the absolute temperature; hence more adsorption of polymers is expected to intervene at lower temperatures. In coagula-

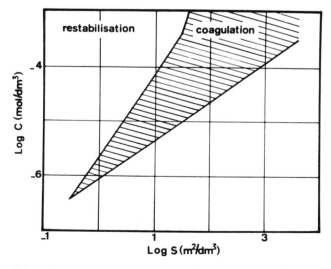

Figure 7 Relation of coagulant dosage to colloidal surface.

tion by adsorption bridging, the process of flocculation is almost concomitant with the coagulation. The relation associated with this sequence is given as follows:

$$-\frac{dC}{dt} = KC^z\theta(1 - \theta)$$

where C is the particle concentration, t the time, K the rate constant, and θ the fraction of the surface covered by polymer segments. Therefore, the maximal efficiency would take place when half of the particle surface is covered.

There does exist some doubt regarding the bridging-coagulation model if one assumes that the dimensions of the colloids compare to those of possible polymeric species intervening in the mechanism of coagulation with inorganic salts of Fe and Al. The scheme seems more adequate for coagulation using synthetic polymers.

Sweep coagulation is a mechanism of coprecipitating during flocculant settling. Fundamental arguments favoring the sweep-coagulation concept are the following: Usually, the best results of the coagulation–flocculation process are obtained in conditions of the lowest solubility of the metal hydroxides derived from the coagulants [e.g., $Al(OH)_3$ or $Fe(OH)_3$]. Furthermore, the coagulation depends above all on the coagulant and water composition rather than on the colloids. In practice, attaining a large floc always favors the water quality. Sweep coagulation is also the basic mechanism involved in coagulation with Ca/Mg salts forming $CaCO_3$ and MgO as insoluble precipitates.

The hydrolysis of aluminum sulfate in distilled water delivers aluminum hydroxide as a fine hydrated colloid with low tendency to precipitate. With a bicarbonate alkalinity which has a better effect than a hydroxyl alkalinity (see Fig. 8), a more agglomerated precipitate is formed and its coagulation value is higher.

The practical observation pointing out that soft waters are usually not easy to coagulate is consistent with the principle of sweep coagulation. The effect of the

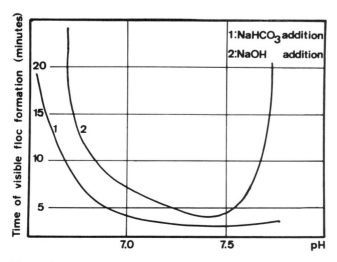

Figure 8   Effect of alkalinity in coagulation–flocculation.

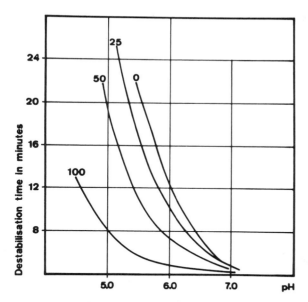

Figure 9   Effect of sulfate ions on destabilization time.

anion associated with the metal salt used as a coagulant has been studied extensively by several authors (8). Generally in increasing the amounts of sulfate, one decreases the coagulation–flocculation time and makes it possible to obtain optimum destabilization of colloids at lower pH values than those in the absence of sulfate ions (Fig. 9). This underscores the fact that these experiments concern the reaction time rather than improvement in water quality through the process. Another approach is that of Packham, which analyzes the coagulant dosage necessary for the coagulation of artificial kaolinite suspensions (9) as a function of pH (final value). A synopsis is given in Fig. 10.

Taking the effects of alkalinity and the anionic part into account, practical design should provide for good capabilities of sweep coagulation: sludge blanket contact, sludge recirculating, and adjustment of the pH after coagulation in order to optimize precipitation.

2.4   Dose-effect Relations in Coagulation

Although no general dosage–concentration relation can be advanced, practical coagulation often corresponds to the following equation:

$$D = K_1[\text{Alk}] + K_2 c^n \qquad (\text{e.g., } D = [\text{Alk}] + 0.33 c^{1.06})$$

Where D is the dosage, [Alk] the starting alkalinity of the water, and c the colloidal charge (e.g., in 10^4 unit charge/m^3).

Studies most often relate to artificially prepared suspensions; however, real suspensions are not homogeneous with respect to the chemical compounds involved (e.g., clay or organics), the dimensions of particles, or their electrical charges. Furthermore, general parameters such as pH or temperature of the water can mod-

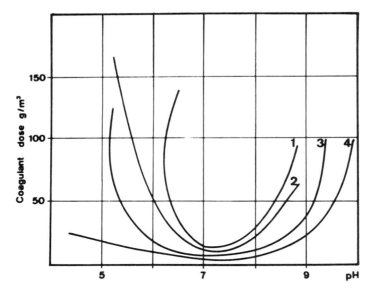

Figure 10   Impact of the anionic part of Al and Fe coagulants. 1, $Al_2(SO_4)_3 \cdot 18H_2O$; 2, $AlCl_3 \cdot 6H_2O$; 3, $FeCl_3 \cdot 6H_2O$; 4, $Fe(SO_4)_3 \cdot 9H_2O$.

ify the colloidal properties. These effects of heterogeneity must be considered over and above those of the change in colloidal properties. To illustrate this, Meuse river water is given as a function of pH in Fig. 11. pH values are adjusted by means of H_2SO_4 or NaOH. The zeta potentials indicated are average values.

A general observation: The suspensions become more polydisperse at lower pH values. These observations regarding the heterogeneity and chemistry of Al and Fe coagulants (Section 3) lead to the conclusion that in real (polydisperse) systems, several mechanisms of coagulation take place simultaneously and that design should provide for the characteristics of at least two different main pathways: adsorption bridging and sweep coagulation (see Fig. 14).

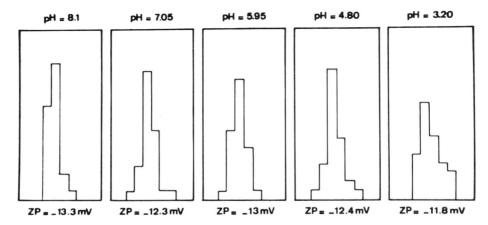

Figure 11   Change in zeta potential as a function of pH (river Meuse).

Table 3 Main Characteristics of the Modes of Destabilization of Colloids

Phenomenon	Electrostatic coagulation	Aggregation or bridging model	Sweep model	Reaction model
Chemical interaction	Absent	Predominant	Secondary	Predominant
Addition of excess	No effect	Restabilization	Favorable	No effect
Zeta potential for optimum aggregration	Near zero	Usually not zero	Approximately zero	Not zero
Relationship between colloid concentration and optimum dosage of coagulant	None	± Proportional	None = excess coagulant required	Stoichiometry
Fraction of surface covered for optimum aggregration	Negligible	50% in general, always 0 < % < 100	± Completely covered	Negligible
Physical properties of aggregates	Dense, poor filtrability, high shear strength	Three-dimensional structured good filtrability, poor shear strength	Hydrated floc, high shear strength	Specific chemical compounds

A summary of some of the principal characteristics of the coagulation process is given in Table 3.

3. ELECTROLYTIC REACTIONS OF COAGULANTS

Except for the seldomly used cationic polyelectrolytes (see Section 3.6, Chapter), the coagulants applied in water treatment are aluminum or iron salts and, to some extent, lime. The dissolution of aluminum salts [e.g., aluminum sulfate ($Al_2(SO_4)_3 \cdot 18H_2O$) or sodium aluminate ($NaA1O_4)_2$] takes place according to several sequential schemes: electrolytic dissociation, hydrolysis and olation or dehydration, equilibrium solubility of precipitated hydroxides, and amphoterism of the latter. When adding aluminum sulfate to water, rapid precipitation of aluminum hydroxide occurs when the solubility equilibrium is reached: $[Al^{3+}][OH^-]^3 = K_s = 10^{-32.7}$. The reaction proceeds according to a pathway involving several fast sequences:

$$Al_2(SO_4)_3 \cdot (18H_2O) = 2Al^{3+} + 3SO_4^{2-} (+18H_2O)$$

3.1 Reactions of Aluminum Salts (Equilibria)

The aluminum ion is known to exist in a hexahydrated form $Al(H_2O)_6^{3+}$. To simplify, the ion is written here as Al^{3+}. The aluminum ion hydrolyses according to the following scheme:

Al^{3+}

$\downarrow$

$AlOH^{2+}$ $Al_2(OH)_2^{4+}$

$\downarrow$ $\uparrow$ $Al_6(OH)_{15}^{3+}$

$\qquad\qquad\qquad\qquad\qquad\qquad\nearrow$

$Al(OH)_2^+$ $x\,Al^{3+} \rightarrow$ $Al_7(OH)_{17}^{4+}$

$\downarrow$ $\searrow$

$\qquad\qquad\qquad\qquad\qquad\qquad\quad Al_8(OH)_{20}^{4+}$

$Al(OH)_3$ $Al_{13}(OH)_{34}^{5+}$

$\downarrow$

$Al(OH)_4^-$

Scheme: Monomeric and polymeric hydrolysis schemes for aluminum ion. Relevant equilibrium constants for the various reaction sequences at 25°C are as follows (10):

$$Al^{3+} + H_2O = Al(OH)^{2+} + H^+ \qquad \log K = -5.03$$
$$2Al^{3+} + 2H_2O = Al_2(OH)_2^{4+} + 2H^+ \qquad \log K = -6.27$$
$$Al^{3+} + 3H_2O = Al(OH)_3(s) + 3H^+ \qquad \log K = -9.1$$
$$Al(OH)_3(s) + H_2O = Al(OH)_4^- + H^+ \qquad \log K = -12.74$$
$$6Al^{3+} + 15H_2O = Al_6(OH)_{15}^{3+} + 15H^+ \qquad \log K = -47$$
$$8Al^{3+} + 10H_2O = Al_8(OH)_{20}^{4+} + 20H^+ \qquad \log K = (?)\,(\text{small})$$
$$7Al^{3+} + 17H_2O = Al_7(OH)_{17} + 17H^+ \qquad \log K = -48.8$$
$$13Al^{3+} + 34H_2O = Al_{13}(OH)_{34}^{5+} + 34H^+ \qquad \log K = -97.6$$

The dissolved species present in equilibrium conditions with solid $Al(OH)_3$ are essentially Al^{3+}, $2Al(OH)^{2+} = Al_2(OH)_2^{4+}$, and $Al(OH)_4^-$, as indicated in the equilibrium diagram in Fig. 12. The isolectric point of $Al(OH)_3$ is nearly at pH 6.

In the present concept, the polymeric species are accepted to be the active molecules intervening through adsorption on the colloidal surfaces with or without secondary bridging. It has also been suggested that in particular the intervention of $Al_6(OH)_{15}^{3+}$ could also be formed by retroreaction from $Al(OH)_3(s)$ (11). The possible intervention of the Al^{3+} ion is of only minor importance and should be neglected, as coagulation can be carried out with sodium aluminate in conditions where the pH is 5 to 8. Coagulation occurs in the zone of optimum precipitation of $Al(OH)_3$, as indicated in Fig. 13. The present general interpretation of the various mechanisms are summarized in Fig. 14.

For the coagulation process as applied in flocculation filtration, low doses (e.g., ≤ 5 g/m^3) of aluminum sulfate are used. The mechanism involved is essentially that of coagulation through adsorption aggregation. The optimum pH zone is between 6 and 7, and rapid mixing on the coagulant dosing (e.g., $G \langle 300$ s^{-1} for less than 30 s) is of the utmost importance in building intermediate polymeric species.

When coagulation–flocculation–settling is used as an integrated process with dosages of 30 to 100 g/m^3 aluminum sulfate, the dominant mechanism gradually

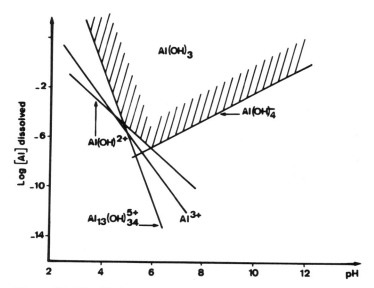

Figure 12   Equilibrium diagram of dissolved aluminum (binary equilibria).

becomes that of sweep coagulation in the optimal pH zone 7 to 7.5. Mixing conditions of the dosing are less relevant in this case (e.g., $G = 100$ to 200 s^{-1} at injection, but back-mixing is important (e.g., $G \leq 100$ s^{-1} with sludge recirculation).

The dosage of aluminum sulfate for optimal coagulation of Meuse river water depends on the initial value of the pH (13) (see Fig. 15), and part of the chemical is used for acidification. This can be substituted for by cheaper sulfuric acid. In the case of the river Meuse at Tailfer, the economic balance becomes optimal in the pH zone 6.7 to 7.1.

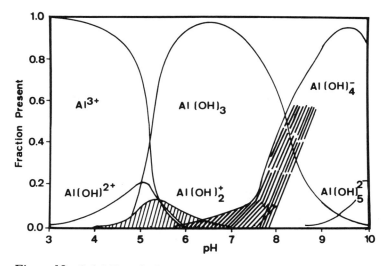

Figure 13   Solubility of Al(OH)$_3$(s). (From Ref. 15.)

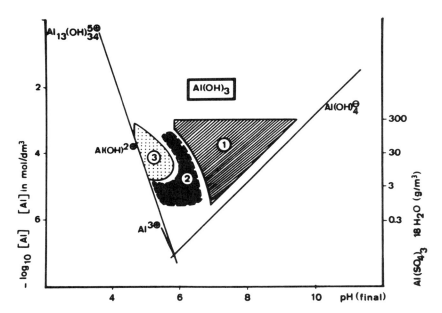

Figure 14 Dosage mechanism relations in coagulation with aluminum sulfate. 1, Zone of sweep-coagulation; 2, zone of adsorption aggregration; 3, zone of reaction coagulation. (From Ref. 12.)

Clay suspensions acquire the electrokinetic properties of aluminum hydroxide, due to adsorption of the precipitating coagulant. The zeta potential of aluminum hydroxide is originally positive, which is due to the existing $Al_6(OH)_{15}^{3+}$ and $Al(OH)^{2+}$ in its structure. Therefore, in case of overdosing, restabilization can occur. On standing in aqueous solution, the ZP of aluminum hydroxide gradually changes to neutral or slightly negative. As a consequence, restabilization is temporary only and plays no significant role if the process as a whole is designed to be operated by sweep coagulation.

3.2 Reactions of Iron Salts (Equilibria)

The behavior of the hexahydrated ferric ion in aqueous solution is similar to that of aluminum. Relevant reactions and equilibrium constants at 25 °C are as follows (7):

$$
\begin{aligned}
Fe^{3+} + H_2O &= Fe(OH)^{2+} + H^+ & \log K &= -2.17 \\
Fe^{3+} + 2H_2O &= Fe(OH)_2^+ + 2H^+ & \log K &= -6.75 \\
Fe^{3+} + 3H_2O &= Fe(OH)_3(s) + 3H^+ & \log K &= -31.1 \\
Fe(OH)_3(s) + H_2O &= Fe(OH)_4^- + H^+ & \log K &= -38 \\
[Fe(OH)_3(s) + 3H_2O &= Fe(OH)_6^- + 3H^+ & \log K &= \ \ ?] \\
2Fe^{3+} + 2H_2O &= Fe_2(OH)_2^{4+} + 2H^+ & \log K &= -2.85
\end{aligned}
$$

Although the polymeric species formed are suspected to be involved in coagulation, their existence remains inconclusive.

The solubility product of ferric hydroxide is equal to

$$[Fe^{3+}] = [OH^-]^3 = K_s = 10^{-38}$$

but ferric hydroxide is subject to hydration, determining an aqueous soluble form:

$$Fe(OH)_3(s) \rightleftharpoons Fe(OH)_3(aq) \qquad K = 2.9 \times 10^{-7}$$

The equilibrium diagram given in Fig. 16 illustrates several differences between iron and aluminum. In the practical zones of pH, the Fe^{3+} ion plays no role as equilibrating species versus preformed $Fe(OH)_3$. The diagram illustrates a broader zone for possible sweep coagulation, especially at lower pH values than for aluminum. At alkaline pH, complex oxides and $Fe(OH)_6^{3-}$ are suspected to interfere in the solubilization of iron (14), but the ratios are seldom established quantitatively. The isoelectric point of $Fe(OH)_3$ approaches pH 8.

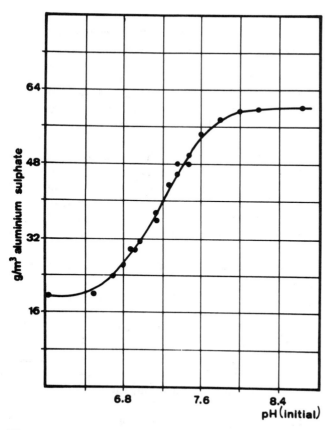

Figure 15   Effect of pH on coagulant dosage for treatment of water of the river Meuse.

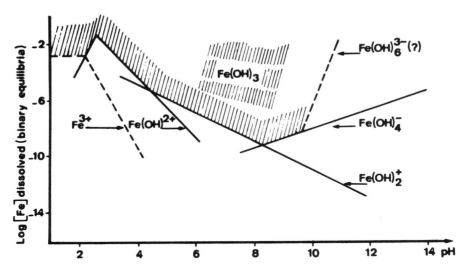

Figure 16   Equilibrium diagram of dissolved iron (binary systems).

The species distribution for iron hydroxide is much more complex than that of aluminum and is concentration dependent. A typical example for concentrations of suspended $Fe(OH)_3$ of 100 g/dm^3 and more (i.e., to the product of hydrolysis of 250 g/dm^3 of $FeCl_3 \cdot 6H_2O$) is given in Fig. 17. However, at a concentration of 10 g/dm^3 as $Fe(OH)_3$, resulting from 25 g/dm^3 $FeCl_3 \cdot 6H_2O$, about 55% of the iron present at equilibrium is in the form of insoluble $Fe(OH)_3$, the other part being in different soluble ionic states, with $Fe(OH)_4^-$ as the most important.

The practical consequence of the solubilities of $Fe(OH)_3$ is that ferric-based coagulation assumes a floc-blanket contact with the incoming water flow. Optional sludge recycling can be advised to improve the precipitation rates. In practical waters, especially in alkaline conditions, the presence of Ca–Mg ions limits the solubility of iron oxides, which means that $Fe(OH)_4^-$ can practically be adopted as the equilibrating species. An investigation of Meuse river water has indicated that ferric chloride intervenes less efficiently than aluminum sulfate in ZP reduction (see Fig. 18).

In summary, the design for coagulation with ferric salts should be based essentially on the concepts of sweep coagulation rather than adsorption bridging. Back recirculation of the sludge, or sludge-blanket contact with the coagulating water, is essential in enabling a reduction in coagulant doses to an average value of 40 to 60 g/m^3 $FeCl_3 \cdot 6H_2O$, which is the case for the river Meuse—one-third to one-half of the dosage is required to obtain reduction of the ZP in a conventional jar test.

The predominant mechanisms of coagulation with iron salts are essentially sweep coagulation and secondary reaction coagulation, with dosages as indicated in Fig. 19. For coagulation followed by flocculation–filtration, the pH is best when higher than 6, with dosages of about 5 g/m^3 $FeCl_3 \cdot 6H_2O$. For coagulation followed by flocculation and settling, the dosages are usually to be set between 30 and 100 g/m^3 at pH values higher than 6. *Overdosing restabilizes the suspension only at acid pH values.*

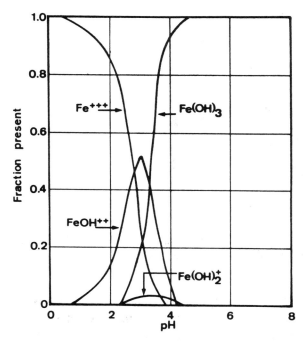

Figure 17   Distribution of Fe species for a 10^{-3} M total concentration. (From Ref. 15.)

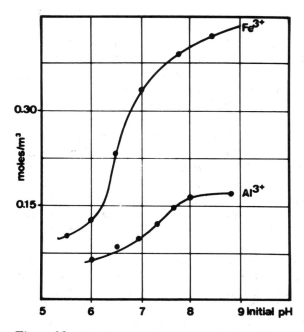

Figure 18   Relative coagulant doses of Al and Fe for a decrease in zeta potential.

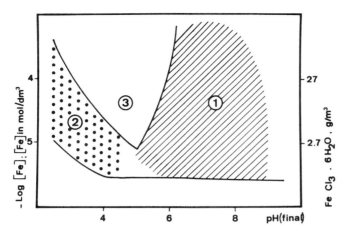

Figure 19   Relation to dosage mechanisms to ferric chloride in coagulation. 1, Zone of sweep coagulation; 2, zone of reaction coagulation; 3, restabilization zone.

When coagulation with iron salts is critical, flash mixing or at least a mixing intensity of $G \geq 300 \text{ s}^{-1}$ is recommended. Ferric floc has a higher shear strength than that of aluminum floc. Also, in flocculation–filtration, there is less breakthrough tendency. In this application, coagulation with ferric salts is generally preferable.

3.3   Coagulation with Lime and Magnesium Salts

Coagulation with lime and related chemicals is based on reaction schemes such as softening, which are discussed in Chapter 15 relative to this process. When using lime CaO or $Ca(OH)_2$, precipitates of $CaCO_3$ (at pH 8.3 to 9.6) and MgO (at pH 9.0 to 11.5) are formed. Colloidal particles can be enmeshed in these precipitates as they are formed. The settling of calcium carbonate can be improved by the addition of small amounts ($\leq 10 \text{ g/m}^3$) of aluminum sulfate or ferric chloride. When magnesium oxide is coprecipitated, addition of other coagulants is not required.

The zeta potential of $CaCO_3$ is negative and that of MgO is positive throughout the entire range of pH in which they exist. On the addition of a sufficient amount of $MgCl_2$, on preexisting $CaCO_3$, the ZP of the latter can be changed from negative to positive. Magnesium oxide is the most active coagulant of both and adsorbs organic products significantly. There is a considerable new interest in the lime treatment in coagulation, particularly in tertiary wastewater treatment (16) or as a preparatory step for the detoxication of raw water to be treated in further biological phases (17) (see Chap. 15 for further details).

Magnesium bicarbonate can be used as a recyclable coagulant, for which the following schemes have been described. The pathway over $Mg(HCO_3)_2$ recycling is more efficient than that over $MgCO_3$. The technical operation involves a double-stage precipitation $CaCO_3$ preceding MgO. The second sludge is then reprocessed. The fractioning of soluble $MgCO_3$ is obtained by filtration after a first stage of reaction of MgO sludge with CO_2, for instance, up to a pH of about 9.6 (18,19).

$$Mg(HCO_3)_2 + Ca(OH)_2 \rightarrow MgCO_3 + CaCO_3 \downarrow + H_2O$$

$$Mg(HCO_3)_2 \xleftarrow[+ H_2O]{+ CO_2}$$

$$\rightarrow MgCO_3 + Ca(OH)_2 \rightarrow \left[\begin{array}{c} CaCO_3 \downarrow \\ + \\ Mg(OH)_2 \end{array} \right]$$

$$MgCO_3 \xrightarrow{+ CO_2}$$

$$MgO + H_2O$$

$$+$$

$$CaCO_3 \xrightarrow{\quad} bleed$$

3.4 Activation with Silica

Activated silica is a negatively charged colloid and can, on precipitation, act as a coagulant by sweep coagulation or in some instances for positively charged systems. The action is, however, poor and the role of activated silica is basically confined to that of a flocculation aid (this is discussed extensively in Chapter 8).

3.5 Use of Polyelectrolytes

Cationic polyelectrolytes can be very efficient as coagulants. The principle of their action is that of adsorption with bridging. There are numerous types, some of which are accepted from the sanitary point of view (see Chapter 19). The most important in water treatment are those of the acrylic type. They are obtained from polyacrylamide as a starting product:

Anionic polymers are formed on hydrolysis:

or by a sequence of methylation–sulfonation:

Cationic polymers can result from quaternization of the amide function or as quaternary polyamines obtained after reduction of the amide function:

$$-\left[\begin{array}{c} CH_2-CH \\ | \\ CO \\ | \\ +NR_3 \end{array}\right]_n \quad ; \quad -\left[\begin{array}{c} CH_2-CH \\ | \\ CO \\ | \\ NH_2 \end{array}\right]_n \xrightarrow{\text{NaOBr}} -\left[\begin{array}{c} CH_2-CH \\ | \\ NH_2 \end{array}\right]_n -$$

$$-\left[\begin{array}{c} CH_2-CH \\ | \\ +NR_3 \end{array}\right]_n -$$

where $R = CH_3$ or C_2H_5.

By adjusting concentration, reaction time, and so on, all reactions can occur between zero and 100%. Manufacturing control can deliver a wide variety of polymers of different molar weights (e.g., 3×10^6 to 10^7 units) and different charge densities. The important class of cationic quaternary ammonium polyelectrolytes has been reviewed extensively in the literature (20).

The charge density of the polyelectrolytes influences the configuration in solution. Increasing charge density at a given molecular weight increases the stretching of the molecule through intramolecular electrostatic repulsion. An increased viscosity of the solution results. At low charge density (e.g., less than 20% of the functional groups ionized), the polymer is reduced to a more-or-less spherical coiled entity. Increasing ionic strength of the solution favors the coiled configuration by decreasing the intramolecular interactions. The selection of a polymer (e.g., molar weight and charge density) always remains a process of trial and error.

Health aspects associated with the use of polyelectrolytes must be considered. Usually, the doses for the use of cationic polyelectrolytes as sole coagulants are too high to make their use economical in drinking water treatment. High-molecular-weight units are more favorable to promote a bridging or aggregation mechanism. The bridging effect is favored at high particle concentration (e.g., more than 10^{14} particles/dm^3). As for river water the total colloidal particle concentration is most often below 10^{12} dm^{-3}; the present role of polyelectrolytes seems to be confined to that of a flocculation aid, sludge blanket stabilizer, filter conditioner, and dehydration agent for sludge compacting.

4. PERSPECTIVES OF COAGULATION

Coagulation is a very ancient process in water treatment technology. Coagulation dates from the early days of recorded history, when various natural materials such as crushed almond beans in Egypt, nuts in India, and alum in China were used to clarify turbid waters. As written in the second book of Kings, Chapter 2, verses 19–22, Elisha healed the waters of a spring in Jericho by treating them with a salt. This must have been in the ninth century A.D.

After the Dark Ages, in 1843, James Simpson used alum experimentally in

England prior to filtration, but this combination was not practiced municipally until after the patent granted to Isaiak Hyatt in 1884 for the use of a coagulant prior to rapid filtration. The first large-scale applications took place in 1885 by the water companies of Somerville and Raritan in New Jersey. The Schulze–Hardy rule dates from 1900. In the early 1920s, it was a generally accepted idea to consider the importance of Al^{3+} or Fe^{3+} present, a pH adjustment being necessary in good practice. The possible importance of an anion (e.g., sulfates) was emphasized.

In 1928, significant progress was made by Sante Mattson (21). Through an electrophoretic investigation of clay suspensions he pointed out that the hydrolysis products of iron and aluminum were more important than the trivalent ions themselves: The greatest effect on the zeta potential is observed in the pH range of hydrolysis of aluminum salts.

During the 1930s the concepts of electrophoretic mobility and ZP were broadened and charge reversal and redispersion were established in case of coagulant overdosage. The idea of precipitation and enmeshment as basic mechanisms in coagulation and the theory of ZP reduction found equal numbers of supporters until the 1950s. It is worthwhile mentioning the discovery of activated silica by Baylis in 1937, the processes of coagulation–flocculation remaining indistinct.

In 1952 came the use of synthetic polyelectrolytes as coagulant aids, and as a consequence, renewed interest in older products based on starch or natural gums. Consequently, new ideas were developed and the late 1960s were the period of development of the new adsorption-bridging (or aggregation) concepts in coagulation. Despite the fact that the protagonists of the precipitation effect based their explanations on sweep coagulation, it was recognized that adsorption was of growing importance, and with the adsorption-mixing concept came the techniques of appropriate flash-mixing with short contacting times.

At the present time, in the treatment of natural waters polydisperse systems, the coagulant dosages generally used correspond primarily to the concepts of sweep coagulation. A renewed interest in online coagulation or coagulation–filtration is part of the present development.

By itself, coagulation does not produce a specific improvement in water quality, but it is essential for the success of integrated processes such as coagulation–flocculation, settling, and coagulation–(flocculation)–filtration. The improvement in water quality relates not only to elimination of turbidity and color, bacteria algae, and viruses (in conjunction with the oxidation process), but also of heavy metals and radioactivity, partial phosphate removal, and taste and odor (TOC) improvement considered in conjunction with the adsorption process.

Although fundamentally a separate process from flocculation, coagulation is in practice followed immediately by flocculation. Practical examples and experience in the use of these processes are given in Chapter 9.

SYMBOLS AND UNITS RELATED TO COAGULATION

c	colloidal charge, units/m^3
e	charge of electron, (1/1850) Coulombs
k_B	Boltzmann constant, 13.8×10^{-24} J/°K
n	number of ions per unit volume
n_i	number of ions

q_i electrical charge of a ion (z_e), Coulombs
x distance from particle surface
z valence number of counterions

C molar concentration, mol/dm^3 or mol/m^3
D dosage, concentration units
°Pt   color, platinate degrees
S particle surface, cm^2 or m^2
T absolute temperature, K
ZP   zeta potential, mV

ϵ dielectric constant, Coulombs/Volt $\times$ m
θ fraction of surface
K reciprocal double-layer thickness, m^{-1}
$\psi(\psi_0, \psi_d)$ electrical potential or Stern potential, mV

[ ]   equilibrium concentrations
[Alk]   alkalinity, g/m^3

REFERENCES

1. F. A. Van Duuren, *Waterworks Eng., 859*, 360 (1967).
2. V. K. La Mer, *Principles and Application of Water Chemistry*, J. Wiley, New York, 1967, p. 246.
3. E. S. Hall, *J. Appl. Chem., 15*, 197 (1965).
4. G. Oskam, *H₂O, 2*, 120 (1969).
5. A. P. Black, *Water Sewage Works*, R 192 (1961).
6. W. Stumm and C. R. O'Melia, *J. AWWA, 60*, 514 (1968).
7. K. M. Yao, *Water Sewage Works, 295*, (1967).
8. A. P. Black and O. Rice, *Ind. Eng. Chem., 25*, 812 (1933).
9. R. F. Packham, *Proc. Soc. Water Treat. Exam., 12*, 15 (1963).
10. A. P. Black and Ch. Chen, *J. AWWA, 59*, 1173 (1967).
11. W. Chamot and B. Stewart, priv. communication, Nalco Chemical Co., Chicago.
12. A. Amirtharajah and K. M. Mills, *Annual Conference Proc., Water for the World Challenge of the 80's*, Part 1 session 1-20 (1980).
13. R. Buydens, W. Masschelein, and G. Fransolet, *Tech. Eau, 250*, 2 (1966).
14. R. De Vleminck, J. Genot, C. Goblet, and W. J. Masschelein, *Trib. Cebedeau*, 439 (1980); 450 (1981); 470 (1983).
15. AWWA-Committee Report, *J. AWWA, 60*, 1280 (1968); *63*, 100 (1971).
16. E. Edelovitch and A. M. Wachs, pg 655, *Proc. 11th IAWPR International Conference*, Cape Town, Pergamon Press, Elmsford, N.Y., 1982.
17. B. Regnier and W. J. Masschelein, *Trib. Cebedeau*, 463–464 (1982).
18. Anon., *Environ. Sci. Technol., 7*, 304 (1973).
19. C. G. Thompson, J. E. Singley, and A. P. Black, *J. AWWA, 64*, 11 (1972).
20. M. F. Hoover, *J. Macromol. Sci. Chem.*, 1327 (1970).
21. S. Mattson, *J. Phys. Chem., 32*, 1532 (1928).

Use of Polymerized Aluminum Flocculants

1. EXISTING PRODUCTS

Prepolymerized aluminum flocculants are obtained by the partial neutralization of an acid aluminum salt. The former exist in two distinct forms at least: products containing 10 to 10.5 wt% Al_2O_3 in the solution of a density of 1.2 to 1.25 kg/L, and the highly concentrated polyaluminum chloride, which contains 18 wt% Al_2O_3 of a solution of density 1.35 to 1.40 kg/L. Of neutral or alkaline pH at the manufacturing phase, on storage the manufactured products become acid (e.g., pH 2.5 for polyhydroxychlorosulfates, 0.7 to 1 for polyhydroxychloride). These values are to be compared to concentrated aluminum sulfate, of which the pH is 2.5.

Obtaining prepolymerized products in the form of a stable solution involves an elaborate manufacturing procedure and close control of the purity of the products reacted. The reagents correspond to the following global formula:

$$nAl(OH)_{1.5}Cl_{1.5} \cdot n(SO_4)_{0.2}$$

They are generally obtained by mixing a hot aqueous solution of aluminum chloride with a solution or suspension of aluminum hydroxide, which in turn can be obtained by precipitation of a solution of aluminum chloride or sulfate by sodium carbonate. The number of sulfate ions present can be controlled by mixing appropriate amounts of aluminum chloride and aluminum sulfate (eventually, sulfuric acid) in the reaction. The molar weight of the prepolymerized products usually ranges from 800 to 900.

Obtaining a highly concentrated, stable solution necessitates the absence of all sulfate ions, thus the use of demineralized water. Furthermore, in water treatment plants the mixing of prepolymerized aluminum flocculants and aluminum sulfate in concentrated form releases an unstable solution that precipitates. This mixing must be avoided.

More recently, a "direct process" for manufacturing polymerized aluminum flocculants by acidification of bauxite has created a new generation of reagents on the European market; they are called PAX. Stability to freezing is improved and the reagents may be stored safely up to less than $-20°C$. In this new generation of flocculants, combinations with silica exist in the range 2 to 15 wt%. Small amounts — less than 1% iron flocculant or high-rate iron containing combinations — can be formulated: up to 12 wt% iron. The product corresponds to a mixed iron-aluminum polychlorosulfate $(FeAlClSO_4)_n$. The density of the chemical increases from 1.25 to 1.4 kg/L with an increase in the silica content of the product. The aluminum concentration of the reagents, expressed as Al_2O_3, ranges from 17 to 24 wt%. The pH value of reagents capable of being stored as liquids is in the range 1 to 1.5.

All manufacturing processes guarantee a stable and reproducible composition with a low content of impurities (e.g., heavy metals) (see Table 1). The most significant elements for quality control are arsenic, cadmium, chromium, mercury, and iron. Because the polyaluminum chloride is free of sulfate ions, there is no increase in the sulfate ions obtained on flocculation of the water. This can be an element for evaluation in cases where the sulfate concentration must necessarily be limited. Mixed iron–aluminum polymeric chlorosulfate coagulants of the new production series are less pure but probably still acceptable for drinking water treatment.

Table 1   Trace Elements of Aluminum Flocculants[a]

	Aluminum sulfate	Polychloro-sulfate (classical)	Polychloride (PAC)	PAX	Polychloro-Fe/Al-sulfate
Ag	7	7.7	0	—	—
As	0.3–14	2.85	0.2	—	—
Ba	8	7.7	0	—	—
Cd	0.08–0.5	3.85	0	0.03	2
Cr	6.5–25	23	0.77	22	2.5
Hg	6.5–1.3	3.4	0.77	0.02	0.02
Ni	6.5–33	7.7	0	0.06	6
Pb	2–10	15.4	1.08	0.06	3
Sb	0.8	0.77	0	—	—
Se	0.015	7.7	0.77	—	—
Cu	70	10	12.3	0.06	1
F	70	77	77	?	?
Fe	400–600	223	190	+	+
Mn	78–250	17	35	+	+
Zn	3.3	154	13.8	0.3	120
Co	0	0	0	—	—
V	10	7.7	3.8	—	—
pH after storage	2.5	2.5	0.7–1	1–1.5	<1

[a]Average values (mg/kg) based on Al_2O_3.

2. PERFORMANCE EVALUATION WITH JAR TESTS

The coagulation value of polyaluminum chloride (PAC) has been compared to that of aluminum sulfate by measuring the amount of coagulant necessary to decrease the ZP of the colloids to approximately zero. The raw water is that of the river Meuse, the pH of which has been changed by addition of HCl or NaOH to broaden the range. At the usual pH values, from 7.5 to 8.5, the effect of PAC is about double of that of aluminum sulfate (Fig. 1).

The flocculation values were measured using the conventional jar test on water of the river Meuse (as in Fig. 1). (The jar test involves mixing at $G \sim 120\ s^{-1}$ for 3 min, followed by 15 min of flocculation at 10 rpm mixing, followed by sedimentation; the transparency of the supernatant water is then measured at 320 nm and compared with that of a standard SiO_2 solution.) The relative turbidities after 5 and 15 min are compared as a function of the coagulant dosing (Fig. 2).

The flocculation value of PAC for removing turbidity is higher than that of aluminum sulfate by a ratio of 1.5 to 2. (We must mention that this experience is carried out by starting with a pH of 8.25 and without activated silica.) The relative results for turbidity removal at final pH 7 are given in Fig. 3, which again indicates that the best yield with PAC is obtained at lower dosing rates than with aluminum sulfate. This conclusion also holds for settled waters at different pH values (Fig. 4).

Filterability of coagulated and flocculated water can be evaluated by the membrane filtration technique (0.45 μm) using a standard volume of 200 mL at a pressure of 0.087 bar. The tests concern river Meuse water treated by the amounts of coagulant necessary to reach a ZP value near zero. The pH of the raw water has been adjusted to reach a representative spectrum of final values. The coagulated water was sampled by jar test after 2 min of rapid mixing $G \sim 120\ s^{-1}$, and the flocculated water, after an additional slow-mixing period of 20 min followed by 45

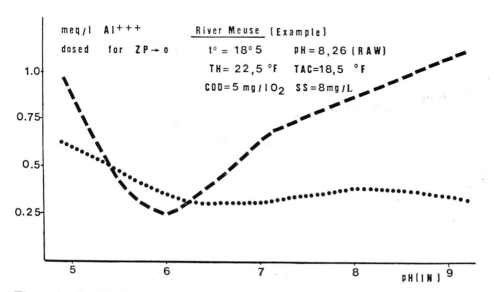

Figure 1 Coagulation effect on zeta potential. Dashed curve, aluminum sulfate; dotted curve, polyaluminum chloride.

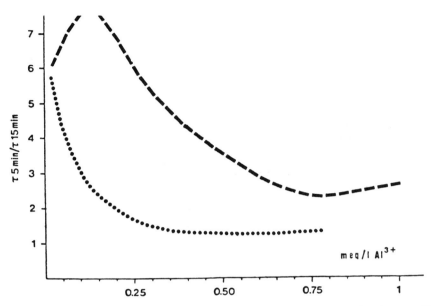

Figure 2   Flocculation efficiency of PAC (dotted curve) versus aluminum sulfate (dashed curve).

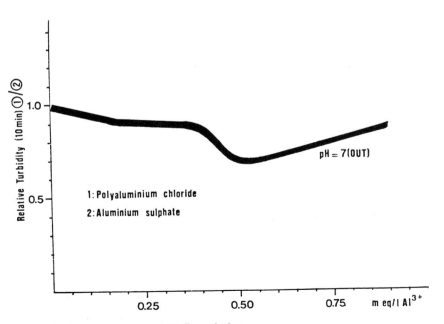

Figure 3   Turbidity removal by flocculation.

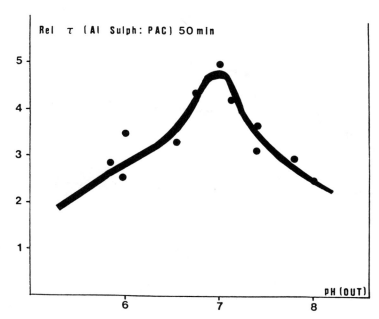

Figure 4   Turbidity removal after settling.

min of rest. The filtrability number (FN) is equal to the time of filtration of the sample, divided by the time required to filtrate the same volume of distilled water in equal conditions. The membrane refiltration time (MRT) is equal to the time required to filtrate the sample volume on a membrane on which the same sample has been freshly passed a first time. The filtration numbers and membrane refiltration times are indicated in Figs. 5 and 6.

The filtration number of coagulated water is lower when using polyaluminum chloride at a pH value above 7, but as far as membrane refiltration time is concerned, the opposite occurs. This means that when using polyaluminum chloride in the coagulation–filtration technique at pH 7 more attention is required at the beginning of the filtration cycle and lessening rapidly during the cycle. During these experiments no activated silica was used.

A residual aluminum concentration in the water when in equilibrium at different values of pH is a parameter of growing importance in view of the stricter quality standards for drinking water, for which a limit of 0.1 mg/L is set at present. The literature values of the equilibrium constants of solid hydrated $Al(OH)_3$ versus dissolved ions are as follows:

$$Al(OH)_3 + H_2O = Al(OH)_4^- + H^+ \qquad pK = 12.7$$
$$Al^{3+} + 3H_2O = Al(OH)_3 + 3H^+ \qquad pK = 9.75$$

and secondarily,

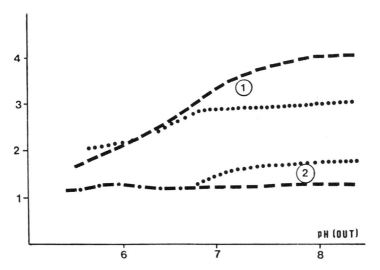

Figure 5   Filtration number as a function of pH value. 1, coagulation; 2, flocculation. Dashed curve, aluminum sulfate; dotted curve, polyaluminum chloride.

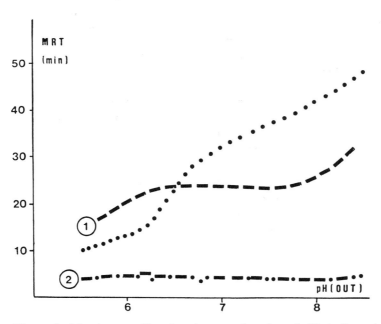

Figure 6   Membrane refiltration time as a function of pH. 1, Coagulation; 2, flocculation. Dashed curve, aluminum sulfate; dotted curve, polyaluminum chloride.

$$Al(OH)_3 + 2H^+ = Al(OH)^{2+} + 2H_2O \qquad pK = 4.41$$

The theoretical solubility is indicated in Fig. 7 as a function of pH. The supernatant of a conventional jar test indicates a slightly higher residual concentration than that expected in theory. The curves for aluminum sulfate and PAC are indicated in Fig. 7 on a comparative basis when water from the river Meuse is coagulated in a jar test up to a PZ approaching zero, followed by flocculation for 20 min and settling for 50 min (all samples are filtered on 0.22-μm Acropor membranes).

This classical jar test can be representative for once-through coagulation–flocculation followed by static settling and to some extent to the flocculation–filtration; however, it is not the case for sludge contact flocculation. To simulate this effect, at the Tailfer plant sludge samples are contacted by slow mixing (for 1 h) with water from the river Meuse to obtain a slurry of 20 vol% solids after 24 h of settling. During the slow mixing, the pH is adjusted to the desired values. Analysis of the supernatant for residual aluminum concentration results in the data given in Fig. 8.

In summary, with raw water samples, the residual aluminum concentrations are higher than those for distilled water or slightly buffered distilled water (10 mg/ L HCO_3^-). This probably corresponds to the effect of the possible existence of solid equilibrium phases other than hydrated aluminum hydroxide when raw waters containing organic material are involved. Some slight differences are observed between aluminum sulfate and polyaluminum chloride (Fig. 7). Although jar tests are hardly significant for evaluation of the performances of sludge-contacting clarifiers,

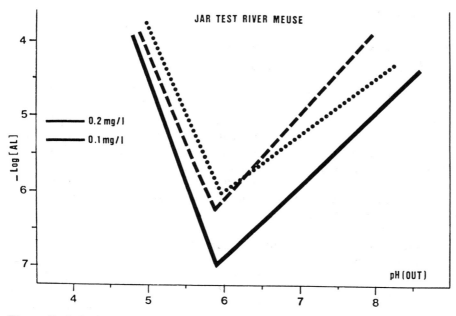

Figure 7 Solubility diagram for aluminum hydroxide (major components). Dashed curve, aluminum sulfate; dotted curve, polyaluminum chloride.

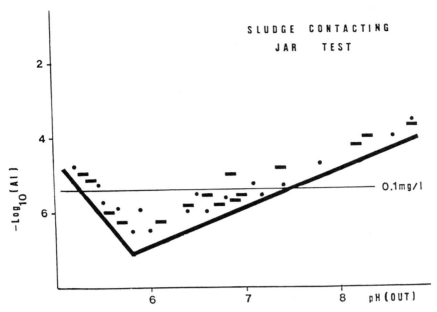

Figure 8   Residual dissolved aluminum in settled water at Tailfer. Dotted indications, PAC; dashed indications, aluminum sulfate.

they do illustrate that a decrease in the maximum admissible concentration of dissolved aluminum in drinking water can narrow the pH zone of application of aluminum-based coagulants.

3.   OPERATIONAL EXPERIENCES AT THE TAILFER PLANT (BRUSSELS)

The Tailfer plant, situated on the river Meuse upstream of Namur, is designed for a full treatment capacity of 3 m^3/s. The plant is a four-line modular design of 0.75 m/s capacity for each modular. It is currently operated at a two-modular capacity, which enables us to study the impact of different treatment schemes in parallel. Basically, the flowsheet for treatment is as follows: pH adjustment with sulfuric acid—chlorination (optional)—preoxidation with chlorine dioxide—coagulation with an aluminum-based coagulant—coagulation–flocculation with activated silica and adsorption with activated carbon followed by flocculation–settling in a Pulsator system working at 5 m/h average overflow velocity. After the Pulsator, pH correction with sodium hydroxide and secondary chlorination is carried out, if required, being completed by secondary horizontal settling with a residence time in the settlers of 3 h. Further treatment consists of classical rapid sand filtration and final ozonization. This scheme was adopted for a two-year period and a full-scale comparative evaluation of polyaluminum chloride versus aluminum sulfate was carried out. At present, because of the stricter standards for residual dissolved aluminum (i.e., 0.1 mg/L instead of 0.2 mg/L), sodium hydroxide is injected after the filtration phase. With river Meuse water, the average pH of "pulsated water" during the reference period was 7.1 with aluminum sulphate and 7.5 with PAC (Fig. 9). These values correspond to optimum coagulation in jar tests and result in equal performance in the treatment plant, as illustrated in Fig. 10 for the dry weight of solids resulting from the double settling stage, in Fig. 11 for the turbidity of filtered water, and in Fig. 12 for light extinction at 254 nm for filtered water.

4. COSTS

The respective costs of coagulants during the reference period (reported to 1990) are 0.40 Belgian francs (BF)/m^3 for aluminum sulfate and 0.60 BF/m^3 for PAC. However, the total costs for the chemical reagents applied in the treatment as a whole are 1.08 BF/m^3 when aluminum sulfate is used and 1.06 BF/m^3 for PAC. The overall costs are first adjusted by the use of less sodium hydroxide for pH correction, and second, by the use of less activated silica as a flocculation aid and less sulfuric acid for eventual initial pH adjustment.

A more careful analysis of the data of Fig. 13 reveals that for the river Meuse the total cost difference can be season dependent. The winter and spring periods are generally more favorable for PAC than are the algae growth periods of early summer. On an average annual basis, the use of PAC requires 20% less aluminum. This is equal to 32% for the reference weeks considered during the first part of the calendar year.

In the Tailfer plant the sludge treatment unit is common to all modules, so for this phase of the treatment, a completely separate evaluation is not possible. However, as treated, thickened, compacted, and transported, the sludge contains on average the following components: the use of 20% aluminum less reduces the coagu-

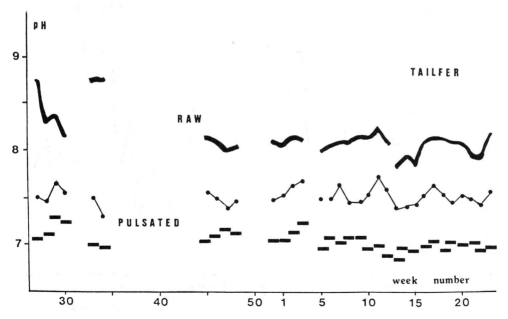

Figure 9 Residual aluminum concentration in settled water. •, PAC; — in bold, aluminum sulfate.

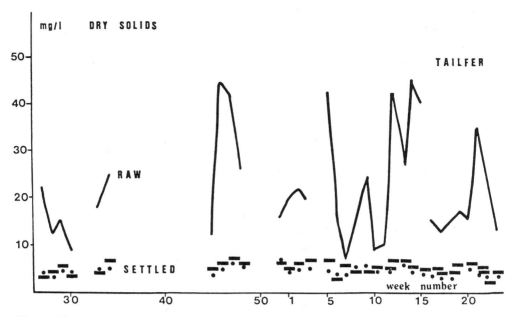

Figure 10   Residual dry solids (mg/L) in settled water. •, PAC; — in bold, aluminum sulfate.

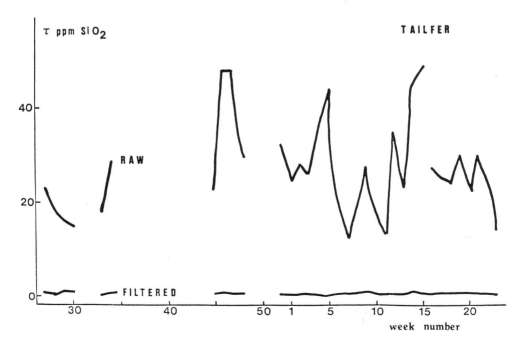

Figure 11   Turbidity of filtered water (PAC versus aluminum sulfate).

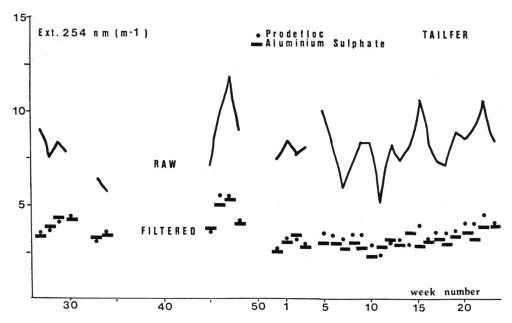

Figure 12   Extinction of filtered water at 254 nm.

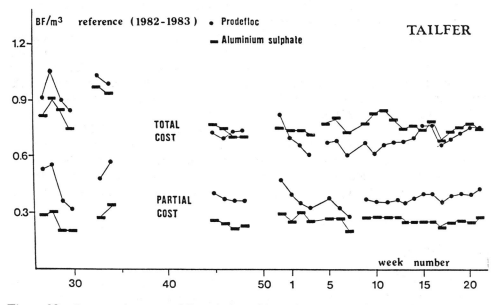

Figure 13   Comparative costs of flocculation with PAC versus aluminum sulfate.

lant residual by 20% and thus also by 20% the amount of lime used in the compaction process. As a consequence, there is the additional benefit of the use of a coagulant that needs fewer aluminum dosings while providing an equal treatment performance. In the case of Tailfer, this additional benefit is 6 to 10% of the cost of sludge transportation and 10% of the cost of the quicklime used in the compaction process (see Chapter 16).

5. CONCLUSIONS AND GUIDELINES

The use of prepolymerized aluminum flocculants presents a high guarantee of quality from the point of view of the content of trace elements due to the elaborate manufacturing process. Their use makes it possible to avoid an increase in sulfate ions. Highly concentrated products simplify the transportation problems and lessen the costs. Preliminary evaluation of performance can be made by jar tests: annulation of the zeta potential of the colloids, turbidity measurements of supernatant, and filtration tests make it possible to evaluate the products on a relative basis, yet the performance in a full-scale plant is superior to that reached in the jar test. Determination of residual aluminum and solubility tests require high skill, adequate stabilization of the glassware, and prefiltration of the samples analyzed. Control of residual aluminum according to the drinking water standards is a challenge in full-scale plant operation given that a pH value in the range 7.4 to 7.5 is required before polyaluminum chloride and aluminum sulfate are truly competitive.

In the once-through flocculation–settling technique, the dosings correspond to sweep coagulation. With low dosings, as in flocculation–filtration, the process is to be associated with secondary bridge coagulation and secondary sweep coagulation-flocculation. Seeing that the dosing rate is 20 to 40% lower with polyaluminum chloride, the sludge density is improved in the once-through processes because the proportion of highly hydrated aluminum hydroxide is lowered.

In sludge contact clarifiers, the conditions are those of sweep coagulation-flocculation. At the Tailfer plant, activated silica is still used as a flocculation aid, but the dosing rate is decreased. Maintaining adequate sludge stability requires readjustment of the sludge bleed rate (e.g., decreasing the bleed by 10% of its conventional value). In the case of Tailfer, this means an increase of concentration factor in the Pulsator by 10% (e.g., from 65% to 75%).

In some cases, flocculation apparently requires reducing the pH level of the raw water (e.g., when high concentrations of humic acids or intensive algae growth occur). These circumstances always require specific preliminary investigations to determine the most adequate conditions. The use of polyaluminum chloride has been proven economically competitive in treating river Meuse water at the Tailfer plant.

REFERENCES

1. J. Y. Bottero, J. E. Poirier, and F. Fiessinger, Study of partially neutralized aqueous aluminum chloride solutions: Identification of aluminum species and relation between the composition of the solutions and their efficiency as a coagulant, *Prog. Water Technol., 12,* 601–612 (1980).

2. Kemira, Technical documentation; Box 902, S-25109 Helsingborg, Sweden.
3. Prodeco, Technical documentation; Via Mozart, 1-20100 Milano, Italy.
4. Rhone-Poulenc S. A., Technical documentation; quai Paul Doumer 25, F-92408 Courbevoie, France.
5. Sachtleben, Technical documentation; Pestalozzistrasse 4, D-4100 Duisburg-Homberg, Germany.

7
Flocculation

1. INTRODUCTION

In this chapter we describe briefly the fundamental principles underlying the flocculation methods used in water treatment, and define general guidelines of application. Practical examples are provided in Chapters 9 and 17.

2. THE FLOCCULATION PROCESS

In the flocculation process, fine dispersed particles are combined into larger agglomerates that can be removed by subsequent processing, such as settling or filtration. Flocculation is determined by interparticle contact, enabling particle growth and particle number decrease. Processes prior to and enabling flocculation are coagulation; chemical or biological oxidation to destroy protected colloids; change in general conditions, such as pH and temperature; seeding and contact with solids (e.g., sludge, filtering material); and the use of auxiliary reagents (i.e., flocculation aids). Coagulation remains by far the most important preparatory step to flocculation.

Flocculation supposes interparticle contact. The function for the repulsion energy between particles is given by the van der Waals equation, in which the repulsive energy increases steeply at short interparticle distances, the repulsive energy term being proportional to r^{-12}. Several mechanisms occur in the flocculation process for which the repulsing energy to interparticle contact must be overcome.

2.1 Perikinetic Flocculation

The first phase generally considered is that of perikinetic flocculation, in which, through the natural random process of thermal agitation, interparticle contacts that occur during Brownian motion promote the coalescence of particles. The energy term involved is $RT/2$ per degree of freedom, hence total energy is equal to $\pm 3RT/2$. According to Fick's second law for diffusion in a homogeneous dispersion (i.e., for particles identical in dimension and chemical composition), the number of collisions due to Brownian motion is equal to

$$I = 8\pi DR_0 n_0^2 \left(\frac{1 + R_0}{\sqrt{2\pi Dt}} \right)$$

in which in practice, after a short time, the term $R_0/\sqrt{2\pi Dt}$ is canceled versus 1 and the simplified expression becomes

$$I = 8\pi DR_0 n_0^2$$

in which R_0 is equal to the radius of interaction of the particles and n_0 is their initial number. D is the diffusion coefficient ($D = k_B T/6\pi\mu R$, where k_B, the Boltzman constant, $= 13.8 \times 10^{-24}$ J/°K and R, the gas constant, $= 8.31$ kJ mol^{-1}°K^{-1}).

As the aggregation probability is equal or proportional to the collision probability, the change in particle number due to perikinetic encounters is expressed as follows:

$$\frac{dn}{dt} = -8\pi DR_0 n^2 = \alpha n^2$$

which is equal to a basic kinetic equation of second order. In its integrated form $(1/n - 1/n_0)$ is a linearly increasing function with time. By generalization of the model to the case of polydisperse solutions according to Hahn (1), we obtain

$$n_k = \frac{n_0}{(1 + t/T)^{\alpha+1}} \left(\frac{t}{T} \right)^{1-\alpha}$$

in which T is the half-lifetime or *flocculation time*:

$$T \simeq \frac{1}{\alpha} \simeq \frac{1}{4\pi DR_0 n_0} \qquad \text{and} \qquad I_{ij} = 4\pi D_{ij} R_{ij} n_i n_j$$

in which n_i and n_j are the numbers of particles of i and j type having a mutual interaction radius R_{ij} and a mutual diffusion constant $D_{ij} \simeq D_i + D_j$.

Owing to the flocculation process itself, initially, homogeneous dispersions rapidly revert to polydisperse systems. Moreover, because of the increase in size, Brownian motion has progressively less effect on particle movement. Also, the magnitude of the energy barrier increases progressively in relation to the cross section of the particles. Consequently, the perikinetic contact probability is also reduced with increased particle dimension.

2.2 Orthokinetic Flocculation

In orthokinetic flocculation, interparticle contact is promoted or induced by velocity gradients created in the bulk of the liquid. Vom Smoluchowski (2) has developed a model for particles of *i* and *j* type coming into contact through the influence of a local velocity gradient dv/dz of constant value and direction (Fig. 1a). Particles collide when the center of particle *i* enters the sphere of influence of particle *j*, the active collision radius being R_{ij} (Fig. 1b). Movement of one particle relative to another is associated with bulk movement of the water. The area through which efficient transit takes place is equal to

$$dA = 2 \sqrt{R_{ij}^2 - z^2}\, dz$$

and the flow through this area (Fig. 1c)

$$dQ = z \frac{dv}{dz} 2 \sqrt{R_{ij}^2 - z^2}\, dz$$

Through the entire sphere the flow is equal to

$$Q = 2 \int_0^{R_{ij}} G \sqrt{R_{ij}^2 - z^2}\, 2z\, dz = \frac{4}{3} G R_{ij}^2$$

Consequently, if particles follow a laminar stream, developed by agitation, the collision probability according to this theory is expressed as follows:

$$J = \frac{4}{3} (R_{ij})^3 n_i n_j \frac{dv}{dz}$$

where dv/dz is equal to the velocity gradient in laminar flow and R_{ij} is the radius of the effective contact sphere. Velocity gradients occur in flocculation systems through mechanical mixing, disturbing flow pathways through sludge blanket or filter layers. The mixing conditions are locally not known and generally turbulent. Camp and Stein (3) therefore assimilate the dv/dz term in the average overall velocity gradient, expressed as $G\ (s^{-1})$.

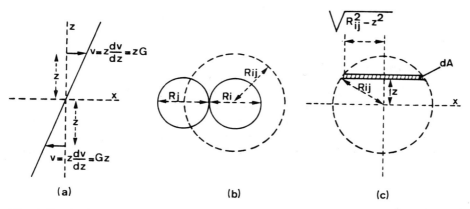

Figure 1   Collision areas in flocculation.

Therefore, the collision probabilities and the agglomeration probabilities are related as follows:

$$\frac{J(\text{orthokinetic})}{I(\text{perikinetic})} \simeq \frac{R_{ij}^2 \mu G}{k_B T}$$

from which at 20°C ($=293$°K), $\mu \simeq 10^{-3}$ kg m^{-1} s^{-1}, and if $R_{ij} \simeq 10^{-6}$ m, the velocity gradient given equal probability that both flocculation mechanisms will occur is approximately

$$\frac{(10^{-6})^2 \times 10^{-3} \times G}{13.8 \times 10^{-24} \times 293} = 1$$

$$G = \frac{293 \times 13.8 \times 10^{-24}}{10^{-15}} \simeq 4 \times 10^{-6}$$

In practical cases, a G value of at least 10 s^{-1} is used.

Consequently, perikinetic flocculation will play only a minor role in practical systems that have a velocity gradient; for example, for a particle 10^{-3} m in size at a G value of 10 s^{-1},

$$\frac{J}{I} \sim \frac{10^{-6} \times 10^{-3} \times 10}{293 \times 13.8 \times 10^{-24}} = 2.5 \times 10^{12}$$

Evolution of the particle number as a function of time in orthokinetic flocculation is expressed as a kinetic equation of first order:

$$\frac{dn}{dt} = -FGn$$

and upon integration,

$$\ln \frac{n}{n_0} = -(FG)t \simeq -kt$$

2.3   Macroscopic Flocculation

An increase in dimension of the individual particle systems takes place very quickly, within 10 s to 1 min of the end of the coagulation phase (Fig. 2). After this phase a more macroscopic flocculation occurs, assumed by Hudson (4) to be described by a system composed of single particles and larger flocs in an "equivalent" number n_F and radius R_F, with flocculation occurring through particle–floc contact according to

$$\frac{dn}{dt} = -\frac{4\pi}{3} n n_F R_F^3 G$$

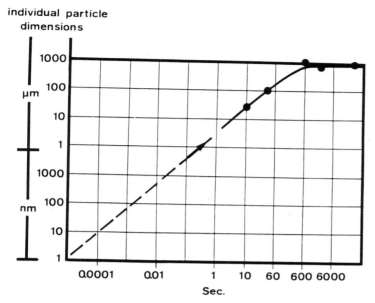

Figure 2 Particle-dimension increase rates.

The basic objections to the foregoing description of orthokinetic flocculation is based on laminar flow and random agitation and a linear velocity gradient. This is likely to occur only over distances comparable to the particle diameter. In a turbulent regime, orthokinetic flocculation has been proposed by Argaman and Kaufman (5) to be based on a diffusion model. The collision rate derived by this approach is expressed as follows:

$$CR_F = 4\pi k R_F^3 n_1 n_F \overline{u^2}$$

in which $\overline{u^2}$ is the mean-square velocity fluctuation and k is a proportionality coefficient expressing the effect of turbulency. Although hypothesized on different bases, all equations are similar in structure; the differences are finally reduced to the proportionality constants, for which experimental values must be determined in each case or assumed empirically.

Finally, the overall decrease in the number of primary particles is the net result of aggregation in the flocculation process, compensated for floc breakup by floc erosion or by splitting through flow and mixing. The global expression follows:

$$\frac{dn_1}{dt} = 4\pi \alpha k R_F^3 n_1 n_F \overline{u^2} + B\frac{R_F^2}{R_1^2} n_F \overline{u^2}$$

and

$$\frac{n_0}{n_1} = \frac{1 + 4\pi \alpha k R_F^3 n_F \overline{u^2} T}{1 + (B R_F^2 n_F \overline{u^2} T / n_0 R_1^2)}$$

in which n_0 and n_1 are the number of single particles at times zero and T, respectively; n_0/n_1 is the *agglomeration rate factor*; α is the efficient fraction of collisions between n_1 and n_F particles resulting in aggregation, supposed constant in any given system; and B is the breakup constant.

3. PARAMETERS FOR FLOCCULATION

Practical parameters that may be approached experimentally are:

1. $\overline{u^2} = PG$, in which P is a performance parameter of the mixing process, called the stirrer performance coefficient for stator stirrers similar to gate paddles (see Chapter 17). According to the observations of Argaman and Kaufman (5), for gate paddles $P \sim 4.6 \times 10^{-6}\,\mathrm{m^2/s}$, and for turbines $P \sim 2.3 \times 10^{-6}\,\mathrm{m^2/s}$.

2. Assuming spherical particles, the total floc volume fraction in a basin is given by

$$V_F = \frac{4}{3}\,\pi n_F R_F^3$$

in which the average floc particle size is directly related to the mean-square velocity:

$$R_F = K/\overline{u^2} = \frac{K}{PG}$$

3. By simplifying and expressing all constant parameters globally in terms of an agglomeration constant k_a and a breakage constant of a floc k_b, the agglomeration factor in a given flocculation system is expressed as follows:

$$\frac{n_0}{n_1} = \frac{1 + k_a GT}{1 + k_b G^2 T}$$

This equation can also be applied for several consecutive basins considered as a function of the basin:

$$\frac{n_0(1)}{n_1(1)},\ \frac{n_1(1)}{n_1(2)},\ \frac{n_1(2)}{n_1(3)},\ \ldots$$

the number (1), (2), and so on, representing, for example, tapered stirring (see Chapter 17). In its differential form

$$\frac{dn_1}{dt} = -k_a n_1 G + k_b n_0 G^2$$

k_b is a constant at a given G value which yields for a batch system (e.g., a test flocculator)

$$n_1 = \frac{k_b}{k_a} n_0 G + n_0\left(1 - \frac{k_b}{k_a}G\right)e^{-k_a GT}$$

and

$$\frac{n_1}{n_0} = \frac{k_b}{k_a} G + \left(1 - \frac{k_b}{k_a} G\right)e^{-k_a GT}$$

or

$$k_a = \frac{1}{GT} \ln \frac{1 - (k_b/k_a)G}{[1/(n_0/n)] - [1/(k_a/k_b G)]}$$

At equilibrium conditions for floc agglomeration versus breakage, the second term in the equation above can be neglected and

$$\left(\frac{n}{n_0}\right)_E = \frac{k_b G}{k_a}$$

Example: Flocculation of a Suspension of Kaolin ($G \simeq 40 \ s^{-1}$)

T (min)	$\frac{n_0}{n}$	T (min)	$\frac{n_0}{n}$
2	2.5	20	13
4	7	23	14
6	8	25	10
7	13	30	12.5
7.5	11	30	15
10	10	33	12
14	15	35	14
16	15.5	35	17.5
17.5	14	40	14.5

The average n_0/n value in this experiment after 14 min of stirring is equal to

$$13.9 = \left(\frac{n_0}{n}\right)_E = \frac{k_a}{k_b G} \quad \text{and} \quad \frac{k_b}{k_a} = \frac{1}{40 \times 13.9} = 1.8 \times 10^{-3} \, s$$

The different values of n_0/n at stirring times that are shorter than that to obtain stationarity enable us to compute the k_a values:

$$k_a = \frac{1}{GT} \ln \frac{1 - (k_b/k_a)G}{[1/(n_0/n)] - [1/(k_a/k_b G)]}$$

T(s)	k_a	
120	2.17×10^{-4}	
240	2.68×10^{-4}	
360	1.99×10^{-4}	Average $k_a = 2.42 \times 10^{-4}$
420	3.11×10^{-4}	
450	2.16×10^{-4}	

Hence at $G = 40\ \text{s}^{-1}$,

$$k_b = \frac{2.42 \times 10^{-4}}{40 \times 13.9} = 4.4 \times 10^{-7}\ \text{s}$$

Taking the average values of k_a and k_b thus obtained, one can calculate the theoretical flocculation curve according to the equation

$$\frac{n_0}{n} = \left[\frac{k_b}{k_a} G + \left(1 - \frac{k_b}{k_a} G \right) e^{-k_a GT} \right]^{-1}$$

The result is given in Fig. 3.

An experimental method for the determination of k_a and k_b has been developed by Bratby et al. (6,7) in which the velocity gradient is estimated by a formula of Te Kippe:

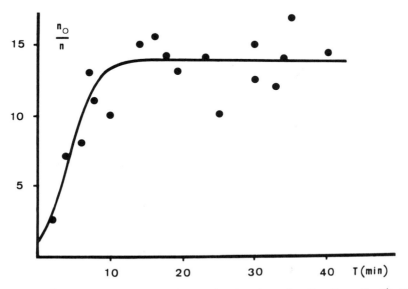

Figure 3 Equilibrium floc agglomeration breakage (kaolin, $G = 40\ \text{s}^{-1}$). Dots, experimental results; curve, calculated average.

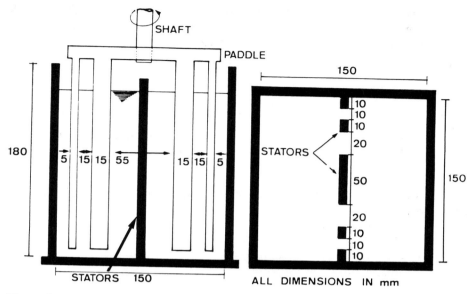

Figure 4 Experimental setup for flocculation test.

$$G(\text{s}^{-1}) = \sqrt{\frac{2\pi NT}{V\mu}} \qquad \text{(see also Chap. 17).}$$

Laboratory experiences carried out by the authors (7) on both turbidity and color removal gave characteristic values of k_a and k_b determined in test systems:

k_a (turbidity) $\simeq 2.5 \times 10^{-4}$
k_b (color) $\simeq 3.09 \times 10^{-4}$

The k_b value may be considered as constant only at a given G value. The relationship k_b versus G is obtained from practical experience as to be approached by

$$G = \exp\!\left(\frac{k_b - k_2}{k_1}\right) \qquad \text{or} \qquad k_b = k_1 \ln G + k_2$$

In experiments using aluminum sulfate as a coagulant flocculant, k_1 and k_2 values to fit the data are obtained as follows:

For turbidity removal: $k_1 \simeq -0.87 \times 10^{-7}$ $k_2 \simeq 7.7 \times 10^{-7}$
For color removal: $k_1 \simeq -3.04 \times 10^{-6}$ $k_2 \simeq 15.2 \times 10^{-6}$

The resulting k_b values are:

For turbidity: $k_b = (7.7 - 0.87 \ln G) \times 10^{-7}\,\text{s}$
For color: $k_b = (15.2 - 3.04 \ln G) \times 10^{-6}\,\text{s}$

The experimental procedure implicates measurement of N_0/N (e.g., through UV absorption or turbidity measurements) on the supernatant as a function of the contact time and for various velocity gradients used.

Exercise.   Suppose that $G = 100 \text{ s}^{-1}$. Then:

Turbidity	Color
$k_a = 2.5 \times 10^{-4}$	$k_a = 3.09 \times 10^{-4}$
$k_b = (7.7 - 0.87 \times 4.6) \times 10^{-7}$	$k_b = (15.2 - 3.04 \times 4.6) \times 10^{-6}$
$\quad 3.69 \times 10^{-7}\,\text{s}$	$\quad 1.2 \times 10^{-6}\,\text{s}$

For flocculation in m consecutive batch-type reactors, the equations can be generalized as follows:

$$\frac{n_0}{n(m)} = \frac{\dfrac{(1 + k_a GT/m)^m}{1 + k_b G^2 T}}{m \sum_{i=0}^{m-1} (1 + k_a GT/m)^i}$$

As a conclusion from Fig. 6, we see that practical polydisperse waters can have quite an unideal behavior.

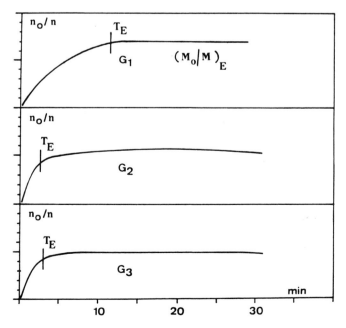

Figure 5   Experimental evaluation of breakage-to-agglomeration ratio.

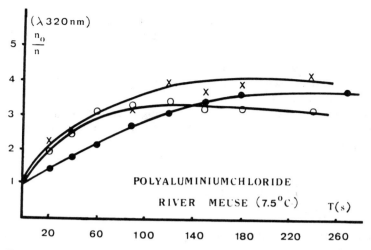

Figure 6 Typical data for water of the river Meuse. •, 20 s^{-1}; ×, 42 s^{-1}; ○, 77 s^{-1}.

4. GUIDELINES FOR FLOCCULATION

All investigations and practical experience reported indicate maximum performance in particle aggregation through flocculation at an average velocity gradient of 40 to 80 s^{-1} when used with a single (completely mixed or plug-flow) vessel. The best performances are obtained at a GT value of $\pm 100,000$ ($>60,000$). In multiple stirred setups (e.g., consecutive baffled basins or tapered stirred vessels) the optimum GT value can be lowered to 50,000 to 60,000. The lowering of the optimal GT value is principally the effect of a lowering of the necessary reaction time rather than that of the recommended G value (40 to 80 s^{-1}). To be consistent, the G value in the GT criterion must be low (e.g., <60 s^{-1}), as "hard flocs" are involved.

Camps number has been developed for flocculation with aluminum sulfate. For dosages up to 50 mg/dm^3, the optimal relation has been found to be

$$(G^*)^{2.8}T = K$$

K being a constant for any given aluminum sulfate concentration (8).

Flocculation is most often performed in vertical up- or downflow chambers which are sometimes incorporated in combined flocculation-clarifiers. The residence time generally selected is about 20 to 30 min. Depth and surface in relation to flow can be set empirically according to the following criteria:

$$\text{Depth} = \tfrac{1}{2}\sqrt[3]{\text{volume}} = \tfrac{1}{2}\sqrt[3]{\text{flow} \times \text{detention time}}$$

$$= \tfrac{1}{2}\sqrt[3]{Qt}$$

$$\text{Surface} = \frac{Qt}{\tfrac{1}{2}\sqrt[3]{Qt}} = 2\sqrt[3]{(Qt)^2}$$

Flocculation time is generally set at 15 to 30 min, or for exceptionally difficult waters, at 60 min. The average velocity gradient should be below 60 s^{-1}; that is a GT value of $<100,000$ s^{-1}. The average velocity gradient should be kept between 20 and 75 s^{-1}. The tip-speed values of the impellers should be between 0.2 and 0.8 m/s so as to prevent floc breakage. The paddle area should be less than 20% of the cross section of the flocculation basin.

The velocity of transport of flocculation water through pipes so as to avoid floc breakage is between 0.15 and 0.2 m/s. In baffled systems a value of 0.4 m/s can be admitted in the ports.

SYMBOLS AND UNITS RELATED TO THE FLOCCULATION PROCESS

k_a	agglomeration constant, dimensionless
k_b	breakage constant, s
k_B	Boltzmann constant, 13.8×10^{-24} J/°K
n_0	particle number
n_F	floc particle number
r	interparticle distance
t	(water) residence time, s
$\overline{u^2}$	mean-square velocity fluctuation, (m/s)^2
v	velocity of particles, m/s
z	reference direction coordinate
A	cross-sectional area of particle interference, m^2
D	diffusion constant, m · s/°K
D_i, D_j	diameter of interference sphere of particles, m
F	kinetic factor in orthokinetic flocculation
G	velocity gradient, s^{-1}
G^*	optimal velocity gradient, s^{-1}
I	collision number (perikinetic)
J	collision number (orthokinetic)
N	number of rotations, s^{-1}
P	performance parameter in mixing, m^2/s
Q	water flow, m^3/s
R	gas constant, 8.31 kJ mol^{-1}°K^{-1}
R_F	floc particle radius, m.
R_i, R_j	collision radius of particles, m
T	absolute temperature, °K
T	flocculation time, s
T	torque input, N · m ($=$ kg m^2 s^{-2})
V	liquid volume, m^3
V_F	floc volume fraction
μ	dynamic viscosity, kg m^{-1} s^{-1}

REFERENCES

1. H. Hahn, *Vom Wasser, 33*, 172 (1966).
2. Vom Smoluchowski, *Z. Phys. Chem., 92*, 155 (1917).
3. T. R. Camp and P. C. Stein, *J. Boston Soc. Civil Eng., 30*, 219 (1943).
4. H. E. Hudson, *J. AWWA, 57*, 885 (1965).
5. Y. Argaman and W. J. Kaufman, *J. Sanit. Eng. Div. ASCE, 96*; *SA$_2$ Proc. Paper 7201*, 223 (1970).
6. J. Bratby, M. W. Miller, and G. V. R. Marais, *Water S.A., 3*, 173 (1977).
7. J. Bratby, *Coagulation and Flocculation*, Upland Press, Croydon, England, 1980.
8. R. Andreu-Villegas and R. D. Letterman, *J. Environ. Eng. Div. ASCE, 102*, 251 (1976).

8
Activated Silica as a Flocculation Aid

1. INTRODUCTION

Activated silica is a glassy product or a silica sol generally presumed to have been isolated by Graham. As early as 1884 he added sodium silicate (water glass) to weak hydrochloric acid and formed a silica sol.

2. USE OF ACTIVATED SILICA

In water treatment, the term *activated silica* implies a suspension of negatively charged colloidal particles, obtained by reaction of a dilute solution (0.5 to 5% SiO_2) of sodium silicate and an activator, usually an acid. Electron microscopy indicates a lack of crystalline forms, the sol being composed of primary spherical particles with a diameter of ± 5 nm. The sol normally attained is of grain size 0.1 to 1.0 μm. The corresponding dry material has a BET surface of ± 250 m^2/g. The electrophoretic mobility, EM (10^{-8} m/Vs), related to the zeta potential (ZP) of the particles, depends on the pH and is distinct for amorphous and crystalline silica. A typical example is shown in Fig. 1.

The use of activated silica in water treatment processes was introduced by Baylis (2), who found that silica in natural waters facilitated the flocculation of water when using aluminum sulfate and to a lesser extent in the case of ferric coagulants. Dissolved silica in natural water is obtained principally by the decomposition of aluminosilicate minerals such as feldspars (3):

$$Na_2CaAl_4Si_8O_{24} + 4CO_2 + 14H_2O = 4H_4Al_2Si_2O_9 + Ca^{2+}$$
$$\text{(andesine)} \qquad + 2Na^+ + 4H_4SiO_4 + 4HCO_3^-$$

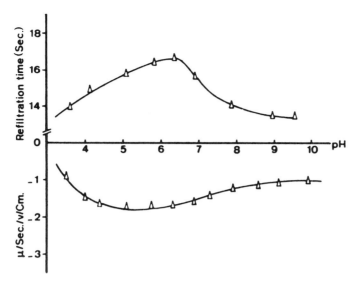

Figure 1 Electrophoretic mobility (EM) of activated silica suspensions as a function of pH. The activation also influences the filterability of the silica. The "solubility" of a gel of silica ranges from 100 to 150 g/m^3, which is equal to 1.65 to 2.5 × 10^{-3} M expressed in SiO$_2$. (From Ref. 1.)

and

$$2Na_2CaAl_4Si_8O_{24} + 6CO_2 + 18H_2O = Na_2Al_8Si_{10}O_{30}(OH)_6 + 2Ca^{2+}$$
$$+ 2Na^+ + 6H_4SiO_4 + 6HCO_3^-$$

Basically, silicic acid appears in the reaction schemes and corresponds to the soluble mononuclear species (as evidenced by direct determination with molybdate) at a solubility value ($pC \sim -2.7$) independent of the pH (Fig. 2).

A simplified solubility diagram according to (3) indicates the scope of stability of soluble and insoluble multimeric and mononuclear species. The accepted equilibrium constants for the dissolution of SiO$_2$ are given in Table 1.

In most flowsheets activated silica is added at the end of the coagulation process or at the very beginning of flocculation (i.e., at the stage of pinpoint floc formation). In this case it acts as a flocculation aid and assists subsequent processes, such as settling and filtration. The major advantages claimed for the use of activated silica, as a flocculation aid, are as follows (4–6):

1. Increased coagulation–flocculation efficiency gives thorough and dense floc particles that are less susceptible to breakdown.
2. Broader pH and temperature ranges promote more effective flocculation.
3. More rapid settling enables smaller constructions at a given flow capacity and thereby lower capital outlay. Color reduction is enhanced significantly.
4. Clearer filter effluents are experienced and a decrease in length of filter runs is sometimes reported; this results from inadequate activation. The best range for

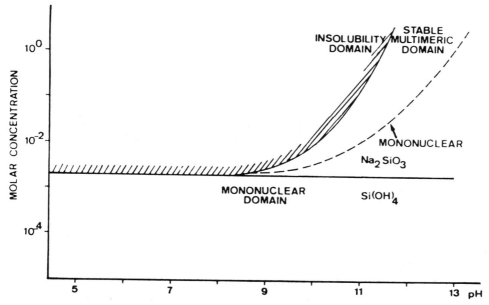

Figure 2 Solubility zones of natural silica.

the ZP of the particles of activated silica gel for optimum filtration is 0 ± 8 mV.

5. Although sometimes claimed, a reduction in coagulant dosage is not in all cases a well-established fact (2,8).

6. Postprecipitation is avoided by the joint use of aluminum sulfate and activated silica, particularly in waters of low alkalinity (e.g., ≤ 1 mEq HCO_3^-/L).

Waters containing sufficient magnesium can be coagulated by the use of lime and silica alone, with the formation of magnesium silicate. Furthermore, the removal of magnesium during the lime-soda softening process can occur at lower pH values by the use of activated silica (9). In silicates, the sodium ions can be replaced by other cations, especially magnesium. Such products, designed as artificial zeolites, are used for their ionic-exchange capacity (9). Activated silica, as such, acts as a cationic exchanger. The removal of heavy metals during flocculation–settling is generally improved by adding activated silica.

Table 1 Acidity Constants of Silicic Acid

Reaction	pK at 25 °C
SiO_2 (quartz) $+ 2H_2O = Si(OH)_4$	3.7
SiO_2(amorph) $+ 2H_2O = Si(OH)_4$	2.7
$Si(OH)_4 = SiO(OH)_3^- + H^+$	9.46
$SiO(OH)_3^- = SiO_2(OH)_2^{2-} + H^+$	12.56
$4Si(OH)_4 = Si_4O_6(OH)_6^{2-} + 2H^+ + 4H_2O$	12.57

3. REAGENTS

Just like silica concentrations in general, activated silica concentrations are often expressed in terms of SiO_2. By this one may not conclude that SiO_2 could be the active species present.

3.1 Preparation

In practice, an activated silica hydrosol is prepared using an activator of sodium silicate. Sodium silicates are manufactured by fusing together sand and sodium carbonate (Solvay soda) to form a vitreous mass which is subsequently dissolved in water. It is most convenient to represent silicates in the form $SiO_2 \cdot xNa_2O$, in which $x > 1$, ranging from 1.65 to 4. For alkaline products, for which $x > 3.4$, partial formation of suspended particles occurs, and above 3.65 the solutions are unstable, making the products less suitable. The commercial sodium silicates fall into two broad classes: the *alkaline grades*, 2:1 by weight SiO_2/Na_2O, and the *neutral grades*, 3.2 to 3.3:1 weight ratio. In fact, both grades exhibit alkaline reaction to different extents.

   Activated silica can be prepared from any of the common sodium silicates, but in practice, only products in the "neutral" range are considered. In the case of alkaline types, activation may sometimes be less successful. Baylis (2) reported that "the active silicate solution was prepared from an old sample of sodium silicate in which the alkali had become partially carbonated, owing to a loose-fitting stopper."

   At present the grades of practical use for the preparation of activated silica are products of specific gravity 1.42 and 1.39. The higher-viscosity form of the sp. gr. 1.42 product makes it less easy to handle, especially for dosing and storage during the winter. Therefore, in practice, the sp. gr. 1.395 product is often preferred. These "neutral" silicate solutions contain mixtures of Na_2SiO_3 and $Na_2Si_2O_5$.

3.2 Activation Process

There is substantial agreement that the activation process starts with monomeric species, generally metasilicic acid (H_2SiO_3), that occur transiently during the activation process. On hydration, metasilicic acid is converted into orthosilicic acid ($H_2SiO_3 + H_2O = H_4SiO_4$).

Table 2   Characteristics of Commercial Silicate Grades

Characteristic	Silicate grade		
	37/40	40/42	40/42
SiO_2 (% wt/wt)	29–29.8	26.6–27	29–29.8
% Na_2O (% wt/wt)	8.8	9.4	8.6–9.0
SiO_2/Na_2O (% wt/wt)	3.3	3.2	3.3–3.4
Specific gravity	1.395	1.42	1.39–1.405
Viscosity (mPas/20°C)	50–100	600–1200	400–600

3.2.1 Monomeric Orthosilicic Acid

Stock solutions of monomeric orthosilicic acid can be prepared by a method proposed by Alexander (10). This is done by converting solutions of metasilicates, $Na_2SiO_3 + 9H_2O$ to their hydrogen form through a cationic-exchange resin (Dowex 50 W-X8). The concentration is best kept under 2×10^{-3} M (solubility of amorphous SiO_2) to prevent polymerization, but concentrations up to 5×10^{-2} M can be prepared transiently. Monomeric meta- or orthosilicic acid reacts completely with molybdic acid within 1 to 2 min (11). Metasilicic acid is a weak divalent acid of which the acidity constants are the following:

$$\frac{[H^+][HSiO_3^-]}{[H_2SiO_3]} ; k_1 = 10^{-9.5}$$

$$\frac{[H^+][SiO_3^{2-}]}{[HSiO_3^-]} ; k_2 = 10^{-12.5}$$

Any acid stronger than metasilicic acid can displace the latter compound from the metasilicate and induce meta- or orthosilicic acid transiently.

When the hydrogen ion concentrations are increased in a dilute solution (1 to 5% SiO_2) of sodium metasilicate, the following consecutive reaction steps occur:

$H_5SiO_4^+$ and general protonated forms of orthosilicic acid are involved in the so-called acid polymerization schemes of orthosilicic acid (12) occurring at a pH value below 4. These schemes are less involved or uninvolved completely in the usual preparation of activated silica for water treatment applications.

3.2.2 Sodium Metasilicate

The preparation of activated silica remains based on the alkaline polymerization of partially neutralized sodium metasilicate. When passed through an acid-regenerated cation-exchange column, dilute solutions of sodium metasilicate produce, on standing, colloidal silica free from salts. This product (i.e., polysilicic acid with a degree of polymerization of 3 to 7 at a pH of 2 to 3) (13) has little or no value in promoting flocculation or the strengthening of flocs. On dilution, depolymerization occurs, partially regenerating the orthosilicic acid. By adding alkali to a suspension of acid-induced polymerization of orthosilicic acid, depolymerization occurs but is followed by rapid repolymerization, releasing an active product (14).

In conclusion, it can be stated that activated silica as used in water treatment is a polymer of partially neutralized sodium metasilicate corresponding to the following scheme:

$$\begin{array}{ccccc}
\text{OH} & \text{ONa} & \text{OH} & \text{OH} \\
| & | & | & | \\
O=\text{Si} & \text{Si} & \text{Si} & \text{Si} & \text{Si} \\
| & | & | & | & | \\
\text{OH} & \text{OH} & \text{OH} & \text{ONa} & \text{OH}
\end{array} \cdots \rightleftharpoons$$

$$\begin{array}{ccccc}
\text{OH} & O^- & \text{OH} & \text{OH} \\
| & | & | & | \\
O=\text{Si} & \text{Si} & \text{Si} & \text{Si} & \text{Si} \\
| & | & | & | & | \\
\text{OH} & \text{OH} & \text{OH} & O^- & \text{OH}
\end{array} \cdots + 2\text{Na}^+$$

Repulsion between the negatively charged groups maintain the polymer in the form of a stretched structure, enabling cross-linking with positively charged molecules.

3.2.3 Polymerization Kinetics

Reversible acid polymerization schemes occur only during accidental excessive acidification or inadequate mixing conditions in the reactor. On growing, the micella of activated silica as used in water treatment processes range between 0.1 and 1.0 μm. As demonstrated by dimerization, the alkaline reaction corresponds to the scheme

$$\text{H}_3\text{SiO}_4^- + \text{H}_4\text{SiO}_4 \longrightarrow \underset{\substack{| \\ \text{OH}}}{\overset{\substack{\text{OH} \\ |}}{\text{HO}-\text{Si}}} - \text{O} - \underset{\substack{| \\ \text{OH}}}{\overset{\substack{\text{OH} \\ |}}{\text{Si}}} - \text{OH} + \text{OH}^-$$

and the dimer further condensates as the monomer, for example, by reaction with an other H_4SiO_4 molecule. The general scheme can be expressed as follows:

$$\text{HO}_{m+1}\text{Si}_m(\text{OH})_{2m}^- + \text{H}_2\text{O}_{n+1}\text{Si}_n(\text{OH})_{2n} \longrightarrow \text{OH}^-$$

$$+ \text{H}_2\text{O}_{m+n+2}\text{Si}_{m+n}(\text{OH})_{2m+2n}$$

or in simplified form as

$$\text{HA}_m^- + \text{H}_2\text{A}_n \longrightarrow \text{H}_2\text{A}_{m+n} + \text{OH}^-$$

where m and n range between 3 and 7. The alkaline polymerization kinetics is expressed by the standard curves shown in Fig. 3.

3.2.4 Fundamental Parameters for Reactor Design

In practice, and for the design of activated silica reactors, the most important parameters involve the following:

Concentration of the metasilicate solution
Nature of the activating agent

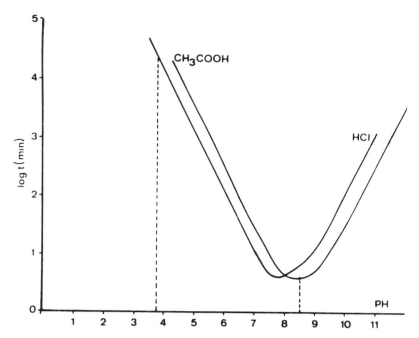

Figure 3 Standard curves for alkaline polymerization of silicic acid (CH_3COOH or HCl used as neutralizing agent).

pH and alkalinity of the activated sol
Aging time before application
Salt content and possible catalysis by salts present

3.3   Activating Agents and Reaction Schemes

In the original Baylis method (2), a 1.5 to 1.8 wt% solution of SiO_2 is neutralized up to 80 to 85% of its alkalinity by sulfuric acid and aged between 5 min and 2 h. The stoichiometry is the following:

$$H_2SO_4 + Na_2O \cdot 3.4SiO_2 \cdot xH_2O = Na_2SO_4 + 3.4SiO_2(act) + (x + 1)H_2O$$

Similarly, with hydrochloric acid, the reaction is

$$2HCl + Na_2O \cdot 3.4SiO_2 \cdot xH_2O = 2NaCl + 3.4SiO_2(act) + (x + 1)H_2O$$

With phosphoric acid it becomes

$$2H_3PO_4 + 3Na_2O \cdot 3.4SiO_2 \cdot xH_2O = 2Na_3PO_4 + 10.2SiO_2(act)$$
$$+ 3(x + 1)H_2O$$

As reported in the literature (4), using aluminum sulfate, N-sol B is obtained as follows:

$$Al_2(SO_4)_3 + 3(Na_2O \cdot 3.4SiO_2 \cdot xH_2O) = 3Na_2SO_4 + 10.2SiO_2(act)$$
$$+ Al_2O_3 \cdot 3xH_2O$$

It is worth mentioning that the aluminum coagulant–flocculant remains available as a secondary product of the activation reaction. Using sodium bicarbonate, N-sol D is obtained by

$$2NaHCO_3 + Na_2O \cdot 3.4SiO_2 \cdot xH_2O \longrightarrow 2Na_2CO_3 + 3.4SiO_2(act)$$
$$+ (x + 1)H_2O$$

In practice, the above-mentioned reactions are only partially completed (see Fig. 4): for example, at neutralization rates of 45% for H_3PO_4, 75% for $Al_2(SO_4)_3$, 85% for HCl, and 100% for $NaHCO_3$. Total reaction with sodium bicarbonate is possible owing to the stability of the sol at these reaction conditions (e.g., fast gelation does not occur).

Using ammonium sulfate, N-sol A is obtained according to the following reaction:

$$(NH_4)_2SO_4 + Na_2O \cdot 3.4SiO_2 \cdot xH_2O = Na_2SO_4 + 2NH_4OH$$
$$+ 3.4SiO_2(act) + (x - 1)H_2O$$

A characteristic of the reaction scheme is that ammonia remains available (e.g., for chloramine formation).

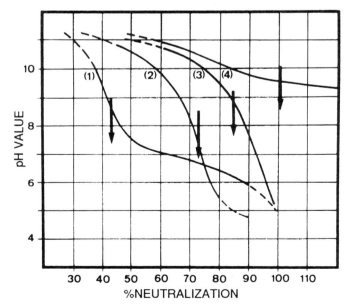

Figure 4 Neutralization rates in silicate activation (usual neutralization rates are indicated by (→)). 1, Phosphoric acid; 2, aluminum sulfate; 3, chlorhydric or sulfuric acid; 4, sodium bicarbonate.

Using dissolved chlorine, N-sol C is obtained through reaction with HCl and HOCl:

$$Cl_2 + H_2O = HCl + HOCl$$

$$HCl + HOCl + Na_2O \cdot 3.4SiO_2 \cdot xH_2O = NaOCl + NaCl + 3.4SiO_2(act) + (x + 1)H_2O$$

However, the exact reaction between silicate and chlorine is pH dependent:

$$Na_2SiO_3 + Cl_2 + H_2O \xrightarrow{\text{pH 10}} NaCl + NaOCl + H_2SiO_3$$

$$3Na_2SiO_3 + 4Cl_2 + 4H_2O \xrightarrow{\text{pH 7.5}} 4NaCl + 2NaOCl + 2HOCl + 3H_2SiO_3$$

$$Na_2SiO_3 + 2Cl_2 + 2H_2O \xrightarrow{\text{pH 5}} 2NaCl + 2HOCl + H_2SiO_3$$

The precise proportion depends on the technical reaction conditions, which will be discussed later. Note that in this method the chlorine remains available in the treatment. On the other hand, in some waters, chlorination may become unnecessarily high. For example, a treatment with 3 g/m^3 SiO$_2$(act) brings about the addition of 1.4 g/m^3 chlorine, originating from the activation.

Using carbon dioxide, carb-sol is obtained:

$$CO_2 + H_2O \longrightarrow H_2CO_3$$

$$H_2CO_3 + Na_2O \cdot 3.4SiO_2 \cdot xH_2O = Na_2CO_3 + 3.4SiO_2(act) + (x + 1)H_2O$$

In the activation process, reaction is achieved between 90 and 100%.

Finally (15), sodium silicofluoride can be used according to the reaction

$$Na_2SiF_6 + 2H_2O = SiO_2 + 4HF + 2NaF$$

and

$$Na_2SiF_6 + 2Na_2SiO_3 = 6NaF + 3SiO_2(act)$$

The silica concentration in the prepared solution is limited to about 2 to 2.75%, owing to the solubility of Na$_2$SiF$_6$. Other characteristics applicable to the method are: maintaining the pH of the activated sol between 5.3 and 7.3 (e.g., +5.8; to apply a SiO$_2$/F ratio between 2.5 and 4.1), and maintaining a silica sol concentration of about 0.3% after an aging period of 2 to 3 min.

4. LABORATORY- AND REDUCED-SCALE PREPARATIONS OF ACTIVATED SILICA SOLS

The repeatability of batch preparations at reduced scales appears to be critical. To obtain a reproducible sol, it is more reliable to proceed by continuous processes such as in plants. The preparation of activated silica at a laboratory scale using a continuous reactor could open up new fields of progress.

4.1 Sulfuric Acid

In the original Baylis method, silicate is activated by sulfuric acid. A 1.5 to 1.8 wt% SiO_2 solution is neutralized at 80 to 85% of its starting alkalinity and aged for 5 min to 2 h. The use of H_2SO_4N is recommended. The residual alkalinity of the suspension is 1200 $\pm$ 50 g/m^3. The solution can be diluted further to a silica content of 0.5 to 1%, thus preventing gelling. The general requirements for the technical generation process must also be observed in the laboratory preparation. Thus aging only appears to be beneficial with silica used as a flocculation aid. Usually, the suspension is best at 25% of the gel induction time or gelling time (9).

4.2 Hydrochloric Acid

The method followed when hydrochloric acid is used is similar to that with sulfuric acid. Walker (16) has formulated a series of recipes for laboratory preparation of activated silica suspensions, starting from 10 wt% SiO_2 stock solutions of sodium metasilicate. For HCl activation, the solution is diluted six- to sevenfold, to reach a 0.66 wt% SiO_2 working solution, to which is added a 0.175 N HCl solution to reduce the alkalinity to 1200 to 1300 g/m^3 (i.e., 83 to 85% neutralization). The sol can be aged for 1 to 1.5 h.

4.3 Sodium Bicarbonate

Equimolecular ratios of sodium bicarbonate to alkali present in the silicate have been recommended to give a sol of optimum activity (5). In the laboratory investigation a 10% SiO_2 solution (25 cm^3 can be mixed with a $NaHCO_3$ solution of 15 g/dm^3 at a ratio of 25 cm^3 SiO_2 solution per 140 cm^3 stock $NaHCO_3$, aged for 1.5 h, and subsequently diluted to 500 cm^3 with distilled water. The final concentration of this sol is then 3 wt%. In our experience, no gel formation occurs under these conditions but a less active sol results. In plant operation, an excess of $NaHCO_3$ is used and for laboratory investigations it is recommended that solutions of 2.5 wt% SiO_2 be mixed as silicate diluted in distilled water with a sodium bicarbonate solution of 40 g $NaHCO_3$/dm^3. This mixture can be used within 1 to 2 h. The final pH should be between 9.2 and 9.8. Seven to eight moles of $NaHCO_3$ are used per mole of SiO_2. The use of distilled water is optional but prevents side effects due to the precipitation of hardness (i.e., calcium and magnesium) in the form of compounds insoluble at pH $\geq$ 9.6.

4.4 Ammonium Sulfate

While stirring, 25 mL of a 10 wt% SiO_2 solution is added to 140 mL of a 10 g/L $(NH_4)_2SO_4$ solution to activate the sol. The mixture is set aside for 1 to 2 h and then made up to 500 mL with distilled water. A sol of about 3 wt% SiO_2 is thus obtained.

4.5 Chlorine

Chlorine-activated sol can be prepared by mixing "chlorine water" (i.e., a saturated solution of gaseous chlorine in distilled water) with a 10 wt% solution of neutral sodium silicate. The silicate solution must be diluted six- to tenfold so as to obtain a mixture of 1 to 1.5 wt% SiO_2. After aging to opalescence for 20 to 40 min, the sol is made up to 500 mL with distilled water so as to obtain a 0.2 wt% SiO_2 sol at a final pH of 7.0 ± 0.2. This sol is claimed to be stable for 2 months. However, for stability the sol is best diluted twice to maintain a 0.1 wt% sol. This process is difficult to carry out, as a high concentration of chlorine water is necessary.

4.6 Aluminum Sulfate

To prepare an aluminum sulfate–activated silica sol, a 10 wt% SiO_2 metasilicate solution is intensely mixed with a 10 wt% $Al_2(SO_4)_3 \cdot 18H_2O$ solution in the proportions 1:6 ($G \geq 200$ s^{-1}). The sol is aged for 1.5 h and then made up to 500 mL with distilled water to obtain a 0.5 wt% SiO_2 sol. The best final pH value is about 7.5 (17). This 25-year-old formula requires adaptation in the light of more recent developments (see Section 6.3), and the possible use of other aluminum-containing flocculants (e.g., WAC or PAC in the silicate activation process) must also be taken into consideration.

4.7 Carbon Dioxide

Carbonic acid facilitates the preparation of an activated sol without introducing additional salts to the treated water. A general accepted practice (16) is to prepare a 1.35 wt% SiO_2 solution of neutral silicate (29% SiO_2), and to bubble carbon dioxide gas through it to obtain discoloration of the bluish shade of the thymolphtalein indicator. After the introduction of the gas, an aging time of about 40 min is required to obtain an opalescent sol. This can be diluted with distilled water—for example, up to 500 mL (i.e., 0.5 wt% SiO_2)—before use.

5. TECHNICAL DESIGN OF ACTIVATION EQUIPMENT AND ITS OPERATION BASED ON LABORATORY TESTS

Measurement of gelling is the first and most important test in the activation process. This measurement is usually carried out by laying a 5-mm-thick film of the sol in a petri dish. The gelling time is considered to be the maturation time of the gel after which when a groove is cut with a knife into the layer of activated gel, it does not coalesce spontaneously.

An alternative technique is to fill 18-mL test tubes with 10 mL of the sol. The gelling time is the time required by the gel not to flow when the tube is tilted to 45 degrees. In this test, care must be taken to prevent the silica from adhering to the inner surface of the tubes as a thin layer just above the sol (i.e., in the zone that will be moistened when the tube is tilted). This precaution is necessary to avoid underestimating the gelling time.

Both tests described above can be completed by the usual ammonium molybdate reaction for dissolved silica, seeing that during aging, more colloidal silica is built up and does not react in this colorimetric test (9). Measurement of the zeta potential can usefully complete the information of colloidal buildup of the silica according to the general data in Fig. 1.

In practice, acid-activated silica should be used between 6 and 25% of its gelling time except for chlorine-activated sols, for which the maturation time is set between 30 and 60% of the gelling time. A 1.75 wt% SiO_2 sol activated with ammonium sulfate is best used within 6% of its gelling time (18). No general guidelines can be given in absolute time values for gelling times considered for use in practical applications, as local conditions such as the salinity of the process water can seriously influence the absolute value of the gelling time.

In most cases, at constant pH and temperature, the gelling time is proportional to the inverse of the square of the concentration of SiO_2, which involves a third-order reactional process. An example is given in Fig. 5.

The gelling time (t) varies as a function of the absolute temperature T according to an equation of the type $T^n t = K$, in which n and K are constants for a given concentration of silica. In the temperature range for water treatment, 22 to 35°C, a threefold variation in the gelling time can be observed and a linear approximation is sufficient for practical use (Fig. 6).

As a rule, the design should be based on the average water temperature to be expected, considering a span of 1.5- to 0.5-fold in gelling time. Possible fast activation of concentrated silica solutions with hot water and subsequent dilution into process waters remains open for investigation. In the bicarbonate-activation system used in the Antwerp waterworks, the activated sol is warmed during winter periods to prevent gradual gelling in the dosing systems.

For acid-activated silica a higher-alkalinity solution requires a longer gelling time than that of a lower residual alkalinity. For instance, according to Fig. 8 and 9, for a 1.5 wt% sol and an alkalinity of 1250 g/m^3 $CaCO_3$, the gelling time is 22 h, while at 1050 g/m^3 the gelling time is 16 h. Consequently, the temperature effect can be compensated through adequate operational control, by neutralization at higher water temperatures and at a lower rate rather than at lower temperature. There is also the possibility of adjusting the SiO_2 concentration.

Another important factor that influences design is the salt content of the process water. This has often been overlooked and makes extrapolations of research results obtained with distilled water questionable for practical applications. This problem was approached by Jackson (5). From his data, a curve may be deduced, expressing

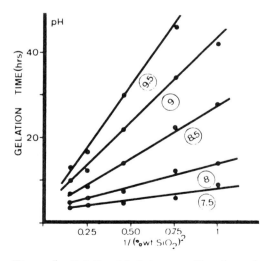

Figure 5 Relationship between gelling time of a sulfuric acid activated sol and SiO_2 concentration (Tailfer water, temperature 22°C, pH as indicated).

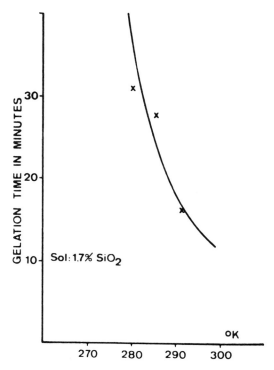

Figure 6   Temperature effect in chlorine-activated silica.

the change in gelling time as a function of the salt content of the process water (Fig. 7). These experiments indicate that in gelling that takes place during maturation, an ionic exchange takes place. Also, divalent ions enhance the gelling and this seems to hold particularly for magnesium ions. For most waters, the gelling time can vary by a factor of 1 to 4, and the design is best based on preliminary investigations with representative samples of the relevant water. Partial gelling is one of the requirements for the action of a silica sol in the treatment; this salt effect can be one of the reasons for some reported cases of failure of activated silica in the treatment of soft waters.

6.   PRACTICAL GENERATION OF ACTIVATED SILICA

Although batch processes for activating silica are feasible, particularly for smaller treatment schemes, the design of activating facilities is best based on a continuous activation using a dosing reactor.

6.1   Common Guidelines for All Methods

1.   The silicate and activator must be prediluted before being mixed into the reactor zone.
2.   At the reactor zone, auxiliary dilution water is necessary to maintain the desired concentration.

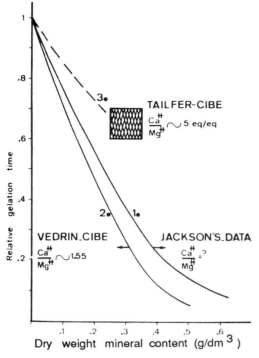

Figure 7   Relative gelling time versus dissolved solids. 1, According to data of Jackson (5); 2, diluted water from the Vedrin plant (CIBE); 3, diluted water from the Tailfer plant (our own experiments at 20°C, temperature and calcium-to-magnesium ratio unknown for the data of Jackson; all sols at 1.5 wt% SiO_2).

3. The mixing power to be released in the reaction zone corresponds to a velocity gradient $\geq 1500 \text{ s}^{-1}$.

4. Auxiliary dilution water can optionally be injected into the reactor zone.

5. Safety measures must provide for the shut-off of reagents should one of them fail. It is recommended that an automatic rinsing phase be supplied should activation failure occur.

6. The injection pipes for introducing chemicals to the reactor space must be located higher than the safety overflow level of the activating mixture and must in no way plunge into the liquid sol.

7. For sol abstraction and contact with the water to be treated, suitably large injectors should be used, made of PVC or brass and capable of a dilution ratio 1 : 10 or greater.

8. The outlet pipe must be located in an outlet zone and bend downward to provide for self-cleaning by the aspiration of air.

9. An overflow warning device must be incorporated.

10. Sampling and waste lines are obligatory.

11. Automation of the activation process is possible if a final pH measurement is taken after dilution using electrodes cleaned by ultrasonics.

12. All reagent flows, including process and dilution waters, must be measured and at least indicated locally. Transmission of data is optional.

The startup procedure recommended is based on the following steps:

1. Position the reactor, the effluent line being directed to a waste line.
2. Set the dilution water(s) to the required flow.
3. Set the silicate flow to the required value.
4. Set the activator flow to the required value.
5. Wait for a twofold turnover rate of the reactor volume.
6. Check the pH value (and/or alkalinity) and make the minor adjustments of reagent flows.
7. Switch over from a waste line to a dosing line.
8. Make all optional checkups — pH, alkalinity, gelling time — and eventually read-just the dosing.

The shutdown procedure recommended is:

1. Position the values to direct the effluent to a waste line.
2. Switch off the activator feed.
3. Switch off the silicate feed.
4. Allow the dilution water in use to rinse the entire system two or three times.
5. Reposition the valves in the dosing position for rinsing.
6. Turn the controls to the waste position and switch off the dilution water.

6.2 Specific Guidelines to Activate Silica with Sulfuric Acid

The maximum concentration that can be treated efficiently with sulfuric acid in the activation process is approximately 5 wt% SiO_2. In practice, the concentration is best kept below 2.5 wt% SiO_2. Consequently, prior to mixing in the reactor zone, the silicate is diluted 10 times. The dilution water of the sulfuric acid is dosed to obtain the desired value of the SiO_2 concentration in the reactor, and the acid is dosed to obtain a residual alkalinity of 1200 ± 100 g/m^3 $CaCO_3$. The corresponding pH values are deduced from Fig. 8.

The standard curves for gelling time as given for water at the Tailfer plant illustrated in Fig. 9 enable us to estimate the approximate gelling time. These evaluations may eventually require correction for the parameters discussed previously. In all instances of the practical pH range of activation, the minimum gelling time is observed at a pH value of about 7. Best results are usually obtained by controlling the design and operation parameters such that the average contact time is between 15 and 20% of the gelling time recorded on the graph in Fig. 9.

During design, the contact time in the reactor volume must be kept between 6 and 25% of the gelling time. To comply with the general guidelines outlined previously, the specific design parameters for the reactors are as follows:

1. To fix the dosing needs of the flocculation in the water treatment according to seasonal demands
2. To design a reactor volume on the basis of an average concentration in the sol of 2 wt% SiO_2 and a residence time of 15 to 20% of the gelling time

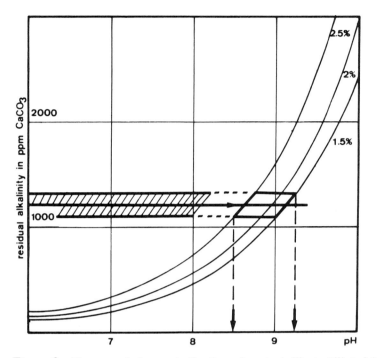

Figure 8   Nomograph for neutralization of neutral silicate (diluted 29% grade) by sulfuric acid (water from the Tailfer plant).

In Fig. 11, an example is given of a reactor able to produce an average quantity of 4 kg/h activated SiO_2. In the sulfuric acid activation of silicate with a continuous flow, both auxiliary maturation and gelling vessel are unnecessary.

6.3   Specific Guidelines to Activate Silica with Aluminum Sulfate

Specific guidelines for activating silica with aluminum sulfate and related compounds are described in patented processes such as the Coventry process (19) and the Norwich process (20). Although the silica produced through aluminum sulfate activation has been described as being less active as a flocculation aid, these processes are of interest since the aluminum salt remains available as a coagulant and is often disposable and stored at the plants. In the Coventry process (19), a 4 wt% SiO_2 silicate solution is mixed with a 10 wt% Al_2O_3 solution of aluminum sulfate in a high-speed mixing vessel. After aging in a dilution to obtain a 0.75 wt% SiO_2-activated solution, the pH should be between 7 and 8, preferably between 7.2 and 7.5.

In the Norwich process (20) marketed by Crossfield, the design bases are as follows:

1.   The effective SiO_2 concentration in the activation process is to be fixed between 0.5 and 5%.
2.   In the activation process, a high level of mixing efficiency in the contact zone must be ensured.

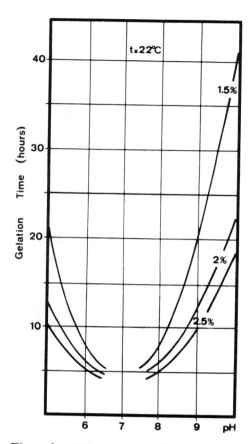

Figure 9   Gelling time of sulfuric acid–activatd SiO$_2$ sols.

3. Double-head dosing pumps are recommended for accurate simultaneous proportioning of the silicate and aluminum sulfate.
4. The final pH should range between 4 and 6.5 and is not considered critical in this commercial application.

In a patent introduced in 1979 (21), silicate solutions were activated with a solution of aluminum polychloride (PAC or WAC) of the general formula $[Al(OH)_xCl_{3-x}]_n$, in which x varies between 1 and 2.5. The process is claimed to build up the active silicoaluminate. The reason for this activation process is considered to be the concentration limits of 1 to 2 wt% SiO$_2$ at a final pH of the activated suspension between 6 and 8. The process produces a less adhesive activated solution: our own lab-scale investigation of the feasibility of silica activation with aluminum sulfate, and PAC or WAC, is illustrated by the typical gelling-time curves reported in Fig. 12.

At low concentrations the "aluminum"-activated sol gels slowly, thus limiting practical activation to the concentration range 1 to 5% except in extreme pH zones. Generally, the gelling time is very short in the pH range for practical activation: 6.0 to 8.5 for aluminum sulfate and 6.5 to 8.5 for PAC-WAC. This must be attributed

Figure 10   Reactor at the Tailfer plant of the Brussels Waterboard.

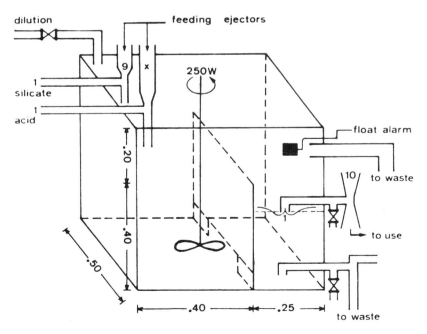

Figure 11   Typical reactor design for activation of silicate with sulfuric acid.

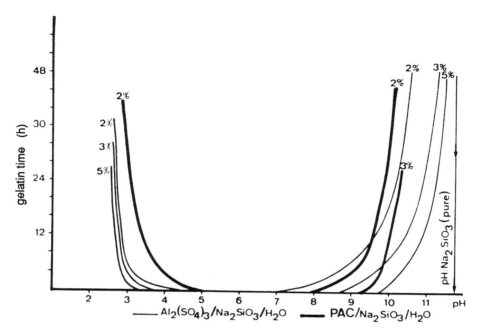

Figure 12 Gelling time curves for activation of silica with aluminum salts.

to the existence of a germination effect due to the colloidal aluminum hydroxide buildup in the system. For practical design, aluminum versus silica concentration relationships can be formulated as a function of the pH at the end of rapid mixing.

The recommended concentration zones for design are indicated in Figs. 13 and 14:

1. Zones ABC indicate extremely fast gelling zones and must be avoided (e.g., to prevent clogging of the apparatus).

2. The recommended safe-design zone is indicated as *abcd*: 0.5 to 1 wt% SiO$_2$ for PAC-WAC activation, and 1 to 2 wt% SiO$_2$ for activation with aluminum sulfate.

3. Intense mixing of prediluted reagents (e.g., each reagent being at 50% of its final concentration) is essential and of leading importance in the case of activation with aluminum sulfate.

4. Maturation is of minor or no importance in these activation techniques; the silica can be dosed after a short mixing time (e.g., 2 to 5 min average residence time in the mixing zone).

5. General design for inlet, overflow, and safety devices are the same as for activation with sulfuric acid.

6. Aluminum hydroxide resulting from the activation process remains available as a reagent in sweep–coagulation–flocculation and can be deduced from the normal dosing rates necessary when this form of activated silica is not applied.

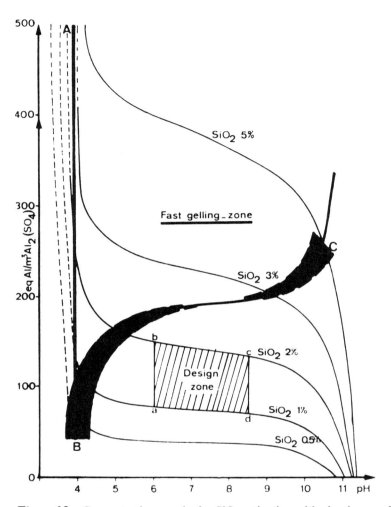

Figure 13   Concentration graphs for SiO_2 activation with aluminum sulfate.

6.4   Chlorine Activation of Silica Solutions

In chlorine activation of silica solutions, the action is based on the neutralization with HCl and HOCl obtained through hydrolysis of dissolved chlorine gas. In practical conditions of pH (± 7.5), about half of the HOCl ($K_a = 3.3 \times 10^{-8}$) is ionized. Thus about 1.5 equivalents of acid are obtained per mole of chlorine used in the process (6). As a rule in practical activation equipment, the global molar ratio Cl_2/SiO_2 is between 0.4 and 0.6. The optimum ratios as a function of the SiO_2 concentration are indicated by line A-B in Fig. 15.

It is recommended that the activation be carried out by mixing chlorine with a dilute solution of neutral silicate, in proportions capable of giving the minimum activation time (e.g., 0.45 mol of chlorine per mole of silicate for a 1.5 wt% SiO_2 solution). However, this requires a solution of ± 8 g Cl_2/L, so the process cannot

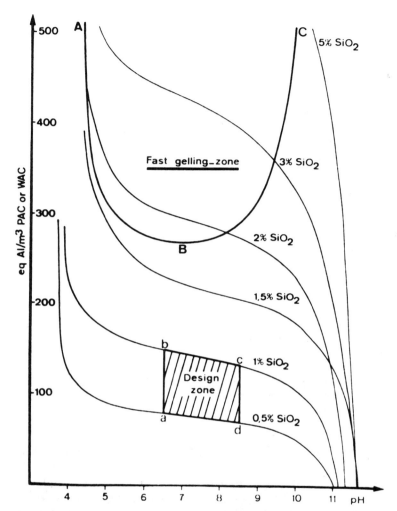

Figure 14 Concentration graphs for SiO_2 activation with PAC-WAC.

be based on batch mixing of chlorine and silicate solutions but is operated on a continuous basis by bubbling gaseous chlorine through a dilute silicate solution.

For safe working conditions the activation contact time should be between 30 and 60% of the gelling time given in Fig. 15. This is equal to between 6 and 12 min for a sol of 1.5% strength. The gelling time (t) is subject to temperature variation ($T = K$) according to the general equation $T^n t = K$, which for this activation process was found to be equal to $T^{16.4} t = 10^{42.7}$. Therefore, when treated surface water is to be used, the reactor volume is best designed for the average temperature of the water. Ideally, locally available groundwater with an almost constant temperature should be used. The total available chlorine necessary in the activation process remains active as hypochlorite in the silica solution injected in the treated water. This means that at doses of 1.5 g SiO_2/m^3 0.8 g active Cl_2/m^3 are also introduced to

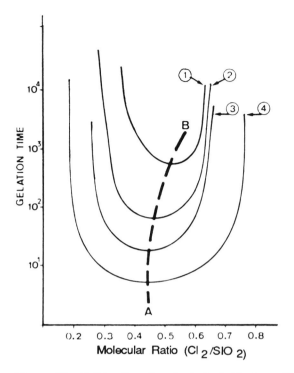

Figure 15 Gelling time (in minutes) of chlorine-activated silica sols. 1, 0.5% SiO₂; 2, 1.0% SiO₂; 3, 1.5% SiO₂; 4, 2.0% SiO₂. Temperature, 17–21°C.

the water. Therefore, in the case of low chlorine demand, this method can lead to superchlorination (22).

Practical equipment consists of a short-time (e.g., 1 to 2 min) contact chamber in which gaseous chlorine is injected in a continuously flowing dilute silicate solution. The overflow of the mixing chamber is directed into a maturation tank on a continuous flow principle, provided with a highly efficient rinsing device by water operated automatically when transfer line clogging occurs. The final pH of the maturated solution should be in the range 7.5 ± 0.5, and the residual alkalinity between 1200 and 1400 g $CaCO_3/m^3$.

6.5 Specific Guidelines to Activate Silica with Sodium Bicarbonate

For the practical design of full-scale processes based on activation of silicate with sodium bicarbonate, the following guidelines can be adopted:

1. Prepare separately in deionized water a solution of sodium bicarbonate containing 40 to 45 g $NaHCO_3/L$ and dilute silicate solution obtained by introducing 1 volume of specific gravity 1.39 product mixed in 17 volumes of deionized water.
2. Add to this mixture 8 volumes of sodium bicarbonate solution and after mixing, allow the solution to mature 1 to 2 h.
3. Then dilute the mixture with 14 volumes of deionized water.

4. An activated sol containing 1 wt% SiO_2 is obtained and is dosed together with aluminum sulfate in the coagulation process stage of the water treatment scheme.

5. The overall molar ratio $SiO_2/NaHCO_3$ equals 192:1571 = 0.12, which means that a considerable excess of the bicarbonate is necessary in this process.

6. The final pH value of the activated sol is between 9.2 and 9.5 and may range to 10.2 for slowly gelling sols. No severe gelling risk occurs through overtreatment with this activator (see Fig. 5 for approximate gelling times). Gelling can become critical at pH $\leq$ 9.

7. In practical applications, the activation is operated in parallel using at least two mixing–activating tanks: for example, one for dosing and another as standby or for use in the activation phase. Furthermore, the activated, diluted sol is abstracted from the bottom of the maturation tank, part of which is measured and dosed into the water. The excess is recirculated at the surface of the maturation tank.

8. To prevent clogging of the dosing and transfer lines in winter, a slight warming of the process waters is recommended (about 20 to 30°C). The lines are doubled in parallel schemes to enable cleaning without interrupting the dosing.

7. MECHANISM OF ACTION OF ACTIVATED SILICA

Little is known of the chemical reactions involved with activated silica used as a flocculation aid in the treatment. The aid usually has minor or no effects when used with coagulants having an iron basis. At more acid pH values, the free silicic acid functions are known to react with ferric salts as follows:

$$Fe^{3+} + H_3SiO_4 + H_2O = FeSiO(OH)_3^{2+} + H_3O^+$$

Generally, silica catalyzes the oxidation of ferrous iron to ferric iron while it delays the hydrolysis of Fe^{3+}. Immobilization of heavy metals in a gellified silica, obtained through precipitation with excess acid, is a method of concentration and stabilization of residuals before disposal. Activated silica is a negatively charged sol with cationic exchange capacity.

Attempts to measure the cationic exchange capacity with methylene blue (MB^+) (23) have failed, owing to the fast gelling induced by mixing of the (MB^+) reagent with active SiO_2 solutions. Although this acceleration in gelling can be considered to correlate with bringing effects in the substitution of MB^+ for Na^+, no definite exchange capacity could be measured as no filtrable liquid could be separated from the gel even at 1.5 wt% SiO_2 concentration.

Among the current filterability tests, the membrane refiltration has been found to be the most useful in ascertaining adequate activation and good filtration-aid properties. An example of a 1.5 vol % solution activated with H_2SO_4 is shown in Fig. 16. The test is carried out by free percolation (at 1 bar) through a 0.4-μm porous membrane of 8.5-cm³ filtrate, starting with a 10-cm³ activated sol. This 8.5-cm³ portion of the primary filtrate is refiltrated upon deposit of the first filtration. The refiltration number is equal to the time required for the first filtration, divided by the time required for the second filtration. In practice, a good activated silica has a refiltration number of less than 2.5, and the material reacts excellently

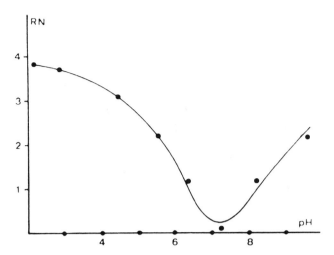

Figure 16   Refiltration number as a function of the pH of a 1.5 wt% SiO$_2$ sol activated with H$_2$SO$_4$.

in treatment if the value is below 1. Obviously, the refiltration number (RN) is only a relative value and must be interpreted in any given condition (e.g., process water composition).

With aluminum coagulants in particular, the action of activated silica probably involves a bridging mechanism of the polymer, with the main coagulant also involving an ionic exchange reaction. The best moment for the addition of activated silica in the treatment scheme is usually after that of the coagulant when a "pin point" floc has already formed. Addition to the raw water has sometimes been suggested, but except for water of very low charge in suspended matter, the results are less satisfactory. The most appropriate procedure should be determined experimentally. Jar test evaluation of the speed of floc formation, residual color, and turbidity after settling are the most appropriate practical tests.

In practice, the density and shear strength of the floc is always increased and good stability of the sludge layer is obtained in upflow settling. Flocculation of aluminum based-flocs is more effective, especially during cold-water periods. Post-precipitation of aluminum hydroxide is avoided, particularly in waters with low alkalinity (e.g., ≤ 1 Eq HCO$_3^-$/m^3). Progress on the removal rates of heavy metals in coagulation–flocculation processes has been beneficial (24), but more specific work on this topic is still required.

8.   CONCLUSIONS

As follows from the literature data and results obtained from experiments or from the authors of the CIBE, the generation of activated silica described in this produces reliable process based on this versatile chemical. Indications of activation are still based primarily on jar tests for flocculation (e.g., speed of floc formation and turbidity of the supernatant water). At appropriate dosing rates, activated silica generally improves flocculation–settling and corrects the filter runs for penetration through turbid waters, particularly during the winter and at the critical periods after back-washing. Even if the entire process is kept under accurate control following the guidelines reported here, the fundamental principles involved in the effects must still be assessed. The most probable theory is likely to be based on the ionic exchange properties of the activated silica:

1. Neutralization must only be partial in activation (e.g., 15% residual alkalinity must be maintained).
2. The active sol has a definite affinity for ions of higher valency (e.g., Mg^{2+}, Al^{3+}).
3. Free silicic acid sols are less active in practice.
4. A negative zeta potential is essential in obtaining an active sol (e.g., -10 to -30 mV).
5. The membrane refiltration test is simple and can confirm adequate activation.

REFERENCES

1. M. Baumann, *Kolloid-Z.*, *162*, 28 (1958).
2. J. R. Baylis, *J. AWWA, 9*, 1355 (1937).
3. J. E. Schenk and W. J. Weber, *J. AWWA, 60*, 199 (1968).
4. L. L. Klinger, *J. AWWA, 47*, 175 (1955).
5. F. R. Jackson, *Water Wastes Eng., 752*, 437 (1958).
6. E. A. Whitlock, *The Society for Water Treatment and Examination* (1954) (special issue, March).
7. J. C. Kane, V. K. La Mer, and H. B. Linford, *J. Am. Chem. Soc., 86*, 3450 (1964).
8. J. R. Baylis, *Water Sewage Works*, p. 61 (Feb. 1937).
9. Ch. A. Black, *J. AWWA, 45*, 1101 (1953).
10. G. B. Alexander, *J. Am. Chem. Soc., 75*, 2887 (1953).
11. G. B. Alexander, W. M. Heston, and R. K. Iler, *J. Phys. Chem., 58*, 453 (1954).
12. Tain an-Pang, *Sci. Sin., 12*, 1312 (1963).
13. G. B. Alexander, *J. Am. Chem. Soc., 76*, 2014 (1954).
14. A. P. Brady, A. G. Brown, and H. Huff, *J. Colloid Sci., 8*, 253 (1953).
15. C. R. Henry, *J. AWWA, 50*, 61 (1958).
16. J. G. Walker, *Water Water Eng., 718*, 534 (1955).
17. R. F. Packham and M. Saunders, *Proc. Soc. Water Treat. Exam. 15*, 255 (1966).
18. R. W. Pitmann and G. W. Wells, *J. AWWA, 60*, 1167 (1968).
19. I. M. E. Aitken, *Effluent Water Treat. J., 1*, 76 (1961).
20. F. M. Perrin and F. Smith, *Publication ref. SS2-11*, Crosfield Chemicals.
21. J. L. Bersillon and Y. Richard, (Degremont), Fr. patent. 79.08.294.
22. V. W. Langworthy, *Water Sewage Works, 104*, 20 (1957).
23. W. Kim, H. F. Ludwig, and W. D. Bishop, *J. AWWA, 57*, 327 (1965).
24. W. J. Masschelein, J. Genot, Cl. Goblet, and B. Regnier, The Elimination of mineral micropollutants, Symposium IAWPR, Brussels; *Water Sci. Technol., 14*, 87–105 (1982).

9

Combined Coagulation–
Flocculation–Clarifying Processes

"No treatment unit is any better than its operator."

1. HISTORICAL BACKGROUND

The concept of sludge blanket clarifiers is more than a century old, but up to the 1950s little fundamental interest was taken in the process (1), although a considerable number of technologies had been developed by that time (2). In 1869, Sillar and Wigner claimed that chemically precipitated sludge has the power of precipitating (3) on recirculating and contacting raw water. The first significant industrial development was that of the Dortmund tank in 1880, which still survives in some wastewater treatment schemes, such as those in the paper industries. In 1982, Archbutt and Deeley used a combined system of two intercommunicating compartments, one for mixing and another for clarification (4). It was understood at that time that intermittent sludge contact with the incoming influent has a favorable effect on precipitation.

These ideas of combining the reacting and clarification zones developed progressively and are the best illustrated by the nested-type Clariflow developed by Walker, shown in Fig. 1. Basically, in this type the flocculation chamber is integrated into the whole construction, but its volume remains separated from the clarifier itself by means of a perforated bottom. By itself the clarifier is of the upflow type, with a separated sludge zone that is not yet integrated in the concept.

At present the accepted concepts for combined process settlers are:

1. The principles of orthokinetic flocculation, applying the velocity gradient concept, underlie the designs.
2. The simultaneous aggregation and breakup mechanisms intervene in flocculation through mixing (velocity gradients).

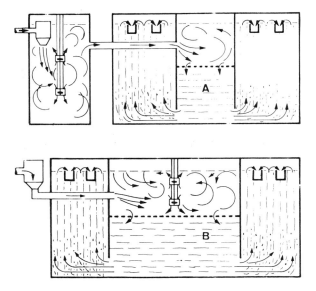

Figure 1   Clariflow units (Walker process): (A) separate type; (B) nested type.

3. Minimum time to initiate floc formation is necessary.
4. Subsequent reactors, especially tapered stirring basins, improve the flocculation.
5. Through sludge contact no definite physicochemical reactions occur that make it possible to reduce the coagulant dosage, but intermittent shutdown of the dosage (e.g., for 10 to 15 min) does not hinder the continuity of operation. Slight permanent overdosing can overcome accidental interruption of chemical feed for a period sufficient to repair the dosage equipment.
6. It is necessary to maintain disinfecting conditions during the flocculation phase when combined with sludge contact or recycling in order to avoid objectionable tastes and odors that occur as a consequence of bacterial development in the reinocultated sludge blanket.
7. The running-in of the process is necessary to constitute the sludge layer and may require several days of operation. This initial period can be shortened by recycling sludge into the inflowing water to be treated.

2.   SUSPENDED SOLIDS CONTACT CLARIFIERS

2.1   Characteristics

Typical characteristics of technologies developed before 1950 are indicated below. Basically, all of the suspended solids-contact clarifiers are to be considered as up-flow clarifiers, for which general guidelines for layer thicknesses are as given in Fig. 2. The suspended solids-contact units are combined flocculators–clarifiers which generally include the following subunits:

A raw water inlet and distribution system
A secondary mixing and reaction zone

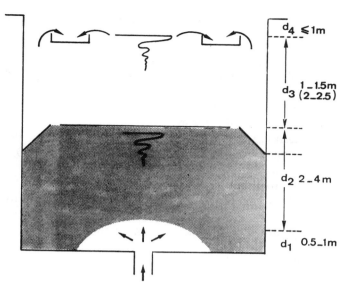

Figure 2 General guideline dimensions in upflow clarifiers.

A sludge concentration zone
A clarification zone with a system for collecting the clarified water

The preferable construction is circular or at least as symmetrical as possible (e.g., square), although for some constructions (e.g., the precipitator), rectangular construction is possible. In that case mixing is usually less satisfactory. Sludge recirculation is intended to favor the flocculation and also sweep coagulation.

2.2 Typical Parameters

Typical general parameters are as follows:

Preferable shape: circular or symmetrical.
Total retention time: 7.2×10^3 to 10.8×10^3 s (2 to 3 h).
Surface loading: usually less than 0.8 mm/s in clarification and less than 1.4 mm/s in softening.
Layer of clarified water: 2 to 3 m above sludge zones.
Sludge concentration zone: inclined at 60° or equipped with scrapers.
Sludge zone residence time: about 3.6×10^3 s.
Sludge bleed: on a continuous basis, 1 to 2% of water flow, temporarily up to 5%.
Sludge concentration: 6 g/dm^3 by dry weight or 10 to 15% in wet volume in coagulation–flocculation and 5 to 10 g/L in weight or 20 to 30 vol % in softening (too high sludge concentration determines the overflow of pinpoint floc). A slurry pool is necessary for correct operation.
Sludge recirculation rate: 3 to 10% of sludge bleed (tends to be increased).
Dynamic range of operation: 50 to 133% of nominal capacity (change in velocity must be progressive and the inlet system must maintain equal distribution at different flow rates).

Characteristics of overflow weirs are at present under reevaluation as compared to the classical concepts. Typical data are:

Clarified water overflow velocity: 0.3×10^{-3} to 0.8×10^{-3} m^3/m$^2 \cdot$ s

  Peripheral overflow weirs should be provided for the entire circumference of circular tanks.

  The acceptable limit of a specific weir is that the sludge layer be located at a depth of $D/7$, D being the diameter of the weir.

  Radial weirs are needed for tanks greater than $D = 15$ m.

  The proper distance of intermediate surface weirs from the wall is about 0.7 times the tank depth.

Sludge overflow velocity: 2×10^{-3} to 4×10^{-3} m^3/m$^2 \cdot$ s

  Sludge flow angle $\geq 52°$. Maximum surface of sludge drainage covers 10 to 20% of total surface.

  Maximum horizontal sludge drainage distance: 10 m (without mechanical scrapers).

  Sludge scraper tip velocity: 0.2 to 0.8 m/s.

2.3 Examples

2.3.1 The Accelator

In the Accelator (see Fig. 3), the raw water containing the necessary dose of reagent is introduced into the primary mixing zone (1) and mixed with an impeller-mixer. The water transits through a secondary mixing zone (2) into a return flow zone (3), where clarification occurs. The sludge is discharged continuously by a regulating valve (4). The total radius is generally equal to the depth of the Accelator. The

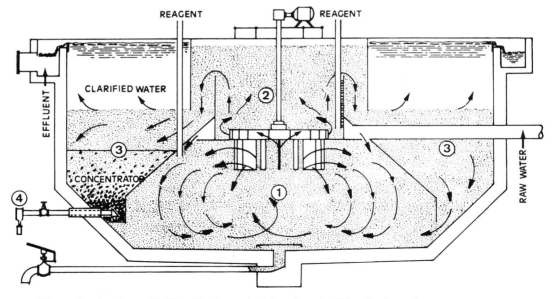

Figure 3 Accelator (Infilco–Hughes patents) or "reactor" (by Cochrane).

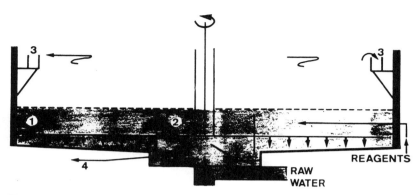

Figure 4   Hydro-Treator (Dorr Co).

residence time in the primary and secondary mixing zones is between 1800 and 3600 s (i.e., 30% of the total detention time. The average velocity gradient in the primary mixing zone (i.e., in 20% of the total volume) is $G \leq 20$ s^{-1}, where the entire concept is based on long back-mixing periods.

2.3.2   *The Hydro-Treator*

In the Hydro-Treator (Fig. 4), the raw water is introduced into the sludge blanket (1) by means of a rotating distribution arm (2). The reagents are added either to the raw water or dispersed into the sludge zone. The basin is flat-bottomed ($<5°$ angle), so rakes are provided on the distribution arm to remove the sludge to the central collecting zone (4). The clarified water is collected by a peripheral weir (3), and the sludge is abstracted through a gate (4). Circular design is inherent in the Hydro-Treator. Total detention time ranges from 3600 to 5400 s. The normal total depth is 4 to 6 m, 30% of it being reserved for the sludge zone (i.e., 1200 to 1800 s). The technology is based on the concept of sweep coagulation and sludge contact flocculation. The design could be improved by separate precoagulation following specific rules of design of coagulation chambers.

2.3.3   *The Liquon Reactor*

In the Liquon reactor (Fig. 5), the raw water is introduced together with the reagents through perforated plates (1) into the coagulation zone, which is mixed by a variable-speed paddle mixer. Sludge is partly recycled via a return port in the sludge concentration zone (3). Mixing on transit from the coagulation zone to the clarification zone is damped by stilling baffles. The sludge bleed is automated with a programmed diaphragm valve.

2.3.4   *The Reactivator*

The Reactivator (Fig. 6) is of circular or square construction, 3 to 30 m in diameter. Raw water and reagents are introduced in a central column (1) with a diameter of about 10% of the basin diameter. The secondary flocculation zone (2) is delimited by reversed conical deflectors. In a similar concept, the Clari-Flocculator of Dorr-Oliver, the deflector is cylindrical. A sludge scraper of 35 to 50 mm/s tip velocity is used to concentrate the sludge in a central outlet zone (4). Other typical data are: surface loading, 7×10^{-4} to 11×10^{-4} m/s; detention time, 3600 to 7200 s; and height of the clarification zone, 2.5 to 3 m.

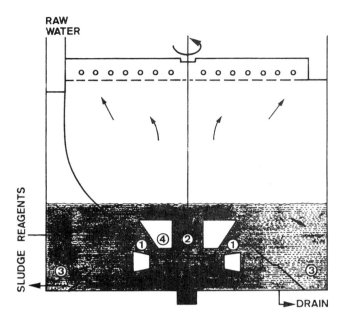

Figure 5 Liquon cylindrical reactor (Cochrane).

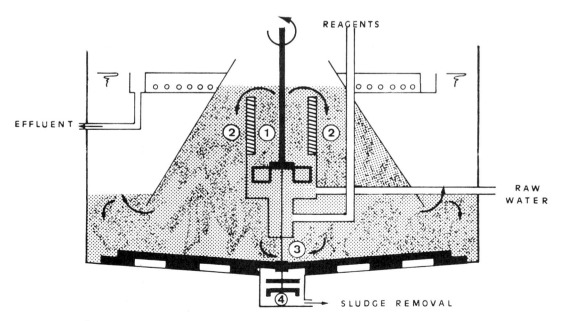

Figure 6 Reactivator (Graver).

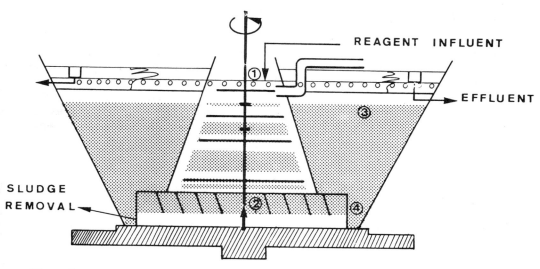

Figure 7 Precipitator (Permutit Co.).

2.3.5 *The Spaulding Precipitator*

The Precipitator (Fig. 7) can be either circular or a rectangular in shape. Basically, the construction is of double conical design, which enables an appropriate and gradual change in water velocity. The mixing conditions are adjusted by paddle mixers in the coagulation–flocculation zone. Reagents and raw water are admitted in the upper part of the central inverted conical or pyramidal space equipped with a mixer (1). At the bottom, swinging baffles (2) complete the mixing for sludge-contact flocculation and prevent rotational movement in the clarification zone. This clarification zone is of the expanding type, as in the hopper tank. Contrary to modern theory (see Section 3), the sludge is collected through a sludge concentrator located at the bottom of the tank (4). The walls are inclined at 60° to horizontal to facilitate sludge collecting. The usual detention time in the clarification zone is 3600 s, and in the coagulation-backmixing zone is 1800 s.

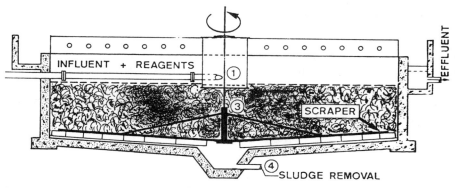

Figure 8 Roto-Rake basin (Graver).

2.3.6 The Roto-Rake Sedimentation Basin

The Roto-Rake basin (Fig. 8) is a typical simplified design from the early 1950s. The reagents and raw water are introduced in a central cylindrical space located at the upper layer (1). Mixing is available to assist the coagulation. There is no specific secondary flocculation zone and the sludge is settled and evacuated from the flat-bottomed tank using scrapers (3). The clarified water is gathered in collectors with orifices. Designs of this general type are not fully in agreement with the concepts of efficient sludge-contact flocculation and are limited at present to preliminary treatment or to concentrating units in sludge dewatering.

3. ADVANCED CONCEPTS APPLYING TO SUSPENDED-SOLIDS CONTACT CLARIFIERS

Orthokinetic flocculation is the basis for sludge-contact systems. In the latter, an equilibrium is installed between floc aggregation and breakage:

$$\frac{dn_1}{dt} = -k_a n_1 G + k_b n_0 G^2$$

wherein orders of magnitude are

$$k_a \simeq 10^{-4}$$

$$k_b(\text{turbidity}) = (7.7 - 0.87 \ln G) \times 10^{-7}\,\text{s}$$
$$k_b(\text{color}) \quad = (15.2 - 3.04 \ln G) \times 10^{-6}\,\text{s}$$

To avoid too much floc breakage:

1. The inlet (jet) velocity must be kept within the range 0.6 to 1 m/s.
2. For paddle mixers the tip velocity is maintained in the range 0.3 to 0.7 m/s; usually, this is equal 0.03 to 0.25 rps.
3. For the disk impeller the tip speed ranges from 0.6 to 1.2 m/s, or usually 0.17 to 0.25 rps, the average velocity gradient being kept below 45 s^{-1}.
4. For helical mixing turbines with $\alpha \leq 35°$ the velocity gradient must be kept below 90 s^{-1} and within the extreme limits of a rotational speed of 2.5 to 25 rps.

Hindered settling occurs during sludge-contact flocculation. An initial period is necessary before flocculant settling. Thus the coagulation phase is best maintained separate from the flocculation and designed according to the specific rules applicable to this process. A short time (less than 30 s) is necessary before settling of flocculated water begins (see Fig. 9). The kinetics of floc settling is expressed as

$$\frac{dh_z}{dt} = k(h_z - h_z, t_\infty)$$

$$h_z - h_0 = (h_{z,\infty} - h_{z,0})(1 - e^{-kt})$$

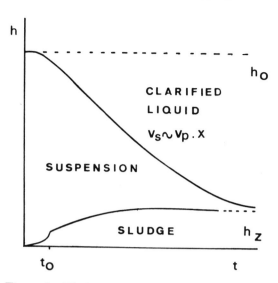

Figure 9   Hindered settling (schematic).

3.1   Upflow Flocculation Settlers

Design criteria and guidelines for upflow flocculator–clarifiers operating with a
sludge layer containing hydrated hydroxide flocs of aluminum or iron: the following
concepts apply to upflow flocculator–clarifiers as well as to the clarification zones
of mechanically scraped, sludge-contact clarifiers discussed in Section 2. Settling
velocities of flocculant particles and sludge can be related to settling velocities of the
individual particles. Temperature correction for particle settling can be computed by
the proportionality $v \sim 1/u$ according to Stokes' law. Typical values of the particle
settling characteristics are given in Table 1.

Table 1   Particle Settling Velocities of Flocculated Material (15°C)

| | | | Conventional data | |
| | | | --- | --- |
Particle type	ρ_s (wet) (kg/dm^3)	v_p (mm/s)	Overflow (mm/s)	Residence time (s)
Clay, sand	2.65	5–10, <7>	28–100	30–540
Crystalline calcium carbonate	1.2	2–3, <2.5>	0.5–3	3600
Softening precipitate	1.15	1.2–1.9, <1.5>	0.8–1.4	<7200
Floc (Al, Fe)			0.5–1.5	<7200
Turbidity	1.005			
Color	1.004	0.7–1.3, <1>		
Algal	1.002	0.6–1.2 <0.9>		
Activated sludge	1.005	1–2, <1.5>	1.4	3600–7200
Organic waste (polyelectrolytes)	1.001	0.3–0.5, <0.4>	0.3	3600–7200
Activated carbon powder	1.2–1.4	2	0.8–1.5	>3600

The experimental equation of Bond (5) is used to evaluate the settling velocity of "sludge particles" v_s as a function of the settling velocity of single particles v_p:

$$v_s = v_p (1 - fS^{2/3})$$

wherein S is the volumetric fraction of solids in the sludge. When expressed in wet volume percentage of settleable solids (vol% after 24 h of sedimentation), $f = 2.78$. The quotient v_s/v_p is assumed independent of temperature changes.

A safe empirical check is to obtain an upflow velocity of less than 0.5 v_p at the separation point of the slurry and the clarified water. The limits for S range from 6 to 15% for most flocculated material, and unless pilot-plant measurements are available, $S \sim 7.5\%$ is to be accepted as an average design value. At the inlet zone of a sludge blanket clarifier, if no sludge recycling is operated into the raw water, $S = 0$ and $v_s = v_p$ at the lower level. Hence at this level,

$$v_s = v_p = \frac{S}{A \ (\text{at } X_l)}$$

(if square, Q/L_l^2). If sludge is recycled, v_s is corrected according to the S value for the inflowing water.

The general accepted criterion for orthokinetic flocculation in a sludge contact clarifier is to consider the factor $G\phi T$, wherein ϕ is the wet volume fraction of the suspended solids (e.g., 5%). For fluidized beds the $G\phi T$ expression has been established according to Ives (6) as

$$\sum_{X_l}^{X_u} G\phi T = \frac{A \ (\text{at } X_l)}{2} \left[\frac{(\rho_s - \rho_l)g}{Qu(2.78)^{9/2}} \right]^{1/2} \times F \left(\frac{X_u}{X_l} \right)$$

The function $F(X_u/X_l)$ for increasing dimensions from the lower (X_l) level to the higher (X_u) level has been defined by numerical integration and can be read from the graph in Fig. 10.

Suggested limits for $G\phi T$ are to be between 60 and 110. The higher the value, the more conservative the design in terms of overcapacity. Hence if a compromise is necessary, it is better to obtain too high a value of $G\phi T$ than too low a value. The sludge residence time equals

$$\frac{\text{volume of the blanket}}{\text{flow of sludge bleed}} = t_s$$

while the equivalent water residence time is

$$\frac{\text{volume of the blanket}}{Q} = t_w$$

We suggest adopting the following criterion for design and operation of sludges of river water coagulated with iron or aluminum salts: $t_s/t_w \times Q\phi T$ between 3000 and 4000. This has been proven to give appropriate sludge consistency for flocculation-clarifying, but further research is desirable in the field. The minimum value for t_w is 1200 s, but up to 3600 s is generally better.

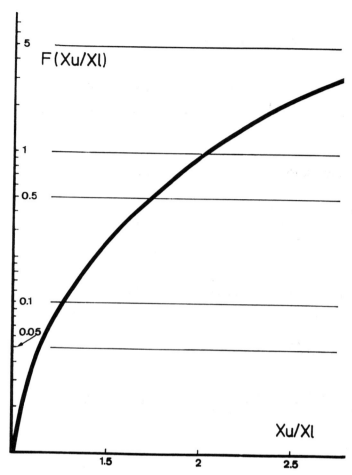

Figure 10 Graphical determination of $F(X_u/X_l)$ from X_u/X_l in expanding upflow clarifiers. (From Refs. 6 and 7.)

Example. River water of $t°$ variable between 10 and 23°C is to be treated at an intake flow $Q = 0.5$ m³/s by coagulation–flocculation with aluminum sulfate at an average (design) dosage of 40 g/m³. The solid impurities in the river average 20 g/m³ in dry weight, and activated carbon powder can be used in the treatment at a maximum dosage of 20 g/m³. Design an inverted pyramidal hopper tank for flocculation–clarifying.

Step 1: $v_p = 0.9$ mm/s at 15°C (see Table 1)

$$v_p \text{ at } 10°C = v_p \text{ at } 15°C \times \frac{\mu \text{ at } 15°C}{\mu \text{ at } 10°C} = 0.9 \times \frac{1.139 \ 10^{-3}}{1.307 \ 10^{-3}}$$

$$= 0.87 \text{ mm/s}$$

Step 2: Maximum admitted value for $v_s = 0.5v_p = 0.435$ mm/s = overflow velocity at upper end of sludge blanket

Step 3: Overflow surface: $\dfrac{0.5 \text{ m}^3/\text{s}}{0.435 \times 10^{-3} \text{ m/s}} = 1150 \text{ m}^2$

Step 4: X_u; $X_u^2 = 1150$; $X_u = 34$ m
(*Note:* This is a large value for X_u but can still be acceptable.)

Step 5: At inlet, $v_s = v_p = \dfrac{Q}{X_i^3} = 0.87 \times 10^{-3}$ m/s $= \dfrac{0.5}{X_i^2}$ $X_i^2 = 575$ m²;

$X_i = 24$ m

Step 6: Check and readjust for the velocity gradient function.

$$\frac{X_u}{X_i} = \frac{34}{24} = 1.42; F\left(\frac{X_u}{X_i}\right) = 0.19 \text{ (Fig. 10)}$$

$$G\phi T = \frac{576}{2}\left[\frac{(1.005 - 0.9997) \times 9.81}{0.5 \times 10^{-3} \times (2.78)^{9/2}}\right]^{1/2} \times 0.10$$

$= 57$, which is slightly below the acceptable limits

If X_i is readjusted to 20 m,

$$\frac{X_u}{X_i} = \frac{34}{20} = 1.7; F\left(\frac{X_u}{X_i}\right) = 0.45$$

$$G\phi T = \frac{400}{2}\left[\frac{(1.005 - 0.9997) \times 9.81}{0.5 \times 10^{-3} \times (2.78)^{9/2}}\right]^{1/2} \times 0.45 = 92 \quad \text{(OK)}$$

Step 7: Sludge blanket:

Dry material:

Settled material	20 g/m³
Activated carbon	20 g/m³
Reagent	5 g/m³
Miscellaneous	5 g/m³
	50 g/m³

$(0.5 \text{ m}^3/\text{s}) \times (50 \text{ g/m}^3) = 25$ g/s

Sludge flow: If maximum 3% bleed: $q_s = 0.015$ m³/s; $t_w \geq 1440$ s;

minimum blanket volume ≥ 720 m³; $t_s = \dfrac{720}{0.015} = 48{,}000$ s

Velocity criterion: $G\phi T \dfrac{t_s}{t_w} = 92 \times \dfrac{48{,}000}{1440} = 3.067 \quad$ (OK)

If $t_w = 3600$ s and the sludge bleed operates at 2.0% ($q_s = 0.01$ m³/s), one has, with sludge blanket volume 1800 m³ and $t_s = 180{,}000$,

$$G\phi T \frac{t_s}{t_w} = 92 \times \frac{180{,}000}{3600} = 4600$$

The velocity criterion is then at the very upper limit of the guidelines for design.

Step 8:   Geometry: problems and options

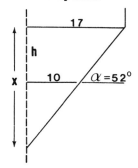

If $\alpha = 52°$, $\tan \alpha = \dfrac{x}{17} \longrightarrow x = 21.75$

$\tan \alpha = 1.28 = \dfrac{21.75 - h}{10} \longrightarrow h = 9\,\text{m}$

Sludge volume: $\frac{1}{3}[(34)^2 \times 21.75 - (20)^2 \times 12.75] = 6680\,\text{m}^3$

Conclusion: The tank is too deep and the sludge volume too overdesigned.

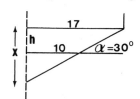

If $\alpha = 30°$, $\tan \alpha = 0.58 = \dfrac{x}{17} \longrightarrow x = 9.8\,\text{m}$

$\tan \alpha = 0.58 = \dfrac{9.8 - h}{10} \longrightarrow h = 4\,\text{m}$

Sludge volume: $\frac{1}{3}[(34)^2 \times 9.8 - (20)^2 \times 5.8] = 3000\,\text{m}^3$

Conclusion: Rather a high sludge volume.

If $\alpha = 15°$, $\tan \alpha = 0.27 - \dfrac{x}{17} \longrightarrow x = 4.6\,\text{m}$

$\tan \alpha = 0.27 = \dfrac{4.6 - h}{10} \longrightarrow h = 2\,\text{m}$

Sludge volume: $= \frac{1}{3}[(34)^2 \times 4.6 - (20)^2 \times 2.6] = 1425\,\text{m}^3$

Conclusion: OK, but the very flat basin requires distribution of the raw water at the bottom by means of a pipe system.

Step 9:   Try to readjust the design on another basis: Subdivide the water flow
in two equal parts and design two tanks.

Start again at point 4.

Step 4': $X_u^2 = \dfrac{1150}{2} = 575 \longrightarrow X_u = 24\,\text{m}$

Step 5':   At inlet: $0.87 \times 10^{-3}\,\text{m/s} = \dfrac{0.25\,\text{m}^3/\text{s}}{X_l^2}$

$X_l^2 = 287\,\text{m}^2 \longrightarrow X_l = 17\,\text{m}$

Step 6': $\dfrac{X_u}{X_l} = \dfrac{24}{17} = 1.4; F\left(\dfrac{X_u}{X_l}\right) = 0.19$

$$G\phi T = \frac{287}{2}\left[\frac{(1.005 - 0.9997) \times 9.81}{0.25 \times 10^{-3} \times (2.78)^{9/2}}\right]^{1/2} \times 0.19 = 39$$

This value is too low; decrease the inlet surface.

$$\text{If } X_l = 12\,m, \frac{X_u}{X_l} = \frac{24}{12} = 2; F\left(\frac{X_u}{X_l}\right) = 0.9$$

$$G\phi T = \frac{144}{2}\left[\frac{(1.005 - 0.9997) \times 9.81}{0.25 \times 10^{-3} \times (2.78)^{9/2}}\right]^{1/2} \times 0.9 = 92 \quad \text{(OK)}$$

Step 7':   For $t_w \geq 2880\,\text{s}$ and 3% sludge bleed, when the sludge volume $\geq$
720 m^3 and $t_s = 96{,}000$, $G\phi T = 92 \times \dfrac{96{,}000}{2880} = 3067$ (OK)

Step 8':   Geometry

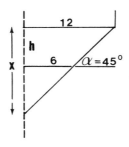

If $\alpha = 45°$, $\tan \alpha = 1 \longrightarrow x = 12\,m\; x - h = 6\,m$

Vol.: $\frac{1}{3}[(24)^2 \times 12 - (12)^2 \times 6] = 2016\,\text{m}^3$

Rather a high value.

$$\text{If } \alpha = 30°, \tan \alpha = 0.58 = \frac{x}{12} \longrightarrow x = 7 \text{ m}$$

$$0.58 = \frac{7 - h}{6}, \qquad h = 2.9 \text{ m}$$

$$\text{Vol.: } \tfrac{1}{3}[(24)^2 \times 7 - (12)^2 \times 4.1] = 1147 \text{ m}^3 \quad \text{(OK)}$$

Conclusion: Geometry factors often determine the construction of upflow clarifiers with a sludge blanket volume much greater than the minimum requirements that would make it possible to flocculate the water adequately. The designs are usually very conservative. Another approach is to increase the permissible overflow velocity, hence to lower the X_u value either by introducing lamella in the clarification zones (see Chapter 10), or by ballasting the floc with heavier material (e.g., microsand).

3.2 Ballasted Floc-Settlers

The design of upflow clarifiers can be improved by sludge recycling. In classical suspended solids-contact clarifiers in which part of the sludge is mechanically recycled in the secondary flocculation zone, the recycling rate is a few percent of sludge production (i.e., 3 to 5%; always less than 10%). The actual tendency is to increase the recycling ratio by injecting part of the sludge bleed (e.g., 20%) into the raw water. This practice changes the inlet conditions. If the average sludge content in the bleed is 10 vol% or $\phi = 0.1$, 20% recycling represents a value of S of 2% at the inlet and ϕ becomes 0.12 in the sludge pool. Then the assumption $v_s = v_p$ at the inlet must be corrected as follows:

$$v_s = v_p(1 - fS^{2/3}) = 1 - 2.78 \times (0.02)^{2/3} \simeq 0.8 v_p$$

For the example given in Section 3.1 (step 5), the inlet zone could then be adapted as follows:

$$\frac{Q}{X_I^2} = \frac{0.5}{X_I^2} = 0.8 \times 0.87 \times 10^{-3} = 0.7 \times 10^{-3} \text{ m/s}$$

Hence $X_I^2 = 714 \text{ m}^2$, $X_I = 27 \text{ m}$.

The design could then be considered further on this basis. The sludge recycling enables us to improve the stability of the flocculation conditions in the sludge layer, to shorten the running-in periods, and to reduce the area of construction, especially at the upper level of the upflow tanks.

Flocs can be ballasted by the introduction of a dense material such as microsand, clay, or diatomaceous earth. In practice, micrograins 0.1 to 1 μm in diameter are used in conjunction with a flocculation aid. The sludge is fractionated in a hydrocyclone and the washed and reconditioned microsand is recirculated into the raw water. The original process, developed in Hungary and known as Cyclofloc, is represented schematically in Fig. 11. Typical surface loadings for the Cyclofloc when using 20 kg of microsand/m^3 sludge issued from aluminum sulfate are given in Fig. 11. The overall residence time is about 1 h.

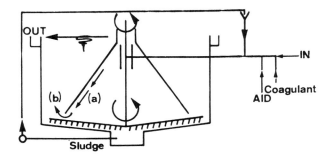

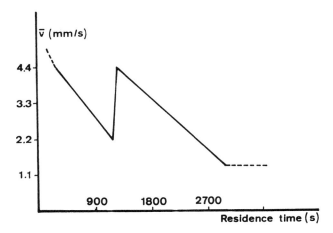

Figure 11   Cyclofloc and its surface loadings.

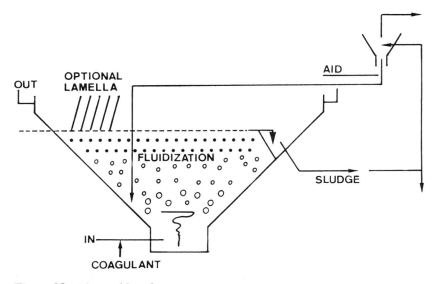

Figure 12   Fluorapid settler.

The dry weight of the solids contained in the floc are increased by a factor of 10 (e.g., to contain 15 to 20 kg/m³ sludge). In the Fluorapid settler (Fig. 12) developed in France (8), stationary conditions of operation rely on a concentration of 15 to 100 kg/m³ sludge (e.g., 50 kg/m³). The standard design also makes use of lamella in the clarification zone. In that case, the overflow velocity can attain an upper limit of 4.5 mm/s, thus reducing the necessary surface by a factor of 8 to 10. The more sand that is added, the less flocculant the settling becomes, and thus the less applicable become the guidelines and criteria above.

3.3   Pulsated Upflow Clarifiers

By distributing the coagulated raw water intermittently or at variable flow into an upflow clarifier, we obtain sequential settling and expansion of the sludge blanket. The original design was patented by the French firm Degrémont under the name Pulsator (see Fig. 13). In the standard design the surface loading remains lower than 0.8 mm/s for flocculated river water. The sludge layer thickness is within the range 2 to 4 m (i.e., 0.5 to 1 h contact time). This is on the conservative side of the sludge-contact flocculation values outlined above. Therefore, the design can easily be uprated to a surface loading of up to 1.4 mm/s. This method is particularly well adapted to the dosing of activated carbon powder incorporated into the sludge blanket, which then acts more or less as in an upflow filter.

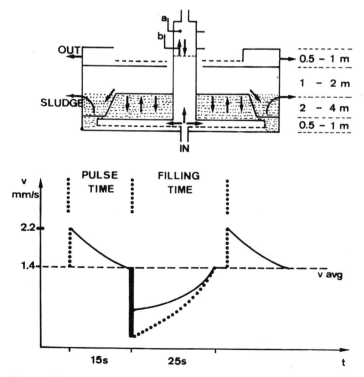

Figure 13   Pulsator.

Normal sludge bleed rates range from 1 to 3% of influent water flow. Raw water inlet is achieved by means of perforated pipes located at the bottom of the tank. The openings are directed toward the bottom, and repartition of the incoming water is improved by deflectors. The tank is operated on a discontinuous inlet cycle, thus preventing compacting and channeling of the floc blanket, which is a problem commonly encountered in upflow settlers. The inlet cycle involves the production of a vacuum in the inlet column, filling the column up to the maximum contact level. At this level the vacuum is shut off and the column is emptied into the settling tank; that is, when the floc blanket expands and sludge overflows into the concentrators, settling takes place during the next phase of filling of the column. Usually, the complete cycle is regulated automatically on an on-off basis for a complete period of 20 to 120 s. The cycle is adjusted empirically to reach a surface loading of about 2.2 mm/s during the pulse time.

The Pulsator can be constructed in any shape and size: square, rectangular, circular, and so on, and for flows ranging from a few cubic meters to several thousand cubic meters per hour. Lamella or tubes can be located in the clarification zone to enable uprating of the overflow velocity (e.g., up to a maximum of 3 mm/s). Stabilization of the sludge layer may then be necessary by lamella elements introduced into the upper part of the sludge layer at a depth of about 0.5 m. This version is called the Superpulsator.

Example of operation conditions (see Fig. 13)

Average surface loading: 1.4 mm/s
$Q = 0.75 \text{ m}^3/\text{s}$
Surface of clarification zone: 540 m²
Incremental surface loading during the pulse time: $2.22 - 1.4 = 0.82$ mm/s
Capacity of the inlet column: e.g., 7 m³ (adjustable)
Filling height of the column: e.g., 0.6 m
Practical rules for pulse time:

$$t_{\text{pulse}} = \frac{72 \times \text{ height in column}}{v_{\text{max}} - v_{\text{avg}}) \times 3600}$$

For example:

$$t_{\text{pulse}} = \frac{72 \times \text{ height in column}}{(2.2 - 1.4) \times 3600} = 15 \text{ s}$$

Practical rules for filling time:

$$t_{\text{filling}} = \frac{72 \times \text{ settling surface} \times \text{ height in column}}{\text{flow of vacuum}}$$

By considering the vacuum flow of 0.28 m³/s (NPT), the filling time becomes, for example,

$$\frac{72 \times 540 \times 0.6}{0.28 \times 3600} = 23 \text{ s}$$

In the short operation cycle considered here, the entire cycle length is 35 to 40 s. Variations up to 120 s are possible in practical equipment by changing the on–off positions of the electrodes in the inlet column and/or the average vacuum-flow capacity of the ventilator.

SYMBOLS AND UNITS RELATED TO COMBINED COAGULATION-FLOCCULATION-CLARIFYING PROCESSES

f	Bond's factor, dimensionless
h_z	height of sludge layer, m
k_a	particle agglomeration constant, dimensionless
k_b	particle breakage constant, s
n_0, n_1	number of particles, dimensionless
q_s	sludge flow, m³/s
t	time, s
$t°$	temperature, °C
t_s	residence time of solids in the sludge, s
t_w	residence time of water in the sludge zone, s
v	settling velocity, m/s or mm/s
v_p	particle settling velocity, mm/s
v_s	sludge settling velocity, mm/s
A	clarification zone area, m²
D	diameter (of weirs), m
F	geometrical function for the sludge layer
G	velocity gradient, s⁻¹
Q	water flow, m³/s
S	concentration of sludge, vol%
T	detention time, s
V	volume, m³
X_l	coordinate of lower layer of sludge, m
X_u	coordinate of upper layer of sludge, m
α	tilting angle of sludge flow zone, degrees
μ	dynamic viscosity, kg m⁻¹ s⁻¹
ρ_l	density of liquid, kg/dm³
ρ_s	density of solids, kg/dm³
θ	wet volume fraction of sludge, dimensionless

REFERENCES

1. F. D. Prager, *Water Sewage Works*, *97*, 143 (1950).
2. AWWA Committee Report, *J. AWWA*, *43*, 263 (1951).
3. U.S. Patent 91373 (1869).
4. Br. Patent 19829 (1892).
5. A. W. Bond, *J. IWES*, *15*, 494 (1961).
6. K. J. Ives, *Proc. Inst. Civil Eng.*, *39*, 243 (1968).
7. A. Amirtharajah, p. 195 in R. L. Sanks, *Water Treatment Plant Design*, 4th ed., Ann Arbor Science, Ann Arbor, Mich., 1982.
8. C. Gomella, *Proc. IWSA Congress*, Brighton, Pergamon Press, Elmsford, N.Y., 1974, p. A1.

10
Lamellar and Tubular Assisted Settling Processes

1. INTRODUCTION

According to the theoretical concepts first developed in 1904 by Allen Hazen, the sedimentation efficiency in settling basins based on nonflocculant settling is assumed to be due to the fact that the velocity of the particle fall is higher than the upflow velocity of the water. The latter is equal to the surface loading: $\overline{S} \geq Q/A$. The surface loading, originally expressed in m/h or $m^3 \ m^{-2} \ h^{-1}$, is also known as the superficial hydraulic charge, the Hazen apparent velocity, the Hazen coefficient, and the overflow velocity (in mm/s). In definitions of static and horizontal-flow sedimentation, the height of the basin is not an element of concern except when it provides a suitable sludge accumulation zone.

In addition to the frictional effects on walls, perturbations observed in practice are density gradients based principally on flow conditions [e.g., if Re > 2000 (Reynolds number) and instability if Fr < 10^{-5} (Froude number)]. The fundamental design criteria of lamellar or tubular settlers are: maintaining laminar (streamline) flow conditions, which provide for sufficient sludge disposal capabilities and area; and appropriate surface loadings within the lamella. Although the technique has been known since the beginning of this century, new interest in its use has arisen due to the work of Culp et al. (1,2).

2. CONSTRUCTION OF LAMELLAR ELEMENTS

Lamellar elements are generally formed by PVC plates starting with flat, waveform, or profiled plates. By combination, various tubular arrangements are obtained: square, rectangular, hexagonal, or profiled (Fig. 1). Furthermore, various combined

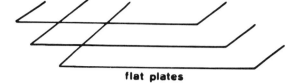

flat plates

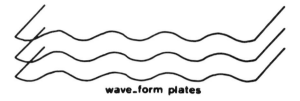

wave_form plates

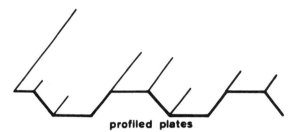

profiled plates

Figure 1   Typical forms of plates.

arrangements make it possible to form packed modules for use in clarification zones (Fig. 2). Waveform plates can be combined for use with low or high levels of hydraulic disturbance and mixing effects (Figs. 3 and 4).

3.   LAMELLAR FLOW CONDITIONS

In every arrangement, very laminar flow conditions must be maintained (e.g., Re $\leq$ 500 and preferably, $\leq$ 200). Considering a temperature of 21°C (i.e., $\nu = 10^{-6}$ m/s), we have Re $= vd_h/\nu$, in which the hydraulic diameter d_h is equal to the surface of the channel section divided by its perimeter, and ν is the average upflow velocity within the channel sections. For square tubes of side w and a v value of 56 $\times$ 10^{-4} m/s (e.g., 20 m/h horizontal velocity in a horizontal-flux settler),

$$d_h = \frac{x^2}{4x} = \frac{x}{4} \qquad \text{Re} = \frac{56 \times 10^{-4}x}{4 \times 10^{-6}} \leq 200$$

and

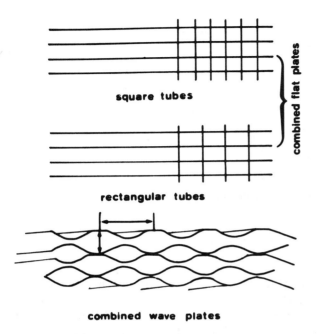

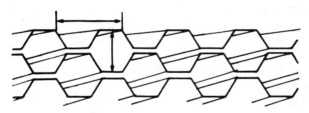

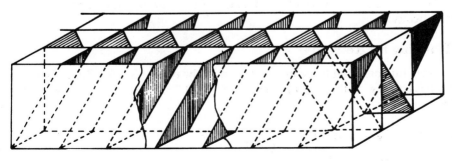

Figure 2   Combined plates to form tubular elements.

Figure 3   Steeply inclined tube modules.

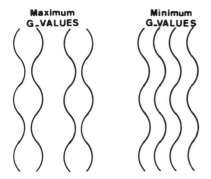

**Maximum
G_VALUES**

**Minimum
G_VALUES**

Figure 4   Mixing effects in wave–plate combination.

$$x \leq \frac{800 \times 10^{-6}}{56 \times 10^{-4}} = 14 \times 10^{-2} \, \text{m} = 0.14 \, \text{m}$$

Considering that the introduction of lamellar or tubular elements in horizontal-flux settling is aimed to increase performance in a given space, the interlamellar distances are on the order of centimeters. For example, a classical horizontal-flow sedimentation tank can be conceived to treat about 0.3 m^3/s at a surface loading of 0.33 $\times$ 10^{-3} m/s. In this case, the necessary settling surface is 0.3 m^3/s: 0.33 $\times$ 10^{-3} m/s = 900 m^2. This surface can be obtained using an arrangement such as the following: width 18 m, length 50 m, height 3 m through which a well-balanced settling tank is obtained with an average detention time of 9 $\times$ 10^3 s and an average horizontal velocity of 5.6 $\times$ 10^{-3} m/s.

   If 20 horizontal plates are inserted in the basin, 21 floors are available for sludge deposition. Assuming that one wishes to maintain the Q/A value of 0.33 $\times$ 10^{-3} m/s, the length of the plates must be of (0.3 m^3/s)/(21 $\times$ 18L) = 0.33 $\times$ 10^{-3}; hence L = 2.4 m. If 5-m-length plates were inserted and the flow maintained at 0.3 m^3/s, the separation efficiency could be improved according to a surface loading of 0.167 $\times$ 10^{-3} m/s.

   As a first conclusion one can state that there is no advantage in equipping the entire length of a classical horizontal-flow settling tank with lamellar or tubular elements. It is sufficient to equip a limited part of the tank in order to maintain laminar flow conditions. However, turbulency of the flow conditions must be considered critically.

3.1   Planning for Re $\leq$ 200

The first general condition to be met in design is always to plan for Re $\leq$ 200. For square tubes (see Fig. 5) this is equal to $vx = 8 \times 10^{-4}$ m^2/s and the surface loading becomes

$$\frac{Q}{A} = \frac{vx^2}{xL} = \frac{vx}{L} = \frac{8 \times 10^{-4}}{L} \qquad \text{(in SI units)}$$

or

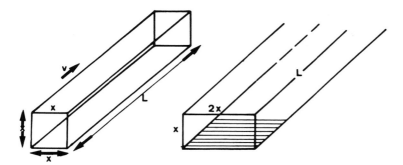

Figure 5   Effects of tube geometry.

$$L = \frac{2.88}{Q/A} \qquad \text{(in conventional units, m/h for } Q/A \text{)}$$

For rectangular tubes (see Fig. 5) of sides x and $2x$, the expressions become

$$d_h = \frac{2x^2}{6x} = \frac{x}{3}$$

$$200 = \frac{vx}{3 \times 10^{-6}} \longrightarrow vx = 6 \times 10^{-4}$$

$$200 = \frac{6 \times 10^{-4}}{Q/A} \quad \text{(in SI units)} \quad \text{and} \quad L = \frac{2.16}{Q/A \text{ (in m/h)}}$$

Generalization of the expression for rectangular sections is as follows:

$$L = \frac{6 \times 10^{-4}}{Q/A} = \frac{12 \times 10^{-4}}{6(Q/A)} = \frac{4(2 + 1) \times 10^{-4}}{2(Q/A)} = \frac{4(p + 1) \times 10^{-4}}{p(Q/A)}$$

in which px is the length and x the width of the rectangular section. The same approach can generally be adopted for curvilinear-rectangular sections obtained by combined wave plates. For hexagonal conduits the relationships become

$$d_h = \frac{2.6x^2}{6x} = \frac{2.6}{6}x = \frac{x}{2.3}$$

$$vx = 4.6 \times 10^{-4} \quad \text{and} \quad L = \frac{4.6 \times 10^{-4}}{Q/A}$$

Owing to the existence of an interstitial zone, the use of circular tubes is less favorable.

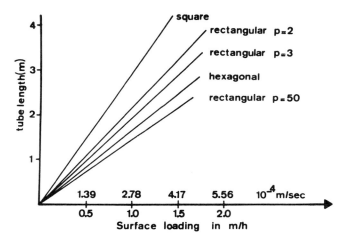

Figure 6   Tube length in laminar settling.

3.2   Planning for Additional Tube Length

The approach in Section 3.1 enables us to compute the general tube length in order to maintain laminar flow conditions, but it also supposes a perfect distribution of the water in the inlet zone without turbulency at the inlet of the lamella area. As this condition is not always satisfied, an additional tube length is often allowed for to stabilize flow at the inlet. The additional length for turbulence at the inlet (L_T) is experimentally related to the Reynolds number as $0.23 \, \mathrm{Re} \, x > L_T > 0.1 \, \mathrm{Re} \, x$ which for $\mathrm{Re} = 200$ equals ($46x > L_T > 20x$, practically an additional length between 50 and 20 times the tubular side, or width x (e.g., in the case of $x = 0.05$ m, L_T is between 1 and 2.5 m to be added to the length for laminar flow). The general formula is expressed as

$$L_T = \frac{5p}{p + 1} \, vx^2$$

To decrease the L_T values to a minimum and consequently shorten the overall length of the lamella zone, the modules are to be placed in zones of stable hydraulic conditions. Therefore, these elements are generally not located in the first third of the basin length in horizontal flow settlers, while in upflow systems or zones they are located midway on (occasionally, three-fourths of) the basin depth starting from the clarification surface.

4.   TILTING ANGLE

The tilting angle is a basic design criterion in the tubular or lamella settling process. In an original experiment carried out by Culp and co-workers (2), they developed a test apparatus to study the effects of tube inclination on settling efficiency. The test is shown schematically in Fig. 7. In this setup, individual settling tubes were inclined at angles of 0, 5, 20, 35, 40, 45, 60, and 90°.

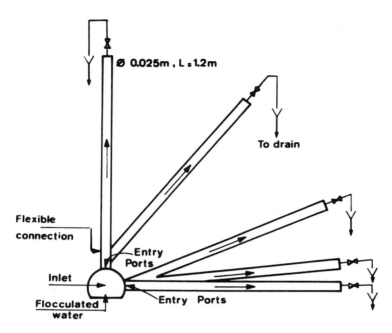

Figure 7   Apparatus for the determination of self-curing effect.

In all the experiments the authors observed an optimum tilting angle for the sedimentation rate of 45 to 60°. At these values and even at high values of surface loadings (e.g., up to 20 m/h), a "self-cleaning" angle was observed, (i.e., a tilting angle at which the deposits flow down to the inlet zone countercurrently to the incoming water stream). In static experiments on the flocculated water of the Tailfer plant we have observed the effects depicted in Fig. 8 (flocculation with aluminum sulfate, all experiments in glassware).

In the design of the tilting angle, the ability to establish a self-cleaning or self-curing effect may depend on several additional factors, such as the rugosity of the construction material, the presence of gases, the water temperature, the presence of bacterial and algal growth on the walls, the nature and composition of the sludge, and the usual settling characteristics of the sludge. Therefore, the tilting angles are fixed according to the following empirical guidelines, which have been proven satisfactory in practice.

4.1   Horizontal Settlers

In specific horizontal flow, settling problems arise with sludge elimination in the tubes or lamella when placed horizontally ($\alpha = 0$). In the system proposed by Culp (3) the lamella are tilted at 5° and sludge is withdrawn by backflushing filtered water countercurrently through the settling system. Even if the system could fit into an existing horizontal flow settler, it suffers the disadvantages of discontinuous operation. The general scheme is indicated in Fig. 9.

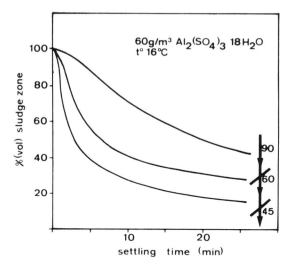

Figure 8 Effect of the tilting angle on settability of flocculated water of the river Meuse.

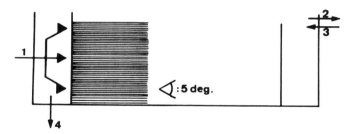

Figure 9 Horizontal settler with slightly tilted plates (5°). 1, Inlet of flocculated water; 2, overflow of settled water; 3, backflow of filtered water; 4, sludge bleed.

4.2 Tilted Plate System

For the continuous drainage of the sludge without backflushing, a 50° inclination of the plates is required in the equistream principle (sludge and water flow moving in the same direction), or a 60° in inclination in the countercurrent principle (e.g., sludge moves downward in upflowing water). Tubes or plates installed in the direction of the basin length can be arranged in two different ways, as is noted in Fig. 10. In the longitudinal arrangement (a), the horizontal surface (A) intervening in the surface loading (Q/A) is equal to $A = n\,(W\lambda \cos \alpha)$, in which n is the number of plates given by $h/x' = h(\cos \alpha)/x$, x being the orthogonal interplate distance. The interplate distance is fixed in order to maintain the laminar flow conditions defined previously. In the transverse arrangement (b), the lamella intake width is equal to $w - \lambda \cos \alpha$ and the individual surface area of each lamella available is equal to $l'\lambda(\cos \alpha)$, in which l' is the horizontal length of the lamella zone. In this arrangement the tilting angle can vary between 50 and 60°, respectively, with $\cos \alpha$ between 0.616 and 0.5. A typical example can be given as follows: $Q = 0.3$ m³/s,

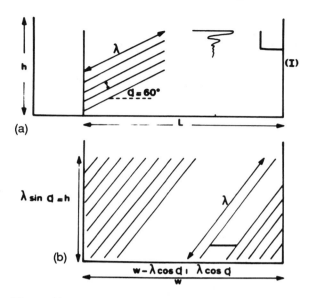

Figure 10   Tilted plate systems in horizontal-flow settlers.

$h = 3$ m, $w = 18$ m, $l = 50$ m; thus classical $Q/A = 1.2$ m/h or 3.3×10^{-4} m/s. For plates of length of 5 m installed in the transverse mode, inclined at 60° and at an interplate distance of 0.05 m, the surface loading becomes

$$h = \lambda \sin \alpha \longrightarrow \lambda = \frac{h}{0.87} = 3.45 \; m$$

(the thickness of the construction material exerts only a secondary influence, which can be neglected)

$$\text{useful width} = 18 - \lambda \cos \alpha = 18 - \frac{3.45}{2} \simeq 16 \text{ m}$$

$$\text{number of complete plates} = \frac{16 \sin \alpha}{0.05(\cos \alpha)} = 275$$

$$\text{surface loading} = \frac{0.3}{275} \times 5 \times 1.7 \sim 1.28 \; 10^{-4} \text{ m/s} \quad (\text{or } 0.46 \text{ m/h})$$

The most important part of the basin (90% of its length) remains available for classical horizontal-flow settling to occur. When operated classically, surface loading in the lamella zone is less than in the complete basin length. Furthermore, the lamella in the inlet zone can exert wave damping in the case where thermally induced, density gradients cause such effects (4). In their longitudinal arrangement, the lamella or tubes are usually suspended in the upper part of the outlet zone of the horizontal-flow settlers. The design criteria for this arrangement, illustrated in Fig. 11, are to be considered as for upflow settlers.

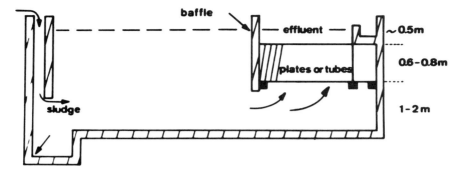

Figure 11 Tubes installed at the outlet of horizontal-flow settlers.

4.3 Lamellar Elements in Upflow Settlers

Lamellar or tubular elements can be installed in upflow settlers either in the clarification zone only, and/or in the flocculation–sludge concentration zone. The plates are to be inclined at least 52° for intermittent sludge withdrawal and 60° for permanent flow of sludge when the water moves upward. Installation in the clarification zone is illustrated in Fig. 12.

The elements are generally tilted in the longest direction of the clarification zone. The useful length of the lamella zone is defined as $l\alpha$ and corresponds to $l - \lambda \cos \alpha$. This defines the number of plates available as

$$n = \frac{l\alpha}{x'} = \frac{l - \lambda \cos \alpha}{x/\sin \alpha}$$

The horizontal surface corresponding to one plate is $\lambda w \cos \alpha$ and the total horizontal settling area is equal to $n\lambda w \cos \alpha = \dfrac{\lambda w \cos \alpha \sin \alpha \, l\alpha}{xl}$, which can be compared to the "classical" area lw by neglecting the thickness of the plates; hence

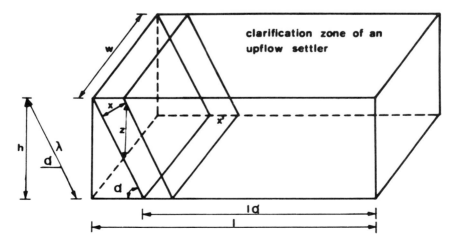

Figure 12 Lamellar clarification zone of an upflow settler.

$$\frac{A_{\text{lamellar}}}{A_{\text{classic}}} = \frac{\lambda \cos \alpha \sin \alpha}{x} \frac{l\alpha}{l}$$

Practical values of x vary from 25 to 200 mm and the A value is increased 3- to 20-fold compared to the customary A value. Also, the laminar flow conditions are maintained at Re < 200 within the lamella, and the zone is designed with additional length for inlet turbulence. The starting point of the design is usually a commercial plate length λ of, for example, 1 m, and in most practical cases of this type, λ/l is maintained between 0.01 and 0.1, with an average value of 0.033.

The lamellar or tubular modules are most conveniently installed in square or rectangular basins but can also be designed for circular basins (5), as shown in Fig. 13. The tube modules are placed in a ring and supported at the outer wall of the basin by radical support members reaching the inner wall, equipped with a baffle to ensure that the entire flow transits through the tubular zone. For basins equipped with radial overflow weirs, these can often be used to hang up the modules. Tubes are tilted as illustrated, from the inner to the outer wall, and the outer length of the segments is on the order of 2 to 2.5 m.

4.4   Steeply Inclined Tubes

Steeply inclined tubes ($\alpha \geq 60°$) can also be suspended in the flocculation zone of an upflow clarifier and improve both flocculation and stability of the sludge layer. Under such conditions, a countercurrent flow pattern between the sludge and the water takes form rapidly. The tubes are inclined steeply (e.g., 60°).

The time available for a particle to settle in the lamella zone of height h is

$$t = \frac{h}{v} = \frac{\lambda \sin \alpha}{v\alpha} \qquad \text{(Fig. 12, 14)}$$

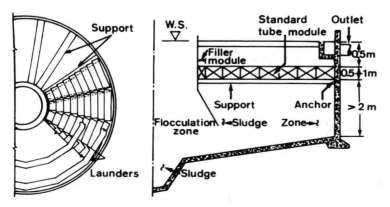

Figure 13   Tube installation in circular basins (schematic).

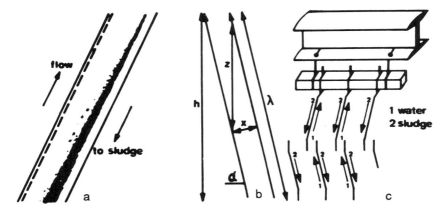

Figure 14   Flocculation within the lamella (schematic): (a) Sludge-water flow patterns; (b) lamella zone velocities; (c) chevron patterns lamella.

However, the velocity within the lamella zone is equal to

$$v\alpha = \frac{Q}{A/\text{lamella zone}} \longrightarrow v\alpha = v\frac{l}{l\alpha} = \frac{v}{1 - \lambda/l\cos\alpha}$$

Within each lamella itself, the available settling time is

$$t = \frac{z}{v(l)} = \frac{x}{v(l)\cos\alpha} = \frac{\lambda\sin\alpha}{v\alpha}$$

$$v(l) = \frac{xv\alpha}{\lambda\sin\alpha\cos\alpha} = \frac{x}{\lambda\sin\alpha\cos\alpha}\frac{v}{1 - \lambda/l\cos\alpha}$$

The critical settling velocity of a sludge [e.g., as given by the formula of Bond (see Chapter 9)], must satisfy the conditions for settling, as a function of the geometrical parameters of the lamella as expressed above. As a guideline it can be recommended to maintain a minimum sludge layer be maintained and the overflow velocity adapted as a function of the temperature and concentration of the suspended matter, as illustrated in Fig. 15. Under the lamella-flocculation area, appropriate arrangements for sludge removal are assumed to be classical.

Example.   Most upflow clarifier designs have been equipped with lamellar-flocculation elements, but one of the first specific developments was the Inka lamella, separated as illustrated in Fig. 16a. It is based on a cocurrent downward flow of both sludge and flocculated water, and classical tilted tubes or lamella equipped with sludge-movement deflecting blades (Fig. 16b). The latter, although promoting local turbulency (favorable for flocculation, critical for clarification), have a general improving effect when placed in the sludge zone (6). The system is related to advanced designs such as the finned-pocket settler (Fig. 18).

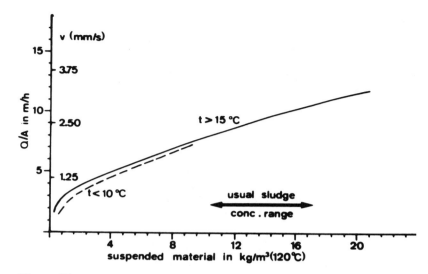

Figure 15   Settling velocity in lamellar settling.

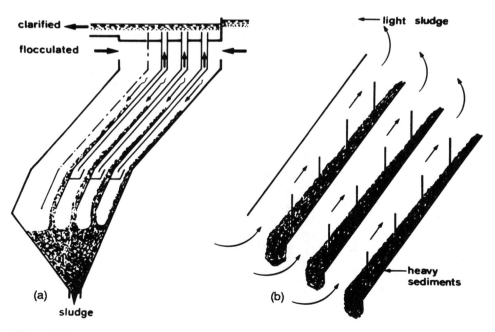

Figure 16   Principles of lamellar flocculation: (a) lamella separator; (b) countercurrent flow of sludge.

5. OXIDATOR PROCESS

In the biological treatment processes (e.g., nitrification), the introduction of vertical elements (Fig. 17) placed in the active zone in which the water is recycled downstream (e.g., in the Oxidator process) can promote the effect by providing a suitable support for oxidation as a pretreatment. Surface loadings remain conventional as for aerobic processes [e.g., between 0.3 and 0.6 m/h (10^{-4} to 2×10^{-4} mm/s)].

6. FINNED CHANNEL SEPARATOR

Further innovation in the lamellar and tubular technologies are on the move. The finned channel separator (7) (see Fig. 18) is a particular arrangement by which horizontal-flow settling can be improved for difficulties due to flow patterns. Moreover, the principle, still in the process of development, calls on centrifugal forces due to the main velocity of the water, supplementary to gravitational forces and drag resistance. In this arrangement, a classical horizontal-flow settler dimensioned according to the usual design parameters is divided into parallel sections and equipped in different zones. The general proportions between the sludge zone and clarification zone remain unchanged (e.g., one-third sludge concentration zone and two-thirds clarification zone. In the clarification zone, a main channel zone delimits active water transfer. It can be subdivided into a laminar flow-through zone comprising an agitated zone and a finned-pocket zone. The cross section of the flow-through zones is about 0.3 to 0.25 of the gross cross section of the settler.

The first design parameter is to obtain a Reynolds number Re $\leq$ 500 for the flow-through zone, so as to maintain overall laminar flow conditions and a cyclone effect at the finned boards. Under comparative overall velocity, this is equal to a velocity for horizontal flow in the flow-through zone of three times the average velocity (e.g., 90 mm/s instead of 30 mm/s). The turbidity improvement rates are claimed to increase the ratio 10 : 1 for standard horizontal settling compared to finned-tube settling.

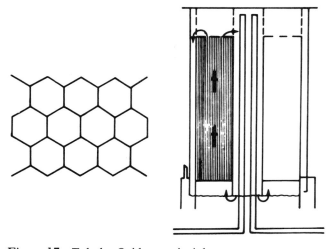

Figure 17   Tubular Oxidator principle.

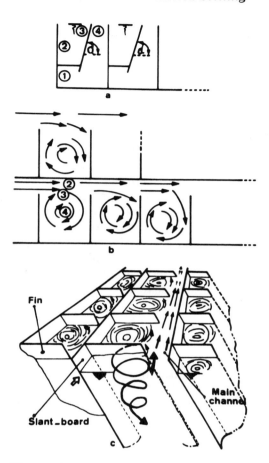

Figure 18 Finned channel settler (schematic): (a) cross section; (b) horizontal view; (c) perspective view. 1, Sludge disposal zone; 2 and 3, flow-through zone; 3, agitated zone; 4, finned-tube zone.

The decrease in turbidity of coagulated-flocculated water as a function of the number of finned tubes (*n*) is expressed by an exponential relationship of the type

$$\tau n = (\tau_o - \tau_p)(1 - k)^n + \tau_p$$

where

τ_o = turbidity at inlet

τ_n = turbidity after *n* fins

τ_p = permanent turbidity after prolonged (e.g., 24 h) static settling (e.g., due to noncoagulated particles)

k = empirical constant depending on the velocity in the flow-through zone

In view of the classical Hazen concept (i.e., the overflow velocity concept), the system is based on a whirling velocity in the finned pocket zone with a local overflow rate of only 20% of the main channel flow. Therefore, gravitational control of the removal of suspended matter into the finned-tube zone is much less critical than in classical horizontal-flow settlers.

7.   MAINTENANCE OF MODULES

Maintenance of lamellar modules may be an important feature in lamellar and tubular assisted-settling processes. Solar irradiation of tubes or plates is best avoided, so as to limit or inhibit algal growth on the plate support. The floc has a tendency to adhere to the upper edges of the tubes. Although it has no significant consequence on water quality, this effect causes the general appearance of the installation to deteriorate. Cleaning can be by means of jet nozzles placed above the settling tubes (in lamellar assisted flocculation), then spraying clear water through a grid installed under the modules when placed in the clarification zone. Facilities must be provided for withdrawal of the elements and periodic cleaning in a basin filled with dilute hydrochloric acid (e.g., 6 wt%). The weight of the units depends on the specific manufacturer. A general indication is for 30 to 40 kg/m^2 per module, and the most common dimensions are 1.5 by 2.5 m.

8.   SUMMARY

The use of lamellar or tubular elements in the settling process supposes in all instances operation under laminar flow conditions: Re < 200 with possible values up to 500 in assisted-flocculation zones. Extra length to compensate for inlet turbulence should be allowed for in the design.

The basic data to be considered relate to the available horizontal (or projected area intervening in the surface loading). Here the plate length λ and the interplate distance x play a dominant role. One should therefore start the design from these values, for which commercially available systems range 0.5 to 1 m for λ and 0.025 to 0.10 m for x. Modular elements can be placed one above the other if necessary.

Typical tilting angles exceed 35° (e.g., 50° for equicurrent sludge and water movement and $\geq 60°$ for countercurrent sludge bleed). Usually, the elements equip less than one-third of the basin length or depth and a sludge accumulation and bleeding zone is required.

The process enables significant increase in surface loading (5 to 10 times) at equal performance rates. Flocculation can be improved by placing the elements in the sludge layer of an upflow settler, but the process has no effect on coagulation, which is to be achieved before the water enters the tubular or lamellar zones.

REFERENCES

1.   S. P. Hansen and G. L. Culp, *J. AWWA*, *59*, 1134 (1967).
2.   G. Culp, S. Hansen, and G. Richardson, *J. AWWA*, *60*, 681 (1968).
3.   G. L. Culp and R. L. Culp, *New Concepts in Water Purification*, Van Nostand-Reinhold, New York, 1974, p. 31.

4. R. R. Hudgins, *Proc. IAWPR Congress*, Toronto, 1980.
5. J. P. Hansen, G. H. Richardson, and A. Hsiung, *Chem. Eng. Prog. Symp. Ser.*, *65*, 207 (1969).
6. Y. Richard, *T.S.M. Eau*, *69*, 113 (1974).
7. N. Tambo et al., *Proc. IAWPR Congress*, Toronto, 1980, p. 409.

11
Testing Methods for Operational Control of Unit Processes

1. INTRODUCTION

Unit processes should be applied individually or in sequences and operated to produce an effluent water within the required standards for any raw water quality entering the treatment premises. Basically, the assessment of water quality as a whole involves measurement of a considerable number of parameters, and this cannot be accomplished on a permanent basis in everyday plant operation. Besides wanting to meet a given particular standard (e.g., in the case of the presence of a specific industrial pollutant), the plant operation relies on a limited number of parameters controlled continuously (e.g., residual concentrations, overall group parameters such as turbidity, pH, bacterial count, zeta potential, and so on, and also on simulation tests such as the jar test, filtrability, and sludge characteristics. All these data require regular updating of correspondence with the more detailed analytical determination of all specific water quality parameters. Although several attempts have been made, there does not yet exist a general integrated scheme correlating water quality and treatment processes as a whole. The purpose of the present chapter is to comment on the scope and limitations of the nonspecific testing procedures applicable to unit processes. For specific analytical methods, appropriate handbooks should be consulted.

2. AUTOMATED WATER QUALITY CONTROL

The reliability of the sensors on continuous operation on raw water or on water undergoing a specific treatment process has proven to limit the available continuous analytical systems to a very reduced number (31): pH, temperature, turbidity, electrical conductivity, dissolved oxygen, oxidation reduction potential, residual ozone

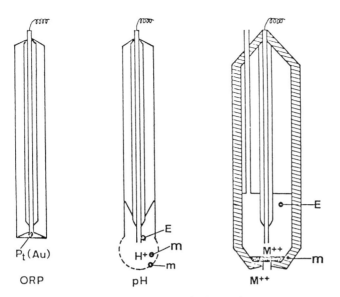

Figure 1   Sensitive potentiometric electrodes.

and residual active oxidants. Sequential analyzers can be used for the control of specific parameters, such as residual coagulant by absorptiometry or specific potentiometric determinations (e.g., fluoride). However, the latter methods are merely intermittent and require regular supervision in the laboratory (1,2).

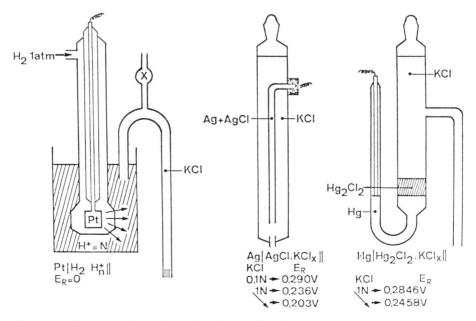

Figure 2   Reference electrodes in direct potentiometry.

2.1 pH Measurement

The continuous measurement of pH is a specific case of the general principles of potentiometry based on the equation of Nernst:

$$E_{25} = E_0 + \frac{0.059}{n} \log[M^{n+}]$$

For measurement of the hydrogen ion concentration, $n = 1$ and $E_0 = 0$ (per definition), whence $E_{25} = -0.59$ pH. In practice, the potential function is measured against the potential of a reference electrode. The so-called normal hydrogen electrode (i.e., an electrode in a medium of pH 0) has a reference potential of zero ($E_R = 0$); however, as a rule, calomel (SCE) or silver/silver chloride reference electrodes are used.

With continuous operation, reliability and good behavior depend on mainte-

Figure 3   Continuous pH measurement of raw water (Tailfer plant).

nance of adequate porosity of membranes, enabling more-or-less specific diffusion of the element to be measured, thus reaching the sensor electrode. Continuous movement of the liquid in the immediate neighborhood of the electrode is essential. A typical setup for measurement and automatic closed-loop correction of the pH of the raw water installed in the Tailfer plant of the Brussels Waterboard to treat river Meuse water is shown in Fig. 3. Sample water is circulated upward along the electrode with an auxiliary sampling pump.

In very difficult cases, one can proceed with ultrasonic cleaning of the electrodes (Fig. 4), making it possible, for example, to automate neutralization of sodium metasilicate to prepare activated silica (see Chapter 8). Therefore, the electrode technology must be adapted to ultrasonic cleaning. Based on practical schemes it is advisable to proceed to weekly calibration of the measurements and a general monthly manual cleaning of the electrodes. In the case of sensors based on a liquid or solid membrane system (e.g., for the measurement of dissolved oxygen), fouling of the membranes or mechanical abrasion by continuous measurement of dissolved oxygen in raw water and wastewater may be a problem.

2.2 Electrical Conductivity Measurement

Continuous measurement of electrical conductivity is based on the electrical resistance recorded on a unit volume at a standard temperature. The conductivity is the reciprocal of the resistivity and for water temperature can be corrected by the following equation:

Figure 4 Setup of pH control with ultrasonic cleaning in activated silica preparation (Tailfer plant).

$$K_{18} = K_t \, [1/1 + 0.0265(t - 18)]$$

In continuous-measurement equipment the electrical resistance of a portion of the liquid is usually measured using a Wheatstone bridge, one branch of which is coupled with a resistance varying with changing water temperature to adjust for the temperature effect. It is important to select equipment with which one can read the results on a linear basis throughout the entire range of measurement. Except for specific cases of industrial effluents, the electrical conductivity provides general information concerning evolution of the mineral content of the water, yet is hardly correlated directly with the operation of treatment schemes.

2.3 Turbidity Measurement

Turbidity is an interesting parameter seeing because it is potentially directly observable by the consumer. The turbidity T is expressed by the ratio I_d (= intensity of the light diffused in a given direction) to the incident light intensity. In "pure suspensions," that is, for a single type of particle equal in shape, weight, dimensions, and chemical constitution, the turbidity is related to the molecular weight of the dispersed colloid. As real systems are to be associated with polydisperse suspensions, turbidity has an empirical value, enabling relative comparisons. Furthermore, the diffusion of the light, which is the underlying principle in turbidity, obeys Bragg's law:

$$n\lambda = 2d \sin \theta$$

For the visual region $\lambda = 400$ to 800 nm and $\theta = 90°$, the average particle diameter is, for first-order diffusion ($n = 1$), $d \sim 200$ to 800 nm. Consequently, turbidity is associated with colloids of a larger size than is color.

The intensity of the diffused light beam depends on the diffusion angle θ, as illustrated in Fig. 5. The diffusion angles most frequently encountered are 25°,

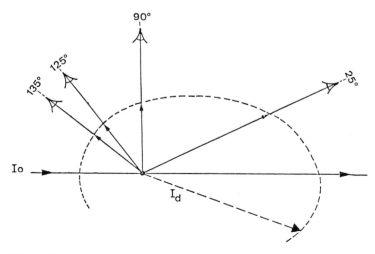

Figure 5   Incidence of the diffusion angle on observed turbidity.

90°, and 135°. When comparing different turbidimeters with the same reference suspension, it is essential for the diffusion angle to be the same for all data compared.

Different units are used to express turbidity. The most important at present are discussed below.

NTU or FTU. In this unit of nephelometry the diffusion angle is 90°. The standard turbidity suspension is prepared with formazine as follows: 5 g of sulfate of hydrazine ($NH_2 \cdot NH_3 \cdot H_2SO_4$) is dissolved in ± 400 cm³ of distilled water and 50 g of hexamethylenetetramine $(CH_2)_6N_4$ is dispersed in another 400 cm³. The two solutions are mixed and diluted to 1 dm³. The suspension is then left to maturate at 20 to 22°C for 48 h. This standard suspension contains 400 NTU (which is equivalent to 1 JTU). The JTU is an empirical unit expressing visual observation of the opacity of a fluid. The unit is more appropriate to concentrated suspensions than to more clear liquids. 1 JTU corresponds to ± 400 FTU or NTU.

The SiO_2 standard corresponds to a suspension of standard silica. Unfortunately, different "standard materials" exist on the market. The most typical, corresponding to the prescriptions of the American Public Health Association (APHA), can be obtained by the dilution of a stabilized suspension containing 200 ppm SiO_2 as marketed by Hellige (877 Stewart Avenue, Garden City, NY). The advantage of the SiO_2 standard for turbidity is that it exists in the form of a opaceous glass, which makes it possible to construct indefinitely stable instrumental standards. These standard glasses, marketed by Sigrist, range down to 1 $SiO_2 \simeq 3.2$ JTU when studied using a Sigrist T 65 turbidimeter (see Fig. 6).

The leading parameters influencing the reliability of turbidimeters for continuous measurement are:

1. The reflection effects of the cell walls.
2. The Corona effect (i.e., the multiple diffusion). When there are high turbidity

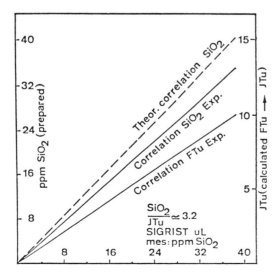

Figure 6   Correlation of JTU (NTU) and SiO_2 turbidity units.

values, linear discrepancies can occur due to the Corona effect, reducing the dynamic range of the instruments.

3. The diffusion angle which influences the measured values (discussed earlier).
4. The optical pathway, which has a determining effect on the reliability of turbidimeters for continuous service.

Single- and double-beam apparatus exist as illustrated in Fig. 7. In the single-beam instrument the measured value can depend on aging of the light source and photocell. These phenomena are eliminated in the double-beam arrangement. A servomotor activates a blend system to produce the same intensity on both light trajectories. Different measuring ranges are available with commercial instruments

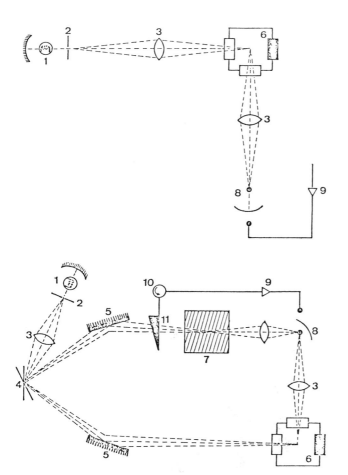

Figure 7   Optical pathways of single- and double-beam turbidimeters. 1, Light source; 2, slit; 3, focusing lens; 4, alternating mirror; 5, fix mirror; 6, measuring cell; 7, standard turbidity; 8, photodetector; 9, transformer-amplifier; 10, electromotor; 11, compensating blend.

fluctuating from 0–0.5 ppm SiO_2 to 0–200 ppm SiO_2. The frequency of adjustment and compensation of the light intensity in both trajectories corresponds to the frequency of oscillation of the mirror and enables the detection of instrumental deficiencies. Most Sigrist instruments are of the double-beam type.

The Hach 1720 A apparatus is illustrated in Fig. 8. It is a single-beam nephelometer calibrated in NTU. The incident light beam is directed vertically into a water column, which is renewed by water upflowing continuously. The general configuration of the apparatus does not make it possible to check easily whether measurements are linear throughout the entire range, but in its version 2100 A for laboratory use, the measurements are linear up to 15 JTU (40 ppm SiO_2).

In the Fischer DRT 1000 instrument (Fig. 9), the instability in light intensity as emitted by the source is automatically compensated through simultaneous measurement of transmitted light. The instrument is designed fundamentally for laboratory use, yet can be equipped with a continuous-flow cell, which, however, remains less suitable for raw water analysis. Except in this case the apparatus is of a good linear correlation up to 15 JTU.

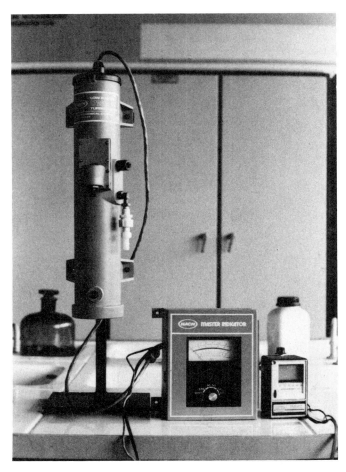

Figure 8   Hach 1720 A turbidimeter.

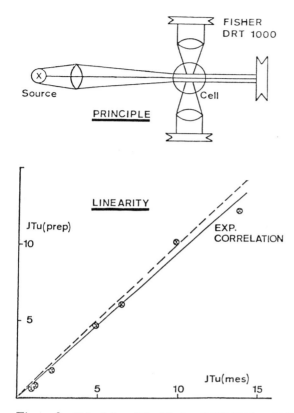

Figure 9   Principles of the Fischer DRT 1000 turbidimeter.

EUR-Control's Mex 2 instrument (Fig. 10) is based on a submersible cell unit. The equipment has the great suppleness inherent in the dip-cell principle and an elaborate electronic and optical system that compensates for the instrumental causes of instability. The concept of optical pathways (modulated comparisons of x-intensities) does not make it possible to obtain very low values (e.g., 0.7 JTU). In this case a high amplification factor is necessary.

2.4   Streaming Current Detection

The streaming current is one of the electrical characteristics of colloids that can be recorded continuously. The principle of measurement is the following: If a colloidal particle is immobilized on the wall of a pipe through which liquid flows, some of the counterions can be mechanically swept by the movement of liquid. This movement can be attained, for example, by a loose-fitting piston driven by a synchronous motor (e.g., at 5 Hz, which is one of the commercial versions).

The relative movement of the counterions to the colloid promotes an electrical current through the annular electrodes inserted in the walls of the plastic pipe (Fig. 11). This current is amplified and redressed to give a measurable output. This "streaming current" is an empirical value without any direct relationship to the fundamental characteristic of the colloids. Its intensity is proportional to the num-

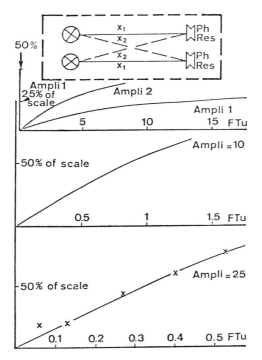

Figure 10   Principles of turbidity measurements with Mex 2.

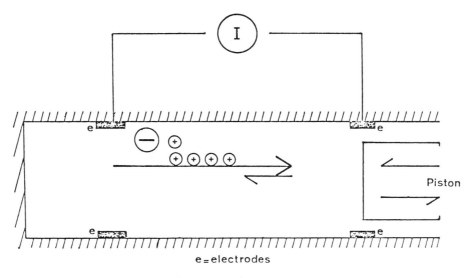

Figure 11   Scheme of a streaming current detector.

ber of counterions swept off from the double layer. Particles that are moved together with their counterions do not contribute to the streaming current because, as a whole, they behave as being electrically neutral.

For homodisperse suspensions and in case of no or only slight changes in pH and ionic strength of the medium, the streaming current can be empirically correlated with the necessary dose of coagulant. In the case of naturally occurring raw waters that are variable and polydisperse, the principle is less reliable. Its application to specific and more concentrated industrial effluents (e.g., painting and coating manufacturing) is promising.

2.5 Amperometric Analysis of Residual Oxidants (3,4)

Amperometric analysis is based on measurement of the diffusion current obtained between a sensor electrode and a reference electrode, both dipped into the same sample water: The current intensity is proportional to the depolarizing elements or product diffusing to the sensor electrode. The basic criteria to be considered in the selection of an apparatus are the current intensity correlated with the concentration of the diffusing element and the absence of residual current intensity (i.e., no current in the absence of the depolarizing compound to be detected). The method is generally applied for residual chlorine and residual ozone.

A widely used electrode couple is Au/Cu. Its reactions are reliable for free chlorine but less so for combined chlorine (e.g., chloramines). On adjunction of KI the electrode couple Au/Ag–AgI releases a depolarization current proportional to those concentrations of oxidants capable of oxidizing iodide into iodine. The general principle of the electrode reactions is illustrated in Fig. 12.

The optimal pH range for the development of good depolarization is between 4

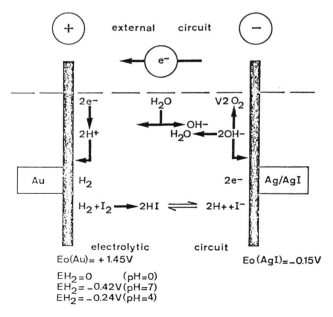

Figure 12 Principles of the amperometric determination of residual active oxidants.

and 5 and is best obtained by the injection of a solution of acetic acid. To develop optimal conditions, the ratio iongram I^- to moles of Cl_2 (active) must be higher than 5. The stock solution for sample water conditioning is of 100 L process water in which are incorporated 80 g of KI and 9 dm^3 of CH_3COOH (conc.). This solution is dosed into the water flow under analysis in a flow-through cell equipped with Au/Ag electrodes so as to obtain a final pH of approximately 4.5.

All forms of active chlorine and oxidants oxidizing iodide to iodine are detected; these are free chlorine, mono- and dichloramine, chlorine dioxide, ozone, and so on. Chlorite and chlorate ions do not interfere. Where the specific detection of ozone dissolved in water is concerned, a similar detector based on Ni/Ag electrodes corresponds to the mechanism indicated in Fig. 14. The same analyzer can also be used for the specific measurement of residual chlorine dioxide in water.

An important feature of analyzers is the fact they are fed with sample water so as to avoid flow-through of bubbles in the cell. See Fig. 15, where a suitable arrangement is shown. It is also important that the electrodes be sufficiently distant from one another to eliminate capacitive effects. In the design developed for these specific analyzers at the Brussels Waterboard, the electrodes are curve blades, with a total surface of 2×30 cm^2 each. The diffusion equilibrium is maintained by a rotating brush, which also cleans the electrodes continuously.

For the analysis of water of drinking water quality, in general, monthly calibration is sufficient and the general maintenance period of the electrodes obtained exceeds 1 year. If high bromide content of the water is possible, the reliability of the Ag/AgCl electrode should be investigated.

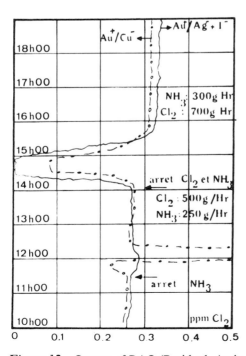

Figure 13   Output of RAO (Residual, Active Oxidants) analyzer.

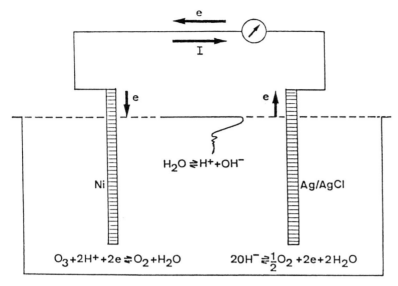

Figure 14   Principle of the amperometric residual ozone analyzer.

2.6   Radioactivity Control

Geiger–Müller counters are available as dip counters and make it possible to survey concentrated radioactive effluents. The sensitivity yet achievable is on the order of 200 pCi/dm^3 or 8 Bq/dm^3; however, this is not appropriate for the survey of drinking water treatment processes. With a continuous flow-through Marinelli cell, one can survey the total γ-activity of a specific γ-emitter. The method requires long counting periods and is better adapted to the control of nondiluted effluents derived from nuclear sites.

2.7   Summary

Automated water quality control does not yet enable plants to operate on a complete closed-loop basis (5,6). The reasons are therefore of different kinds: lack of knowledge of the overall relationship between measurement and reagent dosage; significant delay times in most processes (e.g., coagulation–flocculation), which prevent fast, regulated dose–effect response; and lack of reliability of some sensors. However, some processes can be fully automated (e.g., pH correction, dosing of oxidants, discharge of water with insufficiently low turbidity). The development of computerized "expert systems" is in progress but reaction kinetics is still more accurately known.

3.   TESTING METHODS RELATED TO PARTICULATE SYSTEMS

3.1   Turbidity Removal: Jar Test

In the jar test, samples of raw water are placed in a series of vessels (usually, four to six) and stirred with various doses of chemicals used in the coagulation process.

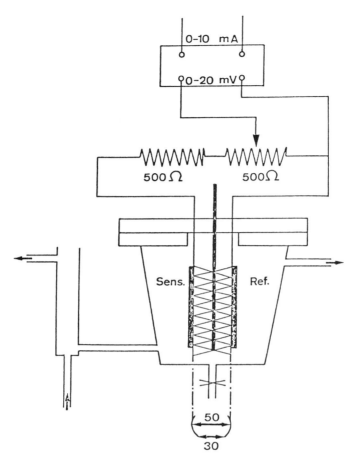

Figure 15   Amperometric measuring cell.

Preliminary addition of oxidants and dosing of acid or alkaline reagents for pH adjustment enables us to simulate as close as possible full-scale conditions in a plant. During the addition of coagulants, a rapid mixing period is used and care must be taken to add the coagulant in a single injection. The reagent is to be introduced into the active mixing zone of the impellers. The first period of stirring, from a lapse time 30 s to a maximum of 2 min, is based on the risk of floc breakage. In conventional techniques the velocity gradient during initial rapid mixing is to be fixed at 200 to 400 s^{-1}. It can be evaluated according to the methods described in Chapter 17.

In average Belgian river waters good general conditions are the following: $G \sim$ 300 s^{-1}, $T \sim$ 30 s (GT value $< 10,000$). To enable comparison at variable dosings, it is essential to employ the same mixing conditions for all samples compared. After the rapid mixing period, mixer speed is set at lower values to be adjusted by trial and error, according to visual observation of floc growth or breakage. As a preliminary guideline, $G \sim$ 25 s^{-1}, $T \sim$ 15 to 30 min can be adopted for lime softening as well as for coagulation. Once again it is essential to maintain comparative conditions for all samples and dosings to be correlated. After the period of slow mixing, the stirrers are stopped and the floc allowed to settle for a period ranging from 15 min to 1 h.

Initial evaluation of the results of a jar test as used in coagulation or softening remains essentially subjective. Most often, the basis used for choosing the optimum chemical dosage is the clarity of the supernatant liquid. The supernatant liquid can be pipetted or siphoned and analyzed for turbidity, color, and so on. Siphoning is a very delicate procedure during which loose flocs can be sucked into the supernatant. Use of a pipette is recommended, with the end bent to 90°. The supernatant must be taken at constant depth (e.g., at 30 mm) to enable comparison of samples.

For manual turbidity measurements the probes are best acidified by HCl to reach a pH of 2.5 (7). Analysis of the residual coagulant concentration in the supernatant of a settled floc suspension provides valuable information. As a sludge blanket is absent in the jar test, this evaluation is often high compared to the needs in the plant. In the very beginning of the flocculation phase there is usually an increase in turbidity, as indicated in Fig. 16. During the process of coagulation–flocculation, particles are aggregated, and as a consequence, turbidity can increase temporarily (2 in Fig. 16). After a variable period of time, flocculation occurs (3 in Fig. 16), ending in settling (4 in Fig. 16) and quality improvement, which means better turbidity in the finished water (5 compared to 1 in Fig. 16). In some cases (e.g., for very turbid or charged waters) settling starts during the flocculation phase (6 in Fig. 16). Automation of the process is attempted by correlating the intermediate turbidity maximum to the final result (see Fig. 16B). The method still requires further investigation, particularly for the case where coagulants other than aluminum sulfate are used.

Visual evaluation of the jar test remains the principal technique of the method, using one of the following observations (8,9):

Speed of floc formation: measurement of the time between the coagulant addition and the appearance of visible floc

Empirical floc-size comparisons: qualitative observation of the size of the floc as it forms, without considering floc strength and density

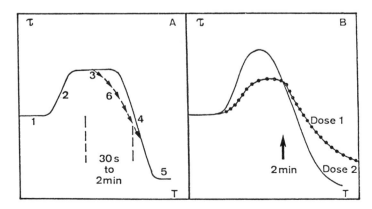

Figure 16 Turbidity during the flocculation evaluated in a jar test.

3.2   Filtrability Tests

Simple filtration tests without absolute significance make it possible to obtain an empirical relative evaluation and comparison of the results in experimental procedures such as the jar test.

The *silting index* (SI) is based on the use of membrane filters of 0.45 μm porosity employed in constant-pressure filtration.

$$SI = \frac{\text{time to filter second 5-mL sample} - \text{time to filter first 5-mL sample}}{\text{time to filter first 1-mL sample}}$$

The *filtrability number* (FN) also uses 0.45-μm membranes.

$$FN = \frac{\text{time to filter 200 mL jar test supernatant}}{\text{time to filter 200 mL distilled water}}$$

In the *membrane refiltration time* (MRT), double-stage filtration is used. A sample is first filtered on a 0.45-μm membrane and the filtrate passed a second time on the same membrane with its preexisting deposit from the first filtration. The MRT is found to correlate adequately with pilot-scale filter experiments, and can be used to optimize the flocculation–filtration process.

In pilot-column filtration tests, as well in full-scale plant operation, the resistance of a floc particle flowing through the filter is indicated by the *floc strength index* (FSI) according to Hudson (10). The FSI is equal to $h\,(ES)^3/l$, in which h is the head loss (in m), ES the effective size of the filter mass (in mm), and l (in m) the penetration depth of the floc or break-through thickness of the filter. For most rapid sand filters, FSI values are in the range 0.2×10^{-9} m. In the pilot-column filtration test (8), part of the sample water leaving the coagulation basin is directed to a small test filter column. By recording the turbidity of the effluent, the chemical dosing can be monitored for the flocculation–filtration practice. As the coagulant dosage increases, the length of the filter run reaches a maximum at optimum dosage. The technique is specifically oriented to flocculation–filtration.

The *break-through index* (BTI) is defined by the FSI $\times$ filtration rate.

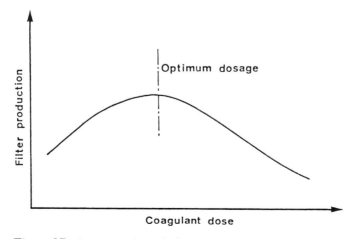

Figure 17   Interpretation of pilot-column filtration.

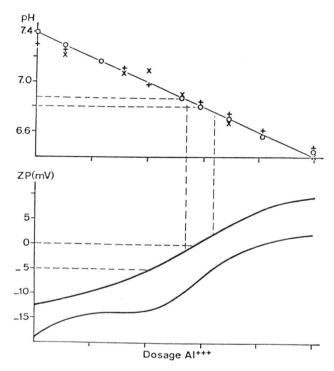

Figure 18   Correlation of pH and ZP in flocculated water. (From Ref. 11.)

3.3   Inorganic Analysis

The concepts of sweep coagulation as a significant or dominant mechanism of coagulation when applied to practical sludge-blanket settling systems make us consider the optimum pH range for precipitation of hydroxides as being optimum for coagulation–flocculation and settling. In the case of aluminum hydroxide, the optimum pH range may thus be fixed between 6.8 and 6.9. This most often also corresponds to a dosing rate by which the ZP of the colloids to be removed approaches zero (11) (see Fig. 18).

In many cases the pH control can be applied to control adequate flocculation. On coagulation with aluminum sulfate the electrical conductivity of the treated water increases due to the sulfate ions, the increase being proportional to the dosage. However, if part of the electrical conductivity of the original water is to be associated with colloids [e.g., concentrated algal suspensions (12)], a breakpoint can occur in the electrical conductivity versus coagulant dosage relationship (see Fig. 19).

3.4   Measurement of the Zeta Potential or
Electrophoretic Mobility (11,13–21)

Although highly criticized, until now the zeta potential (ZP) or electrophoretic mobility (EM) remains the basic principle underlying coagulation practice as applied

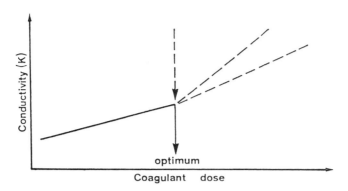

Figure 19   Possible incidence of coagulation on electrical conductivity.

in water purification. The movement of colloids suspended in water when placed in an electrical field is an early recognized fact. The electrophoretic mobility (EM) expresses the velocity of migration of the particles in an electrical field unit. This mobility can be related to the zeta potential (ZP) of the double layer by accepting some theoretical approximation.

Several electrokinetic phenomena that are involved are the following:

Electrophoresis: imposed electrical field causes migration of particles.
Electro-osmosis: imposed electrical field causes migration of solvent.
Sedimentation potential: arises from an imposed movement of charged particles.
Streaming potential: arises from an imposed movement of solvent through capillaries.

Table 1 summarizes some of the characteristics of the different phenomena.

In a simplified model, the double layer of a colloidal particle can be considered as a condenser in which the opposite charges are separated by δ, the very small thickness of the double layer. The potential of that condenser is identified as the zeta potential and expressed as $ZP = 4\pi\sigma\delta/D$, wherein D is the dielectric constant of the medium and σ the charge density of the colloid surface. The application of an exterior potential over a distance L creates a driving force $\delta E/L$ promoting displacement of the counterions. The force opposed to the motion is equal to $\mu\bar{u}/\delta$, where μ is the dynamic viscosity and $\bar{u}$ the velocity of the cloud of counterions.

A stationary state is attained when the driving force is equal to the resisting force, which is expressed as

Table 1   Characteristics of Movement in External Electrical Field

Phenomenon	Liquid	Solids	Potential
Electrophoresis	Motionless	In movement	Imposed
Electro-osmosis	In movement	Motionless	Imposed
Sedimentation potential	Motionless	In movement	Measured
Streaming potential	In movement	Motionless	Measured

$$\bar{u} = \frac{\sigma \delta E}{L\mu} = \frac{ZP \times DE}{4\pi\mu L}$$

Although this equation is a simplification, there is no doubt that for usual values $\bar{u}$ is strictly proportional to E/L. For this reason, the results are generally expressed in terms of electrophoretic mobility.

The electrophoretic mobility is the velocity for a potential gradient of 1 V/cm:

$$\bar{v} = \frac{\bar{u}}{E/L} = EM$$

For most natural waters the electrophoretic mobility is about 10^{-6} m²/V · s at 25°C. On the assumption of spherical particles, this is equal to

$$EM = \frac{ZP \times D}{4\pi\mu} = 7.8 \times 10^{-2} (ZP) \qquad \text{(in mV)}$$

and

$$ZP \text{ (mV)} = -13 \text{ (EM)} \qquad \text{(in } 10^{-6} \text{ m}^2/\text{V} \cdot \text{s)}$$

A typical value of electrophoretic mobility is that of human red blood cells: $-13.1 \pm 0.3 \times 10^{-6}$ m²/V · s

In the case of electro-osmosis the expression of the ZP is

$$ZP = (k) \frac{4\pi\mu}{D} \frac{Ql}{EA}$$

where (k) is the harmonizing constant depending on the units used, Q the flow of the liquid, l capillarity length of the porous tube, and A section of the tube.

In the practice of water treatment control, the following simplifying assumptions are accepted in current determination:

1. Particles are considered as spherical; a shape correction factor is not used.
2. The viscosity and dielectric constant in the double layer are considered to be the same as for the bulk of the liquid.
3. The electrical conductivity of the particles is considered to be lower than that of the liquid so as to avoid privileged passage of the electrical current through the particulate phase.
4. The relative velocity of the movement of the liquid versus the solids is considered to be laminar and inertia negligible.
5. The thickness of the double layer must remain negligible versus the particle radius and the pore size (in the case of diffusion measurements).
6. The relaxation time, which is the delay in the electrophoretic movement due to unsymmetrical deformation of the double layer, is neglected. This hypothesis is acceptable for values of ZP $\leq$ 25 mV.
7. In the electrophoretic methods, based on viewing of particle movement, only the particles moving in the stationary plane of electro-osmotic movement must be taken into account.

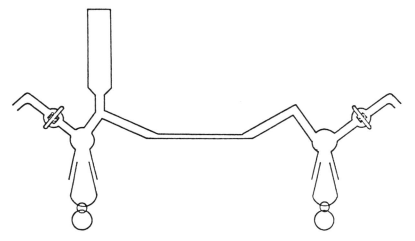

Figure 20   Briggs cell.

The usual techniques for experimental determination of electrophoretic mobility involve a microscopic viewing of particle displacement under the influence of an electric field. Practical techniques for measurement have been reviewed in the literature (18). Basically, the particle examination can be performed either in a flat horizontal cell connected to two electrodes such as the Briggs cell (Briggs cell from T.C. Cooper, P.O. Box 340, Route 1, Fenton, MI 48430) or in tubular cylindrical cells such as the Riddick cell (Zeta Meter Inc., 1722 First Avenue, New York, NY 10028) (Figs. 20 and 21). The Pen-Kem Z-meter is a more recent version based on *simultaneous* measurement on several particles (10 or more). The particle is viewed either by direct microscopic observation (e.g., Briggs cell) or by indirect observation through a stereoscopic microscope (Riddick cell). The optical characteristics of microscopes are summarized in Table 2.

Riddick claims the relative absence of scattering as an advantage of stereoscopic observation, but the Briggs cell allows a wider range of available optics. The cylindrical Riddick cell was found to be better adapted to determinations on fast-settling colloids than the horizontally oriented Briggs cell. The latter can, however, be oriented vertically with horizontal microscopic observation. The Riddick cell, being

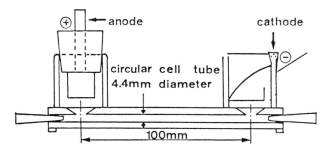

Figure 21   Zeta-Meter cylindrical cell.

Table 2

Cell	Ocular Min.	Ocular Max.	Objective Min.	Objective Max.	Total Min.	Total Max.
Briggs type	5×	25×	10×	40×	50×	1000×
Riddick type	15×	20×	4×	8×	60×	160×

constructed of acrylic material, is easier to carry. Electro-osmosis should be considered in particle viewing. When a dc voltage is used in a closed cell, depolarization of the water molecule takes place. The liquid moves along the wall of the cell to the cathode, while in the center the liquid backflow is directed to the anode. Cancellation of both components takes place in the stationary layers. Measurements of the traveling time of a colloid and the corresponding determination of electrophoretic mobility are to be carried out by observation in the stationary layer of the electro-osmotic movement.

According to the theory of Komagata, the position of the stationary layer can be calculated by following the equation

$$y = \pm \frac{h}{2} \sqrt{\frac{1}{3} \left(1 + \frac{389}{\pi^5 k} \right)}$$

where y is the distance of the stationary layer from the axes of the cell, h the height of the cell, and k the ratio of cell width to cell height (for rectangular cells). For values $k \geq 100$, y becomes $0.211h$ and $0.789h$, corresponding to the data of Smoluchowski. For cylindrical cells the counting "plane" is to be located at a distance of $0.147d$ (diameter) from the wall of the cylindrical tube. In all cases, the counting planes must be situated in the stationary plane of the electroosmotic movement. The rate of particle travel is measured by timing the particle movement through a

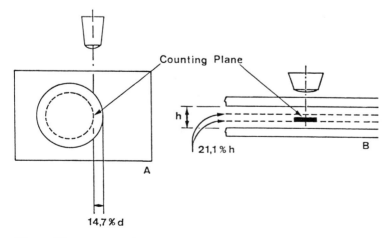

Figure 22 Isoosomotic planes.

few micrometers with the aid of an ocular micrometer. The electrical field is computed from the measured current intensity, the known dimensions of the cell, and the separately determined conductivity of the solution. Most manufacturers provide timing charts, thus eliminating the need for computation.

Electrical current should be maintained constant. The phenomenon of thermal overturn is observed as random movements of the colloids in all directions, making timing impossible. When the specific resistance of the liquid is low (e.g., 100 kΩ/ m), the tendency to thermal overturn is increased. To limit the perturbations due to thermal overturn, the observation time should be limited and the polarity of the electrical field regularly reversed.

The Pen-Kem Z-meter (Fig. 23) illuminates the colloidal particles horizontally through the side of the chamber by means of a 2-mW helium–neon laser. The 7-mm circular beam is compressed in the vertical direction with cylindrical lenses so that most of the laser output power is concentrated within the 5-μm depth of focus of the microscope. The width of the laser beam just fills the microscope field of view. Very few out-of-focus particles are illuminated and a dark-field image is maintained even at a concentration of 100 to 200 ppm. The laser beam is polarized so that the electrical vector is horizontal, thus maximizing the brightness of small particles, as observed at a 90° angle with the microscope. Particles with a high refractive index can be as small as 15×10^{-9} m.

The field strength can be adjusted from zero to 40 V/cm, a current value being 10 V/cm at an applied electrode potential of 100 V. The cell can be used manually by employing a syringe or continuously, with a transfer pump. In this case a 100-mL sample is required. The readout is expressed directly in terms of ZP based on Smoluchowski's equation for spherical particles. For manual operation a 3-mL sample is limiting. The apparatus is based on a rotating-prism technique utilizing an apparent zero movement by compensation of the electrophoretic mobility through the prism velocity. The latter is measured and expressed in terms of mobility or ZP.

In a forced-flow electrophoretic cell (19), mass transport occurs through a porous membrane under the influence of an external electrical field. The membranes are ordinary dialyzing membranes of viscose or cellophane; however, a wide variety

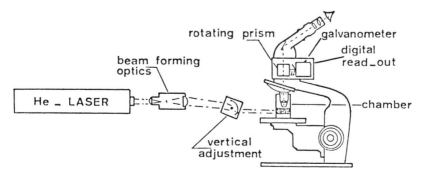

Figure 23   Pen-Kem Z-meter.

of materials can be used. The membranes allow the free passage of electrolytes but not the free flow of water or diffusion of colloids. The entire process is illustrated in Fig. 24. As long as the linear migration velocity (v_m) is equal to or larger than the linear flow of the liquid (v_f), the passage of colloids through the filter remains inhibited.

Analytical determination of the ZP by electrophoretic mass transport has been developed by the Micrometrics Instrument Corporation. The mass transport chamber (Fig. 25) consists basically of a reservoir, sample cell, reversible electrodes, and a shutter isolating the sample cell. To compensate sedimentation by gravity, the cell is rotated at 0.5 rev/s during the test. Basically, the mass transferred through the restricted opening of the sample cell is measured by weight determination after the period of testing. The method is applied to concentrated suspensions (e.g., raw water or wastewater), in which case the transfer time applied is about 300 s. In the technique the ZP is related to EM as follows:

$$ZP = \frac{4\pi}{D} \frac{\mu}{\mu_W} \left(1 - \frac{C}{\rho_s} \right)$$

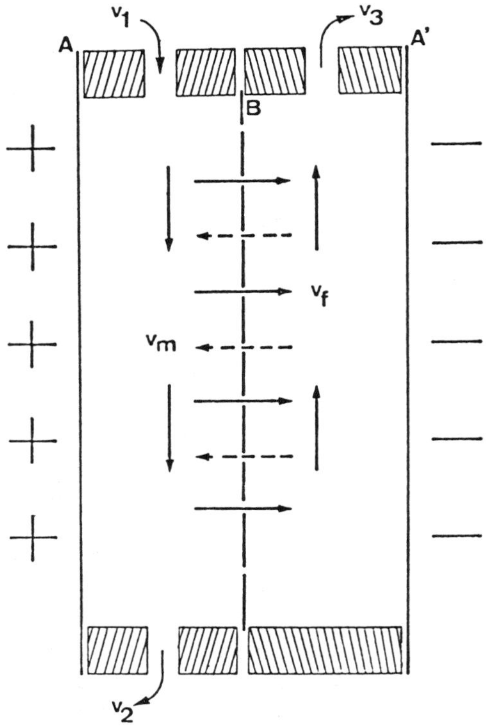

Figure 24 Forced-flow electrophoretic cell. Schematic diagram of a forced-flow electrophoresis cell: A,A′, membranes; B, filter; v_1, raw water inflow; v_2, waste outflow; v_3, purified water outflow; dashed arrows, direction of electrophoretic migration; v_f, linear flow of the liquid; v_m, linear migration velocity. (From Ref. 19.)

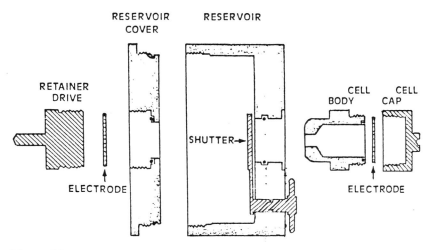

Figure 25 Micrometrics mass transport chamber.

where D is the dielectrical constant of the liquid relative to air, μ/μ_W the relative dynamic viscosity of the suspending liquid to that of water, C the concentration (kg/m^3) of solids in the sample cell, and ρ_s the specific weight of the solids (kg/m^3).

The electrophoretic mobility is evaluated according to the formula

$$\text{EM (m}^2/\text{V} \cdot \text{s)} = \frac{(\Delta w) \times k}{\Omega \times T \times I \times (C/\rho_s)\,(1 - C/\rho_s)\,(\rho_s - \rho_l)}$$

where Δw is the weight charge (kg) of the sample cell after a time of testing T (seconds), Ω the sample electrical resistance (Ω); I the current intensity (amperes); ρ_l the specific weight of the liquid, and k the cell conductivity constant (m^{-1}).

ZP measurement can be standardized according to the following procedure: Prepare stock solutions (A) 1 g/dm^3 of TiO$_2$ colloid (Pen-Kem standard), (B) 1 g/dm^3 KCl, and (C) 1.68 g/dm^3 Na$_4$P$_2$O$_7$, all in deionized water. One can prepare a standard solution by mixing 75 cm^3 (A), 30 cm^3 (B) and 300 cm^3 (C) with deionized water to attain 3 dm^3. The ZP of the suspension is -33 mV.

Ideal coagulation occurs at zero electrophoretic mobility or zeta potential. In practice, the limit of $V = -0.2$ (10^{-6} m/s) (V/ms) or ZP $= -2$ mV can be adopted for adequate coagulation. The real reduction to zero of electrophoretic mobility necessitates large amounts of coagulant. Excess coagulant brings about colloidal charge reversal. Some "residual negative colloidal charges" give place to good results in flocculation–filtration techniques (e.g., ZP $= -3$ mV). The experimental procedure involves the rapid addition of a given quantity of coagulant to the raw water to be treated. After a short period of rapid mixing (± 30 s), an aliquot is transferred to the measuring cell and 5 to 10 particles are viewed. From the observation, one can compute the zeta potential or electrophoretic mobility. The experiment is repeated with various amounts of coagulant and the optimum dosage is selected according to the plant's operational experience.

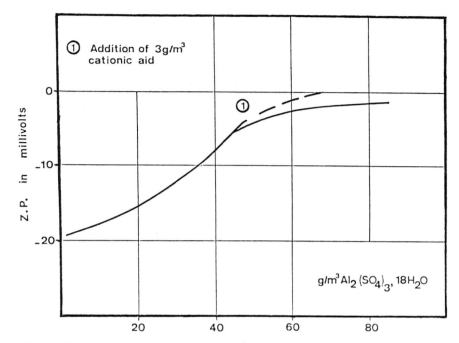

Figure 26 Treatment control by ZP.

3.5 Colloidal Titration

Colloidal particles of opposite electrical charges can virtually neutralize each other and produce neutral agglomerates.

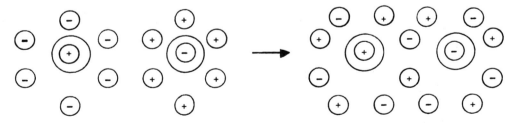

A stoichiometric precipitate can be formed near the isoelectric point. If an appropriate titration agent is found for a colloidal suspension, the reaction can be estimated quantitatively. To obtain this effect, Kawamura and co-workers (22,23) use an excess of positive colloid that is reacted with the natural suspension. The excess is backtitrated with a negative colloid in the presence of Toluidine Blue, which acts as a color indicator at the end of the reaction. The color reaction is based on the phenomenon of *metachromasia*.

Methylglycolchitosan (MGC), in which all the amino groups are fully quaternized, was selected by the authors as a positive polymer:

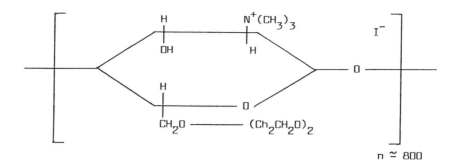

A solution of about 0.001 M quaternary ammonium of MGC is prepared for reaction. Potassium polyvinyl alcohol sulfate is selected as the negative titrant (PVSK), in solution approximately 0.001 M in potassium sulfonate function.

$$\left[\begin{array}{cc} CH_2 & CH \\ & | \\ & O \quad SO_3K \end{array} \right]_{1000}$$

The concentration of the Toluidine Blue indicator is 0.1 wt%.

Progressive neutralization of the colloidal charge during the coagulation can be estimated by gradual increase in the coagulant dose in jar tests followed by colloidal titration. The remaining charge has been correlated with quality parameters such as turbidity or treatment variables such as pH. The following operational relationship has been proposed:

$$D = K_1 A + K_2 C^a$$

where D is the required alum dosage (g/m^3 aluminum sulfate), A the total alkalinity (g/m^3 CaCO$_3$), and C the colloidal charge (10^4 Eq/m^3).

The best approximate relationship found by the authors is

$$D = 1.0A - 0.33C^{1.06}$$

(A is always approximately equal to 1.)

Fresh reagents should always be used, and where the colloidal titration technique is concerned, considerable skill is required. The results are generally less satisfactory with natural complex suspensions than with standard colloidal sols. Protected colloids do not react according the pattern above, and it has been found that detergents may interfere (24). The concentrations of anionic polyelectrolytes, and to some extent, cationic products of sufficiently high molecular weight can be evaluated by colloidal titration according to the above-mentioned technique. The sensitivity rarely attains 1 g/m^3.

3.6 Particle Counting and Sizing

Particle counting was first introduced to the control of water treatment processes with the Coulter counter (8,9). The principle of the instrumentation is illustrated in Fig. 27. The turning rotor forms velocity gradients G promoting collisions and the agglomeration of destabilized particles that facilitate subsequent growth of the floc. The current passes between two platinum electrodes, placed as illustrated in the figure, and is momentarily modified when a particle passes through the calibrated aperture. The change in current is detected by a voltage pulse. As for the change in resistance it is proportional to the volume of the particle passing the aperture. The instrument can be set to different size ranges, generally below 100 μm. The standard limit in water is of 70 μm. Other instruments are similar, but the particle movement through the calibrated aperture is obtained by vacuum. The instruments are generally useful in range 50 to 200 μm.

These techniques are less suited for the analysis of flocculating suspensions because of the principle of calibrated aperture. The addition of an electrolyte to condition the conductivity of the sample may have secondary effects on the suspension. The expansion of multielement optical detectors with computerized interpretation of the zonal output has made possible the development of a new generation of particle analyzers based on the forward diffraction of pulsed light. When a light beam falls on a particle, a diffraction pattern is formed, depending on the size of the particle. Small particles have a larger diffraction angle (Fig. 28).

The light not diffracted by the particle is focused by an appropriate lens to the central zone of the detector. The results are expressed as the percent of total particles having a given size range. Standardization is necessary with appropriate material delivered by the manufacturer (e.g., latex grains or cellulose particles). The dynamic range is equivalent to the particle diameter and can be set at a series of appropriate values: for example, 0–300 μm to 0–600 μm; however, cells for investigation are available up to 1.8 mm.

Particle sizing enables us to optimize the coagulant dosage and mixing condi-

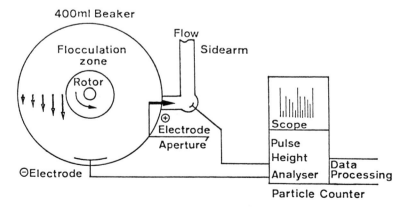

Figure 27   Principle of the Coulter counter.

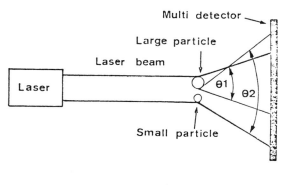

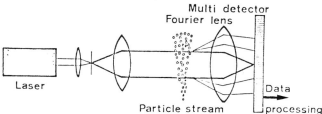

Figure 28 Principle of optical particle counting.

tions in order to remove a given fraction of particles (25). The conditions of sediment transportation in variable-flow periods are also under investigation with this technique.

4. TESTING METHODS RELATED TO SLUDGE EVALUATION

The physical condition of the sludge as obtained by flocculation or observed in sedimentation tanks, including secondary sedimentation, can provide for a measurement of good operation of related unit processes.

4.1 Floc Volume

Floc volume in the sludge layer of upflow settling tanks should reach 10 to 20% wet volume (e.g., ideally 15% after 1 h of settling under quiescent conditions). Too high a sludge concentration causes bulking and consequently channeling along selected passages. Sludge that is too loose causes there to be less sludge-to-liquid contact than that required for sweep coagulation.

With sludge compacting by vacuum filtration (as well as pressurized filtering), a general relationship correlating the vol % of wet sludge volume with the wt % of solids in the compacted cake is expressed as shown in Fig. 29. The relationship holds for sludge layers in flocculation–settling and for softening. The evaluation of flocculation aids and sludge-conditioning products for compacting can also be evaluated by comparison of the easily determined wet sludge volumes. From the data reported in Fig. 29 it appears that preliminary concentration of sludge by settling at levels higher than 30 to 40% wet volume does not significantly improve subsequent filtering performance in sludge compacting.

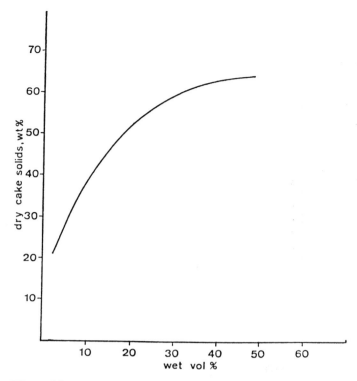

Figure 29 Correlation between wet sludge volume and dry cake content. (From Ref. 26.)

4.2 Sludge Cohesion Index

The *sludge cohesion index* (SCI) is a measure of the expansion ratio of a sludge layer in normal operation of upflow settlers. Besides the physical condition of the sludge, the chemical nature of the latter influences the index measured, as shown by the following experiment. About 50 mL of sludge is poured into a 250-mL volumetric cylinder and allowed to rest for about 10 min (see Fig. 30). A long-necked funnel is introduced into the volumetric vessel to about 100 mm from the bottom. Settled water (with the same pH and temperature as the sludge) is introduced discontinuously through the funnel so that the excess of water flows over the cylinder. As a result of this addition, the sludge layer expands and the additions of water are arranged to maintain a stationary sludge layer at 100, 125, 150, 175, and 200 mL, respectively. Time (T in seconds) is measured at these different stationary sedimentation states as the interval for successive 100-mL additions of water to the sludge layer.

The velocity or surface loading in m/h is equal to

$$v = \frac{3.6A \text{ (mm)}}{T \text{ (s)}} = (\text{SCI})\left(\frac{V}{V_0} - 1\right)$$

in which V is equal to the apparent volume of the sludge [at time intervals of 0 (V_0) to T]. The sludge cohesion index (SCI) is evaluated graphically by analyzing v as a

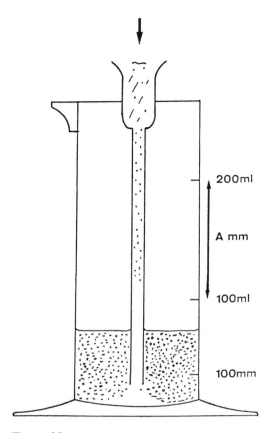

Figure 30   Experimental setup for determination of the SCI value.

function of V. The velocity difference for an increase of V_0 to $2V_0$ corresponds to the SCI value. A good sedimentating floc should have a SCI value between 0.8 and 1.2 m/h (Fig. 31).

4.3   Sludge Settlability

The *settlability of the sludge* has a practical interest for the dimensioning of sedimentation tanks. The testing should be applied in the region of hindered settling (27). A homogenized sludge suspension is introduced by vacuum suction into a cylindrical column of at least 0.1 m in diameter and 1.5 m in height and allowed to settle. The sludge height is recorded as a function of time (Fig. 32). The settling velocity s is deduced from the first part of the curve and determines the maximal surface load as $SL = Q/A$ (Q being the water flow, A the overflow surface). If starting from a sludge concentration S_0, the unit horizontal surface (A_u) that will be necessary to obtain a final concentration S (kg/m^3 dry weight), h_0 being the height of the liquid column, is

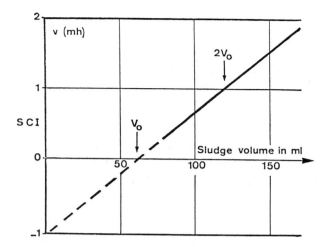

Figure 31   Graphical scheme for determination of the sludge cohesion index.

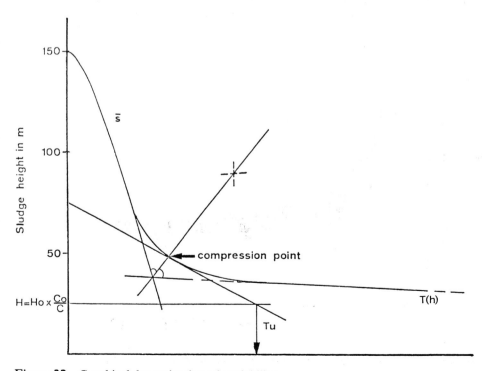

Figure 32   Graphical determination of settlability.

$$A_u = \frac{T_u}{S_0 \times h_0}$$

T_u being defined graphically as indicated in Fig. 32. The horizontal surface loading that will make it possible to thicken a flow Q is then given by

$$SL = A_u \times Q \times C_0$$

In some instruments the sludge limit in a continuous upflow system is detected by a photometer either by a moving detector along a vertical tube through which the liquid flows upward, or by changing the upflow velocity with the detector remaining fixed. This type of sedimeter was on the market during the 1970s (28).

4.4 Sludge Shear Strength

The rotating moment at variable duration of action can be measured with a modified rotating viscosimeter. The general setup is similar to that of the experimental flocculator (see Chapter 17), but the blades and baffles are in plastified grids of 150 cm^2 each; eight blades and four baffles are suitable for analyzing most settled sludges. The floc *shear strength* is indicated on a relative basis by the limit (asymptotic value) of the momentum at infinite agitation time. Generally, a good approximation is obtained after 120 s of agitation time and the values are in the range of 10^{-4} kg $\cdot$ m, while the preformed floc usually attains 5 to 10 $\times$ 10^{-4} kg $\cdot$ m. In the case of insufficiently preformed floc, an initial increasing phase can be observed before the decreasing phase occurs, due to particle fractionating.

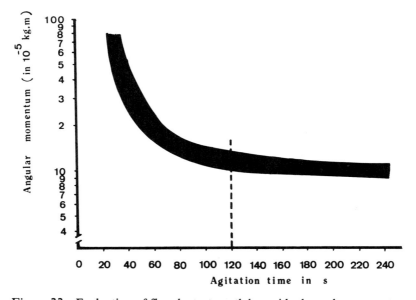

Figure 33 Evaluation of floc shear strength by residual angular momentum.

4.5 Capillary Suction Time

Capillary suction time (CST) is the measure of the potential for dehydration of a sludge in terms of the suction time of water on absorbing pads. The necessary equipment is manufactured commercially by the English firm Triton & Co. based on a development by the Water Research Centre. The principles are illustrated in Fig. 34. Recording of the capillary suction time at a precision of 0.1 s as a function of coagulant dose enables determination of the optimum concentration of the coagulant at which a minimum CST is reached. The results can be correlated with the filtration facility of the sludge and the method applied to the dosing of polyelectrolytes in sludge conditioning.

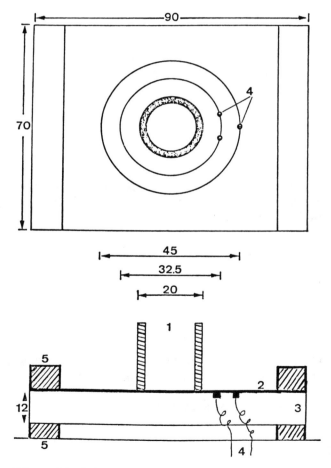

Figure 34 Principle of the CST analyzer (dimensions in mm). 1, Stainless steel feeding cylinder; 2, absorbing pad; 3, isolating acrylic sheet; 4, start–stop contact timer of 0.1 s precision for an electrical resistance of 50 kΩ; 5, stainless steel supports.

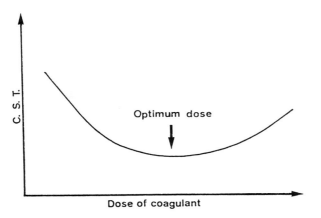

Figure 35 Influence of coagulant dosing (polyelectrolyte) on capillary suction time of sludge (schematic).

4.6 Sludge Filtrability

The *filtrability of sludge* is usually expressed as the specific resistance of filtration (r_s). The latter corresponds to the resistance of a cake laid on a surface of 1 m². By empirical equations developed by Carman and normalized by Coackley (29), the rate of filtration is

$$\frac{\delta V}{\delta t} = \frac{PA^2}{\mu r_s C(V + V_s)}$$

By applying the units as stated, the specific resistances r_s are expressed in s²/kg, but when expressing the pressure P in SI units, the r_s value must then be multiplied by 9.81 and expressed in m/kg as in the equation of Carman:

$\dfrac{\delta V}{\delta t}$ = rate of filtration, m³/s

P = charge of filtration, kg/m²

A = area of the filter, cm²

μ = viscosity of the filtrate, Pa · s

C = concentration of sludge, kg (dry)/m³

V = volume of filtrate, m³

V_s = equivalent volume of sludge, in m³, which would deposit an amount of solids to give a resistance to filtration equal to that of the filtering support

The integration of the preceding equation at constant pressure gives

$$\frac{t}{V} = V\frac{\mu r_s C}{2PA^2} + V_s\frac{\mu r_s C}{2PA^2}$$

of the general form $t/V = bV + a$, where the slope of a plot of t/V versus V is equal to b.

$$b = \frac{\mu r_s C}{2PA^2} \quad \text{or} \quad r_s = \frac{2bPA^2}{\mu C}$$

To evaluate r_s, the apparatus according to Coackley (Fig. 36) has proven to be appropriate when operated at a P value of 0.5 bar positive pressure. Wire disks are available for facultative support of the filtering membrane.

b is evaluated graphically by plotting t/V versus V and r_s computed from this value. The probable error of a single observation of b is $\pm 3\%$, that on A^2 can attain 25% and on μ can be on the order of 5%. Consequently, the standard deviation on the specific resistance must be considered as at least 25%. Measurement of specific resistance is particularly useful to define the best conditions for sludge treatment (30).

4.7 Sludge Compressibility

Sludge compressibility expresses the decrease in the cake permeability when the pressure applied in the filtration is increased. The relation is of the type $r_s = r_{s,o}(\Delta P)^s$, in which s is the sludge compressibility (always less than 1). Highly compressible sludges are characterized by $0.8 < s < 1$ and usually dot not filter well; while materials with $0.3 < s < 0.5$ are constituted by well formed flocs with high shear strength and do filter well. For highly mineral granular sludges s can approach zero. The apparatus and procedure for the determination of sludge compressibility are the same as for the determination of the specific resistance but with appropriate pressure control.

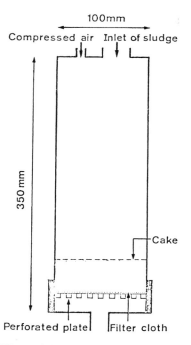

Figure 36   Coackley filter for the determination of r_s.

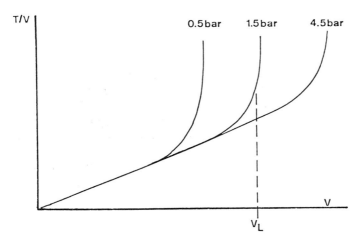

Figure 37 Determination of limiting volume in the dryness test.

4.8 Limiting Dryness

Limiting dryness is the "dryness" obtained on pressure filtering of a sludge during an infinite time. Experimentally, the determination is carried out like for the specific resistance, but the curve of t/V versus V is recorded at different pressures (Fig. 37). The analysis of the incurvation in the graphs enables us to determine the optimal pressure required to reach a given dryness.

SYMBOLS AND UNITS RELATED TO TESTING METHODS

d	diameter (of particles or tubes)
e	electrode
e^-	electron
h	head loss, m
h	height or thickness of an electrophoretic cell, m
h	height, m
k	ratio width to height of an electrophoretic cell, dimensionless
(k)	harmonizing constant for units
l	length, m
m	membrane, porous
n	number of electrons or equivalent charge
n	light diffraction order
r_s	specific resistance to filtration, s^2/kg or m/kg
s	sludge compressibility, dimensionless index
$\bar{s}$	settling velocity, m/s
t	(reference) temperature, °C
t	turbidity
t	time, s
$\bar{u}$	ionic mobility or velocity of ionic motion, m/s
v	velocity of flow or transit, m/s
y	distance of stationary layer, m

A	cross-sectional area, m^2
A	total alkalinity, g/m^3 $CaCO_3$
Au	unit (empirical) surface, m^2
BTI	break-through index, m^2/s
C	concentration, kg/m^3 or mol/m^3
C'	colloidal charge, Coulombs
CST	capillary suction time, s
D	dielectrical constant, $C/V \cdot m$
D	dosage of chemical required, g/m^3
E	electrical potential, V
E	electrolyte
EM	electropheric mobility, $m^2/V \cdot s$
E_R	reference electrical potential, V
ES	effective size, mm
FN	filtrability number, dimensionless
FSI	floc strength index, m
FTU	formazin turbidity unit
G	velocity gradient, s^{-1}
H	height, m
I	light intensity, relative units
I	current intensity, A
JTU	Jackson turbidity unit
K	electrical conductivity, $m\ S\ m^{-1}$
L	length, m
MRT	membrane refiltration time, seconds
(M^{n+})	ion concentration of element M
NTU	nephelometric turbidity unit
ORP	oxidation–reduction potential, V
P	pressure, N
Q	flow, m^3/s
RAO	residual active oxidants, equivalent active chlorine
SCE	standard calomel electrode
SI	silting index, dimensionless
SiO_2	silica turbidity unit
SL	surface loading, m/s or m/h
V	volume, m^3
V_L	limit volume, m^3
W	power, watt
ZP	zeta potential, usually in mV

δ	(elementary) distance (e.g., between microscopic charges)
ΔW	change in weight, kg
θ	diffraction angle, deg
κ	cell conductivity constant, m^{-1}
λ	wavelength, nm
μ	dynamic viscosity, $Pa \cdot s$; $N \cdot s/m^2$
ρ_l	density of liquid, kg/m^3
ρ_s	density of solids, kg/m^3

σ charge density, C/m^2

Ω electrical resistance, Ω

REFERENCES

1. W. Masschelein, *Bull. Anseau, 162,* 3 (1968).
2. W. J. Masschelein, *TMS Eau, 75,* 419 (1980).
3. W. J. Masschelein and G. Fransolet, *Trib. Cebedeau, 32,* 31 (1979).
4. W. J. Masschelein, G. Fransolet, R. Goossens and L. Maes, *Analusis, 7,* 432 (1979).
5. A. Frazer, Special Subject 2, *IWSA Congress,* Kyoto, 1978.
6. W. M. Lewis, *J. IWE, 2,* 137 (1968).
7. J. Bratby, *Coagulation and flocculation,* Upland Press Croydon, England, (1980), p. 272.
8. R. J. Te Kippe and R. K. Ham, *J. AWWA, 62,* 549, 620 (1970).
9. S. A. Hannak, J. M. Cohen, and C. G. Robeck, *J. AWWA, 59,* 1149 (1967).
10. H. E. Hudson, *J. AWWA, 40,* 868 (1948).
11. H. Bernhardt, *Jahrb. vom Wasser, 32,* 193 (1965).
12. J. L. Povoni and W. F. Echelberger, *Water Sewage Works, 118,* 17 (1971).
13. Th. M. Riddick, *Chem. Eng., 26,* 121 (1961); *10* 141 (1961).
14. A. P. Black and A. L. Smith, *J. AWWA, 54,* 926 (1962).
15. F. Edeline, *Trib. Cebedeau, 231,* 72 (1963).
16. A. P. Black and J. V. Walters, *J. AWWA, 56,* 99 (1964).
17. E. L. Beam, S. J. Campbell, and F. R. Anspach, *J. AWWA, 56,* 214 (1964).
18. R. M. Rock and N. C. Burbank, *J. AWWA, 58,* 676 (1966).
19. M. Bier and F. C. Cooper, *Principles and Applications of Water Chemistry*, J. Wiley, New York, 1967, p. 217.
20. G. Schulz, *Water Waste Treat. J., 19,* 121 (1969).
21. R. D. Letterman and R. D. Tanner, *Water Sewage Works,* 62 (1974).
22. S. Kawamura and Y. Tanaka, *Water Sewage Works,* 348 (1966); *Water Sewage Works,* 324 (1967).
23. S. Kawamura, G. P. Hanna, and K. S. Shumate, *J. AWWA, 59,* 1003 (1967).
24. W. J. Masschelein (unpublished).
25. C. H. Tate and R. R. Trussel, *J. AWWA, 70,* 691 (1978).
26. R. J. Calkins and J. T. Novak, *J. AWWA, 65,* 424 (1973).
27. Report Cost 68, supplement to No. 1 of *Tech. Sci. Munic.* (1976).
28. Anon. *Tech. Eau, 327,* 67 (1974).
29. J. D. Swanwick and M. F. Davidson, *Water Waste Treat. J.,* 386 (1961).
30. W. J. Masschelein, J. Genot, C. Goblet, R. De Vleminck, and L. Maes, *Trib. Cebedeau, 439,* 287 (1980).
31. W. J. Masschelein, *Gas/Wasser/Abwasser, 65,* 310 (1985).

12
Adsorption

1. INTRODUCTION

1.1 Adsorption

Adsorption is a term used to describe the existence of a solute concentration (dissolved substance) at the interface between a fluid and a solid higher than that present in the fluid. Adsorption is sometimes classified as *physical adsorption* or *chemisorption*. In physical adsorption the van der Waals forces act between the adsorbed compound and the adsorbing substance. During the process, the heat liberated is between 8 and 21 kJ/mol. These energies fall in the range of diffusion-controlled processes. There is no activation energy, but electrostatic forces can intervene. In chemisorption, there is chemical binding between the adsorbent and the adsorbed substance. This leads to modifications in the molecular structures, so that the liberated heat ranges around 40 to 200 kJ/mol, resulting in activation energy. Consequently, chemisorption takes place more quickly at higher temperatures. Since physical adsorption and chemisorption are both exothermic processes, adsorption is *quantitatively* favored at lower temperatures, according to the principle of Le Chatelier–Van't Hoff.

(*Note*: The term *absorption* means a process through which the absorbed substance does not remain at the surface of the absorbent but diffuses throughout the absorbent's internal material. As the process is merely applicable to the fixation of gases to solids, it will not be considered further here.)

1.2 Adsorbants

Although several materials can be used for adsorption techniques in water treatment—alumina (Al_2O_3), silica gel, Fuller's earth and diatomaceaous earth, molecular sieves, macroporous resins, basic ion-exchange macroporous resins, manganese dioxide, and even "aluminum sulfate flocs" and activated silica—*activated carbon is by far the most adsorbent.*

1.3 Historical Aspects of the Use of Activated Carbon

In 1773, Scheele observed the adsorption of gases, and in 1785, Lowitz observed that of solutions. However, it should be mentioned that a 4000-year-old Sanskrit treatise called "Ousruta Shanghita", stated: "It is good to keep water in copper vessels to expose it to sunlight and to filter it through charcoal."

The specific purifying properties of carbons have been known for thousands of years, but the first commercial application appears to have been in the cane-sugar industry at the end of the eighteenth century (1). Purely by luck it was observed that bone black (bone charcoal), used in the manufacture of boot polish, had a marked decolorizing effect. This led to its widespread use for this purpose in the sugar industry as well as to the beginning of the activated carbon industry.

The use of carbon to dechlorinate water dates back to 1910, when the first municipal filter unit was installed at Reading, England. Brown coal (lignite) was used and it is reported that the plant was working satisfactorily 20 years later. The filters were still charged with the original material; reactivation had not been necessary. It was concluded at that time that as far as dechlorination was concerned, these filters were acting as catalytic reaction promoters rather than as adsorption units.

Activated carbon was introduced in the 1920s and 1930s in Germany for dechlorination, as distinct from carbonized materials such as lignite and coke. Typical large-scale units were installed at Aussig and Dresden and were used at Hamm to remove tastes derived from phenols in Ruhr river water. In the United States, activated carbon was used in 1928 to dechlorinate water in Chicago.

Hard, highly active granular carbons have been developed from coal, and on-site reactivation of this type of carbon has greatly reduced the cost of adsorption as a unit process. Since 1960, new municipal water installations—in particular in Germany, Holland, Denmark, England, and the United States—have proven granular carbon filtration to be practical, economical, and essential in water purification. The modern evolution of ideas in the field of adsorption is given in Table 1.

2. ACTIVATED CARBON

2.1 Preparation

2.1.1 *Materials*

Activated carbon can be prepared from various materials, such as peat, lignite, bituminous coal, charcoal, wood, and coconut shell. Activated carbon made from

Table 1 Use of Activated Carbon

Dechlorination	Since 1930
Taste and odor removal	Since 1955
Organics removal	Since 1970
Biological activated carbon	Since 1976

wood is usually called charcoal. The residues of carbonization are found to have a large pore volume on activation, and therefore a high internal surface area is available for adsorption.

2.1.2 Steam Activation Process

The most widespread process for the preparation of activated carbons used in water treatment is thermal activation. To begin with, the raw material is carbonized to obtain a coke with which steam is reacted in order to increase the pore volume. This is the result of a partial combustion releasing gaseous reaction products:

$$C + H_2O \rightarrow CO + H_2$$

$$C + O_2 \rightarrow CO_2$$

$$C + CO_2 \rightleftharpoons 2CO$$

In the *steam activation process* the reaction takes place between 900 and 1100°C. Control of the activation requires adjustment of temperature, making it possible to diffuse the steam completely into the mass. The pore structure can vary according to the amount of steam and the temperature needed to produce carbon with a given porosity.

2.1.3 Chemical Activation Process

In the chemical activation process, dehydrating products such as $ZnCl_2$ or H_3PO_4 are used to abstract the water from the carbohydrates of the starting material. Carbonization is performed at 400 to 500°C and activation is carried out in the absence of air at 500 to 700°C.

2.1.4 Impurities in Carbons

In the steam activation process, the metals contained in the raw material, unless volatile during the process, will be present in the finished product. To reduce this inorganic content, acid washing of the carbon is carried out. Heavy metals requiring careful consideration are Cu, Pb, Zn, and As. The chemical activation process has its own washing phase. Even so, as far as the metal content is concerned, the purity of such a carbon is usually below that obtained using the steam activation method.

2.2 BET-Surface; Pore Structure

The *internal surface* of an active carbon is usually expressed as the BET surface in m^2/g (see Table 2). This surface can be determined according to the adsorption theory of Brunauer, Emmett, and Teller by measuring the saturation characteristics of the carbon with a single compound, such as nitrogen under pressure. However, there is no guarantee that this entire internal surface is available for the adsorption of organic compounds in water. Activated carbons used in water treatment, have an internal surface between 500 and 1500 m^2/g. In fact, the efficiency of an adsorbent is not necessarily directly related to the internal surface, but good carbons used as granules should attain about 1000 m^2/g.

The *pore structure* seems to be a more important parameter for adsorption than the total internal surface. The pores are conventionally classified according to their average radius:

Table 2   BET Surfaces for Various Materials (m^2/g)

Activated carbon	500–1500
Diatomeous earth	200
MnO_2	100–300
SiO_2 act	250
FeOOH	250–320
Freshly precipitated Fe-hydroxide	300
Aluminum hydroxide floc	50–100
Humic acids from soils	1900
Kaolinite	10–50
Illite	30–80
Montmorillonite	50–150
Calcite ($<2\ \mu$m)	12.5

Source: Some data from Ref. 2.

Macropores	$r \geq 1000$ nm	
Transitional pores	$r = 100$ nm	
Micropores	$r \leq 20$ nm	

A typical pore volume distribution as measured through the adsorption of benzene is given in Fig. 1 according to (3) as well as the saturation volume V_s for different grades. The transitional pores and micropores constitute the most important part of the internal surface ($\sim 95\%$). The macropores can be observed with a scanning electron microscope and evaluated by the penetration of mercury while the total volume is measured by helium or nitrogen penetration. The structure of the micropores is deduced from the adsorption characteristics of water according to

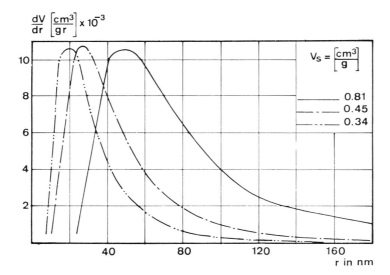

Figure 1   Volume distribution of activated carbon as a function of pore radius.

the Kelvin equation, given below. The macropores are relatively unimportant where adsorption is concerned but are necessary as conduits for rapid diffusion to the micropores.

Two different types of structure are to be distinguished where the pores are concerned: the cylindrical type (e.g., most activated carbons obtained from coconut shells) and the conical type (mineral carbons). At equal pore volume, the latter structure is more efficient for simultaneous adsorption of big and small molecules, while the former is often ineffective when obstruction occurs at the entrance with a larger molecule or a colloidal particle.

The Kelvin equation, which is a more absolute approach to quantifying the adsorption capacity, is given as

$$RT \ln \frac{P}{P_S} = \frac{2\gamma \cos \theta}{r} V_m$$

where V_m is the molar volume in the liquid state, γ the contact angle, and, θ the surface tension of the condensate.

2.3 Adsorption Characteristics

Testing of activated carbons involves the determination of a series of properties (4).

2.3.1 Isotherms

Activated carbon can be used in the form of either powder or granules. In both cases the adsorption characteristics must be considered. The adsorption equilibrium attained in a given case is more or less defined by adsorption isotherms. The Langmuir and Freundlich isotherm equations are the two most widely used models for the description of adsorption equilibria in this field. The Langmuir isotherm model is given by the following formula:

$$q = \frac{q_m b C}{1 + bC}$$

in which q is the amount of compound in g/m^3 or mol/m^3 adsorbed per unit weight of activated carbon in equilibrium with a residual concentration C, q_m is the amount

Figure 2   Pore structure (schematic). 1, Colloid or polymeric molecule; 2, large adsorbed molecules (e.g., detergents, dyes, humic compounds); 3, small adsorbed molecules (e.g., solvents, iodine).

of organic compound per unit weight of carbon forming a complete monolayer on the surface, and b is a constant related to the energy of adsorption. The Langmuir isotherm can be expressed in its normal form (Fig. 3a) or in a linear form (Fig. 3b).

For adsorption on a surface that conforms to all of the requirements inherent in derivation of the Langmuir model, the coefficients also have a defined physical significance. The converse, however, is not necessarily true. If an experimentally determined isotherm agrees with the Langmuir model, this does not necessarily mean that the underlying physical circumstances associated with the derivation of the model have been delineated.

The Freundlich equation is written as follows:

$$q = kC^{1/n}$$

in which k and n are characteristic constants and q and C are as defined above. A linearization of the equation is obtained in log form:

$$\log q = \log K + \frac{1}{n} \log C$$

The value of n is usually less than unity (e.g., between 0.3 and 0.7). Typical graphs are shown in Fig. 4.

An empirical three-parameter isotherm model that combines certain desirable features of the Langmuir and Freundlich models has been employed. The three-parameter model has the form

$$q = \frac{q_m bC}{1 + bC\gamma}$$

where γ is a constant relating to the specific carbon–organic compound system. Although empirical, the equation that generally best fits the experimental data in water treatment is the Freundlich adsorption isotherm.

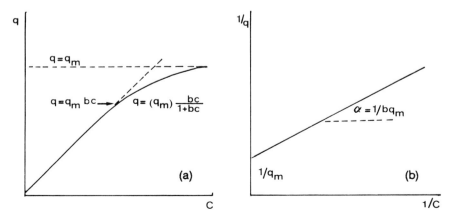

Figure 3 Langmuir adsorption isotherms (schematic).

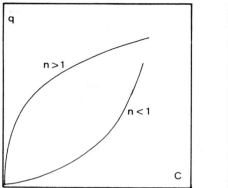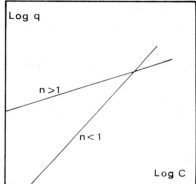

Figure 4 Freundlich adsorption isotherms (schematic).

2.3.3 Isotherm Test

The isotherm test can be carried out by equilibrating definite amounts of carbon with varied concentrations of a product to be adsorbed. For carbon in powder the commercial product is used as such, while granular carbon is best first ground to pass at least 95% on wet sieving on a No. 325 U.S. standard sieve of 44 μm opening size. The fineness of the product should have no influence on the adsorption equilibrium, but the adsorption velocity is reduced with coarse material.

Typical adsorption isotherms are given in Fig. 5. By establishing adsorption isotherms for several substances, both polar and less polar, an overall evaluation

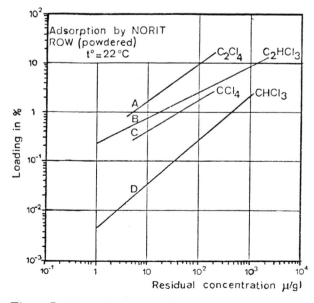

Figure 5 Freundlich isotherms of chlorinated hydrocarbons. (From Ref. 5.)

can be formulated. Substances used are proposed in the sequence phenol, iodine, indole, phenazone, and lauryl sulfate detergent (6). However, the best evaluation is still made with the products to be removed in each particular case.

2.3.3 Interpretation of Adsorption Isotherms

From an isotherm test it can be determined whether or not a particular purification can be performed. A rough estimate of the carbon dosage can be given. A large amount of data can be put in concise form for the evaluation of activated carbons by drawing isotherms (Fig. 6).

The q value, or loading, of carbon 1 is higher than that of carbon 2 over the entire range of concentrations studied. The lower isotherm of carbon 2 compared to carbon 1 indicates less adsorption, although at higher concentrations the adsorption yield is improved.

A carbon described by an isotherm with a steep slope (e.g., 4) is usually better suited for column or filter operation than for use in the form of powder. In contrast, a carbon with a lower $1/n$ value (e.g., 3 compared to 4) reveals less concentration dependence of the adsorption and may be more suitable for batch treatment (e.g., by using activated carbon in powder). For most carbons the value of $1/n$ is between 0.3 and 0.7. Adsorption of a substance is considered to be less efficient if $n < 1$. The adsorption isotherm for a given substance on a given carbon can be used for determination of the dose necessary to reach a desired minimum concentration of the solute.

In its nonlinear form (Fig. 7), the Freundlich isotherm indicates that with an initial concentration of solute of c_1, a quantity m_1 of carbon is necessary to reach a desired equilibrium concentration c_e. If for the starting concentration c_2, the corresponding quantity of carbon to reach the same equilibrium concentration is m_2, then

$$q_e \frac{x_e}{m_e} = \frac{c_1 - c_e}{m_1} = \frac{c_2 - c_e}{m_2}$$

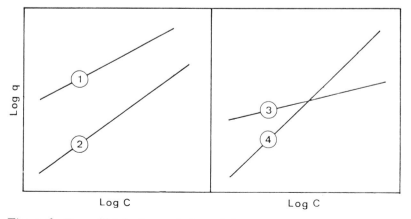

Figure 6 Freundlich isotherms (schematic).

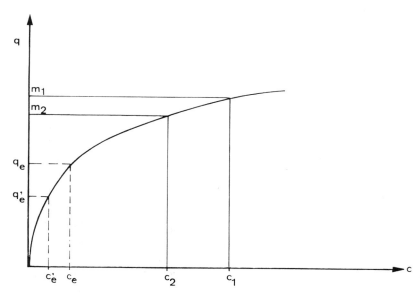

Figure 7 Freundlich isotherm (quantitative evaluation from the nonlinear form).

c_e should be minor in comparison to c_1 and c_2; hence $c_1/m_1 \simeq c_2/m_2$. Thus, as a first approximation: *To obtain the equilibrium concentration of* c_e, the amount of active carbon required is *proportional to the initial concentration of the solute to be eliminated*.

To improve the final quality, which means to attain $c'_e \simeq \frac{1}{2}c_e$ one has

$$x_e = c_1 - c_e$$
$$x'_e = c_1 - c'_e$$

The difference between x and x' is minor, whereas that between q_e/m_e and q'_e/m'_e, is significant, so that to improve the quality (to less than c_e), significantly higher quantities of carbon would be required.

2.3.4 Factors Affecting the Adsorption Isotherm

Temperature. For economic reasons, the temperature at which adsorption is carried out is the process temperature, corresponding to the raw water temperature. As already specified, the temperature has an effect on both the rate of adsorption and the equilibrium concentration. The effect is schematized in Fig. 8 and is often overlooked in experiments done at the temperature of investigation in the laboratory. The pH sometimes has a significant effect on the adsorption characteristics. In most cases the best results are obtained at the lowest pH value. This effect is particularly relevant when acid substances are to be adsorbed.

Competitive Adsorption. Substances simultaneously present can change the adsorption equilibria by mutual interaction. There may even sometimes be elution of a given adsorbed compound. The most typical example of this general phenomenon

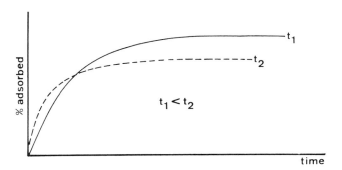

Figure 8   Effect of temperature on adsorption (schematic).

is the displacement of adsorbed phenol with p-nitrophenol in solution. This adsorption–desorption reversibility is of particular importance in view of the biological processes that occur in activated carbon filters (e.g., the filter can be loaded during winter periods of low biological activity and, consequently, desorbed in summer periods). Also in regeneration, chemical desorption can be used (see Section 5.2). In analytical work in which the adsorbed organic products must be recovered, an azeotropic combination of 47 wt% propylene dichloride and 53 wt% methanol extracts at higher rates than do chloroform and methanol (7). The experimental circumstances in which activated carbon testing takes place must correspond very closely to that of practical application. Laboratory tests in artificial media (water, temperature, substances adsorbed) are to be considered as preliminary information only. Each reasonable application for an active carbon should at least specify some indicative isotherms (e.g., for phenol, methylene blue, sodium lauryl sulfonate, and iodine).

Use of the equations of competitive mechanisms in enzyme reactions (30,31) leads to a linear expression of adsorption isotherms (Fig. 9). With different values of C, different values of loading q are obtained. A plot of $-C$ versus q gives at the

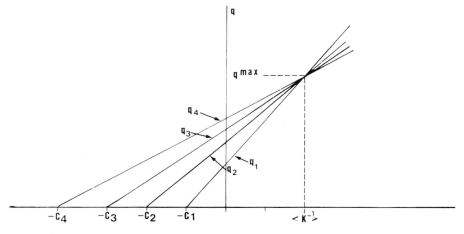

Figure 9   Normalized expression of isotherms.

intersection the average value of K (i.e., the adsorption equilibrium constant) and q^{max} adsorbable per unit of adsorbant.

2.4 Kinetic Aspects of Adsorption

2.4.1 Migration in Solution

The solute must be transported to the surface of the adsorbent. This effect depends on the particle size as well as on agitation of the solution. This phase is usually fast.

2.4.2 Film Diffusion

The solute must diffuse through a boundary layer fixed at the surface of the adsorbing particle. *The velocity of the mass transport through film diffusion is directly proportional to the external surface of the adsorbent, and hence related to the particle size.*

2.4.3 Internal Diffusion or Pore Diffusion

The substance adsorbed at the surface of the particle must be transported to the active adsorbing centers in the pores of the adsorbent (there is diffusion on the particle and in the pores). In the case of processes for which the rate-limiting reaction is adsorption on the exterior surface or transport through an external surface film, the rate is expected to vary, as the reciprocal diameter of the adsorbent particles for a given total mass of adsorbent. (Rate is of first order to surface area, which is inversely proportional to particle diameter. The pore diffusion rate varies as the reciprocal of some higher power of the diameter of the particle.)

Adsorption kinetics relating to the external surface of the particles is fundamental for the use of adsorption characteristics of active carbon. Specific surface area can be defined as being the portion of the total surface area available for adsorption. Thus the amount of adsorption accomplished by a unit mass of solid adsorbent is greater the more finely divided and the more porous is the solid (Fig. 10).

The transport phenomenon is usually considered to be a fast process, but film diffusion can kinetically control the entire operation, which under given conditions of granulometry, carbon type, temperature, and so on, approaches the first-order kinetics of the solute to be adsorbed. Practical mass transfer coefficients in film diffusion can be deduced from the initial adsorption velocities. Values ranging up to 24,000 s/m · mol have been recorded, and most organic products have a film diffusion rate of ≤ 2500 s/m · mol, which means that a contact time of more than 1 h is necessary to decrease the initial concentration by a factor of more than 90%. Adsorption–desorption reversibility can play an important role in the bacterial and biological life in activated carbon. Other general important properties are more specifically related to the comparison of powdered and granular activated carbon.

3. USE OF POWDERED ACTIVATED CARBON

Even though filter beds with granular carbon are employed more and more frequently for the treatment of more and more polluted water, the use of powdered active carbon remains a widespread technique for drinking water treatment. An active carbon dosing device can easily be incorporated in most existing flocculation-filtration equipment and the investment costs are much lower than those required for filtration on granular activated carbon (see Section 4).

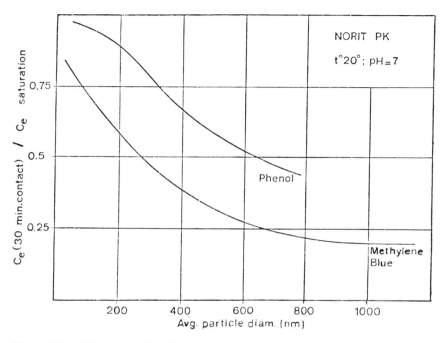

Figure 10   Effect of particle dimension on adsorption rate.

Powdered activated carbon is used for the general improvement of taste and odor and acts as a flocculation aid in the final phase of coagulation by appropriate dosing at this phase, which is at the pinpoint floc stage, in an upflow clarifier (e.g., a Pulsator or similar equipment). The effect of carbon concentration in the sludge is illustrated in Fig. 11.

In the sludge concentration zone of a clarifier the concentration factor can attain 70% (see Chapter 9). Hence when dosing 20 g/m³ raw water, a concentration of 1400 g/m³ is obtained in the sludge, improving the specific weight. Part of this improvement in flocculation must be attributed to the role of activated carbon as a bridging element in the coagulation process (21).

3.1 Dosing

Great progress has been made in dosing equipment and in the transfer procedure for powdered carbon. The flowsheet in Fig. 12 is that of equipment purveyed by the Dutch firm Norit. The carbon is transferred pneumatically from the transport tank. From time to time a pneumatic transfer from storage fills the dosage tank, which is equipped with an automatic weighing device. The pneumatic transfer is carried out from the dosage tank to a mixing tank. The frequency of transfer of a given quantity makes it possible to adjust the dose. A slurry is prepared in the mixing tank, enabling the final transfer of carbon to be accomplished as a "liquid." The concentration of the active carbon in the slurry should not exceed 10 g/L. As most active carbons have alkaline properties, the slurry concentration should not be

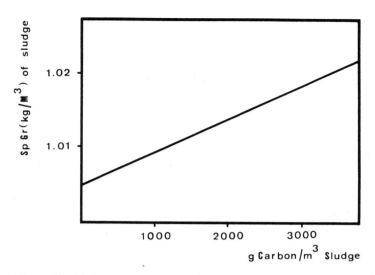

Figure 11   Effect of carbon content on the settling ability of sludge produced from Al-containing coagulants.

too high, to avoid precipitation of calcium carbonate obstructing the dosing lines. Should this happen, passivated hydrochloric acid can be used to clear the obstruction. The inner coating of the dosing and mixing tanks should be protected by chemically resistant enamel (e.g., epoxy). The risk of frictional electricity must also be considered.

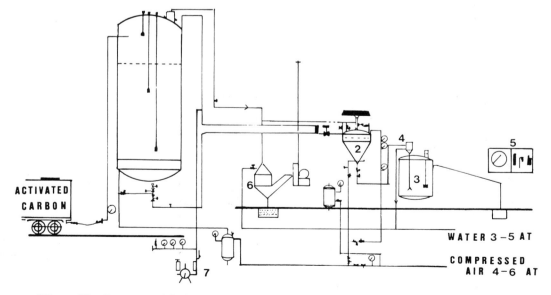

Figure 12   Storage and dosing equipment for powdered activated carbon.

3.2   Particle Size

The particle size of the carbon powder can greatly influence the rates of adsorption. Figure 13 illustrates the grades and size distribution that are normally available. A compromise is to be found between the higher rate of adsorption of fine carbons and the ability of elimination through filtration, or filtrability.

3.3   Contact Time

For the finer grades, a contact time of 30 min is generally sufficient, while for coarse grades 1 h is necessary. To determine the contact time for maximum efficiency, an aliquot of the solution (e.g., 1 L) is to be mixed with a given dose of active carbon (e.g., 20 mg to 1 g) and stirred for 5, 10, 20, 30, 40, 60, and 80 min. Having filtered off the carbon, the residual concentration of the undesired compound can be obtained. From a plot of percent impurity removed versus contact time an assumption can be made as to the required process contact time. Together with the adsorption isotherms for the substances to be eliminated the contact time for the various carbons may be tested on a comparative basis. All tests should be performed at process temperature.

3.4   Specifications

Besides the particle-size distribution and typical adsorption isotherm, the *internal surface* (500 to 800 m^2/g), the *water content* (<10 wt%), and the *apparent density* (400 to 550 g/L) should also be specified. Possible metal elements to be investigated are:

Copper	<	20 ppm
Lead	<	1 ppm
Zinc	<	20 ppm
Arsenic	<	2 ppm

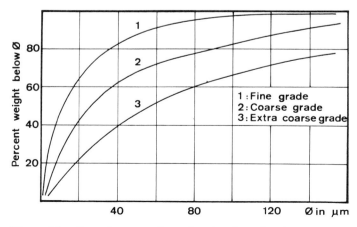

Figure 13   Granulometry of powdered activated carbons.

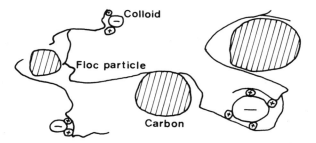

Figure 14 Carbon as a flocculation aid (schematic).

3.5 Bridging Coagulation

The carbon particles dispersed in the water may be considered to be larger than suspended matter. The carbon can be injected either before the filters or during the flocculation phase. Injection during the flocculation phase of treatment can act as a promoter for bridging coagulation, as illustrated in Fig. 14. The application of powdered activated carbon in the flocculation phase can be accomplished ideally when an upflow sedimentation tank is used (e.g., Pulsator or hopper tank), in which the sludge bleed is carried out at an intermediate level. In the sludge zone, which can reach a height of 5 m, there is a permanent provision of active carbon which levels out most momentary irregularities in the water composition. This would otherwise necessitate immediate action where the carbon dose is concerned. Henceforth, it remains possible to adapt the dosage. Moreover, this arrangement has the advantage of ensuring regular renewal of the active carbon through the sludge bleed. Thus there is no possibility of uncontrolled bacteriological fermentation of breakthrough when the carbon is exhausted.

The general observed data for the use of PAC (powdered activation carbon) in the flocculation phase of the river Meuse at Tailfer indicate the distinction between the effect obtained with or without carbon dosing (average 10 g/m³), particularly when the raw water is not heavily charged. The data for chlorated–flocculated and filtered water originating from the river Meuse are listed in Table 3. On an average basis, the TOC of the settled water is improved by 3 to 4 mg/L when using activated carbon.

The TOC value is a sum parameter, resulting from a great variety of compounds. Hence the general trends, if any, depend on the behavior of any particular compound in the mixture. By subdividing the injection of powdered activated carbon into two consecutive phases, a potential increase in the overall removal rate can be obtained (8,9).

Table 3 Quality Improvement with Powdered Activated Carbon (PAC)

Parameter	Raw water	PAC treated and settled
TOC (g/m³)	7 ± 2	4.5 ± 2
Phenol index (mg/m³)	35 ± 30	18 ± 13
MBAS (mg/m³)	48 ± 32	43 ± 31
COD (g/m³)	15 ± 4	8 ± 4

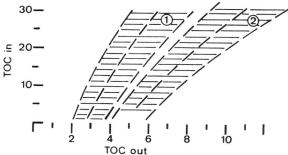

Figure 15 TOC removal with powdered activated carbon as used in the flocculation phase of the Tailfer plant.

4. FILTERING THROUGH GRANULAR CARBONS

The present trend in the application of active carbon to drinking water treatment is to filter through granular activated carbon. This is particularly the case with highly polluted raw water sources. The important advantage of using large amounts of granular activated carbon at all times is a built-in margin of safety regardless of fluctuations in the quality of raw water. If the startup sequence of different carbon filters, operating in parallel, is not reliable, each filter effluent is blended with the others prior to distribution. Moreover, seeing that the simultaneous bursting of all filters is unlikely to occur in these circumstances, this results as an additional safety factor.

4.1 Filtering Properties of Carbon

In addition to adsorption characteristics such as those measured with adsorption isotherms, other important properties for an active carbon to be used in filter beds are the following:

Density and wettability of the carbon
Particle size
Distribution and shape
Resistance to abrasion
Design parameters such as bed expansion and head-loss characteristics
Adsorption kinetics and the possibility of regeneration

4.1.1 Density, Wettability, and Granular Activated Carbon Handling

The bulk density of granular activated carbon is below 0.55 g/mL and usually about 0.4 g/mL. The real density of the material is 2 to 2.1 kg/dm³ for potable water grades. The particle density in water (with the pores of the material saturated) is usually between 1.4 and 1.6. Good activated carbon should contain less than 8%

incombustible ash. The moisture content of the shipped material must be less than 10 wt%.

When immersed in water, granular carbon requires a certain amount of time to hydrate. As a test, the weight percentage of carbon can be determined by recording the amount of carbon that sinks into distilled water within 2 min. A complete wetting-out, which takes place directly in the filter, may need several hours and may have to be completed by a backwash to remove any fine particles of carbon produced during the transport and filling.

Hydrated carbon is corrosive to steel, which means that this can lead to pitting up to 5 mm/year by electrolytic corrosion. AISI 316 grade stainless steel is corrosion resistant. Epoxy coating is also resistant to an activated carbon slurry and protects the steel from corrosion. A minimum total thickness of an alimentary-grade epoxy coating of 400 to 600 μm should be provided. An alternative coating is polyvinyl chloride.

The steel tanks should be equipped with cathodic protection and built 5 mm thicker than necessary for pressure requirements. No particular problems have been encountered with concrete surfaces. Granular carbon can be conveyed either by water or by air. However, on air transport, considerable pipe erosion can arise from "sandblasting effects," also causing great carbon loss in the form of dust. Therefore, hydraulic transport is recommended.

Dry loading of the filters causes problems concerning wetting of the carbon. Activated carbon is available in various quantities: in bulk delivery up to 20 metric tons by dry weight, tote bins of 500 kg net weight, steel or fiber drums up to 100 kg, and paper bags (multiwall or polyethylene-lined) of 12.5 to 50 kg content.

A suitable tank for storage and wetting should be provided for at the site of application. Its dimensions are optional but should allow an intermediate storage of the quantity that is to be introduced to or withdrawn from the filter. Starting with the dry product, the amount of carbon that can be hydrated in 1 m^3 of water and immediately deliver a sufficient slurry for transfer ranges between 100 and 120 kg. The discharge tunnel must be fitted with a vented hood for the collection of dust. Dust and fine particles of carbon can also be washed out on a stainless screen with openings measuring 0.5 mm. Granular carbon slurries can be conveyed by diaphragm pumps, water-jet eductors, and low-speed rubber-lined centrifugal pumps.

Flexible fire nozzles (14 cm) are suitable for intermittent transfer of carbon slurry. Typical injector feeding characteristics are, for example, for the Körting 133207, slurry flow 15 m^3/h (25% carbon) and feed water flow 25 to 30 m^3/h at 8 bar. A window should also be included for observation of transfer to and from the closed tanks.

For each application it is best that the manufacturer furnish all transport characteristics of the various carbons in use. In the absence of specific data, the indications given in Fig. 16 may be considered as guidelines for a 5-cm steel pipe.

On withdrawal of the active carbon from a filter, the material should be washed out on a stainless steel screen with 0.5-mm openings to remove the fine material that is not suitable for reactivation. The allowed concentration range for hydraulic conveyance of carbon ranges between 125 and 375 g/L. Below a linear velocity of 1 m/s the carbon tends to settle, and at more than 2.5 to 3 m/s becomes objectionale for carbon abrasion and pipe blasting.

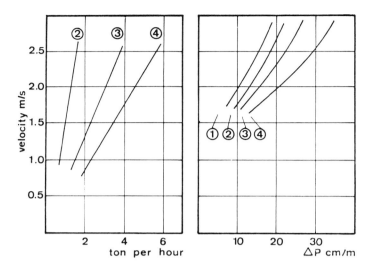

Figure 16 Transport characteristics of granular activated carbon. Carbon transport characteristics (typical example in a 2-in. pipe). 1, Water; 2, 125 g/L; 3, 250 g/L; 4, 375 g/L.

4.1.2 Particle Size, Distribution, and Shape

Granular carbon is available in either broken or extruded form. Extruded carbons are far more uniform in size, due to the method of their manufacture. Pelletized carbons and carbons in briquettes are used less often in drinking water treatment. Particle size and shape are important factors not only for the efficiency of filtration, but also with regard to pressure drop and bed expansion during backwashing.

Typical data related to granular carbon forms are given in Table 4. The particle-size characteristics of a granular carbon can be expressed as a granulometric curve: percent (weight) passing as a function of the screen size opening. The sample is dried at 120°C before being sieved. Typical curves are shown in Fig. 17 (11,12).

Definitions.

Effective size: size corresponding to 10 wt% passing
Diversity: size corresponding to 60 wt% passing
Uniformity index: diversity divided by effective size
Homogeneity: size range between 5 and 95 wt% passing.

Table 4 Granular Characteristics of Activated Carbon

Type	Grain size (mm)	Form	Grain surface	Hardness
Extruded	1–9	Cylindrical	Polished	Excellent
Pelletized	4–9	Spherical	Nearly polished	Satisfactory
Briquetted and broken	0–10	Various	Angulous	Good
Broken	0–10	Various	Angulous	Satisfactory

Source: Ref. 10.

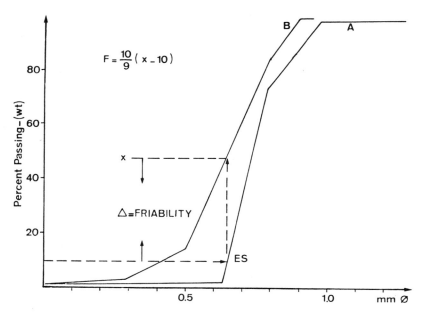

Figure 17 Particle-size distribution of fresh and abraded carbons. (Data from Refs. 11 and 12.)

The effective size of the carbon is an important factor for determination of the length of the filter runs, especially when backwashing is used. In contrast to sand filters, a definite classification of material occurs on backwashing, driving the finer particles to the upper layer. A difference of 3 : 1 in filter run has been observed with particles of an effective size of 0.9 and 0.5 mm, respectively. Consequently, if the water that is to be filtered is heavily charged with suspended particles, that is, if the filtration function is of primary importance in addition to that of adsorption, a sufficient effective size is essential to attain an adequate filter run. In such cases the effective size should range between 0.8 and 0.9 mm. If, by contrast, the water that is to be treated is very clear and the adsorption capacity prevails in the treatment, a finer carbon should be selected. If backwashing is used, a minimum effective size of 0.5 mm must be maintained. Certain carbons, such as the Norit Row 0.8 supra, of which the effective size is 0.7, are used for both applications, by compromise.

Backwashing, conveying, and reactivation involve mechanical handling of the carbon. Granular carbon is known to have a certain mechanical strength. In a conventional test (4), the carbon is submitted to standardized abrasion by stirring and the average diameters, before and after stirring, are compared. If the carbon is submitted to mechanical abrasion or to a sieve test of the granulometric curve, the turbidity of the supernatant gives an impression of the mechanical strength of the carbon. The *friability* is expressed as being the reduction in percentage of carbon of the virgin effective size [$F = \frac{10}{9}(x - 10)$] (see Fig. 17).

The shape as well as the size of the grains is important for pressure drop in the active carbon beds and for the backwash characteristics of the filters. Regularly shaped carbons approach the ideal adsorption behavior and favorable initial head loss. This is the case with extruded carbons such as Norit Row 0.8. Broken carbons are less homogeneous in shape and may therefore lead to more efficient filtration. Pelletized carbons are less favorable for filtration or integrated filtration–adsorption, as they have a decided tendency to clog the filters by falling into parallel layers. For carbons of equal type (e.g., either broken carbons or extruded carbons), the greater the uniformity coefficient, the greater the expansion during backwashing. Practical limits for good carbons require uniformity coefficients ranging between 1.2 and 2.0 and preferably between 1.5 and 1.8.

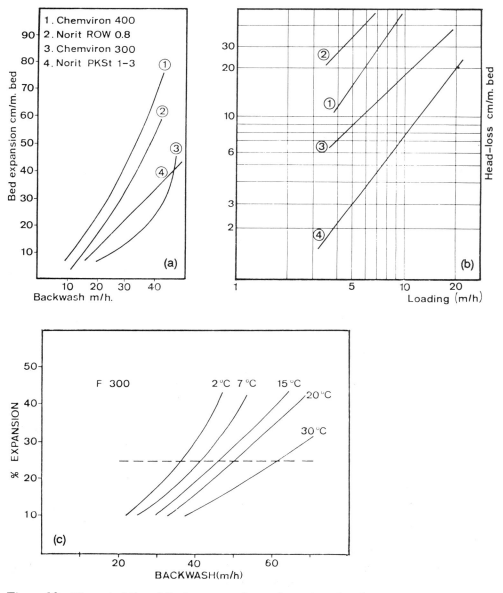

Figure 18   Characteristics of the transport of granular activated carbons.

Typical data for filter bed expansion and initial head loss are indicated in Fig. 18. In practice, bed expansion during backwashing of the filter is kept below 25%. This expansion is sufficient to eliminate gas bubbles (air or CO_2) and to avoid rat-holing. In many cases, an air wash precedes the backwash with water. Simultaneous backwashing with air and water is forbidden, to avoid drawing-over the carbon.

4.2 Adsorption Characteristics in Filter Beds

4.2.1 *Experimental Column Behavior*

Isotherm data do not replace experimental column results giving information on the necessary contact time and residual concentration of a given product in the water to be purified. The ideal and real behavior of the adsorption of a solute contained in a solution passed on a column is indicated in Fig. 19.

One can distinguish two different flow-through patterns in activated carbon filters: the constant break-through pattern and the proportional break-through pattern (Fig. 20). The *constant break-through pattern* (a) is observed for very easily adsorbed substances only: for very small values of $1/n$ as given by the Freundlich isotherm. In the strictly *proportional break-through pattern* (b), the slope of the break-through curve is proportional to the depth (or time) of the bed traveled through by the solute to be adsorbed. This pattern corresponds to less adsorbable solutes (larger $1/n$ values in the Freundlich isotherm).

A criterion sometimes applied to evaluate carbon on a comparative basis is the *half-value thickness*. This is equal to the layer thickness necessary to lower the inflowing concentration of a given compound by 50%. For molasses, some typical values are given in Table 5. Experimental investigations in the treatment of drinking water have indicated that a minimum superficial contact time of 4500 s is necessary in all instances. For the usual carbon grades this corresponds to a layer thickness of 0.6 m (13).

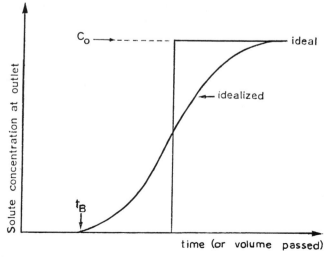

Figure 19   Break-through profile (schematic) of activated carbon filters.

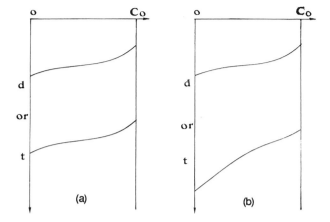

Figure 20   Constant (a) and proportional (b) break-through patterns.

The carbon column should be considered as a multistage process. At the entrance to the column, the carbon comes into contact with the impure liquid, and at the end of the column, with the purified water. Between the entry and the exit is an adsorbed impurity gradient, which moves toward the exit as the carbon is exhausted. Very useful for design are (1) the LUB (length of unused bed) criterion and (2) the BDST (bed-depth service time) equation.

4.2.2   LUB Criterion

At the moment of the first release of solute at the outlet of the filter bed, the total length of the bed can be subdivided (Fig. 21) into a saturated length, L_S, and an unused length, LUB:

$$L_F = L_S + \text{LUB}$$

The displacement of the saturated front is constant with respect to the volume throughout the column or to the time necessary for it to pass:

$$\frac{L_S}{L_F} = \frac{t_S}{t_{st}} = \frac{V_S}{V_{st}}$$

Here "st" represents the steady state or stoichiometric adsorption equilibrium and hence

Table 5   Typical Half-Value for Molasses

Carbon	BET surface (m^2/g)	Half-value (cm)
Hydraffin BK 12	900	12
Norit Row 0.8	1000	8
Chemviron F 300	1050	6
Hydraffin BK 0.5–1.5	400	3

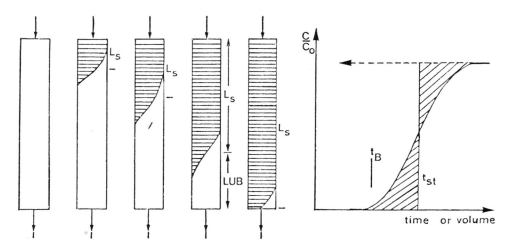

Figure 21 LUB criterion (schematic).

$$\text{LUB} = L_F \times \frac{t_{st} - t_B}{t_{st}}$$

The shape and slope of the adsorption gradient depend on the adsorbability of the impurity and the rate of flow. When designing the adsorption column, one should consider this factor. Data are generally expressed on the basis of the (empty) volume of the filter passed by the water flow. A comparative yield of solute removal can be attained by using a shorter column for substance 2 than for substance 1, as indicated in Fig. 22.

In practical situations, including observation of industrial filters, pilot plants, and laboratory investigations, the following equation is often found to hold (14):

$$\log \frac{C}{C_0} = \alpha \int_0^t Q(C_0 - C)\, dt + \beta = M_t + \beta$$

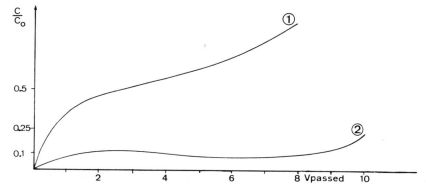

Figure 22 Break-through profiles.

For a given solute to be adsorbed, different carbons can be evaluated by plotting

$$\log \frac{C}{C_0} \text{ vs. } Q_x \frac{\Sigma_i[(C_0 - C) + (C_0 - C)_{i-1}]}{2}$$

which quantifies the increase in leakage C/C_0 as a function of the quantity of adsorbate fixed on the adsorbant.

In practice, break-through curves can be quite different from the ideal S-shaped patterns. An example for the treatment of water from the river Meuse with C 400 (Chemviron) is given in Fig. 23.

4.2.3 BDST Criterion

The contact time in an adsorbing column is defined (15) as the ratio of filter volume (empty) to rate of flow:

$$\text{contact time (h)} = \frac{\text{volume of the empty filter (m}^3)}{\text{flow (m}^3/\text{h})}$$

A design parameter can be determined from the BDST (bed-depth service time) equation, $t = ax = b$, in which t is the time at the break-through, x the bed depth, and b the ordinate intercept in hours. The slope a is a function of the carbon efficiency E_0 (weight impurity adsorbed per volume of carbon), the impurity concentration in the influent, and the surface loading (Q/A):

$$a = \frac{E_0}{C_0(Q/A)}$$

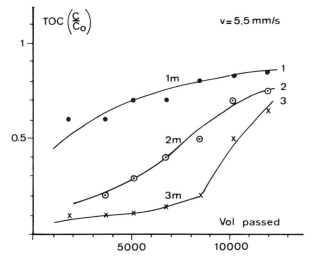

Figure 23   Pilot experiments for TOC removal from river Meuse water (clarified) with granular activated carbon.

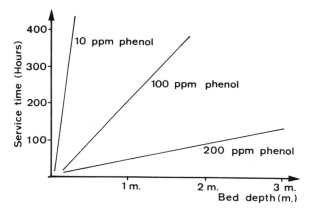

Figure 24   BDST proportionality for phenol.

The ordinate is a function of the adsorption rate of the carbon and of the allowed impurity concentration in the effluent at the break-through. A typical example is shown in Fig. 24. A typical example of BDST curves has been reported by Van der Laan (5) in the case of groundwater contamination with tetrachlorethylenes (see Fig. 25).

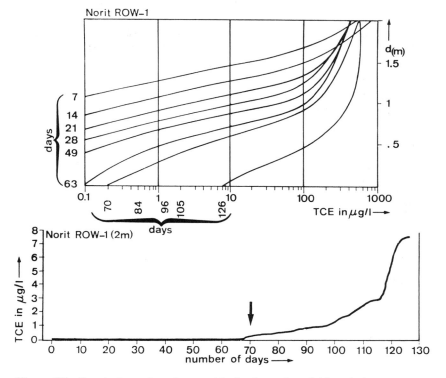

Figure 25   Break-through and unused bed volume for trichlorethylene.

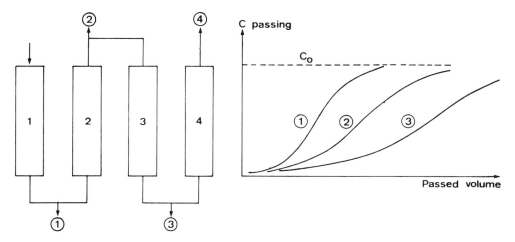

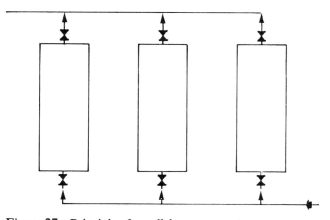

Figure 26   Principles of series arrangement.

4.2.4   Bed Depth

Increasing bed depths may be necessary either for longer-run periods or for higher solute concentrations. Two practical solutions are either an arrangement in series or a subdivision of the total water flow over several substreams.

Series Arrangement.   In the series arrangement, the consecutive columns are saturated one after the other (Fig. 26). Owing to the high cost of valves and pipes, it is seldom that more than three beds are used when working in series. The process is suitable for continuous adsorption and is particularly well adapted for applications where the required bed depth is excessive for a single column. The possibility remains open of using each column individually.

Arrangement in Parallel.   The arrangement in parallel (Fig. 27) does not increase the unit bed depth but is particularly well suited for the changing filters and the regeneration of the adsorbant in an exhausted filter.

Figure 27   Principle of parallel arrangement.

Moving-Bed Technique. The moving-bed technique (Fig. 28) is a typical counter-current technique of modern design, particularly suitable for wastewater treatment. A particular advantage of the process is the elimination of mud balls and gas balls. The gases due to biological activity are eliminated on a continuous basis with the carbon. The device is particularly well suited for continuous operation with recycling of the carbon through equipment for regeneration. The surface loading of the moving-bed column can vary between 1 and 10 m/h. The bed height-to-column diameter is in the proportion 3 : 1. The bed height can be up to 8 to 10 m.

4.3 Design of Activated Carbon Filters

4.3.1 Pressured Carbon Filters

Downflow pressured carbon filters (Fig. 29) are the most classical adsorption filters, existing in essentially two arrangements: monolayer and bilayer filters. In the latter, filtration on sand takes place prior to adsorption. The main reason for using down-flow filters is their suitability for two different purposes: adsorption and removal of suspended matter. The use of horizontal cylindrical closed pressure filters with vertical water flow is quite rare in adsorption on granular carbon. The most wide-spread types in use are vertical pressure filters. Typical units have a diameter of 5 m and a height reaching 8 m. These filters are backwashed in the usual way.

4.3.2 Activated Carbon Rapid Filters of the Open Type

A typical example of this filter is the Aquazur V filter (Fig. 30), in which water flow is equally divided among all the filters provided, in order to treat the whole amount. Standard units are normalized initially with 24 m^2 and capable of attaining 100 m^2.

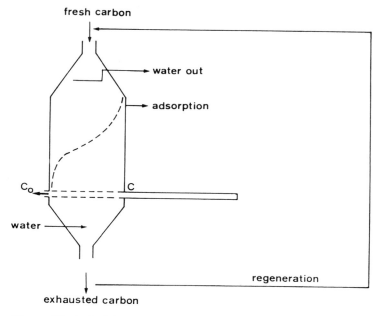

Figure 28   Principles of the moving-bed technique.

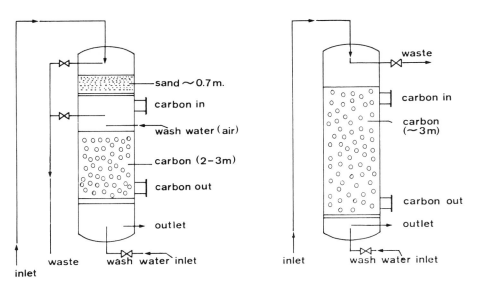

Figure 29   Pressure-activated carbon filters (schematic).

An important characteristic of the filter is the *surface rinse*, which takes place at the end of the backwashing period and during which raw water is introduced in the V-shaped weirs and passed through their apertures horizontally to the surface of the filter medium. A resulting "sweeping off" of the impurities fixed at the surface of the filter medium takes place. The width of standard Aquazur V filters can vary between 3 and 5 m.

4.3.3   Surface-Wash Filters

The first version of the Hardinge automatic backwash filter (Fig. 31) was a segmented filter with divider plates (e.g., polyvinyl-coated steel plates) separating the filter into compartments 8 in. (20 to 30 cm) in width. The filter medium was supported by porous plates (e.g., porous Alundum). In its classical version, the filter medium was a layer of 27 to 29 cm, but later filters have been built to contain up to 1.2 m of granulated carbon. Backwash is carried out segment by segment by

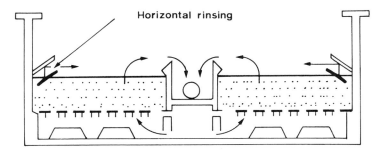

Figure 30   Aquazur® V filter (schematic).

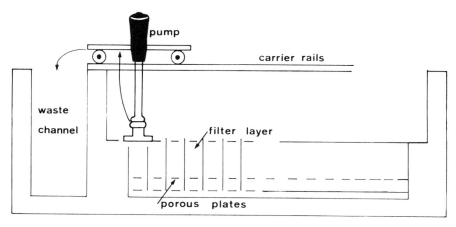

Figure 31 Principles of the Hardinge surface-wash filter.

drawing the wash water from the effluent channel. The backwash expands the filter media in one compartment only and is confined within a cleaner hood equipped with a wash water pump for disposal into the waste channel. The highest wash water operation rate is 40 m/h.

Turbulence within the cleaner hood helps surface washing. During the backwashing of one channel, the remaining segments of the filter continue to filter. The entire backwash mechanism, including hood and pumps, is mounted on carrier rails. It works automatically, either by timing or when the headloss due to suction reaches a predetermined value. In the case of active carbon filters, an eductor can be mounted on the carrier to empty the filter channel by channel. Standard filters of this type vary in surface from 28 to 164 m^2 (16). The principle of surface rinsing has been extended to cleaning of a slow filters capped with granular activated carbon. Illustrated in Fig. 32 is the device used at the Zürich waterworks.

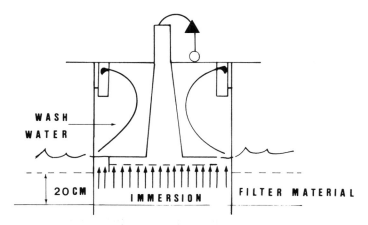

Figure 32 Hood system for filter backwash.

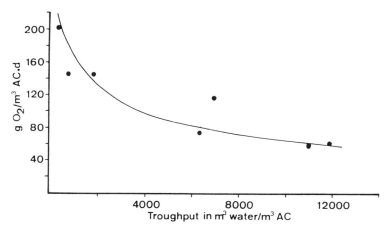

Figure 33   Oxygen uptake from water by activated carbon. (From Ref. 18.)

4.3.4   Emptying the Filter Beds

One must remember that when contained in a closed vessel, activated carbon gradually absorbs the oxygen of the air or consumes it by biological activity in the carbon. Appropriate ventilation or oxygen masks are necessary when entering a closed filter (17).

As for delivery, storage, and filling of the filters, facilities must be provided for emptying the filter and replacing the spent carbon. The usual method of transfer is the eductor. This equipment enables transfer over distances up to 100 m and elevations up to 10 m. A flexible 6-in. line is required. An observation chamber with a pressure-resisting window should be provided in the line.

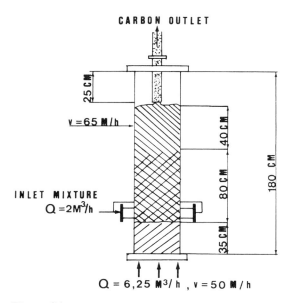

Figure 34   Apparatus for the separation of sand-granular activated carbon mixtures. (From Ref. 19.)

Fractionating of granular activated carbon from the sand when a double-layer activated carbon sand is emptied can be obtained in an upflow scrubber as illustrated in Fig. 34. An intermediate storage tank is required for tranquilization and sedimentation of the carbon before reactivation or discharge into waste. The discharge of supernatant can be facilitated by a pump equipped with a 0.5-mm screen. Ample facilities for rinsing and discharging to waste should be provided in the working areas for the transfer of active carbon. The sewage lines must also be adequately dimensioned (see Fig. 18).

5. REGENERATION OF GRANULAR CARBON

As more molecules are adsorbed, less surface is available for adsorption and the adsorption ability of the carbon is gradually diminished. *Regeneration* involves two consecutive phases: the *desorption* of the matter fixed on the carbon, and *reactivation*, restoring the internal surface and pore structure as much as possible.

5.1 Biological Regeneration

Although biological regeneration has been called into question, it is observed experimentally that under aerobic conditions, bacteria are able to mineralize organic products adsorbed on the carbon:

$$C_xH_yO_z + \left(x + \frac{y}{4} + \frac{z}{2}\right)O_2 = \frac{y}{2}H_2O + xCO_2$$

The biological regeneration is based on two principles:

1. The existence of bacteriological and biological life on the activated carbon, and the possibility of reversible desorption of preadsorbed solutes.
2. That even at higher concentrations of dissolved chlorine (e.g., up to 5 g/m^3), it has been reported that bacteria are not fully eliminated from granular activated carbon filters

This is due to dechlorination effects and is a major drawback for household use of activated carbon filtration. According to the literature (20), the bacterial population on activated carbon attains its maximum level after about 2 months of operation.

The population (*Coliforms* and *Pseudomonas* A) amounts to 10^8 bacteria/cm^3, which is equivalent to a cross section of 0.5 to 1 μm^2 for a 40 μm^2 material surface. The population growth is faster on activated carbon than on nonactivated carbon or sand. Bacteria adhere to the bulk surface, and metabolism is established on macro pores of diameter > 300 nm. The biomass thus formed does not hinder the micropore diffusion rates. According to our experience, with biological activated carbon filtration on a pilot column, to improve groundwater contaminated by phenolic substances (21) at a water temperature of 12°C, only one germ per mm^2 was obtained. Preoxygenation enhances the bioactivity in carbon beds (see Fig. 35). Indirect aeration is achieved in the bulk of the filter mass by injecting hydrogen peroxide into the raw water. No catalytic oxidation effect has yet been observed when decomposing H_2O_2 on activated carbon filters.

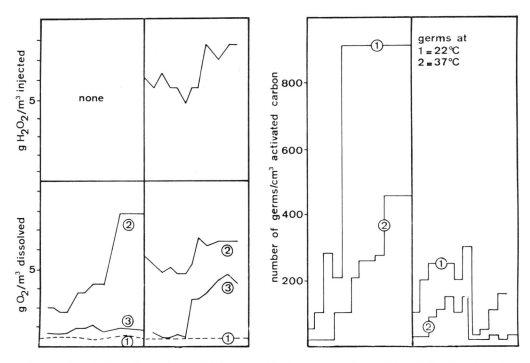

Figure 35   Effect of aeration and injection of hydrogen peroxide on bioactivity of activated carbon. (From Ref. 21.)

Ozone, in particular, seems to enhance the bioactivity in carbon beds, which is probably due, on the one hand, to the oxidation conversion of organic matter to a more readily biodegradable form, and on the other hand, to enriched levels of oxygen in the bed. It is important to note that preoxidation treatment with ozone is not likely to enhance adsorption reactions, as has been suggested in some literature reports. Indeed, such pretreatment tends to convert aromatic systems of compounds to oxygen-rich polar aliphatic systems, which are not as well adsorbed by activated carbon as are the less polar aromatic precursor compounds (22).

Further general items regarding biological activated carbon filtration are the following:

1. It is necessary to balance the C/N/P in the appropriate proportions if not directly available as such (see Chapter 13).
2. The bioactivity is dependent on the temperature, which for surface waters is low during the winter. The design capacity should thus make it possible to operate the filters on an adsorption basis during periods of low bioactivity.
3. An activated carbon effluent can contain between 50 and 45,000 germs cultivating at 22°C according to the season or the available substrate. Post disinfection is necessary in all cases.
4. The carbon dioxide content of the water is increased and the water requires corrections for aggressivity.

The design of biologically operated activated carbon filters relies on experimental pilot investigation. The most accepted criterion is TOC removal, except in the case of very specific pollutants:

$$\Delta(TOC) = (TOC)_0 - (TOC) = kt$$

$$t = \frac{L_F}{V_F}$$

$$N_F = k\epsilon = \Delta(TOC) \frac{V_F}{L_F}$$

where N_F is the specific biochemical yield (g TOC/m³), and L_F and V_F the filter depth and volume flow rate, ϵ the ratio of filter water volume to total filter volume, and $\Delta(TOC)$ the abatement of total organic carbon.

5.2 Chemical Regeneration

Chemical regeneration involves several washings of the spent carbon (e.g., desorption of phenols with caustic soda forming the less adsorbable phenate ion, which is washed out). If successful, the carbon regenerated is directly reusable. The operation is eventually possible without removing the carbon from the adsorbing filter. All adsorbed products are not desorbed in a single operation; therefore, several sequences may be necessary:

Washing with 1 to 2% HCl for 1h
Washing with 10% NaOH for 1 h at 100°C
Washing with 50% alcohol for 3 to 4 h
Vapor treatment for 1 h at 120 to 140°C
Reacidification with 1 to 2% HCl for 10 min

This technique is oriented to industrial applications in which the desorbed substances are recycled into another process. At first sight, the method seems less well suited for drinking water treatment plants.

5.3 Thermal Regeneration

5.3.1 Principles

Thermal regeneration remains the most widespread method for the reactivation of granulated active carbons. In thermal regeneration, water and volatile organic compounds are evaporated and the remaining organic products are carbonized. Reactivation takes place by controlled admission of water in a slightly oxidant atmosphere:

$$2C + \tfrac{3}{2}O_2 \rightarrow CO_2 + CO$$

$$C + CO_2 \rightarrow 2CO$$

$$C + H_2O \rightarrow CO + H_2$$

Thermal reactivation involves heating the carbon in a furnace in a controlled atmosphere in such a way that the organics are volatilized with little combustion of the carbon. The off-gases are burned.

The average energy consumption for the reactivation process may be estimated at ± 11 kJ/kg carbon. Half of this is consumed in the evaporation–carbonization phase and the other half in the degasification–activation phase. The burning of the off-gases involves about 5000 kJ/kg carbon. Several types of furnaces can be used: (1) a multiple-hearth furnace, (2) a rotary kiln, or (3) a fluidized-bed furnace.

5.3.2 *Multiple-Hearth Furnace*

In the multiple-hearth furnace (Fig. 36), the spent carbon is discharged into the top of the regeneration furnace. The carbon is moved downward by scrapers pushing it through the openings in the refractory hearth plates. Burners are currently fed with propane or natural gas. Steam can be injected into the furnace to regulate the temperature and strip off impurities with the off-gases. In the activation, the temperature of the carbon gradually increases to reach 950°C at the last plate. The exhaust gases attain the same temperature. The regeneration capacity ranges be-

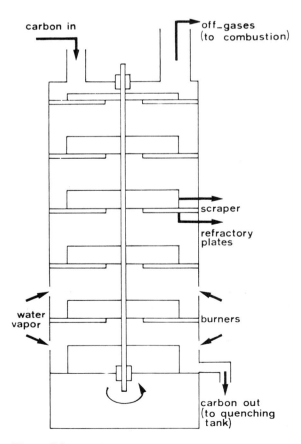

Figure 36 Multiple-hearth furnace (schematic).

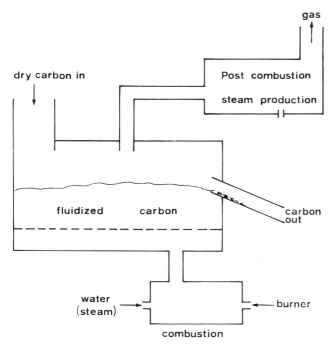

Figure 37 General setup of fluidized-bed furnace system.

tween 20 and 50 kg/m^2. Furnaces have been built for a production capacity of 20 to 500 kg/h. The losses should be between 5 and 10 wt%. Smaller furnaces are constructed using single refractory bricks, which facilitates repairs.

5.3.3 Rotary Kilns

Rotary kilns used for regeneration are of the classical type, with a simple construction and method of operation. The tube is slightly inclined toward the outlet. The burner is placed on the outlet side and can be located either inside or outside the furnace. Local overheating can occur and cause carbon losses (10 to 20 wt%) greater than those encountered in the use of the multiple-hearth furnace.

5.3.4 Fluidized-Bed Furnaces

Fluidized-bed furnaces are the most recent type developed for carbon regeneration. Regenerating gases are produced under a grid on which the spent carbon is fed (Fig. 37). On passage of the gases, the carbon layer expands to 150% of its loose volume. The off-gases are burned as usual. Several types of furnaces of this type are available on the market. The advantages of the process are the following:

1. Residence time is shorter (5 to 15 min) and hence there is less volume of construction.
2. A discontinuous operation is possible.
3. Delay in starting the furnace is much less important than for other types (e.g., ±2 h compared to about 1 day for a multiple-hearth furnace).
4. Thermal homogeneity is excellent in a heated fluidized bed.

5. Losses are minor (~ 5 wt%) compared to other techniques, but reactivation is usually somewhat less than for more static furnaces.

The fluidization technique requires preliminary elimination of material other than carbon, such as sand, which can hinder the fluidization. When a less gradual increase in temperature occurs, the carbon is best dried separately before reactivation, as sudden evaporation of the water contained in the pores may enlarge them excessively.

5.4 Complementary Equipment

After regeneration, the carbon is discharged into a quenching tank (Fig. 38). In this tank a small quantity of carbon is mixed into the water at a temperature of 600 to 900°C (~ 3 L water/kg carbon). The carbon slurry is transferred continuously to the storage tank or filter by means of a pump or eductor. The off-gases of the regeneration furnace are corrosive and generally toxic. They also contain fine carbon particles. Complete oxidation of the gases is carried out in a postcombustion oven, where the temperature is raised to 800°C in the presence of an excess of air. These gases must eventually be cooled by using them to dry the incoming carbon. Water vapor is produced and injected into the regeneration furnace. About 1 kg of vapor is required for 1 kg of carbon.

5.5 Quality of Regenerated Carbon

The original physical characteristics seem to be recovered by regeneration. However, if the carbon has been used as a filtering material as well as for adsorption,

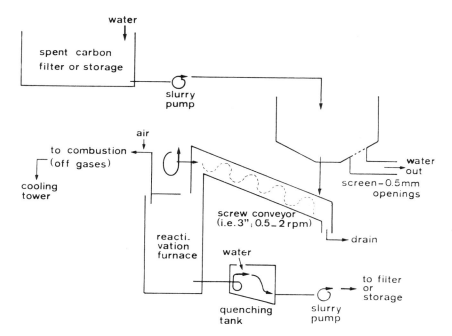

Figure 38   General on-site regeneration sequence.

and where its use for the removal of phenol is involved, its adsorption characteristics are diminished. It remains to carry out a complete and exhaustive comparative investigation of various carbons, before and after regeneration, to determine if the activity sequences of different carbons available on the market are maintained after regeneration. The reactivation process can be questionable, as heavy metals can gradually concentrate in the carbon in the form of oxides. If the amount of FeOx and MnOx exceeds 20 wt%, the total mass should be renewed. These oxides can exert an influence on the weight of the regenerated carbon and must not be overlooked if regeneration is being carried out on a weight contract basis. Spent carbon charged with inorganic products is best acid-washed before recycling.

Too many factors influence the economic balance of regeneration to allow easy formulation of general rules:

Quantity required

Distance from the manufacturer's equipment to the water treatment plant

Possibility of using a cheaper, nonregenerable carbon compared to the costs of regeneration. Each case should be studied taking the following parameters into account (the regenerating equipment being assumed to operate on a continuous basis):

Investment for the building and equipment for regeneration

Cost of activated carbon

Loss of carbon on regeneration

Cost of eventual transport for regeneration by the manufacturer

Operational cost of regeneration

At present one may conclude that the cost of carbon regenerated on the site usually represents 15 to 25% of that of the starting material.

6. USE OF ACTIVATED CARBON FOR DECHLORINATION

Chlorine reacts with carbon according to the overall equation

$$2Cl_2 + C + 2H_2O = 4HCl + CO_2$$

The reaction is often considered to have two consecutive phases: the generation of nascent oxygen,

$$Cl_2 + H_2O \rightarrow 2H^+ + 2Cl^- + O$$

and the reaction of the atomic oxygen adsorbed on the carbon,

$$C_xO_x \rightarrow C + CO \quad \text{or} \quad C_xO_x \rightarrow C + CO_2$$

In a general way the decomposition rate can be described according to an empirical equation of the type

$$\log \frac{C_0}{C} = \frac{F_L}{V_F}$$

in which C_0 and C are the chlorine concentrations at inlet and outlet, F_L the bed length, and V_F the flow rate. This is equal to a first-order reaction principle, which in fact holds only as long as a large excess of carbon is present compared to the chlorine concentration. The surface loading should not exceed 7 m/h. The reaction depends on the size of the carbon grains, pH, temperature, and the concentration of the chlorine. It should be noted that in comparable conditions chloramines decompose less than does chlorine. Dechlorination is usually a side effect of the treatment, although with carbon it may be used intentionally. In the latter case, the water must be free of suspended particles, which are the cause of filter clogging, in which case the service time can be extremely long (e.g., 4 to 6 years for a classical open-type filter, on which water is fed containing less than 10 ppm chlorine). Simultaneous adsorption of large molecules such as those in detergents reduces the dechlorination efficiency. This is not the case with other impurities, such as phenols. However, the ternary system carbon–chlorine adsorbed organics is reactive and extends to the formation of organochlorine compounds. The latter are also formed when high chlorine concentrations are reacted with activated carbon. The bromide content of activated carbons can determine the formation of organobromine compounds.

Dechlorination efficiency decreases after a period of time. One method of characterization is the expression of the dechlorination half-value as the layer thickness for 50% dechlorination. Typical values are given in Fig. 39.

The decomposition of chlorine on activated carbon filters, just like those of chlorine dioxide and hydrogen peroxide, corresponds to an equation of the type Y ($= \%$ chlorine remaining) $= be^{mt}$ from which $t_{1/2}$ (half-lifetime) can be deduced. Typical values for several carbon species are given in Table 6.

An important point is that the process of the decomposition of chlorine dioxide on activated carbon is usually slower than that of the dechlorination. Furthermore, the reactions take place according to a double phase sequence, at least. The chlorite ion may be removed by contact with activated carbon (29). The reaction is a second-order reduction to chloride and undergoes "competition" with dissolved chloride, the activated carbon being a first-order reactant.

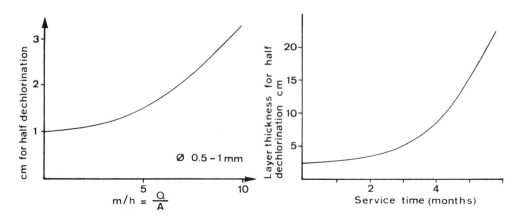

Figure 39 Dechlorination for fresh carbon capped on sand filters.

Table 6 Half-Lifetime (seconds) of Oxidants Dissolved in Water

Carbon	Chlorine	Chlorine dioxide	Hydrogen peroxide
Norit W	4,470	9,000 ⎞	
Norit TNW	3,600	30,000 ⎟	
Norit W52	2,000	18,000 ⎬	10,000
Norit WS52	1,780	63,000 ⎠	
Chemviron 400	2,250	25,000	15,000
Ceca-Te	3,250	85,000	–

Source: Ref. 21.

Hydrogen peroxide decomposes slowly on activated carbon filters and contributes to oxygenation of the filter mass, enabling microbial activity to occur or to be supported. The most probable reaction for chlorine dioxide decomposition is the following:

$$ClO_2 + C + OH^- = Cl^- + CO_2$$

The radical reaction scheme is based on $ClO_2 = ClO + O$ (adsorbed); ClO decomposes in a second stage. A secondary reaction with alkaline carbons cannot be formally excluded. However, no chlorite accumulates in the medium seeing that it is reduced in turn:

$$4ClO_2 + 4NaOH + Ca(OH)_2 + C = 4NaClO_2 + CaCO_3 + 3H_2O$$

$$ClO_2^- + C* = Cl^- + CO_2$$

To decompose ClO_2 efficiently, the carbon surface is best alkaline. The maintenance of an active bactericidal residual in the presence of activated carbon powder is possible with conventional doses of chlorine dioxide during a contact time of 1 h. This may not be obtained with chlorine. However, rates of indirect removal of chlorine dioxide can be high (e.g., 80 to 90%). The mechanisms involved are coprecipitation in $Fe(OH)_3$–MnO_2 matrices and adsorption when complexed with organic compounds (e.g., humic acids).

When present as a consequence of intensive ozonization of manganese-containing water, permanganate is removed according to the following reaction (23,24):

$$4KMnO_4 + 3C + H_2O \rightarrow 4MnO_2 + 2KHCO_3 + K_2CO_3$$

With another form of oxidated manganese, one obtains

$$2Mn_2O_7 + 3C = 4MnO_2 + 3CO_2$$

Seeing that bacterial leakage always occurs through activated carbon filters, this means that a prophylactic post-disinfection is usually necessary. Design should be preceded by an experimental investigation of the problem that is to be resolved, as a great variety of problems can be treated with active carbon — far more than with other unit operations. General guidelines are given in Table 7.

7. WATER QUALITY IMPROVEMENT WITH ACTIVATED CARBON

Activated carbon is certainly not a remedy for all water supply problems; however, it is one of the best techniques for removing dissolved organic products and potentially toxic chemicals, which may also cause taste and odor problems. As a filtration technique, the use of granular carbon is also particularly effective in removing iron, yet this technique limits the adsorption capacity. After filtration on activated carbon, the iron content in the effluent should be lower than 0.03 mg Fe/L. The granular activated carbon is suited for double-layer filtration. As far as removal of organic products is concerned, the oxidability is significantly improved — to the point that one can obtain very low values of the chemical oxygen demand (0 to 2 mg/L) or TOC (<1 mg/L). The phenol content in the mg/L concentration range can be significantly lowered, while detergents are generally adsorbed at a rate of 70 to 80%.

The activated carbon layers usually contain a significant number of bacteria (e.g., 10^5). One result obtained by percolating the water on a carbon filter is the reduction of at least 20% of the BOD_5 if the other general conditions (pH, temperature, nutrients) enable the bacteria to develop. This biological activity is the basis of one of the modern trends in the use of active carbon. Activated carbon eliminates or decreases the number of viruses even though, as is the case in coagulation-flocculation, the sludge remains a source of contamination. The direct removal yield of heavy metals with activated carbon (25,26) often remains marginal; however, when complexed with organic ligands, removal can become efficient.

8. COSTS

The adsorption schemes in treatment are to be considered part of flowsheets as a whole. Representative costs (expressed in U.S. $) for nominal operation based on conditions in the Belgian market are cited in the literature (27).

Table 7   Guidelines for Activated Carbon Filters

	Filtration velocity (m/h)	Bed height (m)	Empty bed retention time (min)	Throughput ratio before regeneration (m³/m³)
Dechlorination	25–30	2	2–4	>1,000,000
Taste and odor removal	20–30	2–3	6–10	~100,000
Organics removal	10–15	2–3	8–15	~25,000
Biological activated carbon	8–12	2–4	15–25	~100,000

Source: Ref. 18.

8.1 Powdered Activated Carbon (0.3 to 0.4 $/kg)

Amortization of dosing equipment: 3×10^{-3} $/m^3 water
Total average per m^3: ≤ 0.005 $/m^3

8.2 Granular Activated Carbon (1 to 1.3 $/kg)

Regeneration costs (twice yearly): 0.5 $/kg
Costs of regeneration losses: 0.25 $/kg
Depreciation costs: $10,000 + $67,000
Total cost per m^3 for a plant of 1000 m^3/h: 0.01 to 0.015 $/m^3

A more detailed estimation, based on practice in Düsseldorf of the regeneration costs of granular activated carbon involving carbon losses, energy consumption, labor, and depreciation costs, has been given in the literature (28).

Evaluation per metric ton (in U.S. $ 1986)

Natural gas (75 m^3/h)	$ 60
Electric power (35 kW)	7
Water (transport)	9
Labor	30
Capital interests	230
Reactivation costs	$330

Carbon losses

Furnace	5%
Transport in the reactivation process	4.5%
Transport in the filters	2%
Total	11.5% = $160

The total costs are consistent with the standard evaluation as given above.

SYMBOLS AND UNITS RELATED TO ADSORPTION

b	Langmuir proportionally constant, dimensionless
c	concentration of dissolved compound, g/m^3 or mol/m^3
h	bed height, m
m	quantity of carbon, g or kg
$1/n$	adsorptivity constant (Freundlich), dimensionless
q	carbon loading, g (or mol) m^{-3} g^{-1} (carbon)
q_m	maximum carbon loading
r	pore radius, nm
t_B	break-through time, s
t_{st}	time to obtain a steady state, s
$t_{1/2}$	half-lifetime, s
x	difference in concentration, g/m^3 or mol/m^3
A	filter area, m^2
BDST	bed depth surface time, dimensionless
BET	BET surface (Brunauer, Emmett, and Teller), m^2/g
C_c	equilibrium concentration, g/m^3 or mol/m^3

C_0	dissolved concentration at starting time, g/m^3 or mol/m^3
E_0	weight compound adsorbed per volume carbon, g/m^3
F	friability, percent or weight fraction
GAC	granular activated carbon
K	Freundlich proportionality constant, dimensionless
L_F	filter length (or depth), m
L_S	saturated length (or depth) of the filter, m
LUB	length of unused bed, m
MBAS	methylene blue active substances, g/m^3 as lauryl sulfate
M_t	integration constant of break-through curves
N_F	biochemical yield of a filter, $g/m^{-3} s^{-1}$ of total organic carbon removed
P	(partial) pressure, $kg\, m^{-2}\, s^{-1}$
PAC	powdered activated carbon
P_s	pressure at saturation, $kg\, m^{-2}\, s^{-1}$
Q	water flow, m^3/s
R	gas constant, $8.3\ J\, mol^{-1}\, °K^{-1}$
T	absolute temperature, °K
TOC	total organic carbon, g/m^3
V_B	volume of the filter mass saturated at break-through time
V_F	filter volume (empty), m^3
V_F	volumic flow rate, m^3/s
V_m	molar volume in liquid state, volume units
β	integration constant of break-through curves, dimensionless
γ	surface tension, $kg\, m^{-1}\, s^{-1}$
γ	Langmuir–Freundlich correlating constant, dimensionless
ϵ	ratio filter water volume to total filter volume, dimensionless
θ	contact angle, deg
Φ	pore diameter, nm

REFERENCES

1. Abstracted from E. Windle Taylor, Presidential address to the 1st IWSA specialised symposium on the use of activated carbon in water treatment, Brussels, 1979, Pergamon Press, and from D. G. Hager, and R. D. Fulker, *Water Treat. Exam., 1,* 41 (1968).

2. U. Forstner and G. T. W. Wittmann, *Metal Pollution in the Aquatic Environment,* Springer-Verlag, Berlin, 1979, pp. 209–214.

3. H. Juntgen, *Veröffentlichungen Lehrstuhls Wasserchemie,* Heft 9, Karlsruhe, Germany, 1975, p. 23.

4. Anon. *J. AWWA, 66,* 672 (1974).

5. J. Van der Laan, *Proc. Specialised Symposium on the Use of Activated Carbon in Water Treatment,* IWSA, Brussels, Pergamon Press, Elmsford, N.Y., 1979, p. 106.

6. C. Gomella, *T.S.M./Eau, 65,* 383 (1970).

7. C. E. Hamilton, *Water Sewage Works,* 422 (1963).

8. H. Sontheimer, pp. 69, *Veröffentlichungen Lehrstuhls Wasserchemie,* Heft 9, Karlsruhe (1975).

9. W. J. Masschelein, *EPA Symposium,* Reston (Washington), 1979.

10. H. Jüntgen, Heft 9, Adsorption, in *Veröffentlichungen Wasserchemie,* Engler Bunte Institut, Universität Karlsruhe, Karlsruhe, Germany, 1975, p. 26.

11. F. Fiessinger and Y. Richard, *T.S.M. /Eau, 10,* 271 (1975).
12. P. Koppe, *Gas-Wasserfach, 107,* 1031 (1966).
13. D. G. Hagh and M. E. Flentje, *J. AWWA, 57,* 1440 (1965).
14. G. Gomella, *T.S.M./Eau, 68,* 151 (1973).
15. R. A. Hutchins, *Ind. Water Eng., 10,* 40 (1973); *Chem. Eng., 20,* 133 (1973).
16. S. Medlar, *Water Sewage Works,* 70 (1975).
17. R. E. Hansen, *J. AWWA, 64,* 176 (1972).
18. H. Sontheimer and B. Frick, *Proc. IWSA Symposium,* Brussels, Pergamon Press, Elmsford, N.Y. 1979, p. 195.
19. B. Stark, Adsorption, in *Veröffentlichungen Wasserchemie,* Engler Bunte Institut, Universität Karlsruhe, Karlsruhe, Germany, 1975, p. 248.
20. D. Van der Kooij, Heft 9, Adsorption, in *Veröffentlichungen Wasserchemie,* Engler Bunte Institut, Universität Karlsruhe, Karlsruhe, Germany, 1975, p. 302.
21. W. J. Masschelein, G. Minon, and R. Goossens, *Tijdschr. Becewa, 49,* 2 (1979).
22. W. J. Weber, *Proc. IWSA Specialised Symposium on the Use of Activated Carbon in Water Treatment,* Brussels, Pergamon Press, Elmsford, N.Y., 1979, p. 7.
23. W. Hopf, *Gas- Wasserfach, 111,* 156 (1970).
24. P. Schenk, *Gas- Wasserfach, 103,* 791 (1962).
25. E. A. Sigworth and S. B. Smith, *J. AWWA, 64,* 386 (1972).
26. A. Montiel, *Proc. IWSA Specialised Symposium on the Use of Activated Carbon in Water Treatment,* Brussels, Pergamon Press, Elmsford, N.Y., 1979, p. 119.
27. W. J. Masschelein, Lecture at the University of Turku, 1981.
28. W. Poggenburg, *Proc. IWSA Specialised Symposium on the Use of Activated Carbon in Water Treatment,* Brussels, Pergamon Press, Elmsford, N.Y., 1979, p. 176.
29. M. Denis, G. Minon and W. J. Masschelein, *Rev. Fr. Hydrol. Appl., 17,* 185 (1986).
30. A. Cornish-Bowden and R. Eisenthal, *Biochem. J., 139,* 721 (1974).
31. R. Eisenthal and A. Cornish-Bowden, *Biochem. J., 139,* 715 (1974).

13
Principles of Microbial Growth and Decay

1. INTRODUCTION

Microbial growth and decay may occur in several important unit processes in water treatment, such as disinfection, repair after disinfection and revival or aftergrowth in the distribution system due to secondary infection, and biological degradation of impurities in natural systems (lakes, ponds, and rivers) as well as in artificial, specially designed systems such as activated sludge treatment or biological filtration processes.

2. GROWTH AND DECAY

Growth is the result of cellular multiplication, which in the case of unicellular organisms leads to an increase in the number of individuals. Decay results from the lysis or intoxication of the cells. In the particular case where growth is inhibited in the environmental system under consideration, without direct killing action, the effect is considered as bacteriostatic in the case of bacteria, for instance. As a first approach to the use of microbial processes, growth and decay are considered to occur on an individual basis for each organism. However, viruses are able to survive and multiply starting with only a piece of an organism, which means that decay kinetics cannot be based on a single particle system alone. Recent indications related to bacteria (plasmids) have suggested a similar scheme even when bacteria are not agglomerated, as in the case of flocculation.

2.1 Cell Number

The most accurate counting of viable cells can be carried out by dispersing the sample on an appropriate culture medium (e.g., a plate-count agar). In this technique only the viable cells, that is, those capable of growth on the medium, are counted. The total count seems to be the most sensitive method for estimating the viable number of bacterial cells. The precision of this technique is $\sigma = \sqrt{N}$, in which σ is the standard deviation, which defines the confidence level of the count, and N is the number of viable cells, or colonies, observed in one sample analysis. Microscopic enumeration in counting chambers does not enable one to obtain a representative result at levels below 5×10^6 cells/mL. Electronic counting (e.g., Coulter or Hiac counting) is more appropriate for particles of larger size (e.g., algae or flocculant particles) than for cells of submicron size. Moreover, the method does not distinguish between viable and dead cells.

2.2 Cell Mass

Besides the *cell number*, the growth characteristics can also be determined by the *cell mass*, expressed as *biomass* in practical systems. However, cell number and cell mass are not necessarily equivalent:

1. The individual cell mass may vary. 1 mg dry bacterial weight can represent between 10^9 and 5×10^9 cells since the cell mass varies during the growth phase.
2. The cell mass determination methods do not distinguish dead cells from viable cells.

The measure of the optical density of a bacterial suspension is expressed as the amount of incident light intensity, attenuated by passage through the sample. In the cases of a bacterial suspension, the light intensity is diminished through scattering. Owing to this effect, the optical density is proportional to the number of scattering particles, which means that to assimilate these to bacteria or viable cells there may be no dead organisms present and all remain assimilable to individual cells (i.e., neither flocculated nor agglomerated in clusters). The wavelength used is a large part of the white-light spectrum of a tungsten lamp of 490 to 600 nm wavelength or, alternatively, a monochromatic light of 600 nm wavelength or 430 nm in a colorless medium such as water of high purity.

The correlation between cell mass and cell number must be determined experimentally for the particular organism, medium, and measuring instrument. The linear zone of the correlation log (I_0/I) versus dry cell weight must also be determined. An accepted guideline is that if the spectrophotometer enables the reading of 0.015 OD unit, the detection limit is equal to 0.01 mg/mL dry cell mass or about 10×10^6 cells/mL. The linear relationship between the OD and the dry cell mass holds strictly for values up to 0.3 mg/mL dry cell only (21). Although no general rule can be given, as a first approximation it is pertinent to consider the individual bacterial dry weight as (1 to 3) $\times 10^{-12}$ g, which is equal to that of 1 mg of ubiquitous bacteria (e.g., *Pseudomonas* species can range 1 mg to 6×10^9 cells, at least in sugar-rich culture media). In natural conditions the dry weight may sometimes attains only 2×10^{-13} g/cell.

Other indirect methods for the estimation of cell mass are the determination of the amount of a particular enzyme, or the overall reaction velocity of a metabolic process such as respiration or fermentation. Greater sensitivity may be obtained by incorporation of radioactive precursors into the cell material.

3. GROWTH RATE CURVES

A microbial population grows by doubling at regular intervals, as bacteria multiply by binary "fission," giving two daughter cells with the same potential of the parent cell (Fig. 1). The total cell number starting from N_0 organisms (in a given volume) evolves as follows:

Generation number (g)	0	1	2	3	. . . g,
Number of cells	N_0	$2N_0$	$4N_0$	$8N_0$	
	$2^0 N_0$	$2^1 N_0$	$2^2 N_0$	$2^3 N_0$	$2^g N_0$

Consequently, the total number of organisms (N_g) present after g generations is $N_g = 2^g N_0$.

In theory, the increase in cell number for a pure culture should be a synchronous process by which the total number increases stepwise. However, small individual differences in metabolism randomize the cell division time or "generation time," so that after several generations the growth can be assimilated to a continuous process. Techniques by which a bacterial population is submitted to cyclic changes, such as temperature, can determine the recurrence of stationary phases followed periodically by growth phases and enable experimental synchronous growth. This technique remains a laboratory method but should not be completely overlooked in periodically variable environmental conditions such as the diurnal–nocturnal cycle. According to the binary scheme for cell multiplication, the total cell number increases as an exponential function of the generation number. As for a given strain in any particular circumstances (nutrients, temperature, pH, oxygen availability, etc.), the time for a cell division, or *generation time*, attains a reasonably constant value and the growth pattern is exponential or logarithmic growth.

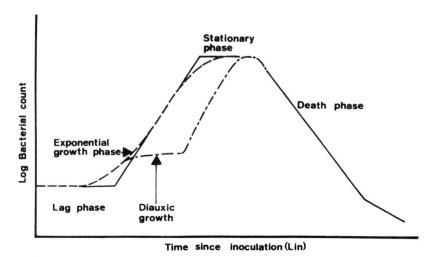

Figure 1   Schematic representation of growth and decay phases.

3.1 Growth Rate Constants

Exponential growth rate constants are defined as follows. The *mean growth rate constant* or *generation rate constant k* is obtained from the binary fission scheme of microbial multiplication:

$$N_g = 2^g N_0 = 2^{kt} N_0$$

Successive generations g of an organism are assumed to occur, under given conditions, at (average) constant time intervals, whence $g = kt$, in which k is designed as the average growth rate constant and expressed as the number of generations per unit time (traditionally per hour). The number of cells present after a time t (i.e., N_t) starting from N_0 at $t = 0$ is equal to

$$N_t = 2^{kt} N_0$$

or

$$\log_2 \frac{N_t}{N_0} = kt$$

where $t = t - t_0$ and

$$k = \frac{\log_2 N_t - \log_2 N_0}{t} = \frac{\log_{10} N_t - \log_{10} N_0}{0.301 t}$$
$$= \frac{3.32 (\log_{10} N_t - \log_{10} N_0)}{t}$$

The time required for a population to double is called the generation time (G) or mean doubling time: $G = 1/k$ (in hours).

The *instantaneous growth-rate constant, r,* is deduced from the first-order kinetical scheme of the logarithmic growth phase:

$$\frac{dN}{dt} = r N_0$$

$$\ln \frac{N_t}{N_0} = rt \qquad \text{or} \qquad \log_{10} \frac{N_t}{N_0} = \frac{r}{2.3} t$$

and

$$r = \frac{(\log_{10} N_t - \log_{10} N_0) \times 2.3}{t}$$
$$= (2.3 \times 0.301) k = 0.69 k$$

By similarity, the instantaneous generation time is expressed as $1/r$, which is equal to $1.45/k$ or $1.45G$. In the literature the two growth rate constants have often been confused, which is of significant numerical importance. Figure 2 illustrates the difference between the two concepts.

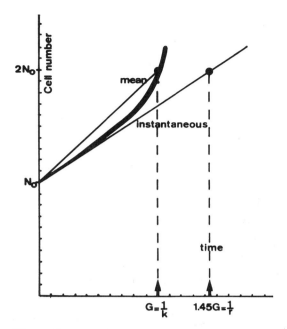

Figure 2 Graphical illustration of the difference in concept of mean and instantaneous growth.

3.2 External Factors

External factors can influence the growth phase significantly, the temperature and the pH value being the most important parameters. Temperature can determinantly influence bacterial growth. A well-known example is the distinction between *E. coli* and coliforms (and related germs), by incubation at 44 and 37°C, respectively. An illustration is given in Fig. 3. Optimum generation times are 0.2 and 0.5 h for *E. coli* and *Aerobacter aerogenes*, respectively. The latter does not grow significantly on Endo medium at temperatures above 42°C.

The temperature characteristics of microbial growth can be described by generalization of the Van't Hoff–Arrhenius concept as applied in chemical kinetics, which, when formulated for the mean growth rate constant, is expressed as follows:

$$k = Ae^{-\mu/RT}$$

or

$$\log_{10} k = -\frac{\mu}{2.3R}\frac{1}{T} + \text{constant}$$

in which μ is analogous to an activation energy. Then plots of $\log k$ as a function of $1/T$, T being in K, make it possible in the linear part to determine the slope from which μ is computed (1). (Application to the instantaneous growth rate is similar in concept.) The total temperature range in which germs can grow is small and extends

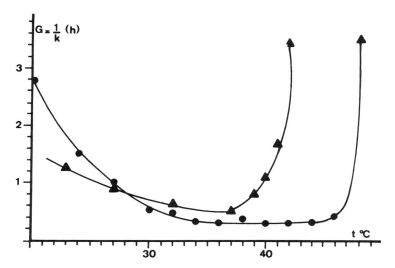

Figure 3   Temperature effect on growth rate (▲: *Aerobacter aerogenes*; ●: *E-Coli*).

from −5 to +80°C. The upper limit is due to the thermolability of essential cellular proteins, including enzymes, while freezing of water intervenes to establish a lower limit.

Three major physiological groups are to be distinguished among the procaryotic microorganisms when considering the temperature range compatible with growth (Table 1). The growth of obligate psychrophilic bacteria is rapidly inhibited at temperatures above 20°C, and in most instances killing occurs under these conditions. Facultative psychrophiles have optimum growth temperatures in the range of mesophiles; however, unlike mesophiles, they can grow slowly at temperatures neighboring 0°C. The existence of cellular proteins of high thermal stability has been established in the case of thermophiles. Up to now there is no satisfactory explanation as to why most mesophiles cease to grow at 20°C. In the case of *Vibrio marinus* species (1) the μ value is not significantly different between the obligate and facultative psychrophilic species: $\mu = 3.9$ kJ/mol. Most microbial cells isolated in the case of aftergrowth in treated waters are mesophilic or thermophilic and only exceptionally psychrophilic, that is, at least in the Belgian climate. A typical example of such ubiquitous strains is given in Fig. 4.

Table 1   Physiological Categories of Bacteria in Terms of Temperature Dependence

	Temperature span (°C)		
Group	Minimum	Optimum	Maximum
Thermophiles	40–45	55–85	60–80
Mesophiles	10–15	20–40	35–47
Psychrophiles			
Obligate	(−5)–(+5)	15–18	19–22
Facultative	(−5)–(+5)	25–30	30–35

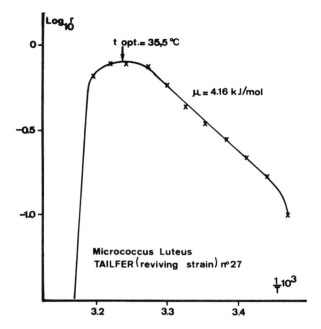

Figure 4   Van't Hoff–Arrhenius plot of a mesophilic aftergrowing strain.

The pH can influence bacterial growth, and the culture medium is one of the adjustable parameters for selectivity. Important examples in water treatment practice are nitrifying bacteria such as *Nitrosomonas* and iron- and manganese-oxidizing bacteria. The first step in the oxidation of ammonia can be represented as follows:

$$NH_4^+ + 1.5O_2 = NO_2^- + H_2O + 2H^+$$

The optimum temperature is 28°C. *Nitrosomonas* are mesophilic bacteria, preferably developing in buffered media containing calcium and magnesium ions; the optimum pH value for nitrification is therefore between 8.2 and 8.6 (see Fig. 5).

The dissolved oxygen content must be higher than 2 mg/L; if not, growth is restrained. The necessary generation time for *Nitrosomonas* bacteria is estimated to be approximately 80 h. Therefore, ammonia removal by nitrification is hardly feasible by biological rapid filtration. Under laboratory conditions the culture rates are improved by the presence of granules of calcium carbonate in the medium, thus buffering the solution at the optimum pH value (2). Further, lime softening with precipitation of $CaCO_3$ is likely to promote nitrification and should be taken into consideration more frequently in wastewater treatment.

In natural systems (e.g., storage or impounding), ammonia removal can be achieved through *Nitrosomonas* development. The process corresponds to a first-order reaction:

$$[NH_4^+] = [NH_4^+]_0 e^{-kt}$$

in which at 20°C, $k = 0.05$ g/m^3 per day. This means that we may consider that a 7-day storage period can remove

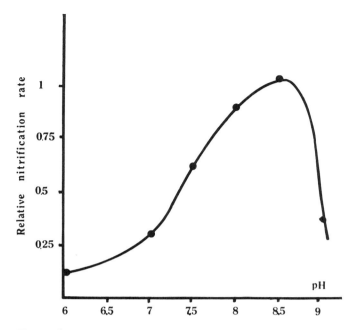

Figure 5   Relative nitrification rate as a function of pH.

$$\frac{[NH_4{}^+]}{[NH_4{}^+]_0} = e^{-0.05 \times 7} = e^{-0.35} = 0.7$$

which corresponds to a possible 70% reduction in the ammonia content during 7-day storage of the raw water. This holds under favorable meteorological conditions and when the carbon sources are not growth limiting.

In bacterial iron and manganese oxidation, the following reactions are implicated:

$$4FeCO_3 + O_2 + 6H_2O = 4Fe(OH)_3 + 4CO_2 + 22.6\,J$$

$$MnCO_3 + H_2O + \tfrac{1}{2}O_2 = MnO(OH)_2 + CO_2 + 12.9\,J$$

$$2FeO + \tfrac{1}{2}O_2 + 3H_2O = 2Fe(OH)_3 + 60.5\,J$$

$$2MnO + O_2 + 2H_2O = 2MnO(OH)_2 + 9.6\,J$$

A first observation is that bacteria capable of oxidizing either iron or manganese must generally oxidize four to six times more manganese than iron to produce the same energy. However, the thermodynamic balance of the bacterial growth is influenced only marginally by the energetics of the inorganic oxidations. Although autotrophic, the normal circumstances for the process to occur is the presence of an available carbon source, which must be more abundant in manganese oxidation than in iron oxidation (3). The bacterial species involved belong to the Chlamydobacteriales order [e.g., *Sphaerotilus* (*Leptothrix*) and *Crenothrix*] and to the Pseudomonadales [e.g., *Gallionella, Siderocapsa,* and *Naumaniella* (siderocapsaceae) and *Thiobacillus ferrooxidans*] (4).

The growth conditions of *Thiobacillus ferrooxidans* are exceptional: the germ is acidophilic and multiplies only poorly at pH > 4. Under laboratory conditions the pH can range from 1.5 to 2, the culture producing sulfuric acid. Even at pH 1 the germ can oxidize ferrous iron and gain the necessary energy for maintenance of growth. *Gallionella* is a typical "gradient organism" living in the boundary layer containing oxygen and anaerobic water. Although autotrophic, it uses small amounts of organic nutrient material to grow. Optimal growth temperature is ±25°C and the pH limits range from 5.8 to 6.6.

The exact generation time of iron- and manganese-oxidizing bacteria is not known but is supposed to be long. The microorganisms live in clusters and are difficult to enumerate. Therefore, cell-mass determination may be more representative, although disturbed by mineral precipitations in which the organisms live encapsulated. The overall compromise in bacterial oxidation of iron and manganese affects the pH value and redox potential as indicated in Fig. 6 (5).

3.3   Initial Lag Phase

The initial lag phase results when bacteria are introduced in a closed medium where the cells must adapt their enzymic set, which is the most appropriate to the environment at the moment. Constitutional enzymes are synthesized all the time, but the possession of a large number of substrate-inducible enzymes facilitates growth on various substrates. A typical case is the genus *Pseudomonas*, of which certain strains can metabolize more than 100 compounds. The *induced-enzyme synthesis* consists of the mobilization of a preexisting genetic code and must not be confused with *enzymic adaptation*, that is, genotypical change or mutation involving an irreversible change. The usual proportion of mutants can be 1 germ in 10^8. The selection of a mutant population can also underlie a lag phase as observed, which is then often

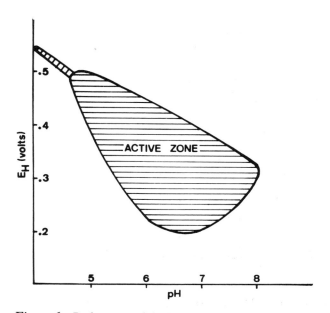

Figure 6   Redox potential–pH zone for bacterial oxidation of iron and manganese.

longer than in the case of induced enzyme synthesis. The lag phase can also occur when bacteria of a preceding stationary or decaying phase are seeded in fresh medium. With the cessation of growth, cells undergo biochemical changes, including enzyme catabolysis. Also, the cells are smaller and recover their size characteristics of the exponential growth phase before divisions occur. Therefore, to determine the lag phase, cells from a preceding exponential growth phase are best taken as inoculum.

The lag phase is best determined by viable counts rather than by cell mass. In this phase, dead or immature cells contribute to the cell mass. As a consequence, the initial lag is often observed in excess when based on cell mass rather than on cell number. Diauxy is sometimes observed in the presence of various nutrients (e.g., glucose and xylose for *E. coli.*). After exhausting the first nutrient (glucose for *E. coli*) an enzymic adaptation is necessary to continue growth when using the second nutrient (see Fig. 1).

3.4 Stationary Phase

Generally, and more particularly in drinking water, the exponential growth phase does not have long-lasting periods, since the concentrations of nutrients quickly reach low levels and restrict anabolism. Usually, the maximum bacterial concentration is limited by the maximal available nutrients in the medium. When an essential nutrient is exhausted, particularly, the (maximum) stationary phase occurs. When this phase occurs, owing to a growth-limiting substrate concentration (S), the stationary phase is observed, regardless of whether viable cell number or cell mass is considered. However, if the growth is limited, due to the accumulation of toxic materials, the culture may be in its stationary phase (or even in the declining phase) in terms of viable cell number, but still be increasing in cell mass. Under such

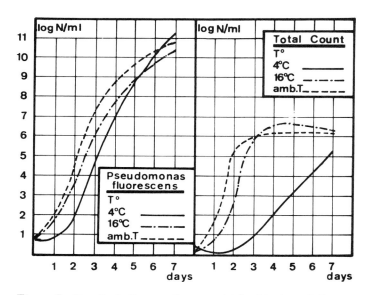

Figure 7   Practical aftergrowth curve obtained by inoculating groundwater into preozonized water and incubating on laboratory scale.

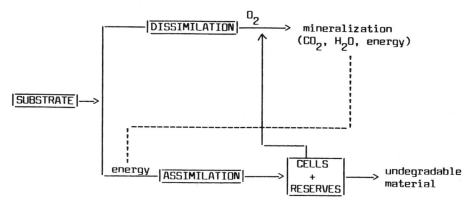

Figure 8 Assimilation–dissimilation scheme.

circumstances, the apparent stationary phase is a statistical phenomenon in which the marginal growth of some members is counterbalanced by the death of other members of the population. In the stationary phase endogeneous respiration can occur, establishing a transfer between assimilation and dissimilation according to the scheme shown in Fig. 8.

The proportion in which cell decomposition occurs depends, at least in continuous-culture systems, on the available amount of substrate. Furthermore, a lethal lag can be superposed on the stationary phase (end of growth), particularly in the disinfection of developing cultures (see Section 3.5).

The growth rate can be expressed as a function of the input rate of nutrient, X being the cell population density (mol/dm^3):

$$\frac{dX}{dt} = -y\frac{dS}{dt} \quad ; \quad -y = \frac{dX}{dS}$$

in which y is the (cell) yield factor with respect to the limiting substrate concentration S. Hence

$$\frac{1}{X}\frac{dX}{dt} = -\frac{y}{X}\frac{dS}{dt}$$

By defining the (instantaneous) growth rate constant as $r = yt_R$, where t_R is the mass transfer rate of the nutrient to the cell mass and by applying the Michaelis–Menten kinetic equation, a steady-state formulation is obtained as follows:

$$r = yt_r^{max}\left(\frac{S}{K + S} - \frac{S_i}{Ki + S_i}\right)$$

where S_i is the intracellular steady-state concentration of the limiting substrate.

Considering S_i as low and back-transportation to the medium as insignificant, we obtain

$$r = r_{\max} \frac{S}{K + S}$$

which is Monod's equation. The relation of r to S, hence the value of K, must be obtained by experiment. The concepts hold better in continuous-culture systems (6,7) than in batch systems, in which dissimilation and resulting back-transportation or equivalent nutrient returning from the cells to the medium can occur more significantly.

3.5 Cell Yield Factor

The cell yield factor, also expressed as growth yield or efficiency of growth, is usually given as $M = yS_L$, in which M is the cell mass concentration and S_L is the concentration of the growth-limiting nutrient consumed. The y value or growth yield is equal to the mass of cell material produced (M) per unit mass of nutrient furnished. By analyzing the carbon content in the dry cell mass and in the nutrient mass, the fraction of nutrient carbon converted to cell carbon can be determined and usually ranges between 20 and 50% for aerobic organisms. The remaining carbon consumed is for energy production (e.g., cellular ATP formation).

The growth rate of microorganisms is not affected by nutrient concentration unless it is fixed at very low values (e.g., a few mg/m^3). In this range the mean growth rate is about proportional to the concentration in substrate, which in this case is growth limiting. It has been shown (6) that diffusion of the nutrient to the biomass does not have a significant influence on the growth rate as a whole. The relationship of growth rates to substrate concentration is different for each strain (7). By selecting a *Pseudomonas* strain capable of metabolizing a large variety of carboneous substrates, one can obtain a group parameter defined as AOC (assimilable organic carbon) (8).

On the basis of a dry cell weight of 2×10^{-10} mg and an average composition of the cellular material as $C_5H_7NO_2P_{0.033}$, 50% of the dry cell weight is carbon. Through the metabolism, 50% of the carbon uptake is for energy production and 50% for incorporation into cellular material. This leads us to assume that per bacterial cell, $\pm 2 \times 10^{-10}$ mg AOC carbon is necessary and for 10^6 cells/mL a concentration of $10^6 \times 2 \times 10^{-10} = 200$ mg/m^3 is required. Since the weight proportion C/N/P is 120 : 14 : 1, the carbon sources are most often the growth-limiting compounds.

The practical determination of AOC is based on the following steps:

1. Isolation and selection of a psychrophilic *Pseudomonas* strain.
2. Determination of the growth-limiting correlation of the cell population density, X, versus carbon content, such as sodium acetate introduced in the water in the concentration range 20 to 300 mg/m^3 [a linear plot of log X versus log (C) helps in further interpolations].
3. The AOC is determined as being the equivalent carbon concentration as sodium acetate, supporting the same aftergrowth level as that supported by the sample water.

The underlying principle of the AOC test is Monod's growth-limiting equation without consideration of dissimilation effects.

3.6 Continuous-Culturing Systems: Chemostat

The maintenance of a microbial population in its exponential growth phase can be obtained for a long period in a continuous culture (e.g., in a system such as shown in Fig. 9). In the culture cell a volume V is maintained while the input corresponds to a flow Q. The culture dilution rate is then $Q/V = D$. The instantaneous growth rate of the population is $dN/dt = rN$ and the rate of cell loss by dilution $dN/dt = -DN$. The overall variation in cell population is equal to $dN/dt = rN - DN = (r - D)N$. In any practical situation, the instantaneous growthrate constant of a given strain cannot exceed a maximal value of r (i.e., r_{max}). The population in the growth chamber becomes stationary when r is equal to D. Under these circumstances, growth is self-regulating and the system is known as a *chemostat*. If the dilution rate is higher than r_{max}, excess outflow occurs and the culture declines. In practice, to obtain a culture in the exponential growth phase, the growth rate is

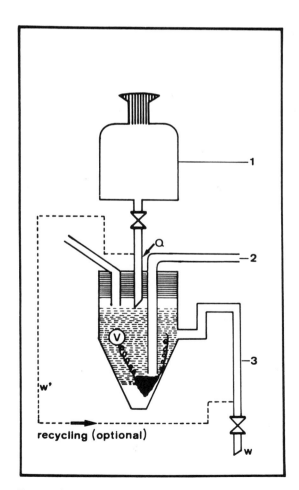

Figure 9   Continuous-culture system. 1, Sterile growth medium; 2, aeration mixing; 3, outlet (possible continuous control).

maintained at a submaximal value determined by a steady-state concentration of limiting nutrient so as to approach a value of r between zero and r_{max} by changing the Q value. Then the mean generation time can be expressed as $G = 1/k = 1/0.69r = V/0.69Q$ and adjusted to a given value by changing Q. The optional return flow W' can aid to run-in the continuous culture or to simulate a recycling process as applied in a full-scale activated sludge treatment.

When the limiting substrate is the energy source used in the chemostat, dissimilation can occur and the uptake rate of limiting nutrient (6) can be decomposed into two terms, one for growth ($t_{R,g}$) and one for maintenance ($t_{R,m}$), and $r = y(t_{R,g} + t_{R,m})$. The specific growth rate r is accepted as a linear function of the specific uptake rate of nutrients used for growth only; then $r = y_{max}t_{R,g}$, the y_{max} or maximum yield factor being defined as $\lim y = y_{max}$ ($t_{R,m} \to \infty$), so that

$$y = y_{max} \frac{t_{R,g}}{t_{R,g} + t_{R,m}} = y_{max} \frac{r}{a + r}$$

in which $a = y_{max}t_{R,m}$, which is called the specific maintenance rate, and

$$a + r = y_{max}(t_{R,g} + t_{R,m}) = y_{max}t_R = y_{max}(S_i - S)D$$

(S being the concentration of limiting nutrient and S_i the same in the feed vessel of the chemostat, and D the dilution rate). From this approach we obtain

$$y = y_{max} \frac{D}{D + a} \quad \text{and} \quad \frac{1}{y} = \frac{1}{y_{max}} + \frac{a}{y_{max}} \frac{1}{D}$$

and $1/y$ plots against $1/D$ will be acceptably linear when the energy metabolization scheme of the culture does not change significantly with dilution (i.e., the limiting substrate is the source of energy).

Remark. To improve the simulation by chemostat of practical water treatment schemes, sludge or solid supports (e.g., glass wool) can be introduced in the culture vessel. A distinction must be made according to the competition between organisms: that is, between the strains growing on supports and the free-living organisms in the culture medium. To simulate significant operational conditions, the vessel can incorporate a thermostat.

3.7 Microbial Interactions

In practice, all forms of microbial interaction described in the kinetics of enzymic reactions (9) can occur between the strains present and those capable of growth, by competition for the available substrates. The definitions of terms applied to microbial interactions and likely to occur in water treatment practice are the following, due to Bungay and Bungay (10).

Neutralism. Coexistence by lack of interaction measurable by bacterial growth rates. Neutralism is possible only at low population densities.

Commensalism. Growth of one member of the bacterial population is favored by the development of another strain that is not influenced by the first. (A typical example of this effect is decomposition by a resistant strain of a product that is toxic for the other.)

Mutualism. The growth of each member benefits from the existence of the other.

Competition. A race occurs for nutrient source and space. [If two organisms have closely overlapping metabolism for a growth-limiting substrate, competition may be severe and result in the dominance of one species according to the competitive elimination principle (11). In a mixed substrate system preferences may be low enough to enable coexistence with occupation of different ecological sites.]

Amensalism. One member changes the environment adversely for the other (e.g., pH changes due to metabolites).

Predation. One organism ingests the other (e.g., amoeba versus bacteria).

Self-inhibition. Auto-regulation of growth occurs as a function of general ecological environment (e.g., space, temperature, etc.) by releasing a growth-limiting cellular factor into the medium.

Synergism. (e.g., joined biochemical synthesis of a metabolite). No real evidence for synergism is yet available, but the observations are possibly explained by symbiotism.

Parasitism. One organism lives depending on and colonizing the other. Typical examples are viruses and phage.

Symbiotism. This is a general term for simultaneous growth of mixed populations with possible mutual growth stimulation. A symbiotic index (SI) can be defined for continuous-culture systems (when using the same medium and conditions in the situations compared):

$$SI = \frac{\text{growth of a species in mixed cultures}}{\text{growth of the same species in pure culture}}$$

Present mathematical modeling of microbial interactions are based on the generalization of Monod's equation (11,12). It applies to continuous-culture systems, for example, for binary interactions: two strains for one limiting substrate, two substrates for one strain (diauxy), and two substrates for two competing strains. The formulations involve interaction coefficients determined experimentally in any particular case by a trial-and-error procedure. The formulations are still not applicable directly to natural systems or treatment processes but serve as the underlying ideas in biological treatment schemes.

In practical systems, in addition to very specific industrial wastewater, the nutrients are always mixed and are individually present at growth-limiting concentrations. They are able to support the growth of a great diversity of organisms. It has been suggested (13) that in such dilute systems the microbial community may differ significantly from that grown in concentrated media. Furthermore, in natural systems, a steady state is not likely to occur. However, a stabilizing effect on population diversity results. In bacterial aftergrowth, with waters that have previously been disinfected, a selection of organisms occurs through competition and the cultures obtained in the first instances are usually due to selected monoculturing organisms. However, the prevailing organism changes with time and general conditions (e.g., temperature, oxygen contents, etc.).

3.8 Oxygen Uptake Rates; Biochemical Oxygen Demand

The concentration of decomposable organic matter in water is often characterized by the biochemical oxygen demand (BOD) test, which associates the oxygen demand with the concentration of organic matter. Among the practical observations one must note that the number of present initially bacteria does not significantly influence the stationary bacterial number after a 5-day incubation at 20°C—that for total counts. One reaches 10^{12} m^{-3}, and up to 10^{14} per m^{-3} for *Pseudomonas* species. In polluted water (e.g., sewage) the stationary level is reached after approximately 1 day. However, in treated water, a period of 3 to 5 days (14) may be necessary, especially at lower temperatures. A count of 20×10^{12} m^{-3} can be associated with an increase in suspended solids of 10 g/m^3 dry mass (up to 20 g/m^3) if it is assumed that bacteria of colloidal size behave as suspended solids and tend to settle in the "bioflocculation." The bioflocculation phenomenon is one of the most important sources of sludge containing organic matter: In a polluted stream, sludge banks or bottom deposits may form even when all primary solids capable of settling have been removed before discharge of the effluents.

In its first stage the deoxygenation can be described according to a kinetic process of the first order:

$$\frac{d(O_2)}{dt} = \frac{d(U - U_t)}{dt} = -d\frac{U_t}{dt} = k'U_t$$

$$U_t = Ue^{-k'U}t = U \times 10^{-kU}t \qquad \text{where } k' = 2.3k$$

$$(O_3) = U(1 - e^{-k't}) = U(1 - 10^{-kt})$$

U is the total first-stage oxygen uptake evidenced by the BOD$_5$ curve and U_t the uptake after time t; both uptake rates are expressed as consumed dissolved oxygen concentrations.

Nitrification is a process that evolutes more slowly than the aerobic degradation of organic matter. Two species are operating: *Nitrosomonas*, which converts ammonium (so-called Kjeldahl ammonia) to nitrite, and *Nitrobacter*, which oxidizes nitrite to nitrate. At 15°C, for instance, the ammonium saturation concentration for *Nitrosomonas* is 0.5 g N/m^3 and the doubling time is 84 h provided that the ammonium and carbonaceous material present is not growth limiting. The data for *Nitrobacter* are not as well established but are considered to be of the same order as for *Nitrosomonas*. For *Nitrobacter agilis*, a G value of ± 24 h has been estimated.

The counts of *Nitrosomas* and *Nitrobacter* in the inflowing sewage are 10^{-4} g/m^3 or, expressed otherwise, $<4 \times 10^5$ counts/m^3. This number is negligible to the population density in a nitrifying (active) sludge (e.g., up to 40 g/m^3 in activated sludge). Consequently, in a batch system such as the BOD test, no nitrification takes place during the first-stage oxygen uptake. The second part of the oxygen uptake versus time, known as the second stage or nitrification stage, corresponds to the part of BOD association with nitrification. In streams polluted by domestic sewage and having a water temperature between 10 and 20°C, the oxygen uptake in the second stage reaches approximately 30 to 35% of that obtained in the first stage. Limiting factors in the nitrification stage are oxygen (>1 g/m^3), temperature (>10°C), and pH (>7; <9). An empirical relationship for changes in the growth rate constant is the following:

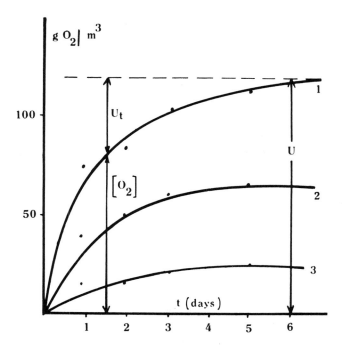

Figure 10   Curve 1, first-stage oxygen uptake by water of the river Senne downstream from Brussels; curve 2, after lime precipitation at pH 11; curve 3, after preozonization and lime precipitation.

$$k_{NO_3^-} \simeq \frac{0.69}{G \text{ (doubling time in days)}}$$

which can be given as

$$k_{NO_3^-} = 0.18e^{0.12(t-15)}[1 - 0.83(7.2 - \text{pH})]$$

which holds in the pH range 6.0 to 7.2. However, in waters less charged than activated sludge, the optimum pH for nitrification can be 8.5 (e.g., in oxygen-free groundwaters containing humic substances and Kjeldahl nitrogen). Nitrification is difficult in such cases. The entire second-stage oxygen uptake is the basis of nitrification in activated sludge treatment, which will be commented upon later.

Extensive investigations of the kinetic constant in first-stage oxygen uptake: $[O_2] = U(1 - 10^{-kt})$ have indicated that $k = 0.1$ at 20°C as an average for all domestic sewages and similar polluted waters. However, the extreme values can be 0.01 for industrial wastes or slowly oxidizable matter or even 0.3 for fresh, easily oxidizable litter. Moreover, the temperature has a direct impact on the deoxygenation constant, which is expressed by the following equation, the reference temperature t_1 chosen being 20°C (15):

$$\frac{k_1}{k_2} = \alpha^{t_1 - t_2}$$

in which α is the thermal constant of bacterial oxygen uptake. By analogy with the (chemical) Arrhenius equation, the following logarithmic correlation has been developed:

$$\log \frac{k_1}{k_2} = \frac{\mu(t_1 - t_2)}{2.3R(t_1 + 273)(t_2 + 273)}$$

in which μ is assimilated to an activation energy and found for natural systems to range from 1750 to 2050 J/mol. However, experimental values for α vary widely from one case to another and extrapolations from measurements, at a given temperature, may be hazardous when the qualitative composition of the water under investigation changes. In practice:

For domestic sewage:	$1.145 > \alpha > 1.065$
For moderately polluted rivers:	$1.047 > \alpha > 1.026$
For activated sludge:	$1.03 \;\; > \alpha > 1.0$
For oxidation ditches:	$1.04 \;\; > \alpha > 1.02$
For oxidation ponds:	$1.07 \;\; > \alpha > 1.03$

In mathematical modeling of river flow, determination of the thermal constant, α, is critical in evaluation of the initial oxygen uptake (16). Natural or artificial ponds can improve the water quality through aerobic oxidation processes operating at the open surface ($O_2 > 2$ mg/L) and anaerobic fermentation in the lower layers. The strict aerobic layer is of <0.3 m thickness and the anaerobic zone of 2.5 to 3 m. The combination of both effects constitutes a *facultative* oxidation pond capable of removing up to 75 to 95% of the BOD of domestic sewage. Empirical guidelines for design are

$$V = 3.5 \times 10^{-5}(\text{BOD}_{20})(IQ)\alpha^{(35-t)}$$

in which V is the volume in m^3 (for a depth of 1.5 m), I the number of equivalent inhabitants, Q the daily flow, and t the temperature in °C. The temperature coefficient can be accepted as 1.08. To avoid odor problems the daily BOD_{20} input should be kept below 2.5 g/m^2 of pond surface.

3.9   Biological Growth Processes in Groundwaters

The aquifer is a complex ecosystem in which bacterial competition occurs in all instances. All of the processes involved in the metabolism of microbial organisms are capable of affecting the groundwater quality: decomposition of organic matter, oxidation of ammonium, reduction of nitrates or sulfates, oxidation of hydrogen sulfide, oxidation of reduced-state iron and manganese, production of methane, and eventually, oxidation of the latter.

    1. The oxidation of ammonium salts corresponds to the nitrification process with the scope and limitations inherent in this transformation. The global reaction scheme is discussed in Section 3.2.

    2. Denitrification in the subsoil can occur according to the following reaction scheme:

$$Ca^{2+} + 2NO_3^- + 3H_2 + C \rightarrow CaCO_3 + 3H_2O + N_2$$

An organic carbon source is necessary as an energy source for the process in which *Pseudomonas* species intervene. The potential accumulation of nitrogen is not desirable, as the permeability may be altered through gas clogging of the pores of the subsoil formations.

3. The reduction of sulfates produces noxious sulfydric acid according to the reaction

$$4H_2 + SO_4^{2-} \rightarrow H_2S + 2H_2O + 2OH^-$$

The species occurring most frequently in sulfate reduction are *Desulfovibrio desulfuricans*, which are ubiquitous. Knowledge of the conditions for optimal development is poor, but what is known is that the germs are strictly anaerobic when cultivated in the laboratory, and traces of ferrous iron favor their development. In natural conditions, the process is usually not observed in waters containing nitrates and therefore enabling denitrification. The energetics of the latter appears to be privileged compared to the sulfate reduction. On slight aeration or introduction of dissolved oxygen in the water, biological sulfate reduction stops and a total count is established at somewhat higher levels than formerly. A typical example is that of groundwater catchment as utilized by the Brussels Waterboard (CIBE) in cretaceous soil.

	Before aeration	After aeration
Desulfovibrio desulfuricans per liter	40–90	—
Plate count (22°C) per milliliter	10–70	50–100
Pseudomonas fluorescens per liter	0	500–1000

Aeration in the groundwater table, or in wells, is a method that can resolve the problems associated with sulfate reduction. The latter also involves corrosion by reaction of H_2S, producing FeS.

4. Oxidation of sulfides under aerobic conditions is caused by the colorless sulfur bacteria of the genus *Beggiatoa*. The principal reactions are

$$2H_2S + O_2 \rightarrow 2S + H_2O$$
$$2S + 2H_2O + 3O_2 \rightarrow 2SO_4^{2-} + 4H^+$$

5. The activity of iron- and manganese-oxidizing bacteria is discussed in Section 3.2.

6. Methane is known to form by fermentation under anaerobic conditions from simple organic acids such as acetic acid, which percolates through peaty soils to reach the water table. The scheme is expressed as follows:

$$Ca(C_2H_3O_2)_2 + H_2O \rightarrow CaCO_3 + CO_2 + 2CH_4$$
$$2H_2 + CO_2 \rightarrow CH_4 + 2H_2O$$

To form methane, extreme reducing conditions prevail (e.g., an E_H value of -0.4 V). Upon changing conditions [e.g., presence of dissolved oxygen at low concentrations (<4 g O_2/m^3)], methane-oxidizing bacteria can cause the formation of methanol.

7. Each of the microbial transformations noted as taking place in groundwater can occur in the distribution system, particularly if anaerobic conditions exist in long-lasting stagnation zones or if the post-disinfection measures are found to be inadequate.

4. DECAY RATE PHENOMENA

4.1 Death Phase

The *death phase* of bacteria takes place at the maximum stationary phase when the medium has been depleted of nutrients or enriched in bactericidal agents. The death rate, just like the growth rate, proceeds exponentially: It sometimes decreases as a result of the consumption of nutrient materials released by dead cells; while a certain autolysis may occur, which is the digestion of the cell material by the enzymes of the cell itself that is the assimilation–dissimilation action. Without multiplication, bacteria indicating fecal pollution cannot survive for long periods (e.g., several days in water free of nutrients), as illustrated in Fig. 11. The overall decay rate is expressed by an exponential function of the type

$$R_t = \frac{N_t}{N_0} e^{\alpha t}$$

in which α is a negative parameter (e.g., -0.3 day^{-1}), for *E. coli*.

In utilizing this decay equation, the germicidal action can be treated as a dy-

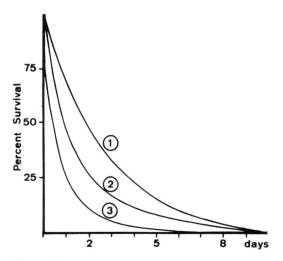

Figure 11   Exponential decay rate curves. 1, *S. faecalis*; 2, *A. aerogenes*; 3, *E. coli* C.

namic time-dependent process, expressed by general kinetic equations in which the extent of killing varies with the nature of the organism, the nature and concentration of the bactericidal agent, and the temperature and characteristics of the medium, such as pH, salinity, and so on. The first attempt to quantify the kinetics of bacterial decay is that of Chick (17). The fundamental equations are known as the Chick's law:

$$N_t = N_0 e^{-kat} \quad \text{or} \quad \ln \frac{N_t}{N_0} = -kat$$

and

$$\log \frac{N_t \times 100}{N_0} = \log(\% \text{ survival}) = \log 100 - \frac{ka}{2.3} t$$

in which N_0 and N_t are the number of cells at the beginning and after t time units, respectively.

Another expression for first-order decay or killing rate is given by the fraction of organisms killed within a unit time t:

$$p = 1 - \frac{N_t + 1}{N_t} \quad \text{or} \quad 1 - p = \frac{N_t + 1}{N_t}$$

in which N_t represents the number of surviving germs in a unit volume after t time units.

$$\frac{N_1}{N_0} \frac{N_2}{N_1} \frac{N_3}{N_2} \cdots \frac{N_t}{N_{t-1}} = \frac{N_t}{N_0} = (1 - p)^t$$

and

$$\ln \frac{N_t}{N_0} = -kat = t \ln(1 - p)$$

$$p = 1 - e^{-ka}$$

4.2 Killing Time

Disinfection results are often indicated by the time (or dose of disinfectant) necessary to kill organisms initially present. The evidence of the killing effect is then stated by the absence of surviving organisms in a tested sample. Examination of the kinetic expressions of the germicidal effect of a disinfectant reveals that the fraction of surviving organisms N/N_0 cannot become zero except at an infinite contact time. Consequently, it is illusive to justify a complete kill in any finite situation. Practically, the discrimination between a small fraction of surviving organisms and zero organisms present depends on the sensitivity of the techniques of investigation. The recording of a result or of a quality requirement (e.g., zero germs) is a frequent practice which, theoretically, cannot be entirely justified on sound statistical grounds. The global efficiency of a germicidal treatment should always be expressed in terms of dosing of disinfectant necessary to kill $x\%$ of the organisms within a time t, or the contact time required in a given situation to kill $x\%$ of the organisms originally present.

4.3 Chick-Watson Law

Chick-Watson law has been completed by Watson to account for the concentration dependence of the bactericidal action of a given disinfectant. The apparent decay rate constant, ka, depends on the concentration of the disinfectant, for example, as

$$\frac{ka}{2.3} = \frac{k}{2.3} C^n$$

Consequently,

$$\log N\% - \log N_0\% - \frac{k}{2.3} C^n t$$

(in which $\log N_0\%$ should normally be 2). n is the concentration coefficient of the disinfectant under consideration. Its value can be obtained from at least two survival curves. At a given percent survival rate,

$$C_1^n t_1 = C_2^n t_2$$

$$n \log C_1 + \log t_1 = n \log C_2 + \log t_2$$

$$n = \frac{\log (t_2/t_2)}{\log (C_1/C_2)}$$

For example, if a given strain of bacteria in water is killed up to 90%, either by 0.2 ppm chlorine after 4 min or by 0.1 ppm chlorine after 10 min, one obtains

$$n = \frac{\log (10/4)}{\log (0.2/0.1)} = \frac{\log 2.5}{\log 2} = 1.33$$

The disinfection time (e.g., the time required for a 99.5% killing) is related to the concentration of the disinfectant. This relation is linear on a log–log correlation for a particular organism capable of promoting waterborne diseases (Fig. 12). There are numerous examples of good conformity of bacterial decay expressed by Chick-Watson law. However, a number of deviations are observed, as indicated in Fig. 13.

4.4 Multihit Theory

In multihit theory, a number of interactions n to a single sensitive target of an entity are considered necessary to completely kill or inactivate the organisms. If n hits are necessary for a killing action, all organisms with $n - 1$ or fewer injuries survive. If the probability of a single interaction within a time t is equal to a, the chance of n equal hits is given by the nth term of Poisson's series:

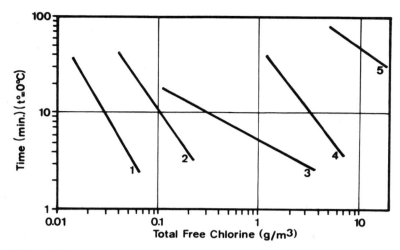

Figure 12 Exponential decay rate as a function of concentration of disinfectant. 1, *E. coli*, pH 7; 2, *E. coli,* pH 8.5; 3, Coxsackie, A_2, pH 7; 4, Coxsackie, A_2, pH 9; 5, *Histolytica,* pH 8.

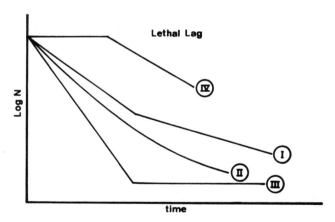

Figure 13 Typical deviations from Chick-Watson law. *Type I:* If the plot of log *N* against *t* consists of two distinct segments, the most attractive interpretation is that of the composite survival of two populations of organisms having different specific resistance to the germicidal agent. *Type II:* If the graph is a smooth curve, it means either gradual inactivation of the germs or consumption of the disinfectant. *Type III:* Agreement with Chick-Watson law is good except that a limited number of organisms survive. This phenomenon often occurs when bacteria are sheltered by suspended matter. *Type IV:* There is a lethal lag phase. This may be the result of a preparatory step before effective disinfection [e.g., mixing of the disinfectant to the water, penetration in the cell, chemical transformation of the disinfectant when introduced into the water (hydrolysis)]. An alternative explanation is that of the multi-site and multihit killing.

$$P_n = \frac{(at)^n}{n!} e^{-at}$$

The fraction of organisms that have received fewer than n hits and which survive is equal to

$$\frac{N}{N_0} = P_0 + P_1 + P_2 + \cdots + P_{n-1}$$

$$= -e^{-at} \sum_{i=0}^{i=(n-1)} \frac{(at)^i}{i!}$$

$n = 1$; $N/N_0 = e^{-at} = e^{-kat}$. This would indicate an equation structured as Chick's law by assimilating a to ka.

$n = 2$: $N/N_0 = (1 + kat)e^{-kat}$
$n = 3$: $N/N_0 = [1 + kat + \frac{1}{2}(kat)^2]e^{-kat}$

The drawback of this approach is that it is indeterminate, as both k and n vary independently and have to be deduced from a single experimental measurement. By generalization of the multihit equations above, Wei and Chang (18) have developed a multi-Poisson distribution model to enable an interpolative interpretation by superposition of the calculated graph to the observed decay rate curve (see Fig. 14).

A typical example of the multihit interpretation is the change in apparent decay rate constant as a function of the n-hit number. Among the experimental methods for achieving a suspension of single-cell spores or cysts, ultrasonication is the most useful to assess the multihit number or n value.

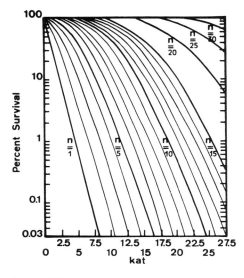

Figure 14   Generalized multihit killing graph.

4.5 Multisite Theory

In the multisite or multitarget concept, a number of vital centers m must each be hit once to kill the organism. For first-order kinetics, proceeding to every hit, the fraction of the total number of sites that is hit within time t is equal to $1 - e^{-kat}$, and the probability of m sites to be hit within time t is

$$P_t = (1 - e^{-kat})^m$$

and the fraction surviving is given by

$$1 - P_t = \frac{N}{N_0} = a - (1 - e^{-kat})^m$$

Binomial expansion of the mortality probability function gives

$$P_t = (1 - e^{-kat})^m = 1 - me^{-kat} + \frac{m(m-1)}{2} e^{-kat^2} - \cdots$$

$$\simeq 1 - me^{-kat}$$

and

$$\frac{N}{N_0} = me^{-kat} \qquad \text{or} \qquad \log \frac{N}{N_0} = \log m - \frac{kat}{2.3}$$

Most interesting in this approach is the independence of the m values in the linear portion of the decay curves—which means that the curves are parallel. The site number m is deduced graphically, independent of the ka value corresponding to the linear portion of the decay curve. Moreover, the m value can be considered in theory as independent of the disinfectant applied to a given strain under constant conditions.

However, the m value can result from global appreciation: number of sites to be inhibited, and number of organisms living in clusters and to be entirely devitalized, to deliver a zero count. This phenomenon may be of particular importance in the analysis of membrane filtration methods applied to flocculated waters, in which bacteria may be agglutinated and counted as single individuals. However, interpretations on the basis of Chick's law remain useful in the generalized decay rate law, according to Morris: The rate of devitalization is assumed to be expressed by an equation of the following general form:

$$\frac{-d \ln N}{dt} = -(F)kc \frac{N}{N_0}$$

in which (F) is again a function of N/N_0, that is, a dimensionless function such that $F^*(N \to N_0) = 1$.

Consequently, seeing that kc is inversely proportional to the time for a given fractional killing rate at a selected time (e.g., t_r, which is equal to the time required

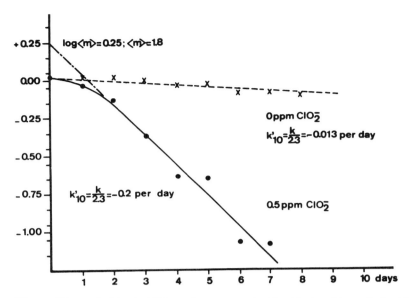

Figure 15   Lethal lag in killing of enterococci with chlorite. (From Refs. 19 and 20.)

to render $N_t/N_0 = 1/e$), kc can be substituted by x/t_r, and $\log(N/N_0)$, as a function of t/t_r, should express a universal function of F. Substituting t_R for t and $1/e$ for N/N_0 in this case, one obtains

$$\frac{1}{e} = 1 - (1 - e^{-kctR})^n$$

in which

$$kct_R = -\ln\left[1 - \left(1 - \frac{1}{e}\right)^{1/m}\right]$$

and

$$\frac{N}{N_0} = 1 - \left\{1 - e^{t/tR}\ln\left[1 - \left(1 - \frac{1}{e}\right)^{1/m}\right]\right\}^m$$

in which m is the number of sites to be hit in the multiple-target-single-hit killing action. Graphs for interpolation are given in Fig. 16.

4.6   Germicidal Action

Germicidal action corresponds to an endothermic reaction that is favored by an increase in temperature. The effect is expressed in terms of variations in the specific killing rate, constant k, according to the Arrhenius equation: $k = Ae^{-E/RT}$, in which

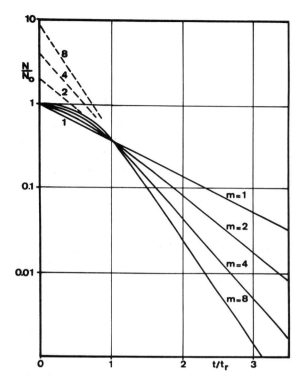

Figure 16   Dimensionless multitarget/single-hit killing graphs.

E is the activation energy and A a temperature-independent constant. The equation is also expressed as

$$\log k = \log A - \frac{E}{2.3RT}$$

and

$$\frac{d \log k}{dT} = \frac{E}{2.3RT^2}$$

Seeing that practical water temperatures do not change within an extended range (e.g., between 0 and 25°C), one can approximate the fractional change in k per degree as $E/2.3RT^2$, or, as a percentage, by $100E/2.3RT^2$. In the temperature range considered, $RT^2 \simeq 39.5$ kJ/°C. Most activation energies for bacterial killing range from $E \simeq 1.9$ to 3 kJ, which amounts to a 5 to 8% change in rate per degree. Also, by integration,

$$\log_{10} \frac{k_1}{k_2} = \frac{E(T_2 - T_1)}{2.3RT_1T_2}$$

and for a small (e.g., 1°) difference,

$$\log \frac{k_1}{k_2} \simeq \frac{E}{2.3RT_1T_2}$$

4.7   Effect of pH and Salt Content on Survival of Bacteria

Besides the effects of pH, causing chemical changes in the germicidal agent, acidification or alkalinization usually hastens the death rate of bacteria present in water. Therefore, all survival experiments are best carried out on the water that is to be distributed. Compared to the intracellular fluid of the bacteria, distilled water is hypotonic, while a concentrated saline solution is hypertonic. Effects on cell membranes may alter the results. If the experiments cannot be carried out with real water, the solutions used should preferably be isotonic (e.g., 9 g NaCl/L).

It may be necessary to buffer the pH of the water. It must be noted that some buffers can interfere with disinfectants (e.g., organic buffers with chlorine, chlorine dioxide or ozone, phosphate buffers with silver ions, and perborate buffers with chlorine dioxide or ozone). Therefore, the interpretation of bactericidal results should always emphasize these external parameters.

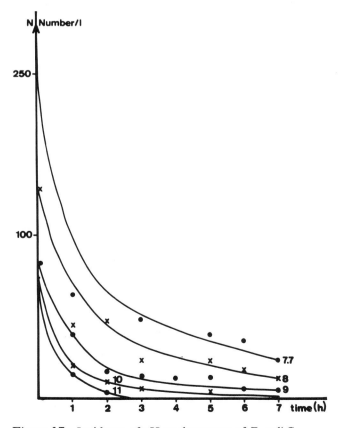

Figure 17   Incidence of pH on decay rate of *E. coli* C.

The introduction of lime to the water, which increases the pH, was experimented with as an operational method for disinfection in treatment plants. For water with a pH of 11 to 11.5, 2 h was necessary to kill the known pathogenic bacteria; at a pH of 10.5, up to 11.4 h of contact time was required. These pH values necessitate a final acidification (Fig. 17).

Until now the foregoing method has been limited to emergency use in critical circumstances where the usual disinfectants are insufficient. To use the method, sufficient contact time must be provided in the design of the treatment plant. This application can play a secondary role in water-softening processes.

4.8   Action of Disinfectants on Living Cells

Most chemical disinfectants used in drinking water treatment are oxidizing agents, (e.g., chlorine, chlorine dioxide, ozone, permanganate, etc.). Earlier, the action was supposed to be that of nascent oxygen released by the oxidant (e.g., $HOCl = HCl + O$ or $O_3 = O_2 + O$). *General oxidation* of the organisms was an alternative to nascent oxygen theory. The oxidation characteristics are determined by the oxidation-reduction potentials (ORP), and this concept was long the basis for interpretation of the disinfection efficiency. However, in practical efficiency sequences, diffusion of the active oxidants can intervene predominantly (e.g., $KMnO_4$ is not a very efficient disinfectant, although its E_H value is high). Therefore, the diffusion characteristics of the disinfectant can play an important, even dominant role in the disinfecting efficiency.

Table 2   Oxidation–Reduction Potentials of Disinfecting Agents

Reaction	Potential at 25°C $E°$ (V)
$F_2 + 2e + 2F^-$	2.87
$O_3 + 2H^+ + 2e = O_2 + H_2O$	2.07
$H_2O_2 + H^+ + H_2O + 2e = 4H_2O$ (acidic)	1.76
$MnO_4^- + 4H^+ + 3e = MnO_2 + 2H_2O$	1.68
$HClO_2 + 3H^+ + 4e = Cl^- + 2H_2O$	1.57
$MnO_4^- + 8H^+ + 5e = Mn^{2+} + 4H_2O$	1.49
$HOCl + H^+ + 2e = Cl^- + H_2O$	1.49
$Cl_2 + 2e = 2Cl^-$	1.36
$HOBr + H^+ + 2e = Br^- + H_2O$	1.33
$O_3 + H_2O + 2e = O_2 + 2OH^-$	1.24
$ClO_2(gas) + e = ClO_2^-$	1.15
$Br_2 + 2e = 2Br^-$	1.07
$HOI + H^+ + 2e = I^- + H_2O$	0.99
$ClO_2(aq) + e = ClO_2^-$	0.95
$ClO^- + 2H_2O + 2e = Cl^- + 2OH^-$	0.9
$H_2O_2 + H_3O^+ + 2e = 4H_2O$(basic)	0.87
$ClO_2^- + 2H_2O + 4e = Cl^- + 4OH^-$	0.78
$OBr^- H_2O + 2e = Br^- + 2OH^-$	0.70
$I_2 + 2e = 2I^-$	0.54
$I_3^- + 2e = 3I^-$	0.53
$OI^- + H_2O + 2e = I^- + 2OH^-$	0.49

4.9   Practical Used Disinfectants

The disinfectants most commonly used at present are chlorine, chlorine dioxide, and ozone. In some instances several others are also used, such as chloramines, potassium permanganate, silver ions, bromine, and iodine as well as UV (254 nm) irradiation and ultrasonic cell disintegration. The latter two, physical means are used either alone or simultaneously with the usual reagents to upgrade their activity.

For the various disinfectants used in practice, the sequence of decreasing activity in terms of efficiency is at present as follows: O_3, ClO_2, Cl_2 (HOCl, NaOCl, OCl^-), Br_2 and NH_2Br, NH_2Cl–$NHCl_2$, UV rays, I_2, $KMnO_4$, H_2O_2, and Ag^+. (H_2O_2 is a newcomer. Not highly active by itself, it may be a significant disinfectant in association with other agents. More research must be undertaken to establish the feasibility of its use.) The sequence as given here differs from the oxidizing power sequence and results partly from the diffusion capability of the oxidant enabling cell penetration. Guidelines for design of the units to apply the above-mentioned disinfectants are outlined in the individual chapters concerning each reagent and process.

SYMBOLS AND UNITS RELATED TO MICROBIAL GROWTH AND DECAY

a	specific maintenance rate, t^{-1} (in bacterial growth)
a	interaction probability in chemical disinfection, dimensionless
g	generation number, dimensionless
k	mean growth rate constant or generation rate constant, t^{-1}, generally h^{-1}
k	corrected decay rate constant, t^{-1}
ka	apparent decay rate constant, t^{-1}
kc	generalized decay rate constant in multitarget killing, t^{-1}
k,k'	oxygen uptake rate constants, t^{-1}
m	number of vital centers, dimensionless
n	concentration factor of a disinfectant, dimensionless
r	instantaneous growth rate constant, t^{-1}, generally h^{-1}
r_{max}	maximal (possible) instantaneous growth rate constant, t^{-1}
t_r	time necessary to lower the bacterial count in the ratio $1/e$, time units
t_R	mass transfer rate of nutrient, mt^{-1}
$t_{R,g}$	mass transfer of nutrient for growth, mt^{-1}
$t_{R,m}$	mass transfer rate of nutrient for maintenance, mt^{-1}
u,u_t	oxygen uptake; in concentration units (e.g., mol/dm^3)
w'	return flow in a chemostat, selected flow units
y	yield factor (cell mass with respect to nutrient mass consumed), dimensionless
y_{max}	maximal yield factor, dimensionless
A	temperature constant in bacterial killing, dimensionless
AOC	assimilable organic carbon, units as carbon concentration
BOD	biochemical oxygen demand; oxygen concentration (e.g., g/m^3)
C	molar concentration of chemical, mol/L
D	dilution rate, t^{-1}

E	activation energy in bacterial killing, J/mol
E_H	oxidation–reduction potential versus normal hydrogen electrode, V or mV
G	generation time, t, generally h
I	equivalents inhabitant, dimensionless
K	Monod or Michaelis constant, dimensionless
M	cell-mass concentration (e.g., g/L)
N	bacterial count, dimensionless
OD	optical density, dimensionless
P	fraction of organisms killed per unit time, dimensionless
P_n	nth term of Poisson's series, dimensionless
P_t	probability of hit, dimensionless
Q	flow, (selected flow units) m^3/s, L/s, L/min
R_t	fraction of surviving cells at time t, dimensionless
S	substrate or nutrient concentration, concentration
S_i	intracellular steady-state concentration of nutrient, concentration
SI	symbiotic index, dimensionless
S_L	concentration of limiting nutrient, concentration units
T	absolute temperature, K
V	volume, dm^3 or m^3
X	cell population density, in concentration, generally mol/dm^3
α	thermal constant of bacterial oxygen uptake, t^{-1}
μ	activation energy, J/mol, kJ/mol

REFERENCES

1. F. J. Hanus and R. Y. Morita, *J. Bacteriol.*, *95*, 736 (1968).
2. H. Seppanen and M. Wunderlich, *Ann. Bot. Fenn.*, *7*, 58 (1970).
3. H. Seppanen, *Nordisk Hydrologisk Konference*, Aalborg, Vol. II, 1974, p. 420.
4. R. E. Buchanan and N. E. Gebbons, Eds., *Bergey's Manual of Determinative Bacteriology*, 8th ed. Williams & Wilkins, Baltimore, 1974.
5. R. S. Wolfe, in *Principles and Applications in Aquatic Microbiology*, H. Heukelekian and N. C. Dondero, Eds. Wiley, New York, 1964, p. 82.
6. N. Van Uden, *Ann. Rev. Microbiol.*, *23*, 1536 (1969).
7. M. L. Bungay, *Adv. Appl. Microbiol.*, *10*, 269 (1968).
8. D. Van der Kooij, *H₂O*, *12*, 269 (1979).
9. A. L. Lehninger, Chap. 8 in *Biochimie*, Flammarion, Paris, 1977.
10. H. R. Bungay and M. L. Bungay, *Adv. Appl. Microbiol.*, *10*, 269 (1968).
11. H. Yoon, G. Klinzing, and H. W. Blanch, *Bioengineering*, *19*, 1193 (1977).
12. N. Van Uden, *Ann. Rev. Microbiol.*, *23*, 473 (1969).
13. E. Stumm-Zollinger, *Appl. Microbiol.*, *14*, 654 (1966).
14. W. J. Masschelein, G. Fransolet, and E. Debacker, *Eau Quebec*, *13*, 289 (1980).
15. T. R. Camp, *Water and Its Impurities*, Reinhold, New York, 1963, p. 247.
16. E. W. Moore, *Sewage Works J.*, *13*, 561 (1941).
17. H. Chick, *J. Hyg. (Cambridge)*, *8*, 92 (1908).
18. J. H. Wei and S. L. Chang, in *Disinfection Water and Wastewater*, J. D. Johnson, Ed., Ann Arbor Science, Ann Arbor, Mich., 1975, p. 11.
19. W. J. Masschelein, in *Oxidationsverfahren in der Trinkwasseraufbereitung*, University Karlsruhe, Karlsruhe, Germany, 1979, p. 484.
20. W. J. Masschelein, *Water S.A.*, *6*, 116 (1980).
21. J. C. Senez, *Microbiologie générale*, Doin, Paris, 1968, p. 201.

$$14$$

Filtration Powders and Diatomaceous Earths

1. INTRODUCTION

Filter powders are largely used in the food industry, but their use in the treatment of drinking water is still limited. One of the reasons, therefore, is that they are more appropriate for limited flows smaller than what is required for the supply of potable water. The filtration method as used is a consequence of World War II, during which diatomite or diatomaceous earth was used to filtrate water for the troops. It has been applied successfully for the purification of water in swimming pools, in which only a part of the water is treated and where a temporary failure of the filtration system is sometimes accepted. This is completely opposite to the reliability required for the purification of potable water. We should also mention that in building installations for filtration with powders, certain valuable principles are not always applied. The recommendations for good design are treated in Section 5. At present, a thousand installations for the treatment of potable water using this method exist in the United States.

Although versatile in operation and to some extent labor intensive, filtration with powders and diatomaceous earths is very successful for the removal of parasites, animalcules, nematode eggs, and larvae. The city of New York is undertaking a major pilot-plant investigation to update and optimize the process by possible automation. Changes in the concept and operation can be anticipated upon release of the results of this extensive investigation.

2. POWDERS AND POWDER FILTRATION

The concept *powder* relates to the grain size of a material: the diameter of powders is always smaller than 150 μm. Larger diameters are described as granules. The technical use of powders in the preparation of potable water supposes:

The choice and characterization of a filter powder, sometimes also called a filter aid

The choice of a support layer

Operational principles and parameters for design

Evaluation of costs and benefits

Powder filtration is used primarily in two different circumstances: for more accurate treatment of surface water, already purified by conventional techniques; and, for direct treatment of raw water from reserve storage, to ensure a supply during peak periods. In this case, diatomite filtration is generally used for economic reasons, because the use of such peak stations, based on diatomaceous earth filtration, is less costly than conventional methods. Both methods are in use at the Brussels Waterboard (CIBE): at Yvoir-Champale and Spontin, respectively.

2.1 Filter Powders

1. Diatomaceous earth or diatomite is nearly pure SiO_2 in the form of skeletons of dead diatoms. In a sufficient pure condition, the material is chemically inert. Among other advantages, it does not consume chlorine and does not give a bad taste. The powder has an inner porosity of approximately 100 to 200 m^2/g BET surface. The granulometry is discussed later, but only material with a grain size smaller than 100 μm and with average diameters between 10 an 50 μm is suitable. Because of the inner porosity, the powder is light; in other words, it has minimal weight per liter (e.g., 0.4 kg/L). Owing to the principles of depth filtration, the powder can also withstand hydraulic variations. Consequently, diatomite is by far the principal filter powder used in the treatment of potable water.

2. Microsand, with a comparable grain size differs from diatomite in being of greater weight per liter (approximately 1.5 in place of 0.4). Thus it provides a far more compact filter layer than diatomite and a considerably shorter filtration cycle. The use of microsand to treat water remains limited to swimming pools and to the acceleration of floc formation (by the seeding effects in precipitation).

3. The use of asbestos fibers is avoided for general hygienic reasons (i.e., possible carcinogenic properties of chrysolite when handled under dry conditions).

4. Cellulose fibers present a relatively high degree of hydration: 80 to 90% humidity. Swelling of the filter material takes place on the adsorption of water. Consequently, the specific resistance (see Chapter 11) is higher than that of diatomite. These powders are hygienically safe; however, for economic reasons (250 BF/kg, 1990) they are less well adapted to the treatment of potable water. Yet the material serves as an ideal bottom layer in aiding floc formation. The residue is flammable and thus appropriate for sludge treatment by combustion.

5. Wood powder has the same advantages as cellulose fibers but reacts with chlorine and is not advisable for drinking water treatment. Impurities of lignin are suspect because of the bad taste and possible formation of chlorine compounds. However, wood powder can be used as an additive during the filtration of sludge on vacuum drums.

6. Magnesite as an adduct can considerably improve diatomaceous earth filtration during iron and manganese removal, among other uses (1).

7. Gyrolithe, burned calcium silicate, has shown encouraging results, but its use remains experimental (2).

8. Pumice stone or comparable artificially expanded granite (e.g., perlite) is a possible competitor for diatomite. The material presents a weight per liter 20 to 39% lower than that of diatomite (i.e., 0.3 kg/L). This parameter can be adjusted at any value according to the conditions of preparation of the material. The specific resistance can be compared with that of diatomite. The compression during the filtration is lower than that of diatomite.

9. Carbon powder, as activated or nonactivated carbon, is also taken into account for filtration. Nonactivated carbon is somewhat coarser than diatomite or perlite (i.e., with an average diameter of 80 to 150 μm). The material is specially adapted for the filtration of alkaline liquids that can dissolve silicium dioxide. It is used less in the preparation of potable water. Carbon powder of a 10- to 140-μm granulometry is not used directly because of break-through of the filter septa. Moreover, the thickness of the layer of filtration powder is insufficient to reach a kinetically sufficient adsorption efficiency, considering the short period of contact with the water.

10. Calcite or calcium carbonate in powder is less adequate since the layer porosity greatly decreases with increasing layer thickness.

11. Because of their grain size (i.e., 10 to 50 μm), absorbing powders such as Fuller's earth, bentonite, and kaolinite do not permit filtration. They can eventually be used as filter aids.

12. Slime can certainly be considered as the first filtration material in powdered form. The filtration is progressively refined, owing to the fact that smaller parts are retained on the layer. Here it concerns a typical surface filtration.

13. The operation of different filtration powders can be improved by aids (e.g., polyelectrolytes or coagulant aids). When using chlorine, their efficiency can decrease. The use of iron or aluminum salts as coagulants may increase the efficiency of the filtration but induce rapid clogging of the filtering layer.

2.2 Support Layer

Filter septa must be constructed of noncorrosive material. Rigid as well as flexible structures can be used, but the former are preferable where water quality is concerned. The risk of distortion and rupture of the filter cake is then smaller. The various materials used therefore present both advantages and drawbacks.

1. Porous ceramic, aluminum silicate, or silicium carbide are appropriate but in general too fragile.

2. Sintered metals made of stainless steel or Iconel with pore openings of 3 to 165 μm allow sufficient elimination. However, the risk of clogging is very high.

3. Perforated stainless steel plates are not sufficient since the smallest sieve opening is approximately 75 μm. They are really good support layers for the cloth filters that comprise some filter septa.

4. Metal cylinders surrounded with Monel thread make it possible to fabricate filtering candles with a sieve opening of 5 μm and higher. Generally, the opening between the threads ranges from 40 to 50 μm. Thus it is the filter powder itself that determines the membrane-type porosity of the support layer, while the filtering candle acts as a sieve.

5. Tissues in stainless steel wiring (AISI 136 quality is required for water chlorination) are available with mechanical openings down to 5 μm. The most commonly

used for flexible filtering candles are the linen type, with warp and weft of the same thread thickness. The inner support is assured by a helical spring. For cloth filters in stainless steel, we usually employ the satin mode with reinforced warp. Because of the flexibility of the tissues, changes of 1 to 10 can occur in the openings. Silting is also liable to occur in the tissue.

6. Porous plastics are presently available in PE, PTFE, PVC, and PU. Pore openings can be as low as a few micrometers. Thin layers (approximately 10 μm thick) are available. Until now, their practical use in the filtration of potable water has been limited.

7. Fabrics are often used together with support layers in metal. The most commonly used fabrics together with disk filters are propylene, polyesters, and polyamides such as nylon. Further data on these fabrics are available in the literature (4). Some information is also given in Chapter 16.

8. Paper is not used on an industrial scale in drinking water treatment plants. However, on a domestic scale, it is used more and more commonly, combined, for instance, with active carbon powders (e.g., for dechlorination). Normally, the openings of the pores are smaller than 1 μm — in any case, below 5 μm. Precipitation of minute amounts of iron hydroxides, even at acceptable concentrations in water, often hinders continuous operation of the process.

9. Filtering candles made of agglomerated diatomite are also available on the market. Here we are concerned with hollow cylinders that can be disinfected in an autoclave. The normal working pressure is limited to 25 to 30 kg/m^2 to avoid breakage. The possible flow per filter remains lower than 3×10^{-4} m^3/s, with loading of approximately 1.4 mm/s.

3. QUALITY OBJECTIVES OF FILTRATION USING DIATOMACEOUS EARTHS

The origin of the use of diatomite filtration for potable water is the elimination of pathogenic germs and parasitic organisms in water for armies during World War II. Unlike sand filters, cysts of amoebae such an *Entamoeba histolytica* are also sufficiently well removed by this technique (5). Viruses and bacteriophage f2 are eliminated with a yield of 75 to 95%. The yield can be increased by pretreating the powder with polyelectrolytes (6). The elimination seems to occur during filtration with a powder layer rather than by the addition of powder to raw water, so-called "body feed" (7). Coliform bacteria are also retained sufficiently. This effect is improved by the presence of iron, owing to simultaneous flocculation.

Higher-structured organisms such as crustaceae and larvae are also eliminated: for example, *Copepoda* and *Diaptomus* (8). Colloids causing color or turbidity problems are generally eliminated with a yield of 50 to 60% and 95 to 98%, respectively. Diatomite is more efficient than microsand for this purpose, showing better adsorbtion. Here also, the effect is increased by the preparation of powders with polyelectrolytes. Algae and unicellular organisms are eliminated directly by diatomite filtration (10), but at the expense of a rapid increase in head loss.

Removal of iron and manganese is satisfactory. Concentrations below 0.05 to 0.1 ppm are reached in a single filtration stage. This effect is improved by the addition of polyphosphates for manganese removal, for example. As an alternative, potassium permanganate can increase the velocity of precipitation of manganese in

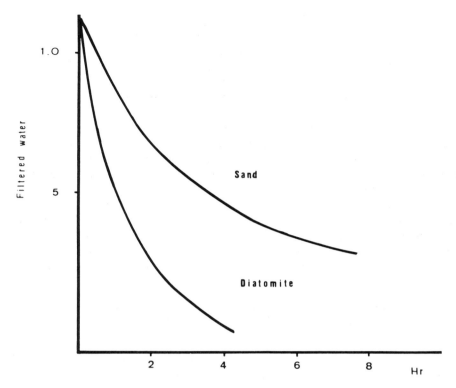

Figure 1   Improvement in turbidity by filtration on diatomaceous earth (units JTU).

solution, due to oxidation by the classical mechanisms to form manganese dioxide (11). When the pH of the filtrated water is between 7.4 and 7.9, the yield for manganese removal is 80 to 85%.

Softening by precipitation with lime can be very efficient in conjunction with postfiltration on diatomaceous earth. Indeed, we must take into account an accelerated increase in the loss of head and a corresponding shortened filter cycle (1). Thanks to the formation of a deposit of MgO in the case of a pH higher than 9.6, the elimination of bacteria is also promoted (5). Radioisotopes are also partially adsorbed by filtration on diatomite (12).

4. NATURE, COMPOSITION, AND CHARACTERISTICS OF DIATOMITE

4.1 Nature of Diatomite

Diatomaceous earth, diatomite, and even Kieselguhr are fossilized rocks consisting of dead diatomaceae occurring in most clay or lime deposits. To render their exploitation profitable, the deposits must be of sufficient purity and the geological formations of sufficient thickness.

Diatomaceae belong to the vegetable world, in particular to the Pheophyceta or brown algae. Chlorophyll is masked by the phycopheine they contain. The latter is soluble in soft water. Diatomaceae show all the properties of unicellular organisms: nucleus, protoplasm, and so on. The skeleton consists of nearly pure silica; nevertheless, the membranes of the walls of the living organisms consist of cellulose. Like greens, diatomaceae can be fed with minerals in the presence of sunlight. Alive, diatomaceae are thus to be found at depths where sunlight still penetrates.

Diatomaceae live as well in fresh water as in salt water. A concentration of 1.25% and more in salt strongly promotes their growth. At 3.25% salt content, optimum development of "sea diatomaceae" takes place. The species that live in lakes and pools containing sludge (i.e., the so-called "Sapropel"), furnish nonhomogeneous deposits and are taken into consideration less for exploitation as a filter powder since they contain a certain level of impurities. It is estimated that approximately 7.2% of the surface of all the seas in occupied by diatomaceae. The Sargasso Sea is an intensive area of cultivation, where there are up to 200,000 cells per column of water of 1 m^2. A layer of fossils can be exploited from a thickness of 5 to 10 m. The winning area of Lompoc (California) has a layer thicker than 400 m.

Diatomite include skeletons and chips from a great number of species (at least 200). The principal ones are needle- and disk-shaped:

Needle-shaped: Nitzschia, Synedra, Thallasionema, and Thallasiothrix
Disk-shaped: Actinoptychus, Arachnoidiscus, Aulacodiscus, Auliscus, and *Coscinodiscus*

The percentage of the needle-shaped variety fluctuates between 15 and 85%.

By microscopic analysis one can determine the diversity of the various types of diatomites. Such analysis takes place after washing in concentrated sulfuric acid and rinsing with water. Then a drop of slurry is put on a plate and covered with a slide. Another method consists of evaporating the water, pouring out the sample in cedar oil, and analyzing the residue by immersion. The magnification recommended is from 150 to 200. From the literature data (13) it appears that the filtration velocity decreases when the percentage of disk-shaped species increases.

4.2   Composition of Diatomite

The SiO$_2$ content can vary between 85 and 95%. The filter aids used by the food industry are in general calcined products. By calcination, one decomposes the remaining organic material, immobilizes the carbonates, and evaporates the water. Besides SiO$_2$, diatomite usually contains a few inorganic oxides. Typical analysis for well-known trademarks are given in Table 1. The pH at the surface is approximately 7 to 8 but can reach 10. The loss of volatile compounds is determined by heating at 980 to 1000°C. Among the impurities, arsenic alone appears to be considered from the viewpoint of health, but its maximum content remains below 10 mg As/kg powder.

4.3   Characteristics of Diatomite

4.3.1   Physical Properties

The general physical properties of diatomite are the following for fresh powders:

Table 1 Weight Percent Composition of Commercial Diatomaceous Earths

Product	SiO_2	Al_2O_3	Fe_2O_3	TiO_2	P_2O_5	CaO	MgO	$K_2O + Na_2O$	H_2O	Calcination loss
Celite 535 (Iceland)	89.8	0.8	2.8	0.37	2.1	6.9	0.6	2.3	–	0.1
Celite 545 (Spain)	86.4	2.6	0.3	0.05	0.36	1.1	0.35	2.7	–	0.16
Filtercel (raw, U.S.)	85.2	3.8	1.5	0.2	–	0.5	0.6	1.2	4	3
Supercel (U.S.)	91	4.1	1.6	0.2	–	0.6	0.6	1	0.5	0.4
Primisil 741 (France)	90.7	3.6	1.78	0.5	0.05	0.55	0.4	0.95	–	0.15
Sil-flo 443 (perlite)	76	14	0.06	–	–	0.44	0.04	7.5	–	1.9
Codaflow PR-H-4 (perlite)	74.4	15.35	0.43	–	–	0.44	1.53	4.26	0.43	2.11

Melting point: 1500 to 1600°C (if not pure, possibility of 1200 to 1300°C)
Softening point: 1100°C
Allotropic change: 870 to 875°C
Light refraction index: 1.44 to 1.46
Specific heat capacity: 1 kJ/kg
Density: 2.35 kg/L
Apparent density
　　　Wet: 310 to 380 kg/m³
　　　Dry: 180 to 270 kg/m³
Outer surface: 1 to 2 m²/g
Porosity: 75 to 85%

For determination of the *apparent density* of the dry powder, or the weight per liter, one proceeds as follows. The volume of a given weight of powder is determined after being placed in a volumetric cylinder under standard vibration conditions such as ASTM D 2854-70 or DIN 53 194.

The wet *density* is determined as follows: 1 g of sample is placed in a centrifugation tube and moistened with distilled water. After having obtained a suspension, the tube is filled to 14 mL in total volume and centrifugation is carried out at 30 rev/s for 300 s. The volume of deposit is to be read with a precision of 0.05 mL.

4.3.2　Electrical Properties

Diatomite has a negative zeta potential with values included between 25 and 50 mV (14,15) (Fig. 2). By the addition of less than 0.1 g/m³ cationic polyelectrolytes to the water to filter, there is eventually a charge reversal and a positive charge at the

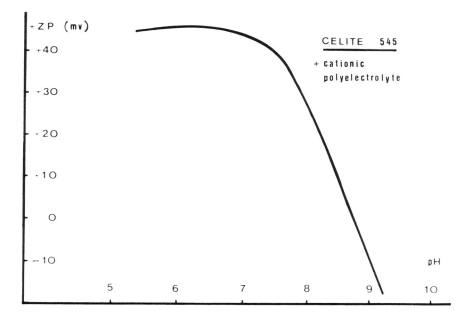

Figure 2　Zeta potential of diatomaceous earth.

coated surface. As an alternative, one can pour out 5 g of diatomite in a 250-mL solution of polyelectrolyte at a concentration of 20 g/m³. After 15 min of mixing, the maximum adsorption is reached.

A positively charged filter mass has the property of improving the quality of the water filtered by retaining colloidal particles through the mechanism of mutual coagulation. This effect appears in the elimination of color, bacteria, and viruses (6). The filter aid can reach a zeta potential of up to + 43 mV (15).

More conventional coagulants such as aluminum or iron salts were also described as possible aids in changing the electrical loading of natural diatomite (16). This effect can be used during floc filtration when, because of an excessive dosage, not all the added aluminum or iron salt has flocculated and settled. The elimination of the positively charged floc then occurs on the negatively loaded diatomite. With continuous dosage, an inversion of charge can occur on the filter cake. An improvement in the quality of the water filtrated, obtained by the use of electrically inversed powders, is also due partially to hydration of the settled cations, among which are aluminum and iron salts. This results in a decrease in permeability.

4.3.3 Granulometry

The size of the diatomite grains can play a very important part in filtration: It determines the initial permeability of the filter layer and therefore also improvements in quality at the beginning of the filtration cycle. Determination of the grain size for fine powders causes difficulties, mostly because of the phenomenon of agglutination. Therefore, the determination is done using a settling test for the analysis of soils (e.g., ASTM method D 422, alternatives 39, 63, and 61 T).

A sample is poured into water. Agglomeration of the particles is avoided by the addition of dispersion means such as sodium silicate or sodium hexamatophosphate. To obtain complete dispersion, air is blown through the solution for 5 min at a pressure of 3.5 kg/m². The method limits degradation of the powder grains as contrasted with that resulting from mechanical agitation. After dispersion, the density of the suspension is measured with a hydrometric balance. By the application of Stokes' law, the dimensions of the parts are determined as a function of the relative settling velocity.

For the most representative determination, concentrations of 20 to 30 kg/m³ are employed. Typical hindered sedimentation appears from 40 to 50 kg/m³ and higher. The result is expressed, as in screen tests, as a "percent weight thinner than" (see Fig. 3) as a function of the log of the diameters of the particles. Particle size determination using a Coulter counter permits more accurate determination of grain size below 100 μm. Here we also use a means of dispersion for instance, CTAB (cetyltrimethylammonium bromide), to avoid interference by agglutination.

Dry sieving gives rise to overevaluation of dimensions. Wet sieving remains an acceptable method for the practical choice of powders; however, it is less accurate than the techniques mentioned earlier. However, in industry, sieves are generally available. In contrast to the techniques described earlier, no use is made of dispersion means, and thus no preconditioning of the surface of the grains is necessary.

For the wet sieving method, one lets water seep through a thin layer of diatomite. This is the best approach under the conditions of industry. This method is also applicable on powders of sawdust and cellulose. The data given in Fig. 3 are typical for cellulose powders that can be used for water filtration. Screens with an

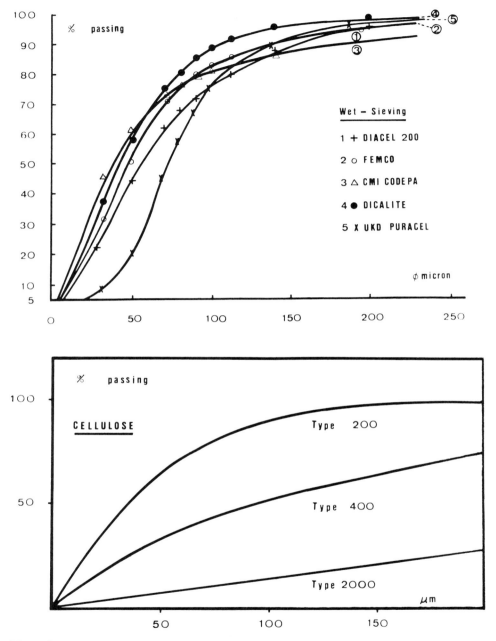

Figure 3   Granulometric curves of diatomaceous earths and cellulose powders.

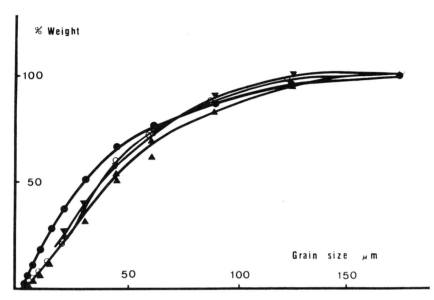

Figure 4 Granulometry as determined by laser diffusion. ▲, Dicalite 4200 + Puracel 100 S; ▼, Celite 535; ●, Coda Flow PRH4; ○, Celite 545.

air-jet device (such as the Engelsmann type) abrade the grains of diatomaceae. Laser-light diffusion based on forward diffusion is, at present, the most reliable way to determine the granulometry of filtration powders (Fig. 4).

4.3.4 Permeability

The permeability of a layer diatomite plays a fundamental part in filtration. In relation to this, one may cite the Kozeny–Carmen function (17):

$$\frac{p^3}{(1 - p)^2}$$

where p is the volumetric porosity. From tests it appears that pressure modifications between 500 to 70,000 kg/m^2 do not involve a significant change in the porosity of diatomite. It is thus a practically incompressible filter material, in contrast to other means, such as cellulose.

Hutto (17) has determined a porosity function using the formula X/L (i.e., the specific weight of diatomite divided by the thickness of the layer of filter material).

For wood powder, the relative decrease in porosity on the surface of the filter layer is due to the settling of "flocculated" powder on the surface. This interpretation and the different values of porosity mentioned in the literature (17) cause us to be more attentive to the hydration of the filter powders that were used in the Spontin plant of the Brussels Waterboard.

As a test, a given weight (e.g., 4 g of powder) has been poured into distilled water and hydrated for one night. Then the suspension was filtrated on a filter membrane in glass fiber at 5000 kg/m^2 vacuum "just sucked dry." The wet sample was then weighed and the result expressed as follows:

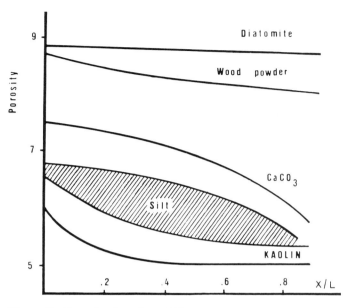

Figure 5   Porosity function of filter aids.

$$\% \text{ humidity } = \frac{(\text{wet weight } - \text{ dry weight}) \times 100}{\text{dry weight}}$$

Ideally, this percentage represents the hydration water or the absorbed water by the filling up of the inner pores. On the other hand, the specific resistance of a filter cake is also measured (on a relative basis): On 4 g of powder we filtered 1 L of raw water from the river Meuse (Tailfer plant) at a positive pressure of 500 kg/m². A clear difference between the behavior of diatomite and other means is evidenced by this determination (see Fig. 6).

The practical "permeability" of the filter layer (K_3) can be defined by the head loss (H), in meters, divided by the flow (Q) in m³/s · m²and the weight of powder per surface of layer (W) in kg/m². The full comparison according to Baumann also takes makeup into account:

$$H = K_3 Q W + H_2 \qquad (H_2 \text{ is due to makeup})$$

Initially, only K_3 is significant as a physical characteristic of the material. Typical values for standard diatomite are located between 0.6 and 1.2×10^{-2} s/kg. For finer powders, one can obtain values up to 3×10^{-2} s/kg. These powders are nearly unusable for the treatment of potable water (16). To become acceptable, deposits on the surface of the grains can sometimes increase the value of K_3. According to this point of view, one must also distinguish the effect of different polyelectrolytes.

4.3.5   Adsorption

Actually, no representative test has been developed for the adsorbing properties of diatomite. The bleaching of sugar solution is sometimes suggested by producers.

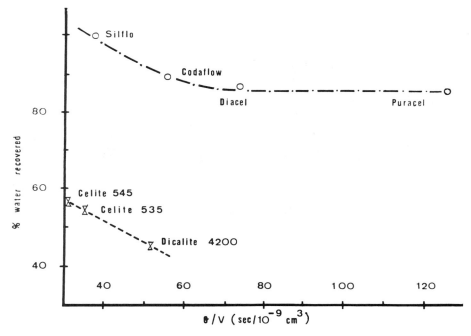

Figure 6   Influence of water retention on specific resistance.

5. PRACTICAL REALIZATION OF DIATOMITE FILTRATION

Diatomite filters exist as both pressure and vacuum filters. The former are used up to a pressure difference of 15 to 20 g/mm^2 and the latter at 4 to 7 g/mm^2.

5.1 Pressure Filters

Pressure filters can consist of candle or plate filters. For candle filters, cylindrical elements are hung vertically into the water. The filtration occurs from outside the candle toward the inside. Here, as for other filters, it is most important that the support material of the septa not become too deformed under the influence of pressure. Therefore, rigid candles surrounded by Monel threads are preferred. Fabrics of stainless steel on a spring structure are less reliable. The former are usually backwashed with air (so-called "air bumps") from the inside of the candle to the outside. The powder settled is separated in this way, followed by backwashing with water. For supple septa in the form of a candle, the powders can sometimes be separated by shaking the surface (Fig. 7).

Plate filters for diatomite filtration under pressure consist of classical press filters with square or circular frames. The diatomite is first coated on the filter cloths. After completion of the cycle by increase of the head loss, cleaning can take place in different ways:

1. Dry cleaning through mechanical scraping of the filter cake
2. Wet cleaning through rinsing of the exhausted filter cake (usually occurs on circular plates submitted to a rotating movement during rinsing in a vessel filled with water)

Figure 7   Cleaning supple septa by shaking.

3.   Wet cleaning through backwashing with pressure water, although less used in
     this case

5.2   Vacuum Filters

Vacuum filters in a closed operation are comparable to the pressure filters mentioned above. The only difference is that the water is sucked under vacuum through
the filter cloth and septa, not pushed through them. A possible difference is the
open design, where the filter elements remain visible. The septa can also maintain a
flat structure and thus be easier to clean and control (12). Rotating drum filters
under vacuum are very efficient for diatomite filtration. However, their use seems
to be limited to sludge treatment rather than to drinking water filtration.

5.3   Filtration Properties

The support layer of the septum has to satisfy a certain number of requirements.
Beyond the necessary mechanical resistance, proper porosity, and corrosion stability, one must assure a homogeneous deposition of the filter powders over the total
surface of the septa. The "hydraulically equal" distribution of powder over all the
septa is an important element of a good support material. As a guideline for septa,
it is recommended that the loss of loading of a pure septum at the maximum filter
velocity not exceed 50 mm column of water. Mainly for vacuum filtration, the
accumulation of gas bubbles on the septum must be avoided. Electrolytic corrosion
is one of the most common difficulties with metallic septa.

The thickness of the layer of the diatomite on the pure septum may not exceed the distance between the septa (18). Most filters are built with a distance between the pure septa of approximately 2 to 2.5 cm (1 in.) (8). For a fresh layer of powder settled on the septum, the head loss may not exceed 35 mbar for a maximum flow of water. The practical thickness of layer for the filter material can thus vary between 1.2 and 3.5 mm. The minimum quantity of diatomite necessary is therefore approximately 500 g/m² filter surface (19). As a guide, 1 to 1.5 kg/m² can be recommended (20). As a possible maximum value, one may consider 3 kg/m² (1). To avoid irregular layer structure, it is necessary to wet the diatomite properly before dipping.

As classical filtration velocity, one can accept 0.7 mm/s (2.5 m/h) as an average. Variations between 0.3 and 12.5 mm/s are mentioned in the literature (21). The powder filtration can thereby be considered as a more-or-less conventional rapid filtration. The fundamental differences are layer thickness and grain size. The filtration mechanism is more one of surface filtration than of depth filtration. The hydraulic variations therefore have an important role in the normal working of this type of filter. In the plants of the Brussels Waterboard, the filtration velocities are the following:

Lienne: 0.56 mm/s (2 m/h)
Yvoir–Champale: 1.4 mm/s (5 m/h)

6. GUIDELINES FOR THE DESIGN OF DIATOMITE FILTRATION AND FILTRATION WITH OTHER POWDERS

It is not sufficient to employ a good powder and proper filter for diatomite filtration. The design must also be adequate.

6.1 Plant Operations

Diatomite filtration requires regular cleaning periods. If the production has to remain continuous, each plant will thus have to dispose of a minimum of two parallel filters or a parallel series of several filters. For good operational control, we must be guided by the following (18):

1. Measurement of the head loss for each filter or series of filters separately
2. Use of a flow meter per filter with, if possible, a totalizer
3. Use of a meter for the quantity of diatomite or other filter powder introduced in the system
4. If possible, constant control of the quality of influent and effluent

6.2 Filtration Process

For filtration of raw water, the average length of a filtration cycle is between 10 and 12 h. For the final filtration of already sand-filtered water or for treatment of swimming pool water, one may expect a filtration cycle of 24 to 48 h with a head loss below 1 kg/cm². Cleaning and restoring to service requires 20 to 30 min. During the period, the production of the unity concerned is interrupted.

Backwashing with water is very short in duration, 2 to 3 min. Normally, the consumption of wash water remains under 1% of production. For the design of the filters, one must also take into account separation of the settled powder through the "air-bumping" or "autopact" technique. The filter vessel must resist an overpressure 200% higher than the operational pressure, to accept the pressure surges for the separation of the powders. This method, possible only for pressure filtration, make it possible to economize filter powder.

6.3 Hydraulic Factors

The good running of the filtration depends on a certain number of hydraulic factors. These principally concern the inclusion of air, breakdown of the powders, and mixing conditions for the preparation of the suspension.

1. Diatomite must be made wet in advance. This operation occurs in a tank provided with a slow mixer (e.g., $G \leq 20 \text{ s}^{-1}$, slower than 4 rpm). One may also use rooms with spatter plates. The concentration of the suspension is generally maintained below 15 wt%. The first or beginning of coating or "makeup" is operated at approximately 500 to 750 g/m^2. Obviously, wetting before the first coating of the septa is done with pure water. For the supplement, or "body feed," raw water is sometimes used. To maintain diatomite in suspension, it is necessary to have a minimum flow velocity of 0.07 m/s. Values used in practice can be four to five times higher.

2. The transfer of the suspension must occur without suction or inclusion of air; if not, it can present important drawbacks. If there is pump cavitation, the hydraulic shock causes the powder to fall from the septa. This generally induces an irregular covering of the septa later. Erosion of the powders can also occur through dispersion by air (22). When air is settled on the septa or is enclosed in the powder layer, the first head loss is too high. Also, due to pressure changing, a part of the powder can break loose from the supporting septa during gas exhaust.

3. The transport pipes for diatomite in suspension must be as short as possible, and straight, to avoid undesirable zones of deposit or abrasion through friction. In this context, the following pipe diameters are to be considered as indicative: 5 mm for 15 mL/s, 25 mm for 50 mL/s, and 50 mm for 90 mL/s. For the transport of diatomite in suspension, centrifugal pumps are recommended and are used up to a value of 5 to 50% of the nominal flow capacity. According to this method, one can avoid the abrasion of the powder.

4. The addition of diatomite to raw water preferably occurs by dosage of a suspension of the powder through a pipe bend of 90 to 45° and a pipe length of 5 to 6 m upstream of the filter inlet.

5. Typical examples of practical arrangements are discussed later, including the Yvoir–Champale and Spontin (Lienne) plants of the Brussels Waterboard.

6. With diatomite filters placed for security or more accurate purification, short-circuiting in the pipes is to be avoided between incoming and outcoming water. This is naturally even more severe in connective with filtration of raw water. Tightness of the valves isolating the filters when mounted in series or in parallel is absolutely necessary. The water that comes out during powder filling may not be mixed with the filtered water. It must be either discharged or brought back to a drainage basin.

7. Potentially, the sludge of the filtration contains only matters coming from raw water and silica from diatomite. Thus it concerns primarily suspended solids rather than pollutants. The sludge from diatomite filtration drains well and is perfectly adapted for landfill.

Technically, it is of the utmost importance that the cleaning or detaching of powders by shaking supple septa be very complete. If not, there is the possibility that the septa will gradually clog at some sites, and consequently that during the following cycle a regular recoating is not obtained. (This complication occurs regularly, especially with supple candle filters.) Discharge into public sewers can lead to obstruction. Therefore, the sludge concentration must be maintained lower than 2 kg/m^3. The limit for the transport velocity without significant sedimentation is approximately 25 mm/s in this case.

All classical means of drying and compacting, centrifugation, press filters, vacuum box filters, and so on, are appropriate for treatment of sludge. According to the literature (23), the following technical uses are also possible: filling for roofs; ceramic, brick, or asphalt products; and abrasives.

7. OPERATIONAL PARAMETERS

The practical data mentioned in this chapter are based on experience acquired in the Yvoir–Champale (Figs. 8 and 9) and Spontin-Lienne (Figs. 10 and 11) plants of the Brussels Waterboard. The operation of the plants is discussed below.

Figure 8   Battery of diatomaceous earth filters at Yvoir–Champale.

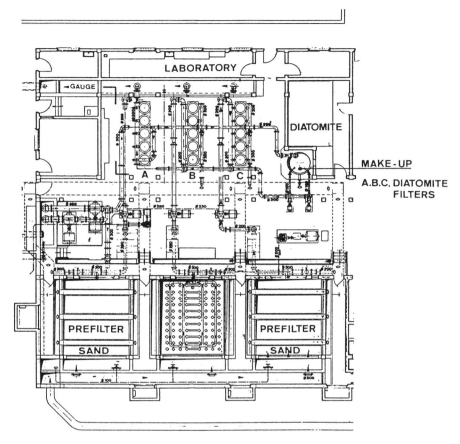

Figure 9   Schematic of the Yvoir plant.

7.1   HEAD LOSS

During the filtration of raw water, as in Lienne on supple septa in metallic fabric, the head loss must be limited to 10 g/mm^2. If not, deformation of septa can occur. This leads to irregular coating of the filter powder after a few cycles. For a given layer of filter medium, the head loss increases in a quadratic manner as a function of the time of operation or the volume filtered. Experimental curves illustrate this evolution (see Fig. 12) for a coating of 1 kg/m^2 septum. For filter velocities up to 0.7 to 0.8 mm/s, the operation time decreases more or less linearly, while for higher filter velocities, this tendency rapidly levels out. In other words, the volume of water produced per filter surface is a function of head loss (Fig. 13).

By the addition of filter medium (body feed), clogging of the filter layer can be slowed down. In the meantime, the head loss becomes a linear function of the time (see Fig. 14). An optimum yield is difficult to reach in practice when the quality of the raw water is subject to variations.

Figure 10   Diatomaceous filters at Spontin–Lienne.

7.2   Autopack–Recoat

The autopack–recoat system (24) is worth special attention. This is used nearly every day in the Yvoir plant for a finer filtration after rapid sand filtration. For the autopack portion, compressed air is brought into the filter (valve 16 in Fig. 16a, the water supply being interrupted). Gradually, a situation like that shown in Fig. 16b evolves. The valve is opened and by the sudden expansion of air inside the candle, powder is flushed down from the septa. After this operation, the powder can be either withdrawn or recoated on the septa.

In the autopack–recoat system, the use of filter powders has been considerably extended (see Fig. 17). After the operation, the freshly filtered water must be discharged because of turbidity. Another typical example from Yvoir is shown in Fig. 18. During washing of the primary sand filters (2.2 mm/s) in this station, an autopack step is operated simultaneously for peak consumption. This makes it possible to use the powder longer.

7.3   Septa Characteristics

The two types of filters operated at the Brussels Waterboard differ from each other by the pressure the septa can resist without deformation: 10 g/mm^2 (supple) and 30 g/mm^2 (rigid). The concentration of the suspension coated on the septa must remain

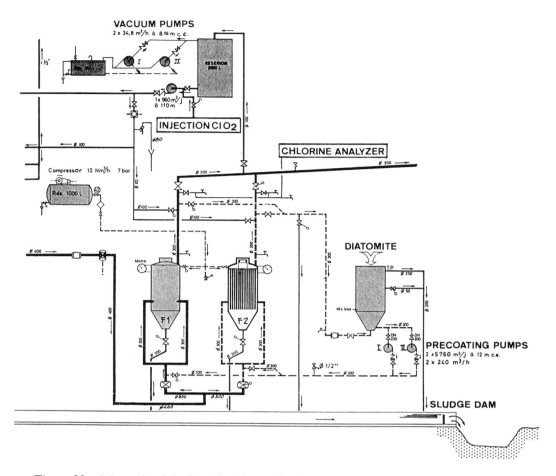

Figure 11   Schematic of the Spontin–Lienne plant.

under 15 wt% to allow complete hydration. It is best to work with a maximum concentration of 25 to 30 kg/m^3 water. A normal septa coating rate of 500 g/m^2 is acceptable. It is very important that the layer of coating be of uniform thickness, with a variation of no more than about 10 wt%. Sometimes this causes problems with supple septa, especially when iron and manganese are present in the water.

Bridging of the zones between septa is to be avoided. When this condition is not fullfilled, the flow is hindered and cracks appear in the filter layer. When cracks in the filter layer cannot be avoided, it is best to use the system with continuous powder feed (so-called "body feed"). This is done, preferably, by an extrahigh dosage of a diatomite suspension with a concentration of 2.5 to 5% in the water to be filtered. In this case one must take care that there is very good mixing of the powder in the raw water. This dosage could be 100 g of diatomite per hour per square meter of filter surface. This occurs more frequently with supple septa than with rigid ones.

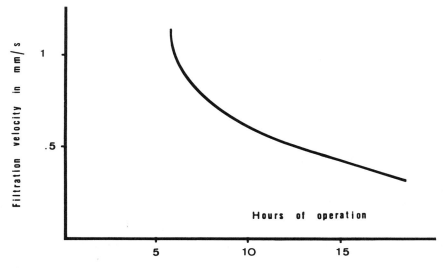

Figure 12   Relationship filtration velocity versus length of cycle.

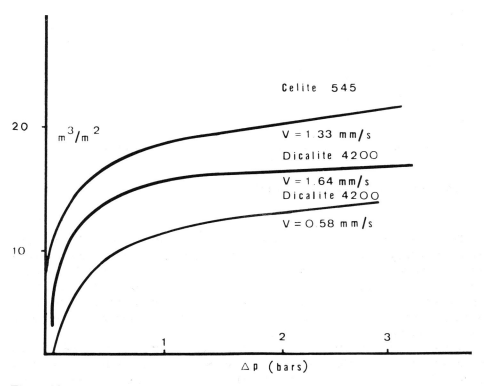

Figure 13   Production per unit surface as a function of head loss.

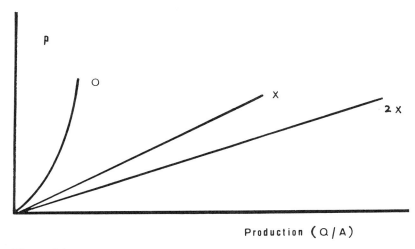

Figure 14   Effect of body feed (schematic head loss in ordinate). (From Ref. 16.)

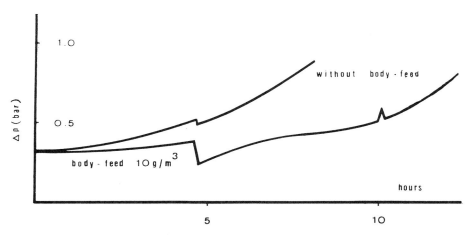

Figure 15   Effect of body feed (Spontin–Lienne).

The septa must be cleaned at least once a year. For the filtration of raw water, maintenance may be required every 2 to 3 months. At the Brussels Waterboard, it is done after each production campaign at the annual peak period. First, the septa are dry-brushed mechanically. Sometimes further mechanical cleaning by brushing in water is necessary. This operation must be carried out thoroughly, especially for rigid septa surrounded with Monel wires. (Rinsing with water under pressure seems to be efficient only for septa with a filter cloth support layer.)

Septa supported by anticorrosive material, especially AISI 316 steel, can be treated with 2 to 3 wt% HCl to eliminate less soluble salts. Afterward, it is necessary to rinse thoroughly with water. It is better to finish cleaning by immersing in a solution of diluted bleach water (e.g., 10 g/m^3 active chlorine for approximately 30 min). Afterward, the septa are kept dry.

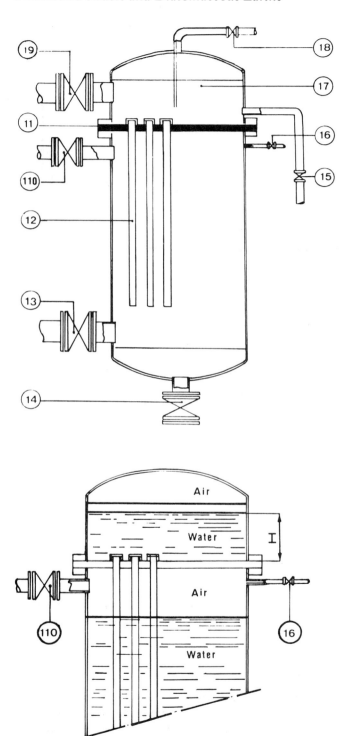

Figure 16 Principle of autopack–recoat operation.

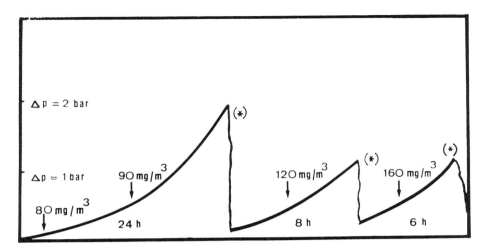

Figure 17 Effect of autopack–recoat on the filtration cycle. $v = 0.7$ mm/s; raw water: 2300 mg/m³ dry solids; (*), autopact–recoat.

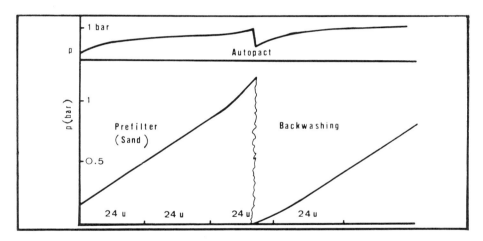

Figure 18 Combination of autopack–recoat with backwashing of prefilters.

(During filtration it is recommended that a minimum concentration of residual chlorine of 0.1 g/m³ be present in the water to avoid formation of a slime layer through bacterial growth on the septa.)

8. ESTIMATION OF COSTS

Like filter powders, but contrary to other filter media, diatomite is always used to exhaustion and replaced. In the field of potable water, there seems to be no description of a case of regeneration of dirty diatomite. However, this process, which is theoretically possible, seems to be subject to problems in practice because of the diminishing grain size. The first cost factor in this filtration process is thus the price of the powder. On the Belgian market, the price varies between 9 and 13 BF/kg according to the point of origin and the quantities ordered. "White diatomite" is approximately 10% more expensive than the standard quality. The latter is sufficient for the filtration of potable water.

For final treatment of prefiltered water at Yvoir–Champale, the necessary quantity varies between 1 and 2 g/m^3 water produced, according to the length of the cycle. This is equal to 0.015 to 0.025 BF/m^3. For the treatment of raw water, the price of the filter powder depends more directly on the quality of the raw water. For treatment of the water at the Lienne quarry, which has a concentration of suspended solids between 5 and 15 g/m^3, the quantity of diatomite used varies between 15 and 30 g/m^3 of treated water, or 0.20 to 0.40 BF/m^3 (all 1989 values).

Extrahigh dosages lengthen the filter cycle, but the cost of the powder consumed does not change and staff time is reduced. [As a point of comparison, the price of reagent for a total treatment scheme (coagulation, disinfection, adsorption on carbon powder, and ozonization) such as that at the Tailfer plant (river Meuse) varies between 0.8 and 1.2 BF/m^3.] The cost of diatomite is thus important, so at the Brussels Waterboard the technique is reserved for "peak plants" or emergency use.

The first dosage (precoat) of a filter aid in cellulose has as a consequence lengthening of the filter cycle through an in-depth filtration effect. The ratio of 100 g of cellulose per square meter of septum at 1 kg/cm^2 pressure gives a layer thickness of approximately 1 mm. The price is high (i.e., ±200 BF/kg. This material, which significantly lengthens the filter cycle and prevents blocking of the septa, is better suited for industrial filtration than for large volumes such as those required in drinking water distribution.

For normal surface loadings of approximately 5 m/h, the global price of the diatomite filter installation is approximately 50% of that for conventional rapid sand filtration. Based on normal industrial capacity, the price approaches 0.3 BF/m^3 (the minimum price for investment is about 1000 BF/m^3 per day production capacity). These costs would naturally increase considerably if the plant were not in permanent operation.

Automated diatomite filtration is mentioned in the literature. Our experience has been that the method is not sufficiently reliable for completely automated water treatment. (Based on the 5-day work week) at the Brussels Waterboard, for Belgium this factor would be approximately 0.5 BF/m^3 for theoretical full-capacity staffing. During further filtration of previously treated water, this parameter can be calculated at half of the value cited above. Building larger plants would also decrease this cost factor by 50%.

Maximum energy costs, based on those required to pump water at 1 to 3 bar pressure, are in the range 0.03 to 0.1 BF/m^3. Maintenance and general costs are marginal when calculated on the basis of industrial capacity (approximately 0.05 BF/m^3). They depend strongly on the division of the work of the maintenance staff on the premises.

9. CONCLUSIONS AND RECOMMENDATIONS

Among powder filtration techniques for the preparation of potable water, diatomite is predominant. In the United States, about a thousand plants use diatomite filtration for the treatment of potable water. In Europe, this method is used less.

The filters could be either pressure filters or vacuum filters. The latter have the advantage of allowing direct visual observation of the behavior of the septa during operation. In Europe, practical experience with pressure filters is less well developed, but candle filters are used more than plate filters. The latter are more preferable for sludge thickening than for water filtration.

Choice of the appropriate powder is made empirically by its use in the industry. The determination of grain size requires a considerable amount of work as well as special evaluation equipment. The wet screen method furnishes nearly representative results.

The septa for candle filters can be rigid or flexible. The former are satisfactory up to an operational pressure of 30 g/mm², the latter, only up to 10 g/mm². Filtration velocities are in the same range as for conventional rapid sand filtration. Because of the structure of candle filters, the ground surface occupied is far smaller than for sand filters. The cost is only about 50% of that for sand filtration. Operational costs are quite high and diatomite filtration plants generally require constant manual control.

In conclusion, we can say that the technique is more acceptable for "peak stations" than for a permanently working plant; however, the latter is possible. There are large plants (in permanent operation) in the United States.

Diatomite filtration normally achieves the following improvements in quality:

1. Parasites, larvae, macroscopic organisms, and algae are eliminated.
2. The bacterial population is decreased by 50% when chlorine is not used. Complete elimination by the use of 0.1 ppm free chlorine is possible. Viruses are eliminated by at least 50%.
3. The concentration of solids and suspended matter is reduced to 100 mg/m³ or less.
4. Iron and manganese are decreased under the limit of 80 to 50 mg/m³, but this occurs at the expense of the length of operation of the powders and septa.
5. Diatomite filtration guards against excess flocs in coagulated water. The method is also applicable in a coagulation–filtration system.
6. Dissolved heavy metals are removed by absorption only at the beginning of the cycle, as the thickness of the layer is not sufficient to stabilize this effect.
7. Organic compounds in solution are not adsorbed or are adsorbed only partially. The chemical oxygen consumption and UV extinction remain nearly unchanged by diatomite filtration, at least for dissolved products.
8. It is better to disinfect the water to avoid biological effects with possible gas formation in the filter. The presence of gas in the powder layer hinders normal working of the filters. The formation of biological slime can damage the structure of the septa.

SYMBOLS AND UNITS RELATED TO FILTRATION POWDERS AND DIATOMACEOUS EARTHS

p volumetric porosity, dimensionless either in percent or as a ratio
w weight of powder per coated surface, kg/m²

BET   surface of Brunauer, Emmett, and Teller, m²/g
H head loss, m

H_2 head loss due to makeup, m
K_3 permeability of the layer on body feed, s/kg
L layer thickness, m
Q water flow, m^3/s
Q/A surface loading, m/s
X specific weight of diatomite, kg/m^3
X/L porosity function, kg/m^4
ZP   zeta potential, mV

Δp head loss of filters, m or bar

REFERENCES

1. G. J. Coogan, *J. AWWA, 54,* 1507 (1982).
2. F. B. Hutto, *Chem. Eng. Prog., 53,* 328 (1957).
3. S. Norman, *Pulverkohle* (Lezing Postakademiale kursus), 1979, Delft.
4. D. B. Purchas, *Filtr. & Sep., 6,* 465 (1965).
5. R. W. McIndoe, *Water Waste Eng.,* 48 (1969).
6. Th. S. Brown, J. F. Malina, and B. D. Moore, *J. AWWA, 66,* 98 (1974).
7. Th. S. Brown, J. F. Malina, and B. D. Moore, *J. AWWA, 66,* 735 (1974).
8. S. Syrotynski, *J. AWWA, 59,* 867 (1967).
9. G. R. Bell, *Water Water Eng., 72,* 482 (1968).
10. Anon., *Effluent Water Treat. J.,* 235 (1963).
11. F. J. Costabile and Ch. H. Perron, *J. AWWA, 63,* 230 (1971).
12. H. N. Armbrust, *Water Sewage Works,* 197 (1960).
13. A. B. Cummins, *Ind. Eng. Chem., 34,* 403 (1942).
14. E. R. Baumann, J. L. Cleasby and P. E. Morgan, *Water Sewage Works,* 331 (1964).
15. Ch. S. Oulman and E. R. Baumann, *J. AWWA, 56,* 1047 (1964).
16. Ch. S. Oulman, D. E. Burns and E. R. Baumann, *J. AWWA, 56,* 1233 (1964).
17. F. B. Hutto, *Chem. Eng. Prog. 53,* 328 (1957).
18. G. R. Bell, *J. AWWA, 54,* 1241 (1962).
19. Anon., *J. AWWA, 57,* 157 (1965).
20. M. E. Depauw, *Trib. Cebedeau, 343–344,* 329 (1972).
21. F. J. Costabile and Ch. H. Perron, *J. AWWA, 63,* 230 (1971).
22. J. H. Dillingham and E. R. Baumann, *J. AWWA, 56,* 793 (1964).
23. Anon., *J. AWWA, 62,* 507 (1970).
24. M. Costello, *Swimming Pool Age, 6,* 66 (1966).

Softening and Mineralization

"And we have also pools of which some do strain water out of salt and others do turn fresh water into salt."

— Lord Francis Bacon, *The New Atlantis*, (1623)

1. DEFINITIONS

The calcium and magnesium content determines the hardness of water. The concentration of these ions is expressed in one of the following ways:

French degrees	10 ppm $CaCO_3$
German degrees	10 ppm CaO
British definition	10 ppm $Ca(OH)_2$
ppm hardness	1 ppm $CaCO_3$
English degrees	1 grain $CaCO_3$: (0.0648 g) per gallon (4543 L) = 14 ppm
U.S. units	14.26 ppm $CaCO_3$
	1 grain $CaCO_3$: (0.0648 g) per cubic foot (28.317 L) = 2.29 ppm

These expressions are based on an equivalent amount of calcium carbonate even when the compounds present are not calcium salts but magnesium salts. For example, 10 ppm $Mg(HCO_3)_2$ represents 0.0685 mM or 0.137 mEq/L Mg^{2+}. Then the equivalent of calcium concentration is also 0.137 mEq/L or 6.85 ppm $CaCO_3$ or 0.685 Fr. hardness.

Temporary hardness is due to bicarbonates or carbonates only, while *permanent hardness* also implies ions other than those from carbonic acid (e.g., sulfates and chlorides).

2. IMPORTANCE OF MINERALIZATION
     OF DRINKING WATER

Drinking water is considered to be consumed at 2 to 2.5 L/day per person. The calcium needs per day per person range from 600 to 1000 mg. Of this total, 200 mg/day is taken in directly as calcium ion through the intestine. The contribution from drinking water can range between 10 mg/L (25 ppm $CaCO_3$ hardness) and 120 ppm (300 ppm $CaCO_3$ hardness) (1), ranging from marginal to complete needs. Also, evidence has been produced that calcium is probably less well absorbed from water than from other foodstuffs.

The daily need for magnesium ranges from 300 to 500 mg. Standard contents in foodstuffs are much lower than for calcium; examples (in mg per 100 g food) are cheese, 40; milk, 11; fish, 25; eggs, 11; vegetables, 20; and meat, 25. This means that the contribution to the magnesium needs provided by water is significant, and moreover, it appears that the magnesium ion is well absorbed. Therefore, if drinking water is softened, the process should maintain the magnesium concentration high enough to contribute to the average daily intake of this significant element.

The intake of increased amounts of sodium (e.g., in ion-exchanged water or soda-softened water; see below) has been analyzed critically in relation to the risk of increased blood pressure. In a normal food regime daily intake as NaCl is considered as 3 g/day (equivalent to 1.2 g expressed as Na). In Europe (1), the normal food intake brings about 3.5 g Na/day. A strongly restricted diet represents a daily intake of 0.9 g Na. In Europe, the maximum admitted concentration of sodium in drinking water is 150 mg/L. The exchange of 300 ppm $CaCO_3$ (6 mEq/L calcium) can introduce 138 mg/L sodium; hence care must be taken not to exceed the maximum allowable concentration.

3. POTENTIAL ADVANTAGES OF SOFTENING

Problems such as corrosion and bacterial development often occur with water softeners installed in private houses. These are due to lack of maintenance and could be avoided or lessened if the softening were centralized by the water authority. Italy has established legally binding conditions for the household treatment of water (2).

Calcium and magnesium salts of soaps (e.g., fatty acids) are nearly insoluble in water, so precipitation of these salts, particularly on textiles, may occur during washing. The problem is less important when synthetic detergents are used.

Scale formation in boilers is another drawback of hard water. The bicarbonate ion decomposes as the temperature rises and forms carbonates:

$$2HCO_3^- \rightarrow CO_3^{2-} = H_2O = CO_2$$

Since the carbonate of calcium is only slightly soluble, precipitation occurs:

$$Ca^{2+} + CO_3^{2-} \rightarrow CaCO_3 \downarrow$$

If large amounts of magnesium are also present, insoluble magnesium hydroxide can be precipitated:

$$Mg^{2+} + 2OH^- \rightarrow Mg(OH)_2 = MgO = H_2O$$

If present, other anions can also precipitate (e.g., calcium and magnesium silicates, which are almost insoluble):

$$Ca^{2+} + SiO_3^{2-} \rightarrow CaSiO_3$$

$$Mg^{2+} + SiO_3^{2-} \rightarrow MgSiO_3$$

The normal flow of the water is hindered by clogging of the pipes due to these scale deposits, while heat conduction in boilers is also reduced. Water containing a total hardness of 1 mEq $CaCO_3$/L is considered "soft"; while a content above 5 mEq/L is called "hard water."

Certain secondary benefits of lime softening are:

1. A bactericidal effect, due to the high pH reached in some processes
2. Elimination of iron, particularly the ferrous form found in well waters
3. Partial elimination of organic products, by coprecipitation
4. Reduction in concentrations of trace elements, such as Hg, Pb, and Zn, by incorporation in the crystals when crystallization method is used

The softening of drinking water is unnecessary from a hygienic point of view; soft waters are even suspected to promote cardiovascular diseases (3). However, for most uses, hard water is less desirable, specifically for laundering.

4. CALCIUM CARBONATE EQUILIBRIA

Waters oversaturated with calcium carbonate tend to precipitate $CaCO_3$; when undersaturated, they tend to dissolve calcium carbonate and are, eventually, corrosive. There exist numerous indexes and graphical methods of evaluating the equilibrium conditions versus $CaCO_3$ (4,5). Only a few of the fundamentals are outlined here.

The total concentration of carbonic species present in water involves carbonic acid (or dissolved carbon dioxide), bicarbonate, and carbonate: H_2CO_3, HCO_3^-, and CO_3^{2-}, respectively. The concentrations of these species are interrelated by equilibrium reactions:

$$H_2CO_3 \rightleftharpoons H^+ + HCO_3^-$$

$$K_1 = \frac{[H^+][HCO_3^-]}{[H_2CO_3]}$$

$$HCO_3^- \rightleftharpoons H^+ + CO_3^{2-}$$

$$K_2 = \frac{[H^+][CO_3^{2-}]}{[HCO_3^-]}$$

$$K_1 = 4.07 \times 10^{-7} \qquad pK_1 = 6.39 \quad (20°C)$$

$$K_2 = 4.17 \times 10^{-11} \qquad pK_2 = 10.38 \quad (20°C)$$

The dissolution of carbon dioxide in water corresponds to an equilibrium characterized by

$$CO_2 + H_2O \rightleftharpoons H_2CO_3 \qquad \text{(in which } H_2CO_3 \text{ represents the sum of dissolved} \\ CO_2 \text{ and } H_2CO_3)$$

$$K = \frac{[H_2CO_3]}{[CO_2]}$$

$$K = 4 \times 10^{-2} \qquad pK = 1.41 \quad (20°C)$$

Hence at ordinary temperature about 4% of the total "acid CO_2" exists in the form of carbonic acid.

The alkalinity of the water is represented by the sum of HCO_3^-, CO_3^{2-}, and OH^- expressed in conventional units (e.g., ppm $CaCO_3$). *At the usual pH values for drinking water, the OH^- concentration is not significant compared to the HCO_3^- value.* The concentration of the hydroxyl ion is related to the pH as expressed by the dissociation constant of water:

$$H_2O = H^+ + OH^-$$

$$K_w = 6.76 \times 10^{-15} \qquad pK_w = 14.17 \quad (20°C)$$

Calcium carbonate dissolves in water according to the following dissociation:

$$CaCO_3 \rightleftharpoons Ca^{2+} + CO_3^{2-}$$

The solubility constant is equal to the product of the equilibrium concentrations of calcium and carbonate ions.

However, solid calcium carbonate can exist in at least three forms, which have slightly different solubility constants: calcite, aragonite, and vaterite.

$$K_{s,ca} = 3.55 \times 10^{-9} \qquad pK_{s,ca} = 8.45 \quad (20°C)$$

$$K_{s,ar} = 4.90 \times 10^{-9} \qquad pK_{s,ar} = 8.31 \quad (20°C)$$

$$K_{s,va} = 1.35 \times 10^{-8} \qquad pK_{s,va} = 7.87 \quad (20°C)$$

The effect of temperature on the equilibrium constants must be considered if the water temperature is significantly different from 20°C (6,7). Suggested formulas are (T in kelvin):

$$pK_1 = 356.3094 + 0.06091964T - \frac{21{,}834.37}{T} - 126.8339 \log T + \frac{1{,}684{,}915}{T^2}$$

$$pK_2 = 107.8871 + 0.03252849T - \frac{5151.79}{T} - 38.92561 \log_{10} T + \frac{563{,}713.9}{T^2}$$

$$pK_w = \frac{4471}{T} + 0.01706T - 6.0875$$

$$pK_{s,ca} = 171.9065 + 0.077993T - \frac{2839.319}{T} - 71.595 \log T$$

$$pK_{s,ar} = 171.9773 + 0.077993T - \frac{2903.293}{T} - 71.595 \log T$$

$$pK_{s,va} = 172.1295 + 0.077993T - \frac{3074.688}{T} - 71.595 \log T$$

For approximations, the graphs in Fig. 1 can be used. For the equilibrium $CO_2 + H_2O \rightleftharpoons H_2CO_3$, the pK value varies linearly with temperature in the range of approximately 1.2 at 5°C and approximately 2 at 70°C.

4.1 Effect of Ionic Strength

The above-mentioned "thermodynamic" equilibrium constants are values applicable at infinite dilutions of a single salt. To account for real conditions a correction for the ionic strength of the water needs to be used. Operational equilibrium constants can be deduced from the thermodynamic constants by introducing an appropriate factor for correction of the salt content of the water, called the *ionic strength factor*. The ionic strength is equal to

$$\mu = \tfrac{1}{2}(\Sigma\, C_1 V_1^2 + \Sigma\, C_2 V_2^2)$$

in which $V_1 = 1$, $V_2 = 2$, and C_1 and C_2 are the respective concentrations (ion g/L) of mono- and divalent ions in water. If a fully worked-out ionic balance for the water is not available, reasonable approximations are given by

$$\mu = 4H - A$$

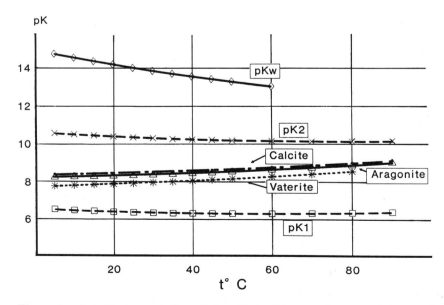

Figure 1 Graphical correlation of the effect of temperature on calcium carbonate equilibria.

in which H is the total hardness in moles per liter and A the alkalinity in equivalents per liter.

The ionic strength of the solution determines an approximate activity factor of each ion:

$$-\log f = 0.5 V^2 \sqrt{\mu} \qquad (\text{e.g., } V = 2; -\log f = 2\sqrt{\mu})$$

The operational equilibrium constants K° are obtained by correcting the equilibrium constants for an activity factor:

$$K_S^{\circ} = f_{Ca}[Ca^{2+}] f_{CO_3}[CO_3^{2-}] = K_S = [Ca^{2+}][CO_3^{2-}]$$

Hence

$$K_S^{\circ} = K_S f_{Ca^{2+}} f_{CO_3^{2-}}$$

$$pK_S^{\circ} = pK_S - \log f_{Ca^{2+}} - \log f_{CO_3^{2-}} = pK_S - 2\sqrt{\mu} - 2\sqrt{\mu}$$

$$pK_S^{\circ} = pK_S - 4\sqrt{\mu}$$

Similarly,

$$K_w^{\circ} = K f_{H^+} f_{OH^-}$$

$$K_w^{\circ} = pK_w - \log f_{OH^-} = pK_w - 0.5\sqrt{\mu} - 0.5\sqrt{\mu}$$

$$pK_w^{\circ} = pK - \sqrt{\mu}$$

Also,

$$K_1^{\circ} = K_1 f_{H^+} f_{HCO_3^{2-}}$$

$$pK_1^{\circ} = pK_1 - \sqrt{\mu}$$

And

$$K_2^{\circ} = K_2 = \frac{f_{H^+} f_{CO_3^{2-}}}{f_{HCO_3^-}} = K_2 f_{CO_3^{2-}}$$

$$pK_2^{\circ} = pK_2 - 2\sqrt{\mu}$$

Calculation of the pH_S value is corrected by taking as a basis the difference in the equilibrium constants corresponding to the operational constants of K_2 and K_S. Consequently,

$$pH_S = (pK_2^{\circ} - pK_S^{\circ}) + pCa + pAlk$$

4.2 Effects of Heterogeneity

The equilibrium constants given above are related to homogeneous solutions of ions. However, when precipitation or dissolution of hardness is concerned, the system is heterogeneous and solid–liquid contact must be considered. An approach

(8) has been published that can be handled on the basis of the equilibrium constants.

immersed solid $\overset{K_S^*}{\rightleftharpoons}$ equilibrated solid

particle size d — molar size d

molar size   — molecular surface s

An approximation is for the relationship between particle size and molecular surface. Suppose that 1 mol of finely powdered solid consists of N uniform particles of equal size. Than the surface of a single particle corresponds to $S = kd^2$ and the volume is $V = ld^3$. The molar volume is $v = M/\rho$ and

$$s = NS = \frac{N\alpha V}{d} = \frac{M\alpha}{\rho d}$$

where $\alpha = k/l$ is the empirical shape factor of the particles ($=1$ if perfect spheres); or, the diameter/length ratio in cylinders, ρ the density of the solid, M the formula molar weight of the solid, and $\bar{\gamma}$ the mean Gibbs energy of the solid–liquid interface.

$$\Delta G = -RT \ln K_{S,o} \pm 2RT \ln 55 = \tfrac{2}{3}\bar{\gamma}$$

$$= \frac{2M}{3} \frac{\alpha\bar{\gamma}}{d} = RT \ln \frac{K_S(s,d)}{K_S(s = 0; d = \infty)}$$

The equations become

$$\log K_S^* \text{ (immersed solid)} = \log K_{S,o}^*(d = \infty) + \frac{0.2895 M\alpha\bar{\gamma}}{RT} d^{-1}$$

$$\log K_S^* \text{ (equilibrated solid)} = \log K_{S,o}^*(s = 0) + \frac{0.2895\bar{\gamma}}{RT} s$$

An experimental approach to estimating $\bar{\gamma}$ is to set $\log K_S^*$ as measured versus $s(\mathrm{m}^2)$.

The mean Gibbs energy can also be estimated by a semitheoretical method. Assuming that 1 mol of a coarse crystalline solid composed of $A^+ + B^-$ is immersed in aqueous medium, when the large crystal is cut into pieces, s increases, as there is a gain in the Gibbs energy. At the end of the pulverization hydrated ions are formed with total molar surface

$$s = 4N_a(r_{A^+}^2 + r_{B^-}^2)$$

where N_a is Avogadro's number and r_{A^+} and r_{B^-} are the ion radius of A^+ and B^-, respectively.

For the aqueous phase, the molar fraction is related to the molarity by a factor of about 55 (1000 : 18); therefore, uncertainty exists as to the calculated Gibbs energy (about $\pm 2RT \ln 55$). Hence

$$\Delta G - RT \ln K_{S,o} \pm 2RT \ln 55$$

However, the uncertainty is small as long as $K_{S,o}$ is large. The $\bar{\gamma}$ value can then be computed as

$$\bar{\gamma} = \frac{3(-RT \ln K_{S,o} \pm 2 \ln 55)}{8\pi N_a(r_{A^+}^2 + r_{B^-}^2)}$$

An agreement between measured and calculated values is acceptable on the basis of this hypothesis (8), as the maximum uncertainty is 15 to 20% of the value of the Gibbs energy.

As far as calcium carbonate is concerned, K_S^* (immersed solid) can range from 3.2×10^{-6} to 3.2×10^{-7} (respectively, $pK_S^* = 5.5$ to 6.5) instead of the pK_S value in homogeneous solution of 7.9 to 8.5. The increase in "apparent K_S value" in the presence of solid calcium carbonate illustrates the presence of "colloidal and unprecipitated solid" calcium carbonate species when a given water is equilibrated versus an excess of solid crystalline calcium carbonate.

5. EVALUATION OF CALCIUM CARBONATE AGGRESSIVITY

A water that has a CO_2 ($+ H_2CO_3$) content in excess of the equilibrium condition has a tendency to dissolve $CaCO_3$ up to the point of equilibrium. Again numerous graphs for evaluation of the extend of the calcium carbonate aggressivity have been published (5,6,8).

5.1 Tillmans Equilibrium Curve

The first systematic approach is due to Tillmans (9). The graph corresponds to the equilibrium

$$H_2CO_3 \rightleftharpoons HCO_3^- + H^+ \qquad (H^+) = K_1 \frac{|CO_2| + |H_2CO_3|}{|HCO_3^-|}$$

in which CO_3^{2-} is supposed to be absent (i.e., pH lower than 8.3). For the facility $|CO_2| + |H_2CO_3|$ is represented as CO_2 concentration. Fundamentally, the equilibrium corresponds to the dissolution of $CaCO_3$ by $CO_2 + H_2O$ into $Ca(HCO_3)_2$. Hence 2 mol of HCO_3^- are in equilibrium with 1 mol of CO_2. This proportion determines the iso-CO_2 equilibrium line in Fig. 2.

All waters with CO_2 in excess versus the Tillmans equilibrium line are potentially aggressive to $CaCO_3$. Compositions with higher HCO_3^- (alkalinity) than the equilibrium line are potentially scaling. Iso-pH lines can be constructed to evaluate the balance aggressivity versus scaling for a given HCO_3^- content; the concentration of CO_2 being determined by analysis:

$$pH = pK_1 - \log |CO_2| + \log |HCO_3^-|$$

A simplified formula to compute iso-pH values is given by

$$pH = 7 - \log \frac{3 \times CO_2 \ (mg/L)}{0.61 \times Alk \ (in \ ppm \ CaCO_3)}$$

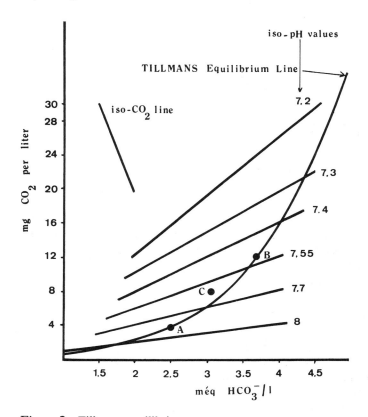

Figure 2   Tillmans equilibrium curve.

5.2   Langelier Saturation Index

The Tillmans graph provides an initial approach, but more detailed development of the subject is described by Langelier (10,11). The equations are valid in the pH range 4.5 to 10.3, based on the following development:

$$K_S = |Ca^{2+}| \times |CO_3^{2-}|$$

$$|Alk| = 2|CO_3^{2-}| + |HCO_3^-| + |OH^-| \qquad \text{(OH}^- \text{ can be neglected in normal conditions)}$$

$$|CO_3^{2-}| = \frac{K_2 |HCO_3^-|}{|H^+|} = \frac{|Alk| - |HCO_3^-|}{2}$$

$$|HCO_3^-| = \frac{|Alk|}{1 + 2K_2/|H^+|}$$

$$\frac{|Alk|}{2} = \frac{|HCO_3^-|}{(K_2/|H^+|) + \frac{1}{2}}$$

Hence

$$|CO_3^{2-}| = \frac{K_2}{|H^+|} \frac{|Alk|}{1 + 2K_2/|H^+|}$$

At equilibrium versus $CaCO_2$, $|H^+| = |H_s^+|$ and thus

$$K_S = |Ca^{2+}| \frac{|K_2|}{|H_s^+|} \frac{|Alk|}{1 + 2K_2/|H_s^+|}$$

The term $2K_2/|H_s^2|$ is very small and is neglected. Then

$$pH_S = pK_2 - pK_S + p|Ca^{2+}| + p|Alk|$$

which is the fundamental equation of Langelier, *in which the calcium concentration* $|Ca^{2+}|$ *is expressed in moles per liter, and the alkalinity concentration* $|Alk|$ *is given in equivalents per liter.*

    The Langelier saturation index (I) is defined by

$$I = pH \text{ as existing (measured)} - pH_S$$

If I is negative, the water is aggressive; if I is positive, the water is scaling or precipitative. Although the SI is not a quantitative expression, it is generally considered that the water at the tap should have an SI value of $+0.2$. Spontaneous precipitation of $CaCO_3$ occurs in water at SI values of approximately 2.5 and higher but can be induced by nucleation.

5.3 Effect of Water Temperature and Magnesium Concentration

The effects of water temperature on the aggressive or precipitative properties of a given water composition can be evaluated by Langelier equations by using appropriately corrected equilibrium constants. However, in boiler feedwater the dominant precipitate found in the scaling is often magnesium hydroxide-oxide associated with silicates and, eventually, corrosion products. The phenomenon has been studied extensively (12) and found to be in line with the solubility product of $Mg(OH)_2$; $pK_S = 11.26$ (25°C), but precipitation can occur when the solubility product exceeds 9.2 (13). Precipitation of magnesium hydroxide is a first-order process of the type

$$n = n_0 \exp\left(-\frac{L}{Gt}\right)$$

where n, n_0 is the particle population density ($mL \cdot \mu m$), L the particle size (μm), G the growth rate ($\mu m/min$), and t the reaction time.

    The "birth rate" is defined as $B° = n°G = k_N G^i$, where k_N is a kinetic constant and i a kinetic order. The value of i has been found to be dependent on pH (13):

$$i = 8100[OH^-] - 0.43 \qquad [OH^-] \text{ expressed in mol/L}$$

or

$$i = 0.1621[\text{Alk}] - 0.43 \qquad [\text{Alk}] \text{ expressed in ppm CaCO}_3$$

For solutions of $MgCl_2$ treated with an excess of 20% NaOH versus stoichiometry a birth rate has been determined as $B° = 2.1 \times 10^3/\text{mL} \cdot \text{min}$, which indicates a fast precipitation in such oversaturated conditions. Equilibrium lines have been determined from practical data by Larson (12) and are reproduced in Fig. 3. If the magnesium concentration is in excess of the equilibrium concentration for a given temperature of the water source, precipitation of $Mg(OH)_2$–MgO will occur in boiler systems with or without problem, depending on their design and operation as a whole.

According to Langelier (11), the stability index is the extreme value of the saturation index (SI), at which precipitation necessarily occurs. This index is pH dependent; at +0.1 SI the tendency for scaling by $CaCO_3$ will be three times as great at pH 8.2 than at pH 9.2, given the solubility of $Ca(OH)_2$.

5.4 Indicative Rules

Practical indicative rules of thumb for solubilities at 18°C are (in mEq/L):

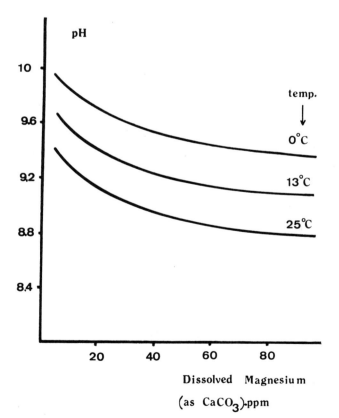

Figure 3 "Practical" equilibrium ratios for Mg in distribution water. (From Ref. 12.)

$Ca(OH)_2$	34.9
$CaCO_3$	0.3
$Ca_3(PO_4)_2$	Nearly insoluble
$Ca(Mg)SiO_3$	Nearly insoluble
$Mg(OH)_2$	0.29
$MgCO_3$	2.62
$Mg_3(PO_3)_2$	Nearly insoluble

Equations for the precipitation equilibria as a function of increasing temperatures (Fig. 4) indicate that at low alkalinity an equilibrated water can be aggressive at increased pH values. Water with low alkalinity represents a greater risk of aggressivity at increased temperature than do waters with higher alkalinity.

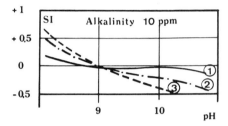

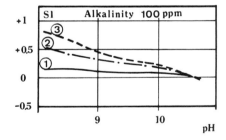

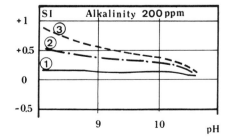

Figure 4   Impact of water temperature on $CaCO_3$ stability. Heating limits: 1, $13°C \rightarrow 25°C$; 2, $13°C \rightarrow 50°C$; 3, $13°C \rightarrow 75°C$. (From Ref. 12.)

5.5 Sources of CO_2

Carbon dioxide is one of the equilibrating substances involved in carbonate equilibria. It results from various sources:

1. Indirect increase through equilibration among carbonic acid species on dissolution of minerals
2. Acidification of the water with reagents (acids, aluminum sulfate, exchange resins, etc.)
3. Equilibration versus air
4. Bacterial decomposition of organic compounds in the aerobic mode
5. Chemical oxidation processes oxidizing organic compounds
6. Artificial injection of CO_2 into water

   Sources of CO_2 as used in recarbonation (see below) are:

1. Aeration with a blower or compressor, saturating the water with CO_2 by passing an excess of air
2. Evaporators of liquefied CO_2 resulting from bacteriological metabolism
3. Underwater burners fed with propane gas (use of CO_2 resulting from the combustion of fuels is questionable where public health is concerned)
4. CO_2 issued from calcination of chalk rocks

   The mean origin of CO_2 dissolved in water is the exchange with air or the atmosphere (14). As long as the gas pressure remains low (e.g., lower than 1 atm), the dissolution obeys Henry's law, relating $P_\gamma = H \times x_\gamma$, where P_γ is the partial pressure of the gas considered, H the Henry constant, and x_γ the molar fraction of the gas dissolved in the liquid. (The expression as a weight fraction can be obtained by the formula

$$\frac{x_\gamma}{1 - x_\gamma} \frac{M_\gamma}{M_S}$$

wherein M_γ is the molar weight of the dissolved gas and M_S the molar weight of the solvent; for water $M_\gamma/M_S = 44/18 = 2.44$).

Henry's constant is often expressed in atm or in mmHg. Conversion is $H(\text{mmHg}) = H(\text{atm}) \times 760$. The CO_2 content in open air is usually in the range 0.033 vol%, or 3.3×10^{-4} atm. Henry constants as a function of temperature are as follows:

$t(°C)$	0	5	10	15	20	25	30	35	40
$H(\text{atm})(\times 10^{-3})$	0.728	0.876	1.04	1.22	1.42	1.64	1.86	2.09	2.33
$H(\text{mmHg})(\times 10^{-5})$	(5.53)	(6.66)	(7.60)	(9.27)	10.81	(12.46)	14.06	(15.88)	17.69

For example, at 10°C and $P_{CO_2} = 3.3 \times 10^{-4}$ atm one has

$$x_{CO_2} = \frac{3.3 \times 10^{-4}}{1.04 \times 10^{+3}} = 3.17 \times 10^{-7}$$

and

$$\text{mg } CO_2/L = \frac{3.17 \times 10^{-7}}{1} \frac{44}{18} \times 1000 \times 1000 = 0.774$$

The diffusion of CO_2 in water and in porous materials is fast, since the diffusion constant is $D = 2 \times 10^{-7}$ m/s (20°C).

6. CHEMICAL SOFTENING

Chemical softening can be obtained either by precipitation or by crystallization. Furthermore, the processes are distinguished on the basis of the reagents used.

6.1 Lime-Soda Process

During the lime-soda process, the hardness is precipitated in the form of $CaCO_3$ and $Mg(OH)_2$ with the addition of lime [CaO–$Ca(OH)_2$] and soda ash (Na_2CO_3). The reactions involved in the process can be given schematically as shown in Table 1. The quantities of lime and soda needed to remove a given amount of hardness can be calculated from the data in Table 1.

To Remove Carbonate Hardness:
Reaction B or C–D: (1 + 2) mEq $Ca(OH)_2$/mEq

To Remove Alkalinity:
Reaction B or C–D or G: 1 mEq $Ca(OH)_2$/ mEq

To Remove Permanent Hardness:
As calcium hardness, reaction H or I: 1 mEq Na_2CO_3/mEq
As magnesium hardness, reaction E–F or F–I: 1 mEq $Ca(OH)_2$ and 1 mEq Na_2CO_3/ mEq

To Remove CO_2:
Reaction A: 1 mEq $Ca(OH)_2$ per 22 mg/L

The usual method is based on an excess-lime treatment. For the precipitation of $Mg(OH)_2$ the pH must be higher than 9 (usually between 10 and 10.5). The process

Table 1

| | | mEq added | | mEq |
| | | | | hardness |
Reaction		$Ca(OH)_2$	Na_2CO_3	removed
A.	$Ca(OH)_2 + CO_2 \rightarrow CaCO_3 + H_2O$	1	—	0
B.	$Ca(OH)_2 + Ca(HCO_3)_2 \rightarrow 2CaCO_3 + 2H_2O$	1	—	1
(C.	$Mg(HCO_3)_2 + Ca(OH)_2 \rightarrow MgCO_3 + CaCO_3 + 2H_2O$	1	—	0
)D.	$MgCO_3 + Ca(OH)_2 \rightarrow Mg(OH)_2 + CaCO_3$	1	—	1
(E.	$MgCl_2 + Ca(OH)_2 \rightarrow CaCl_2 + Mg(OH)_2$	1	—	0
(F.	$MgSO_4 + Ca(OH)_2 \rightarrow Mg(OH)_2 + CaSO_4$			
G.	$2NaHCO_3 + Ca(OH)_2 \rightarrow CaCO_3 + Na_2CO_3 + 2H_2O$	1	(+1)	0
(H.	$CaCl_2 + Na_2CO_3 \rightarrow CaCO_3 + Na_2SO_4$			
(I.	$CaSO_4 + Na_2CO_3 \rightarrow CaCO_3 + Na_2SO_4$			

involves a dosing of lime. The completeness of the reaction is influenced by the catalytic effect of the slurries, but new, more reactive types of lime become available on the market (see Chapter 19). To promote a good settling of the precipitates, a small dose of $Al_2(SO_4)_3 \cdot 18H_2O$ or sodium aluminate is often added (e.g., <5 ppm). Recirculation of the sludge also promotes the precipitation.

The process is generally incomplete and there is a possibility of postprecipitation of calcium carbonate, especially if the water contains excess alkalinity after the treatment. To avoid this complication the softened water must be treated either with sulfuric acid or with $CO_2 : 2CO_2 + Ca(OH)_2 = Ca(HCO_3)_2$. The latter reaction is called *recarbonation*. The addition of polyphosphates (e.g., sodium hexametaphosphate) in small amounts (0.5 to 5 mg/L) can prevent postprecipitation.

In practice, softening does not need to be complete. If permanent hardness can be allowed to remain in the treated water, the use of soda ash can be omitted. This technique is called *partial softening*. With "excess lime" treatment, after a minimum hardness has been reached (about pH 10), the hardness can increase again by dissolving calcium hydroxide. An example is given in Fig. 5.

Split treatment consists of adding an excess of lime to part of the water. After precipitation, the softened substream is mixed with a fraction of the unsoftened water. In this way the CO_2 of the untreated substream recarbonates the water. The proper proportions of the substream to be treated versus that to be bypassed can be calculated from an analysis of the raw water.

By expressing all the concentrations in equivalent mg/L $CaCO_3$ except CO_2 (in mg CO_2/L), the following equations can be applied to calculate the substreams:

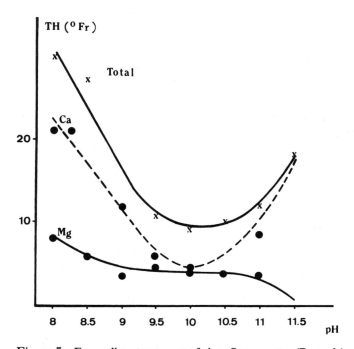

Figure 5   Excess lime treatment of river Senne water (Brussels).

percent bypass

$$= \frac{100[(\text{primary OH alkalinity}) + (\text{desired bicarbonate alkalinity})]}{\text{primary OH alkalinity} + \text{raw total alkalinity} + \text{raw } CO_2 \times 2.27}$$

(The "primary" characteristics correspond to the treated effluent, and the "raw" characteristics to the untreated substream.) The magnesium hardness of the effluent can be computed from the formula

magnesium hardness

$$= \frac{(\% \text{ bypass} \times \text{raw Mg hardness}) + (100\% \text{ bypass}) \times \text{Mg primary hardness}}{100}$$

6.2 Caustic Soda Process

Caustic soda (available as a 50 wt% solution) is easier to handle than are lime and soda ash. The reactions are shown in Table 2. The use of sodium hydroxide must be considered on a comparative economic basis. Sodium hydroxide precipitates both carbonate and noncarbonate hardness, and there is less sludge than with lime. Below 6°C the reaction velocity with lime decreases strongly, while that with caustic soda is quite independent of temperatures between 1 and 22°C. The global environmental impact is in favor of lime since sodium hydroxide usually results from synthetic procedures which also deliver residues.

6.3 Softening with Sodium Phosphate

When reacting with calcium and magnesium ions, sodium phosphate produces insoluble phosphates; for example,

$$3Ca(HCO_3)_2 + 2Na_3PO_4 \rightarrow Ca_3(PO_4)_2 + 6NaHCO_3$$

$$3CaSO_4 + 2Na_3PO_4 \rightarrow Ca_3(PO_4)_2 + 3Na_2SO_4$$

The high cost of phosphate makes the process less suitable for drinking water treatment, but the reactions can be technically applied in conjunction with the lime-soda process to produce a water with a very low residual hardness. Lime-phosphate softening is often used industrially.

Table 2

Reaction	mEq NaOH/mEq neutralized or removed
$CO_2 + 2NaOH \rightarrow Na_2CO_3 + H_2O$	1
$Ca(HCO_3)_2 + 2NaOH \rightarrow CaCO_3 + Na_2CO_3 + 2H_2O$	1
$Mg(HCO_3)_2 + 2NaOH \rightarrow MgCO_3 + Na_2CO_3 + 2H_2O$	1
$MgCO_3 + 2NaOH \rightarrow Mg(OH)_2 + Na_2CO_3$	1
$MgCl_2 + 2NaOH \rightarrow Mg(OH)_2 + 2NaCl$	1
$CaCl_2 + Na_2CO_3 \rightarrow CaCO_3 + 2NaCl$	x

6.4 Technical Execution of Lime Softening

6.4.1 *Precipitation*

The softening process may be carried out in conventional flocculator–clarifiers, but usually the operation is performed at higher surface loadings. A rapid-mixing basin ($G > 100$ s^{-1}, preferably 300 s^{-1}) is necessary and 5 to 10 min is sufficient for the residence time in the mixing zone. Floc growth and precipitation take place within 40 to 60 min. Since the process is facilitated by contact with preformed calcium carbonate, the contact clarifiers (e.g., Pulsators) with sludge recirculation (e.g., accelator, Hydrotreator, Reactivator, precipitator) are particularly well suited to the backmixing concept (see Chapter 9). Classical surface loading is kept under 5 m/h. In horizontal flow-sedimentation basins used in lime softening, the retention time should be between 2 and 4 h, with a horizontal flow of 18 m/h. The sludge is drawn off as a 5 to 15% slurry. Scrapers are used to facilitate sludge removal.

6.4.2 *Crystallization*

The calcium carbonate formed in the softening process can be precipitated either as flocs or as crystals. The formation of these crystals is promoted by crystallization nuclei of calcium carbonate. Classically, the dimensions of the inoculation grains are between 0.2 and 1 mm. It is essential that the grains remain in suspension and do not grow together to form a solid mass. This process proceeds much faster than floc growth, so that surface loadings of 50 to 100 m/h are possible. The method eliminates primarily the calcium hardness, but some magnesium is also removed as $CaMg(CO_3)_2$ and $CaCO_3 \cdot MgCO_3$ built in the calcium carbonate crystallization lattice. The Gyractor (Fig. 6) is a typical form of traditional commercial equipment used in the crystallization technique. The total height of the apparatus can attain 5 m, with a crystallization zone of about 2 m.

Much recent progress has been made in the technology of crystallization as a result of research and applications in the Netherlands. From this research the following facts are now known, which can determine new progress:

1. The rate of softening can be very high; crystallization on nucleation material occurs within seconds (15).

2. Well-calibrated sand is necessary as nucleation material, starting from 0.1 to 0.3 mm and growing to 0.2 to 0.6 mm. At the size of 0.4 to 0.6 mm on crystallization, they need to be removed to prevent clogging. These guide values depend also on the surface loading.

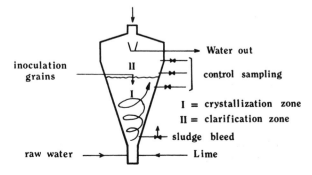

Figure 6 Schematic of the Gyractor crystallizator.

3. A steady state must be obtained which corresponds to the formula

$$Q_1 = \frac{d^3}{D^3} Q_2$$

where Q_1 and Q_2 are the flows of fresh nucleation material and the pellets withdrawn from the reactor, respectively; and d and D are the diameters of the seed grains and the material as withdrawn, respectively.

4. The upflow-velocity value of the water needed to maintain the nucleation material in fluidized condition is given in Fig. 7. At higher water temperature (e.g., 20 to 25°C) the pellets can attain 1.1 to 1.2 mm in size, while in colder periods (e.g., less than 10°C) the size is best when under 0.8 mm (see Fig. 7).

5. The contact zone can be conical (Spiractor, Gyractor) or cylindrical. However, the design of the inlet structure is of utmost importance to prevent clogging. A cyclone operated by the raw water flow is at present the preferred design.

6. The total grain surface per unit reactor volume, S in mm^{-1}, called the *specific pellet surface*, is influenced by three principal factors: grain size, upflow velocity of the water, and water temperature. The specific surface is determined by the formula $S = 6(1 - p)/d$, where p is the volumetric porosity of the material and d is the grain diameter. The variations are illustrated in Figs. 8 to 10, according to the literature data (17).

7. The process kinetics has been described by the following equation according to (18):

$$-\frac{d|Ca^{2+}|}{dt} = k_T \times S \times \left[|Ca^{2+}| \times |CO_3^{2-}| - \frac{K_s}{f^2} \right]$$

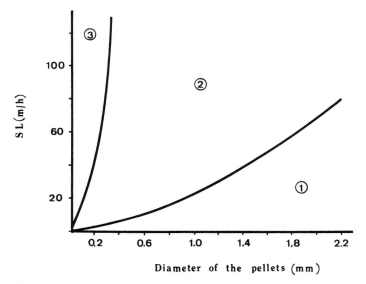

Figure 7   Possible surface loading in a pellet reactor as a function of grain size. 1, Compact layer; 2, design zone; 3, washing-out zone. (From Ref. 16.)

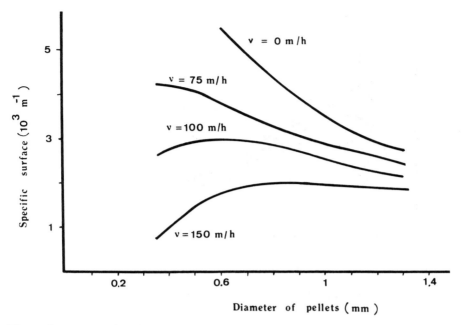

Figure 8   Impact of grain size on the specific surface.

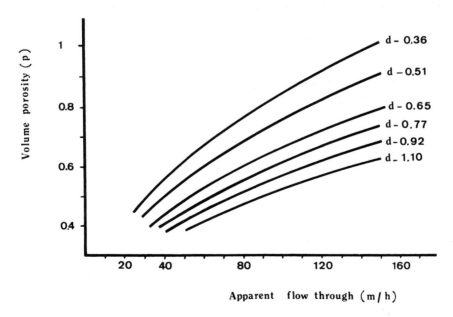

Figure 9   Volume porosity, depending on specific surface.

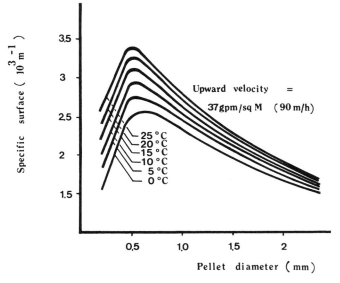

Figure 10   Impact of water temperature on specific surface.

where $|Ca^{2+}| \times |CO_3^{2-}| - K_s$ is the oversaturation in mol^2L^{-2}; S the specific surface area in m^2/m^3 or m^{-1}, k_T the (total) rate constant of precipitation of $CaCO_3$ ($L \, mol^{-1} \, t^{-1}$), and f the activity coefficient.

8. The hydraulic process parameters of a plug-flow fluidized-bed crystallization have been reviewed by Wilms (19).

Headloss: In the transition zone between laminar and turbulent flow,

$$5 < Re < 100 \qquad Re = \frac{1}{1-p}\frac{vd}{v}$$

$$H = 130 \times \frac{v^{0.8}}{g}\frac{(1-p)^{1.8}}{p^3}\frac{v^{1.2}}{d^{1.8}}L$$

$$H_{max} = (1-p)L\frac{\rho_p - \rho_w}{\rho_w}$$

Porosity:

$$\frac{p^3}{(1-p)^{0.8}} = 130 \times \frac{v^{0.8}}{g}\frac{\rho_w}{\rho_p - \rho_w}\frac{v^{1.2}}{d^{1.8}}$$

Bed height:

$$E = \frac{L}{L_0} = \frac{1-p_0}{1-p}$$

Velocity:

$$v_{min}^{1.2} = \frac{p_0^3}{1 - p_0} \frac{\rho_p - \rho_w}{\rho_w} \frac{d^{1.8}}{0.8} \frac{g}{130}$$

$$v_{max} = \frac{1}{10} \times \frac{g^{0.8}}{v^{0.6}} \frac{(\rho_p - \rho_w)^{0.8}}{\rho_w^{0.8}} d^{1.4} \qquad \left(1 < \text{Re} < 50, \text{Re} = \frac{vd}{v}\right)$$

Velocity gradient:

$$G = \left(\frac{W}{\rho_w v}\right)^{1/2} = \left(\frac{H\rho_w g v}{\rho_w v L}\right)^{1/2} = \left[\frac{\rho_p - \rho_w}{\rho_w} (1 - p) g \frac{v}{v}\right]^{1/2}$$

Symbols and units:

Re = Reynolds number, dimensionless
H = hydraulic loss, m H_2O
v = kinematic viscosity, m^2/s
g = gravity acceleration, m/s^2
v = velocity, m/s
G = velocity gradient, s^{-1}
W = energy dissipation, W/m^3
E = expansion of bed, dimensionless
p = porosity, dimensionless
p_0 = porosity of fixed bed, dimensionless
L = bed height, m
L_0 = height of fixed bed, m
d = diameter of pellets, m
ρ_p = specific gravity of pellets, kg/m^3
ρ_w = specific gravity of water, kg/m^3
ρ_p for sand grains   2740 kg/m^3
ρ_p for pellets ($CaCO_3$)   2840 kg/m^3

Preliminary Rules of Thumb for Design.
1. At least two reactors must be installed in a parallel hydraulic scheme to guarantee flexibility of operation.
2. The surface loading is in the range 50 to 125 m/h with caustic soda dosing and 60 to 100 m/h with dosing of milk of lime.
3. The height of the nucleation grain layer when at rest is in the range of 2 m in the case of cylindrical reactors and 3 to 3.5 m in the case of conical reactors. The expansion rate when fluidized is on the order of 1.5 to 2.
4. The total heights of the reactors are 5 to 10 m for reactors operated with caustic soda and 8 to 10 m when operated with milk of lime (reactors up to 14 to 15 m in height exist).
5. Characteristics of the pellets and inlet–outlet mass balance are as described above.

6.4.3   Recarbonation

The lime softening processes are usually completed by rapid sand filtration. Recarbonation is carried out prior to this operation. Therefore, if carbon dioxide is used, various techniques may be employed (see also Section 5.5). A few guidelines for the contact chambers in recarbonation:

Detention time: 15 to 30 min.
Depth: 3 to 5 m.
The chamber must be ventilated or remain open to the atmosphere.

Other recarbonation techniques are sulfuric acid dosing or high-rate chlorination. In the latter case the hydrochloric acid formed may be sufficient to redissolve enough $CaCO_3$ to obtain minimum alkalinity.

6.4.4   Disposal of Sludge Resulting from Lime Softening Processes

The sludge abstracted from concentrators can be transferred with centrifugal pumps, preferably of the open impeller type. The use of sludge drying beds or lagoons operated on an alternate basis (fill-and-let-dry basins) are the usual methods for sludge disposal of lime-softening sludge. The sludge removed from softeners seldom exceeds a concentration of 10 wt% dry mass. On drying it gradually concentrates to 50 wt%, occupying 15 to 20% of its initial volume. Vacuum filtration is an alternative method of delivering sludge that may be suitable for agricultural purposes. The sludge can be burned to prepare quick lime suitable for reuse. Rotary kilns are frequently used for this purpose. The process is generally unsuitable when there is much magnesium in the sludge. The water-softening sludge resulting from precipitation or crystallization techniques is generally suitable for agricultural use. Crystallization pellets are appropriate as starting product to manufacture construction materials or as product for livestock.

7.   SECONDARY BENEFITS OF CHEMICAL SOFTENING

7.1   Removal of Heavy Metals

A large number of heavy metals can be removed or lowered in concentration due to the reduction in solubility of their oxyanions, including hydroxy carbonates at increasing pH. This is the case, for example, for Zn, Ni, Cu, Co, Cd, Hg, Te, Cr, Ag, Mn, and Fe. An example is illustrated in Fig. 11 (initial concentrations in metal ions: 1 mg/L). It is worth noting that to be efficient, pH values of 9 and higher must be obtained. This puts into question the necessity of maintaining a minimum magnesium content in the drinking water. Also, water with too high a pH is less well suited for cooking dough and pastalike materials.

7.2   Removal of Organic Compounds by Chemical Softening

Addition of lime has been described as increasing the removal of BOD and improving the sludge conditioning reported in earlier literature (21,22). Indicative data are illustrated in Fig. 12. About half of the BOD_5 can be removed by lime precipitation (line 2 versus line 1). If ozone is also used (line 3), the removal rate can be 80 to 85%. However, the pH must be 10 and higher to obtain this result, which is in agreement with literature data (23).

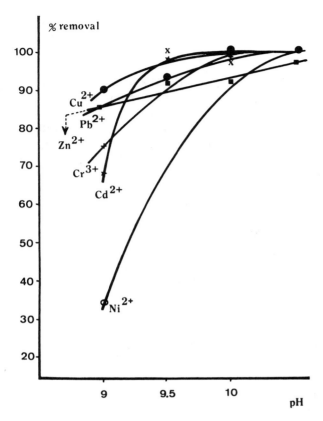

Figure 11 Coprecipitation of heavy metals in chemical softening. ●, Cu^{2+}; ■, Zn^{2+}; x, Pb^{2+}; +, Cr^{3+}; ★, Cd^{2+}; ○, Ni^{2+}. (From Ref. 20.)

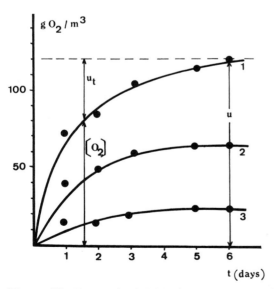

Figure 12 Removal of BOD from wastewater by precipitation with lime, river Senne, Brussels. (From Ref. 20.)

7.3 Disinfection

Excess-lime treatment of water was reported very early as a method of removal of coliforms as well as other enteric pathogens (24). Typical results obtained in the laboratory are illustrated in Fig. 13. To summarize: High pH values speed up the effect, but the necessary reaction time ranges up to several hours. For more ubiquitous bacteria such as wild strains of *Pseudomonas fluorescens*, the behavior on lime treatment may be hazardous, as illustrated in Fig. 14. In all instances high pH values and long reaction times are necessary to obtain a significant amount of decay. In conclusion: For the purpose of disinfection, excess-lime treatment is only an emergency technology for use in the absence of other disinfectants.

7.4 Phosphate and Ammonia Removal

Calcium orthophosphate can be coprecipitated on lime softening. The maximum result is obtained at pH 9 to 10 (see Fig. 15). Under the same conditions, the removal of ammonia is only partial (20) (i.e., 10 to 30%). Removal can also result from partial stripping at alkaline pH values.

7.5 Potential Recovery of $Mg(OH)_2$–MgO

Magnesium hydroxide-oxide has a coagulation value as described in the literature (25,26) (see also Chapter 5). The principle of the method is based on the following equation:

$$Mg(OH)_2 + 2CO_2 = Mg(HCO_3)_2$$

$$Mg(HCO_3)_2 + 2H_2O \xrightarrow[\text{air}]{34-45\,^{\circ}C} MgCO_3 \cdot 3H_2O + CO_2$$

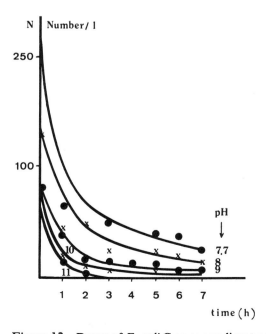

Figure 13 Decay of *E. coli* C on excess-lime treatment. (From Ref. 20.)

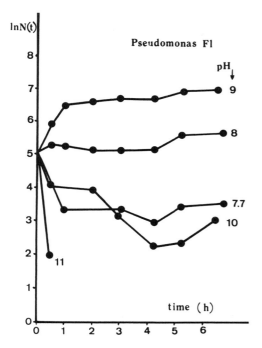

Figure 14 Decay of *Ps. fluorescens* (wild strains) on excess-lime treatment. (From Ref. 20.)

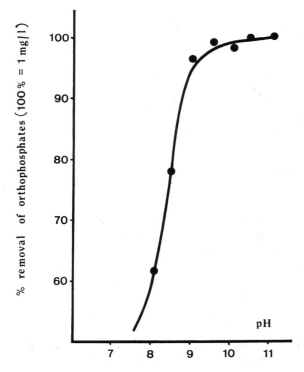

Figure 15 Removal of dissolved orthophosphate with lime treatment.

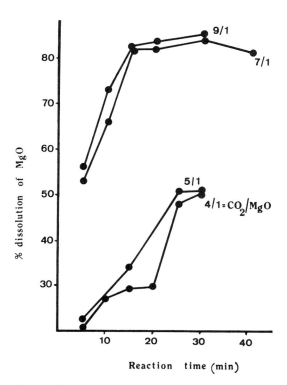

Figure 16   Recovery of precipitated $Mg(OH)_2$–MgO with CO_2.

Experimental conditions are poorly detailed in literature. In the laboratory, dissolution of precipitated $Mg(OH)_2$ in lime-softening sludges is obtained by bubbling a mixture of CO_2 and air. The molar proportion of CO_2 to MgO is important, as shown in Fig. 16. In all instances an excess of CO_2 is required. The percent recovery can range up to 80% when a molar CO_2/MgO ratio of 7 : 1 is contacted for 15 min (20).

8.   WATER TREATMENT BY ION EXCHANGE

8.1   Ion Exchangers

Ion exchange is defined as the *reversible* interchange of structural ions of materials called ion exchangers with ions in solutions contacted with the ion exchangers. General reviews have been published on the subject as applicable to water treatment. *Cationic exchangers exchange cations and thus contain negatively charged groups in their structure. Consequently, contrary to the definitions of the charge of cations and cationic polyelectrolytes, cationic resins are negatively charged cationic exchangers.* The reverse are the anionic exchangers containing positively charged groups that combine reversibly with various negative ions.

In 1833, Fuchs observed that "when certain clays are treated with lime, they release potassium and sodium." In 1848, evidence of exchange of calcium versus ammonium ions in soils was produced by Thompson and Way (29). The reversibility of the process was fully assessed around 1850 (29). Manganese oxide as a nonstoichiometric compound is composed of a mixture of MnO and MnO_2 approaching the structure

$$\left[O{=}Mn \underset{\diagdown_O}{\overset{\diagup^{O}}{}} \right]^{2-} Mn^{2+}$$

in which Mn^{2+} can be exchanged with Fe^{2+}, for example.

Around the beginning of this century ion-exchange properties were assessed for several natural minerals, including bentonite clays, glauconite (greensand), and conditioned silica gels. The first synthetic ionic exchangers were the synthetic zeolites (from the Greek *zein-lithos*, meaning "boiling stone"), patented in 1906. They were composed of a fused mixture of SiO_2 + $Al_2O_3(H_2O)$. Later, the product was prepared by a gelling process of silicate and aluminate, the gel later being dried and crushed to grain size. These first synthetic exchangers had very slow exchange kinetics and could eventually deteriorate by losing the silicate portion by dissolution in water.

8.2 Synthetic Exchange Resins

Current commercial ion exchangers are, at present, synthetic resins in which functional groups act as exchangers. Effective ionic exchangers must:

1. Contain ions in their own structure
2. Be insoluble in water under the operational conditions (temperature, acidity, basicity, etc.)
3. Have a porous structure enabling diffusion of the ions throughout the structure

During the 1930s it was discovered that sulfonated organic products were capable of cationic exchange (28). At that time sulfonated phenolic resins were obtained by condensation of *m*-phenolsulfonic acid and formol.

A fundamental discovery was made in 1945 by d'Alelio, who patented a sulfonated polystyrene polymer capable of cationic exchange.

Sulfonated coal and sulfonated phenolic resins deteriorate in the presence of dissolved chlorine. Obtaining and maintaining a reproducible porous structure was a problem with sulfonated polystyrene *homo*polymers. The resins, designated here by "Res," are actually synthetic *reticulated* polymers, usually resulting from copolymerization of styrene and divinylbenzene.

Functional groups are substituted into the benzene rings (e.g., on sulfonation or ammoniation), thus providing the exchanger groups.

Advanced technologies (30) of copolymerization have made it possible to improve the pore structure to obtain equally spaced cross-links determining an "isoporous structure" (i.e., a material with micropores of uniform size). In "macroporous" resins a spongelike structure is obtained with a network of tightly cross-linked molecules with intermolecular holes that are larger than the single molecular size.

Polyacrylic resins (31) macroporous resins can allow efficient reversible adsorption of organic material, which is a more recent development in ionic exchange. The most common functional groups for cationic exchangers are the carboxyl group (COOH) (weak acid resins) and the sulfonic group ($-SO_3H$) (strong acid resins). Anionic exchangers are the weak basic imino group ($-NRH_2^+/OH^-$) and the strong basic quaternary ammonium group ($-NR_3^+/OH^-$). If the proton of the acid groups ($-COOH$) or ($-SO_3H$) is replaced by a sodium ion (e.g., $-COONa$, $-SO_3Na$), the resin is defined as being in its sodium form. General characteristics of resins applicable at present for water softening are given in Table 3. *Granulometry* is determined by wet sieving.

Table 3 General Characteristics of Cation-Exchange Resins

Appearance	White to reddish-brown spheres	
% Divinylbenzene	8–14	
Density	1.3	
Liter weight	800–880 kg/m^3	
Moisture content (native)	30–50 wt %	
Sieve analysis (example)	Diameter (mm)	Passing (%)
	<0.2	0.5
	0.22	3
	0.42	14
	0.63	42.2
	0.84	36.3
	1.19	4
Effective size (mm)	0.4–0.6	
Uniformity coefficient	1.5–1.8	
Void volume	30–40 (vol %)	

8.3 The Exchange Process

The *exchange process* can be represented as follows:

$$Res^-H^+ + Na^+ = Res^- + H^+$$

$$2Res^-Na^+ + Ca^{2+} = Res_2^{2-}Ca^{2+} + 2Na^+$$

Charged ions are exchanged; consequently, the exchange capacity of an ion exchanger is expressed by the number of charges (i.e., equivalents of ions) necessary to maintain the electroneutrality of a given quantity (volume or weight). This capacity is determined analytically by the equivalents of ions exchangeable when the reaction is completed.

The design of ion-exchange processes is based on the ion-exchange equilibria defined by the equilibrium concentration quotients, Q:

$$\frac{[ResNa^+][H^+]}{[ResH^+][Na^+]} = Q \ ResH^+ \rightarrow ResNa^+$$

$$\frac{[Res_2Ca^{2+}][Na^+]^2}{[ResNa^+]^2[Ca^{2+}]} = Q \ ResNa^+ \rightarrow Res_2Ca^{2+}$$

The quotients are not thermodynamic equilibrium constants, as they vary with the composition of the exchanger phase. They are selectivity coefficients, characterizing the relative affinity of the exchanger for specific ions. For practical cases of displacement of the sodium ion by bivalent ions, the values of Q are on the order of 20 to 40 and the usual sequence is

$$Mg^{2+} \simeq Zn^{2+} < Cu^{2+} < Co^{2+} < Ca^{2+} < Sr^{2+} < Ba^{2+}$$

The affinity of exchangers for bivalent ions in dilute solution is much higher than for monovalent ions. Therefore, Ca^{2+} in a dilute solution will be practically completely adsorbed on a $ResNa^+$ exchanger.

However, an exhausted exchanger of the $ResCa^{2+}$ form can be restored in the $ResNa^+$ form by exchange with a concentrated NaCl solution (e.g., seawater). Calcium and magnesium present in seawater can increase the volume of water required for regeneration. Exchange reactions are slower than electrolytic reactions and are widely controlled by diffusion rates into the boundary layer adjacent to the exchanger or within the pores of the material.

8.4   Exchange Capacity

The exchange capacity of an ion exchanger is expressed in mEq/g or per mL of exchanger (see Table 4). Alternative units can be g $CaCO_3$/g or mL of exchanger. The *total exchange capacity* is not obtained in practice. Moreover, an oversaturated resin can lose part of its reversibility of exchange. Therefore, a preliminary test of the *breakthrough capacity* is always necessary.

For regeneration, an excess of the regenerating ion is always required; it can range 150 to 200 times the theoretical stoichiometric amount. The exchange capacity is increased at higher temperatures. Seasonal variation between minimum and maximum exchange capacity may range to 10% of the nominal capacity.

8.4.1   Testing Procedures Applicable to Ion Exchangers
Tests give only an indication of possible obtainable results.

8.4.2   Definitions (32)

Wet weight: weight of a bulk volume of resins saturated by an absorbed liquid (water), free of adherent droplets of liquid.

Table 4   Exchange Capacity of Ion Exchangers

Substance	Functional Group	Total exchange capacity (approximate)		
$Fe_2O_3 \cdot xH_2O$	—	4	mEq/g	
MnO_x	—	15	mEq/g	
Montmorillonite	Na^+ silicate	0.8 –1	mEq/g	
Kaolinite	Na^+ silicate	0.02–0.1	mEq/g	
Illite	Na^+ silicate	0.2 –0.4	mEq/g	
Greensand	Al silicate	0.12–0.14	mEq/mL	
Improved greensand	Al silicate	0.15–0.24	mEq/mL	
Zeolite	Na^+ silicate	0.15–0.8	mEq/mL	
Synthetic zeolite	Na^+/Al silicate	0.6 –0.7	mEq/mL	
Sulfonated coal	SO_3H aromatic	0.3 –0.36	mEq/mL	
Phenolic resins	Phenol SO_3H(Na)	6 –9	mEq/mL	
Polystyrene resins	(Copolymers)			
Weak acid	COOH, COONa	8 –10	mEq/g	(4 –5 mEq/mL)
Strong acid	SO_3H, SO_3Na	4.5	mEq/g	(1.6–1.9 mEq/mL)
Weak basic	RH_2N^+	4.5–5.5	mEq/g	(2 –2.5 mEq/mL)
Strong basic	R_3N^+	3.5–4.5	mEq/g	(1.2–1.4 mEq/mL)

Resin volume: volume of the resin itself as determined by pycnometric techniques.

Standard state: 1 g of resin in the dry state in the H^+ or OH^- form for cationic and anionic resins, respectively. The "dry state" is obtained by drying over phosphorus pentoxide up to constant weight. Oven drying is not recommended and should be limited to 105 to 110°C (33).

Settled volume: volume of a resin mass in exchange conditions. Several methods are proposed. The simplest and the one generally used is to agitate the resin suspension in a graduated tube and to record the first settling volume (i.e., the volume after proportional settling).

8.4.3   *Determination of Total Exchange Capacity*

The total exchange capacity is defined at 20°C and must be determined at 20 ± 5°C. The determination is achieved either by a batch technique or in a testing column containing a given volume of ion exchanger in the standard state. Preliminary hydration with pure water is the first step of test procedure. Very detailed descriptions of standard test columns are indicated in the literature (33), as well as mathematical models for break-through profiles (34).

Simplifying, the test is carried out in a column of diameter/height ratio between $\frac{1}{20}$ and $\frac{1}{15}$. The column is filled with wetted resin at two-thirds of its height. For resins used in the sodium form the exchanger is stabilized in the column by percolating a solution of 50 to 70 g NaCl/L and using a quantity of 0.4 kg NaCl/L of wetted resin. The column is rinsed with distilled water and must be maintained permanently wetted.

The water under investigation is passed through the column at a flow of 5 to 10 column volumes per hour. If W is the volume of the water exchanged for S (milli)equivalents per liter by an ion exchanger of volume V in the column, the total exchange capacity is

$$T = \frac{W \times S}{V}$$

The break-through capacity corresponds to

$$B = \frac{W \times (S \times S')}{V}$$

where S' is the average ionic content in the effluent (e.g., $Ca^{2+} + Mg^{2+}$ in the case of softening).

8.5   Use of Ion Exchangers

Apart from rare cases, such as the reduction of sulfates, nitrate removal, or desalinization, total demineralization of drinking water is seldom undertaken. Softening with substitution of Ca/Mg ions by sodium ions is more frequent, particularly in municipal supplies, and in Europe, particularly on a household scale.

The *softening process* does not alter the alkalinity of the water:

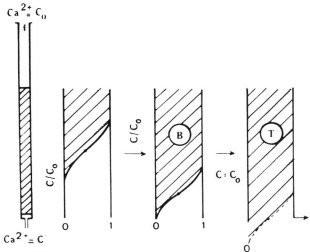

Break-through profiles of exchange. columns

B : Break-through Capacity

T : Total Exchange Capacity

Figure 17 Break-through profiles of exchange columns. B, break-through capacity; T, total exchange capacity.

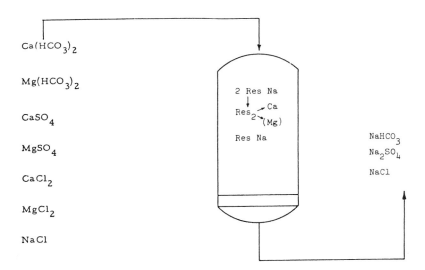

Although the process has been developed in open filters, the most common method remains the downflow pressure filter, particularly on a household scale.

In the *decarbonatation process*, exchange on a $ResH^+$ basis and using acids (e.g., HCl or H_2SO_4) for regeneration, decarbonation of the water can be obtained according to the following scheme:

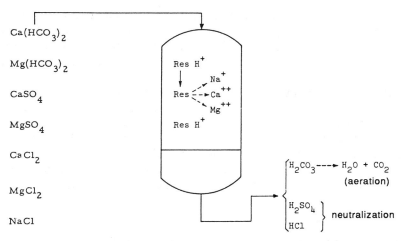

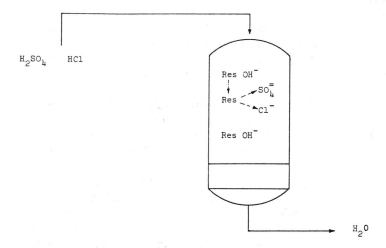

The neutralization of the acids H_2SO_4 and HCl can eventually be obtained by subsequent passage through an anion exchanger (e.g., ResOH$^-$).

8.5.1 Removal of Silica and CO₂

The removal of dissolved silica (silicic acids) as well as CO_2 (carbonic acid) requires strong basic or anionic resins. The application is important for waters to be used in high-pressure boilers. The same principles hold for the removal of carbonic acid. As a *strong basic resin* is also necessary, the latter must be regenerated with sodium hydroxide (e.g., a 4% solution). An aeration step is to be provided to dissipate most of the CO_2 and thus to spare the regeneration solution.

8.6 Use of Macroporous Resins (35)

Strong basic anion-exchange resins in the chloride form [e.g., Lewatit MP 500 A (Bayer Chemical Co.) or similar from other manufacturers] can be used for the removal of humic acids. A full-scale plant is in operation at the Fuhrberg plant of the Hannover Waterworks (2.500 m³/h). The best resin has a macroporous structure that is favorable for the adsorption of larger molecules. Sulfate ions interfere since owing to their higher exchange coefficient, sulfates replace chlorides. In Fuhrberg water, containing 130 mg/L SO_4^{2-}, a throughput of about 300 bed volumes can be reached. With the resin in the sulfate form, about 50% of the humic acids is still removed up to 5000-bed-volume transit.

Possible throughput is up to 50 bed volumes per hour (four filters of 12.5 m³ resin each), with a total cycle of 4 days between regeneration. Removal of the humic acids ranges from 85% (beginning of the filtration cycle) to 55% (end of the cycle), with an initial dissolved organic carbon input of 6.5 mg/L. Nitrification of the dissolved nitrogen compounds sets in gradually. Elution of nitrite must be considered. Regeneration of the exchange filters is obtained by the transit of two bed volumes of a solution containing 100 g/L NaCl and 20 g/L NaOH.

8.7   Operational Sequences

A single-phase strong cationic resin in its sodium form is the most used technology for softening. Regeneration is possible by passing a suitable solution downward (e.g., NaCl 25 g/L; complete regeneration may necessitate 5 mol NaCl per mole of Ca/Mg salt exchanged); the regeneration contact time is in the range 20 to 45 min; the volume of brine is between 0.5 and 1.5 times the volume of the resin. Often a less concentrated brine solution is used (e.g., 10%). The regeneration is followed by washing with clear water. Part of this wash water (36) can be recovered as the so-called "compound" and used subsequently in regeneration preliminary to transit of the brine solution.

Solid conditioned NaCl combined with adequate rinsing devices can be used for household applications according to the instructions of the equipment manufacturer (e.g., washing machines). Filtered seawater can be used for regeneration, but its Ca/Mg content must be considered. The flow of the regeneration liquid must be increased accordingly (i.e., by 60 to 80%).

A cationic exchanger followed by a strong basic exchanger is the simplest design of a flowsheet for complete demineralization. The exact sequence of the different exchangers must be considered as a function of the water to be treated: The classical sequence is a strong cationic exchanger–decarbonation–deaeration–a strong anionic exchanger. However, if the water is heavily charged with anions necessitating a strong anionic exchanger, the preferred sequence can become strong cationic exchanger–weak anionic exchanger–decarbonation–deaeration–strong anionic exchanger.

Also, a "top layer" of a weak acid-exchanging resin can remove the calcium-magnesium ions, and a subsequent strong acid exchanger, the remaining alkalinoterrous ions together with the neutral salts (K, Na) (31). Mixed beds can also be used: mixed anionic and cationic resins separated hydraulically for regeneration.

8.8   Guidelines for Design

General data on softening with cationic exchangers are given as follows (28) (manufacturers should be consulted for more specific information):

Exchange capacity: 700 to 1250 kg/m³
Bed depth: 0.6 to 1 m

Surface loading: 10 to 16 m/h (exceptionally, 50)
Backwash speed: 12 m/h
Regeneration time: 25 to 45 min
Rinse flow:
  Fast 20 to 40 m/h
  Slow 8 to 15 m/h
Rinse volume: 3 to 5 m³/m³ resin

Ion exchange can be operated with technologies similar to the ones applied in rapid sand filtration. The bed depth must be at least 0.75 m, to avoid premature break-through. Usually, it is not higher than 2 m, to avoid excessive head loss. The usual surface loading reaches 6 to 15 m/h, but higher velocities (e.g., 20 to 40 m/h) are possible. The process has the following components:

1. The fixative exchanger cycle (e.g., Ca^{2+}, Mg^{2+} retention); the velocity depends on the water flow (e.g., 5 to 10 filter volumes/h).
2. Expansion by countercurrent washing (elimination of deposits, colloidal precipitates, air pockets, or channels); an expansion zone of 50 to 100% is required.
3. The rinse, which is to replace the excess of regenerating solution by a downward water flow; the volume of the rinsing solution is kept between 2.5 and 5 times the volume of the exchanger.

The entire process is best controlled automatically (e.g., by measurement of the electrical conductivity of the effluents). The surface loading of the ionic exchange filters influences both the exchange efficiency obtained and the head loss, which is also dependent of the water temperature. The exchange efficiency is largely independent of the water flow since the ion-exchange process is fast (Fig. 18). The limits of 16 to 40 m³/h resin are recommended to maintain steady exchange conditions.

As a conclusion, the head loss (Fig. 19) remains very low in all instances as long

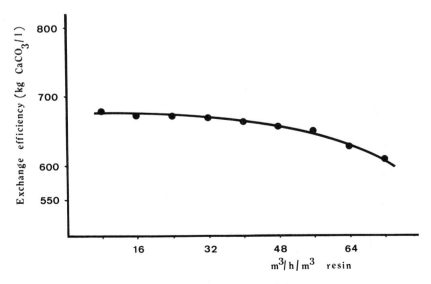

Figure 18 Incidence of flow rate on exchange efficiency.

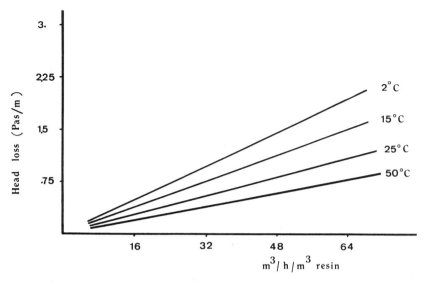

Figure 19   Head loss in resin beds as a function of flow rate and water temperature.

as ion exchange is the sole process occurring (i.e., as long as no suspended matter is filtered by the exchanger). As the bed expansion on countercurrent regeneration is high (Fig. 20), with considerable risk of resin loss, an indicative sequence that is generally preferred is described below.

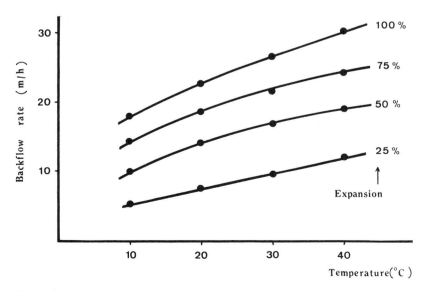

Figure 20   Backwash expansion of cation-exchange resins.

8.9 Design Rules of Ion-Exchange Softeners

An applicable general formula is

$$E = \frac{K \times B \times Q \times H}{G \times T}$$

where
  E = volume of the exchange material (m^3)
  K = safety factor for irreversible adsorption (generally, 0.75 to 0.8)
  B = break-through capacity (kg CaCO$_3$/m^3)
  Q = volume of water to be treated between regeneration (m^3)
  H = hardness of the raw water (kg/m^3)
  G = hardness removed between two regenerations (kg/m^3)
  T = total exchange capacity of the resin (kg CaCO$_3$/m^3)

Design Checklist:

Water flow
Softening requirements
Time between regenerations
Layout of the units
Resin exchange capacity
Hydraulic characteristics of the resin
Underdrain and backwash system
Brine makeup and storage

In general:

Layer thickness ranges: 50 to 150 cm.
Backwash expansion: 80 to 100% of bed depth is taken into consideration.
Resin volume range of softeners: 1.5 to 120 m^3. Vertical softeners operated under
  pressure may have diameters up to 3.6 m (minimum 25 cm). (For household or
  limited uses, pressure-type systems of smaller volume are available.)
Water inlet: similar to that for rapid sand filters.
Brine inlet: same as for water inlet if downflow regeneration.
Underdrain system: must be resistant to corrosion by brine and to aggressive softened waters. Homogeneous distribution of fluids must be obtained by a mani-
  fold pipe distribution system. The underdrain can be constructed on the basis
  of built-in strainer in a concrete floor or as a composite gravel underdrain. The
  minimum layer thickness is of 25 to 30 cm composed, for example, of several
  sizes of gravel (e.g., 10 to 15 cm 0.3 to 0.6 mm in size; completed with 10 to 15
  cm of 10 to 20 mm).
Brine makeup and storage: experience has proven that the withdrawal of brine from
  the bottom of the storage tank with water feed at the top is the preferred dosing
  form. Wet storage is preferable and the brine is best made up immediately after
  delivery or even during unloading of the transport units themselves (boat,
  trucks, etc.) (for small uses, continuous-flow-through units with solid conditioned tablets releasing the necessary brine on flow-through of the regeneration
  water).

Materials:

Generally, stainless steel, except for the high-titanium grades, are not acceptable because of "chloride corrosion."

Suitable epoxy coatings are recommended for both steel and concrete surfaces.

Foodstuff-grade plastics on the basis of PE, PTFE, PP, and PVC are recommended.

Ebonite-vitrified equipment performs well but is outside the acceptable price range.

Operational control: similar to that of rapid sand filtration except that the measurement of electrical conductivity of inlet and outlet water and of regeneration and rinsing waters is the determinant.

8.10   Problems Associated with Ion-Exchange Softening

8.10.1   *Spent Brine*

Disposal of spent brine or exhausted regeneration liquids can involve considerable problems. The volume of wastewaters usually represents 3 to 4% (range 1.5 to 7) of the production capacity. Moreover, the waste brine composition may contain 35 to 100% dissolved solids. Direct uncontrolled discharge in a river may be harmful to fish life. Where dilution in a river is allowed, it must be controlled so that the preestablished level of salinity (e.g., 1000 μS/cm) will not be exceeded. Ocean disposal of the spent brine is undoubtedly the most acceptable solution to the problem.

8.10.2   *Bacteria*

Bacterial growth in the resins, particularly in household applications of the process, is common. Some of the strains may even be pathogenic (37).

Bacterial Results in Softening Exchangers.   (Results in percent; total 143.)

Achromobacter	8.4	*Aerobacter*	3.5	*Alcaligenes*	5.5
Aspergillus	5.5	*Bacillus*	26	*Brevibacterium*	4.2
Candida	11	*Clostridium*	17.5	*Corynebacterium*	8.4
Cryptococcus	3.5	*Escherichia*	8.4	*Flavobacterium*	9
Gestrichum	4.2	*Micrococcus*	5.5	*Mycobacterium*	17
Paecilomyces	5.5	*Paracolobactrum*	5.5	*Penicillium*	4.2
Pseudomonsa	6.3	*Rhodotorula*	5.5	*Sarcina*	7.7
Serratia	4.2	*Staphylococcus*	23	*Streptococcus*	7.7
Streptomyces	3.5	*Coccus* (gram neg.)	8.4	*Diplococcus* (gram neg.)	8.4

For large-scale plants, appropriate measures to prevent biological fouling of resins are disinfection of the resin beds. This operation can be carried out with formol, sodium chlorite, sulfites, or quaternary ammonium salts.

8.10.3   *Organic Substances*

The presence of organic substances in the water (particularly humic acids) can hinder the exchange by fouling the ion-exchange material. This complication is more pronounced for strong basic anionic exchangers. These can act as a pretreatment to protect the subsequent exchangers.

8.10.4 Precipitates

Iron, manganese, and aluminum precipitating in the resin beds can seriously affect the exchange capacity. Ferrous iron can be flushed to the waste during regeneration. A rejuvenation process is based on the reduction in situ of the ferric deposits by the use of $NaHSO_3$ in the regenerating liquid. Occasional treatment of strong cationic resins with HCl can improve resin lifetime when a problem occurs with manganese fouling. Preliminary tests or advice of the manufacturer of the resin are required to be able to use these techniques.

8.10.5 Aggressivity

The bicarbonate content of the water is unchanged but the calcium is eliminated; in most cases this renders the water aggressive according to Langelier:

$$I = \text{pH} - \text{pH}_S$$

$$\text{pH}_S = (pK_2 - pK_S) + \text{pCa} + \text{pAlk}$$

The possible aggressivity of the resulting water must always be taken into consideration.

8.10.6 Sodium Content

With strong cationic resins in their sodium form, in the softened water the calcium ions are exchanged for sodium ions. Although the level at which the sodium may be present is believed to be harmless, it still remains necessary to pay attention to possible effects of an increase in sodium.

Example. Several operating municipal plants have been described in the United States (28). In Europe, the municipal Tournai plant (Belgium) is a typical example (36).

Example of Tournai (Belgium).   Open type-filter exchangers:
Bed expansion at 6 to 12 m/h, starting at high velocity and then slowing down; total time, for example, 8 min
Downflow regeneration at a speed of 2 to 3 m/h
Downflow rinsing at a speed of 5 to 10 m/h

The entire regeneration cycle can last up to 1 h, the different phases being in the range of 1 for bed expansion, 2.5 for regeneration, and 1 for rinsing.

9.   MINERALIZATION AND STABILIZATION OF WATER

Starting from an equilibrated water, an aggressive water can result due to several causes:

1. Hydrolysis of aluminum sulfate, converting bicarbonate to carbon dioxide:

$$Al_2(SO_4)_3 + 6HCO_3^- \rightarrow 3SO_4^{2-} + 2Al(OH)_3 + 6CO_2$$

2. Fermentation: for example, for glucose

$$C_6H_{12}O_6 + 6O_2 \rightarrow 6CO_2 + H_2O$$

3. Softening, particularly by ionic exchange in closed pressure filters, removes the calcium content but leaves the carbonic acid content unchanged

4. Treatment of water with acids for auxiliary reasons such as optimization of coagulation
5. Natural waters with low mineralization
6. Mixing of two or several waters [the Tillmans graph is not linear in correlating CO_2 versus HCO_3^- (e.g., in mixing waters A and B in Fig. 2 in equal proportions, a composition C results which is aggressive); consequently, the water service must take care of the problem as a whole and compute the equilibrium of the final product]

9.1   Lime-Softened Waters

Lime-softened waters can require recarbonation as described above. The American Water Works Association have defined a goal standard of minimum 80 to 100 mg/L hardness as $CaCO_3$. Most European legislations converge to a minimum required hardness of 54 to 60 mg/L calcium (135 to 150 ppm as $CaCO_3$), but also, a minimum in magnesium concentration is often required [e.g., 6 mg/L (or 25 ppm as $CaCO_3$)].

The major problem associated with lime softening with or without recarbonation is post-precipitation, as has been known since the 1950s (38). The problem is associated with the level of supersaturation and only partial post-precipitation occurs. Data as reported are, for example, that waters of 24 ppm supersaturation versus calcium bicarbonate equilibrium lose about 2 ppm hardness on 24 to 48 h of storage. A supersaturated water of 30 ppm hardness can lose up to 15 ppm on longer storage (e.g., 15 ppm precipitated). Moreover, if in the storage reservoirs, the water transportation mains or household equipment scale deposits exist, they can act as seeds for further precipitation of calcium carbonate. The particle size of magnesium oxide seeds plays an important role in the speed of deposition or settling; at 75 μm or less, deposition occurs within minutes.

The zeta potential or electrophoretic mobility of the colloidal $CaCO_3$ or $Mg(OH)_2$–MgO particles also plays a very important role (39). At alkaline pH values such as those that occur in softened water, it is remarkable that calcium carbonate colloids are negatively charged, while magnesium hydroxide-oxide colloids appear to be positive. Typical values are given in Fig. 21. Hence the complementarity of Ca/Mg can play an important role in lime precipitation through coagulation by mutual neutralization of colloids (see Chapters 5 and 10). It is generally recommended that a filtration stage be included after lime treatment.

9.2   Use of Polyphosphates

The nomenclature is as follows:

$$P_2O_5 + H_2O \rightarrow 2HPO_3$$

Metaphosphoric acid

$\xrightarrow{\text{heating}}$

$\begin{cases} [NaPO_3] & \begin{array}{l}\text{orthophosphate} \\ \text{sodium phosphate}\end{array} \\ [(NaPO_3)_2] & \text{sodium di-metaphosphate} \\ (NaPO_3)_x & \text{sodium } x\text{-metaphosphate} \end{cases}$

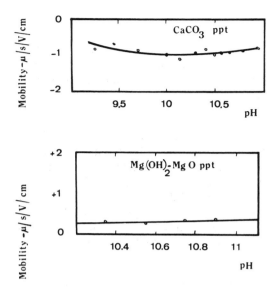

Figure 21 Electrophoretic mobility of $CaCO_3$–$Mg(OH)_2$–MgO precipitates. (From Ref. 39.)

Simple metaphosphates have a great tendency to polymerization

$$P_2O_5 + 2H_2O \rightarrow H_4P_2O_7$$
Pyrophosphoric acid

$$P_2O_5 + 3H_2O \rightarrow 2H_3PO_4$$
Orthophosphoric acid

$Na_4P_2O_7$ tetrasodium pyrophosphate
$Na_2H_2P_2O_7$ disodium dihydrogen pyrophosphate
Na_3PO_4 trisodium phosphate
Na_2HPO_4 disodium phosphate
NaH_2PO_4 monosodium phosphate

The action of polyphosphates (e.g., so-called sodium hexametaphosphate) is based on partial covering of the crystals of calcium carbonate with polyphosphate (40). The Ca/P ratio is about 300 : 1. In most drinking waters this indicates the use of 1 to 2 mg/L of sodium hexametaphosphate.

On hydrolysis, orthophosphate is formed:

$$(NaPO_3)_6 + 6H_2O \rightarrow 6NaH_2PO_4 = 6H_2O + 6NaPO_3$$

On a general medical basis, as well as a consequence of the possibility of bacterial aftergrowth, the systematic use of polyphosphates in drinking water treatment remains questionable. A temporary use of polyphosphates for the stabilization of distribution mains and galvanized household lines can be recommended, preferably in conjunction with silicates. Mixed formulations on a phosphate basis can be used to prevent corrosion in older distribution systems, for example, by dosing 3 mg/L expressed as PO_4^{2-} of an equal-weight solution of orthophosphate and hexametaphosphate.

9.3   Correction of Aggressivity

Although the Langelier index cannot be correlated immediately with the quantity of potential precipitate or dissolving of material, it may generally be recommended that I be higher than $+0.3$ and never lower than -0.2. If the water is aggressive, several corrective methods can be employed.

9.3.1   Aeration

The equilibrium concentration of dissolved CO_2 versus air is about 1 mg CO_2/L. Consequently, by spraying the water, it is possible to diminish the amount of dissolved CO_2. The process has the advantage of making it possible to saturate the water with oxygen. In practice, by spray aeration alone, it is not possible to obtain a water containing less than 3 to 5 mg CO_2/L. The equilibrium implicates an alkalinity of at least 100 to 120 ppm (see the Tillmans curve). The process is self-limited and does not alter the hardness of the water. Aeration is indirectly involved in the dispersal of ozonated air in water during ozonization as a final stage of treatment.

9.3.2   Filtration Through Granulated Marble

The analytical determination of the aggressivity or precipitativity of a water requires equilibration versus $CaCO_3$ and subsequent determination of its change in composition. From a theoretical standpoint, filtration through marble should be a very elegant method for the correction of aggressivity: The equilibrium cannot be surpassed, and if the filters are sufficiently large, changes in the carbon dioxide content of the raw water do not necessitate direct adjustment. The precipitation of iron and manganese on the marble grains is inherent in the process and thus inactivates the material. Consequently, backwashing becomes necessary.

The operation can be carried out in open and closed filters. However, in the latter case, the necessity of renewing or adding to the material from time to time gives an advantage to the open filters. The filters are usually dimensioned to call for supplementary filling twice a year. The kinetics of the exchange has been studied extensively by Tillmans and co-workers. Their conclusion was that the finer the material, the faster the exchange (e.g., grades of 0.5 to 1 mm, 1 to 2 mm, and 2 to 3 mm); relative exchange times are 1, 2.3, and 3.7. To exchange 1 m^3/h, quantities ranging from 200 to 250 kg $CaCO_3$ are required. The material also tends to cake, and the water to be exchanged must be appropriately prefiltered to avoid clogging of the exchange material.

9.3.3   Filtration on Magno or Neutralite

By heating dolomite at 700°C one obtains magno:

$$CaCO_3 \cdot MgCO_3 \xrightarrow{\;700°C\;} CaCO_3MgO + CO_2$$

A typical composition of the material is (in weight percent)

$CaCO_3$	72
MgO	22
$MgCO_3$	3
$Fe_2O_3 \cdot Al_2O_3 \cdot SiO_2$	3

The magno or neutralite particles react with the CO_2 of the water 5 to 10 times quicker than does marble. The principal reaction is

$$3CO_2 + CaCO_3MgO + 2H_2O = Ca(HCO_3)_2 + Mg(HCO_3)_2$$

It has been established indeed that an analytical ratio CaO/MgO in the material is optimal in practice. The velocity of exchange depends on the water temperature and the following empirical formula can be handled:

$$v_x = (1 + 0.085t) \times v_t$$

where v_x and v_t are the acceptable velocities at temperature x and reference temperature t, generally chosen as 12°C.

The process can be operated in classical open filters of the same design as used for rapid sand filters (e.g., 1- to 1.5-m filter-medium layer). Synthetic magnos exist in regularly shaped hard materials with a head loss of about 30 cm/m and supporting backwashing by water or air and water. Another advantage of this material is that it is not influenced by the iron and manganese content of the water up to a concentration of 1 mg/L. Iron is removed simultaneously. Classical grain size is in the range 0.5 to 2 mm, but coarse grades range from 2.5 to 5 mm.

The exchange rate is directly proportional to the active surface, that is, the surface of the grains and the thickness of the active part of the filter layer. Before operation, the product is best washed to limit the hydroxide alkalinity of the water during the first periods of operation. After this stabilization phase, the reaction with CO_2 is practically stoichiometric. The necessary exchange capacity is best determined experimentally during the design.

A first approximation for design is to consider the process to be described as a mass transfer system:

$$-\frac{dC}{dt} = k \times C$$

and

$$k = \frac{\ln C_0 - \ln C}{t}$$

where C and C_0 are the concentrations of dissolved CO_2. However, as the process is strongly temperature dependent, the activation energy must be considered:

$$k = K \exp\left(-\frac{A}{RT}\right)$$

where
$\quad k =$ velocity constant (s^{-1})
$\quad A =$ activation energy (J)
$\quad R =$ universal gas constant (8.31 J mol^{-1} K^{-1})
$\quad T =$ absolute temperature (K)
$\quad K =$ steric factor (in the range of temperatures for drinking water, e.g., 5 to 25°C, the steric factor is approximately 0.185 per degree)

9.3.4 Injection of Alkali

Alkali dissolved in water can neutralize aggressive CO_2. The most used products are lime and sodium hydroxide. The reactions are

$$2CO_2 + Ca(OH)_2 = Ca(HCO_3)_2$$

$$CO_2 + NaOH = NaHCO_3$$

The use of lime increases the hardness of the water; that of sodium hydroxide alters only the alkalinity. The process must be controlled to avoid any excess of reagent that would initiate softening or post-precipitation.

The theoretical calculation involves a step-by-step approach in which successive compositions are introduced in the equations as was done by Langelier. The entire problem is actually dealt with by computer, but the operational procedures remains very empirical. The attainment of a given pH is usually a sufficient monitoring parameter confirmed by analytical determination of the aggressivity or precipitativity on solid calcium carbonate.

If lime is used, it is recommended that it be before a filtration stage. This also holds for caustic soda, but with careful dosing a clear water reservoir ensures sufficient safety against post-precipitation. Minimum hardness is necessary when water is transported by iron pipes (even protected) to the consumers. As a guideline, the total hardness should be at least around 50 ppm, preferably 120 ppm, with an equilibrium pH > 7.

To dissolve calcium when using lime, it is necessary to introduce a sufficient amount of CO_2:

$$CaO + H_2O \rightleftharpoons Ca(OH)_2 \xrightarrow{\;2CO_2\;} Ca(HCO_3)_2$$

If the increased hardness is applied before the coagulation–flocculation phase, the lime can be added directly to the raw water and the desired dose is then dissolved by bubbling CO_2 through porous pipes into the water. The excess lime is eliminated during the subsequent coagulation–flocculation process.

9.3.5 Use of Sodium Bicarbonate

For reduced uses, increased hardness and alkalinity can be obtained by employing sodium bicarbonate and calcium sulfate, which are to be dosed and incorporated separately in the water.

9.4 Mineralization and Corrosiveness of Water

Most waters are capable of dissolving a certain quantity of metals. A great variety of materials have been used in existing distribution systems: cast iron species; steel, both with different coating systems or without coating; polyvinyl chloride; polyethylene; asbestos–cement; various cement coating systems; copper; lead; brass; galvanized steel; welding materials; and so on. Some aspects of "bacteriological corrosion" through sulfate-reducing bacteria are described in Chapter 12.

A KIWA group (Netherlands) has published a report (42) in which a compromise for water composition that would preserve the best of the above-mentioned materials is proposed as a guideline (reference temperature is 10°C):

TIC (total inorganic carbon) (CO_2 + HCO_3^- + CO_3^{2-}) higher than 2 mmol/L
Langelier index between -0.2 and $+0.3$

$$\text{Permanent ions (in mmol/L)} = \frac{|CL^-| + 2\,|SO_4^{2-}|}{\text{TIC}} < 1$$

pH generally not lower than 7.8 and lower than 8.3, but a practical lower value to
be considered is given as

$$0.38\,|TIC| + 1.5\,|SO_4^{2-}| + 5.3$$

The highest of both values, 7.8 or the computed value, must be taken into consideration. The highest pH value of 8.3 is defined to limit the loss of zinc from galvanized steel and brass and to prevent post-precipitation of calcium carbonate. The lowest value concerns the plombosolvency (>7.8) and the variable lowest value the dissolution of copper. Although in such conditions, in most cases the maximum admissible concentrations in lead and copper are not exceeded, the problem may need local or, occasionally, wider attention as a function of the water temperature and water residence time, for example.

A minimum TIC value of 2 mmol/L is necessary to prevent pitting corrosion of copper. The lowest value of I, -0.2, is necessary to prevent the liberation of chrysolite parts of asbestos–cement pipes, and the highest value, $+0.3$, is to limit the scaling in heating equipment. However, this point still needs further attention. The combination of TIC and chloride sulfate is important for corrosion of steel and cast iron and the chloride content concerns the stability of brass and welding materials. These tentative guidelines are now tested on a large number of distribution systems.

SYMBOLS AND UNITS RELATED TO SOFTENING AND MINERALIZATION

f	activity factor, dimensionless
p	volumic porosity, dimensionless
A	alkalinity, total, equivalents per liter
$B°$	birth rate of precipitates, number/mL · min
H	total hardness, mol/L as $CaCO_3$
I	saturation index, dimensionless
Re	Reynolds number, dimensionless
SI	Langelier saturation index, dimensionless
$V_1, V_2, \ldots$	valence of ions, number
μ	ionic strength, concentration units (e.g., mol/L or mEq/L)

REFERENCES

1. R. Amavis, W. J. Hunter, and J. G. P. M. Smeets, *Hardness of Drinking Water and Public Health*, Pergamon Press for EEC, Elmsford, N.Y., 1976.
2. Circolare 30 Ottobre 1989, No. 28, *Gazz. Uff. Repub. Ital.* (Mar. 11, 1989).

3. R. Masironi, Z. Pisa, and D. Clayton, *Bull. WHO, 57*(2), 291 (1979).
4. *Corrosion Control by Deposition of CaCO₃-Films*, AWWA, Denver, Colo., 1978.
5. Method 2330, Calcium carbonate saturation, *Standard Methods APHA-AWWA-WPCF*, 17th Ed., Washington, D.C., 1989.
6. Report of task group on CaCO₃ saturation indexes, *J. AWWA, 82*, 71 (1990).
7. G. Dorange, A. Marchand, and M. Le Guyader, *Rev. Sci. Eau, 3*, 261 (1990).
8. P. W. Schindler, Chap. 9 in *Equilibrium Concepts in Natural Water Systems*, Adv. Chem. Ser. 67, American Chemical Society, Washington, D.C., 1967.
9. J. Tillmans and O. Heublein, *Gesund. Ing., 35*, 669 (1912).
10. W. F. Langelier, *J. AWWA, 28*, 1500 (1936).
11. W. F. Langelier, *J. AWWA, 38*, 169 (1946).
12. T. E. Larson, *J. AWWA, 43*, 649 (1951).
13. B. Dabir, R. W. Peters, and J. D. Stevens, *Ind. Eng. Chem. Fundam., 21*, 298 (1982).
14. J. F. Thomas and R. Rhodes Trussel, *J. AWWA, 62*, 185 (1970).
15. A. Graveland, *H₂O, 20*, 291 (1987).
16. KIWA, *Communication 102,* 1990.
17. J. C. Van Dyk and D. Wilms, *Dutch Experiences with Pellet Reactors*, TI, Antwerp, 1990.
18. H. N. Wiechers, P. Sturrock, and G. v. R. Marais, *Water Res., 9*, 835 (1975).
19. D. Wilms, *H₂O, 22*, 628 (1989).
20. B. Regnier and W. J. Masschelein, *Trib. Cebedeau, 35*, 259 (1982).
21. T. J. Tofflemire and L. J. Hetling, *J. WPCF, 45*, 210 (1973).
22. R. C. Faro and N. L. Nemerow, *Water Sewage Works*, Oct. 1976, p. 82.
23. M. J. Semmens and G. A. Hohenstein, *Proc. Water Conference on Environmental Engineering*, ASCE, Minneapolis, July 1982.
24. E. Wattie and C. W. Chambers, *J. AWWA, 35*, 709 (1943).
25. C. G. Thompson, J. E. Singley, and A. P. Black, *J. AWWA, 64*, 11 (1972).
26. N. Martin Helin, Jr., *Env. Sci. Technol., 7*, 304 (1973).
27. H. A. Alsentzer, *J. AWWA, 56*, 742 (1963).
28. E. Bowers, Ion-exchange softening, Chap. 10, in *Water Quality and Treatment*, AWWA, McGraw-Hill, New York, 1971.
29. Y. I. Gladel, *Rev. Inst. Fr. Petrole (IFP), 12*, 936 (1957).
30. Anon., *Water Water Eng., 69*, 16 (1965).
31. Anon., *Water Serv., 950*, 141 (1975).
32. H. P. Gregor, K. M. Held, and J. Bellin, *Ann. Chem., 23*, 620 (1951).
33. Cation exchanger test procedures, AWWA standard, *J. AWWA, 41*, 451 (1949).
34. Y. I. Gladel, *Rev. Inst. Fr. Petrole (IFP), 12*, 864 (1957).
35. W. Kölle, *EPA 570/9-84-005*; CCMS-112, EPA, Office of Drinking Water, Washington, D.C.
36. F. Lefevre, *Techn. Eau., 161*, 31 (1960).
37. J. M. Stamm, W. E. Engelhard, and J. E. Parsons, *Appl. Microbiol., 18*, 376 (1969).
38. H. O. Hartung, *J. AWWA, 48*, 1523 (1956).
39. A. P. Black and R. F. Christman, *J. AWWA, 53*, 737 (1961).
40. G. Corsaro, H. S. Ritter, W. Hrubik, and H. L. Stephens, *J. AWWA, 48*, 683 (1956).
41. F. Brummel, *Gesund. Ing., 83*, 65 (1962).
42. Anon., *Optimale Samenstelling van drinkwater*, KIWA report 100.

16

Treatment Processes of Aluminum-Based Sludge

"Sludge disposal is a challenge to the drinking water service."

1. SLUDGE COMPOSITION AND QUANTITY

In the accumulating zones of static sedimentation tanks or horizontal flow settlers, the sludge can attain a dry solids concentration between 100 and 200 kg/m^3. At these concentrations the sludge is practically solid. On normal abstraction, dilution occurs and the concentration of the removed material ranges from 5 to 15 kg/m^3 dry weight. When the basins are completely emptied, a concentration of 40 to 50 kg/m^3 can be obtained.

In the sludge zones of dynamic settling tanks such as upflow clarifiers or combined flocculators–clarifiers with sludge blanket recirculation, a stationary sludge concentration of 5 to 15 kg/m^3 dry weight is maintained. In this case, one obtains a wet volume fraction of 7 to 20%. In all cases the sludge bleed flow is below 5% of the nominal water flow of the plant, and in most circumstances the bleed ranges from 1 to 2% of the water flow. The sludge bleed can be either continuous or intermittent. The velocity of 0.75 m/s or more is ideal to prevent settling of the concentrated sludge in the pipelines.

The blended sludge from settling tanks of 5 to 15 kg/m^3 dry mass contains 98 to 99% water. The density of the abstracted sludge is always below 1.1 kg/dm^3 (e.g., 1.006 kg/dm^3). When obtained by filter backwash, the total sludge flow usually represents 3 to 8% of the nominal flow of the plant; however, the backwash is intermittent. During the first third of the washing phase, the approximate waste-solids content reaches an average of 0.5 to 1 kg/m^3 dry solids. This value, then, sharply decreases to reach practically zero at the end of the washing period. A value of 0.3 to 0.4 kg/m^3 can be accepted as an average concentration of dry solids in the backwash water of the filters.

On an average basis, the sludge content expressed in grams of dry material per cubic meter of treated water reaches the following amounts:

Matter from raw water: 10 to 50 g/m³ (e.g., 20 g/m³)
Hydrated aluminum hydroxide: 10 to 20 g/m³
Inorganic compounds (iron, manganese, activated silica): 5 g/m³
Activated carbon (if used): less than 20 g/m³
Miscellaneous and organic: less than 20% of total

As a general rule, the composition of the dry material contained in the sludge when abstracted from the treatment, whether activated carbon is used or not, is as follows:

	Without carbon	*With carbon*
Material extracted from raw water	57%	51%
Coagulant residue	33%	30%
Carbon	—	12%
Miscellaneous	10%	7%

The coagulant part [i.e., $Al(OH)_3$ hydrated] includes the heavy metals as removed from the raw water source, especially iron and manganese. Their contribution to the sludge depends on the characteristics of the raw water source. It usually attains less than 10% of the total coagulant residue. It has been reported that the volatile solids can range between 20 and 35% of the total dry solids (1). Experimental data from the Tailfer plant treating the water of the river Meuse are illustrated in Fig. 1.

Organic compounds heated in the presence of air, and originating from the raw water, are burned at temperatures below 400 to 450°C. Temperatures up to 550 to 600°C are necessary to burn activated carbon powder used in the treatment and finally, incorporated in the sludge. When operated in a continuous manner, the

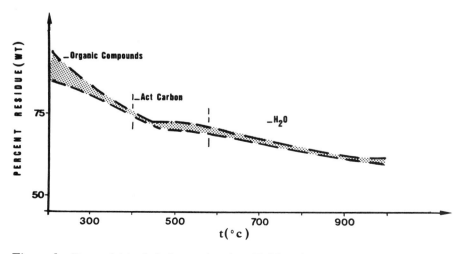

Figure 1 Dry weight of sludge on heating (Tailfer plant, Brussels Waterboard). Typical examples: initial concentration in dry weight, 30 to 40%.

differential determination of organic compounds and activated carbon is obtained with a slight systematic error (10 to 15%) by overestimating the organic compounds versus the activated carbon residue. At the Tailfer plant, organic and volatile substances are below 15% of the total solids. When polyelectrolytes are used in the treatment, the inorganic coagulant dose is decreased in relation to increase in the relative portion of volatile organic compounds (1).

Although sludge composition is claimed to be widely variable (1) as a function of the season, river loading, type of chemicals used, and so on, for every plant an average composition can be given. As an example, long-range averages such as obtained during operation at the Tailfer plant of the Brussels Waterworks are given in Table 1. The loss in weight situated between 400 and 620°C can be associated with the activated carbon and nonvolatile but combustible organic compounds.

Remark. Ferric sludge solids vary and the total solids content of iron sludges in sedimentation tanks is usually higher than that of aluminum sludges (1).

The wash water of filters represents 5 to 10 times the volume being handled by the settling tanks. Most of the processes involve compacting by settling before chemical handling. Therefore, it should be noted that upon dilution of the hydroxide sludge from the settling tanks, with filter wash water, this sludge rapidly decreases to its original volume. Hence there exists no pertinent objection to dilution if it takes place before the subsequent thickening process. This involves the construction of an appropriate flow-regulating capacity, such as a reservoir, to maintain a constant total sludge flow without sudden changes. For continuous operation of the presedimentation tanks (e.g., thickening with classical scraped clarifiers) in the sludge treatment scheme, the possible instantaneous flow variations should be maintained below 5% of the total sludge flow.

2. SLUDGE HANDLING TECHNIQUES

The following processes can be applied to the treatment of sludge obtained in drinking water plants: direct discharge, direct recycling, lagooning, sludge compacting and disposal , and coagulant recovery. Coagulant recovery is the unique method that offers some economical favorable perspectives. Sludge compacting is obtained by heating, freezing, sand-bed drying, vacuum filtration, centrifugation, and pressure filtration.

2.1 Direct Discharge

Direct discharge of sludge to rivers or water bodies is contradictory with pollution control directives. By treating more and more river water for drinking water purposes, treatment plants have emerged as a large potential source of wastewater. Discharge to sanitary sewers is an alterative; however, waste treatment plants should be capable of treating an additional amount of sewage. In the digestion processes of sewage plants, particularly in anaerobic digestion on sand drying beds, gelatinous precipitates of aluminum hydroxide can hinder the dewatering. Except in very local conditions in which the sludge issued from drinking water plants is of secondary importance in comparison to the capacity of the wastewater treatment facilities, discharge to the sanitary sewers is generally not a satisfactory solution to the sludge-handling problem.

Table 1 Composition of the Average Sludge from the Tailfer Plant (River Meuse, Belgium)

Treated water (m^3)	Year	Dry Solids from Meuse water (tons)	Alum residual $Al(OH)_3$ (tons)	Activated carbon (tons)	Lime $Ca(OH)_2$ (tons)	Total dry solids (tons)	Compacted sludge transported (tons)	Water content (%)
40,110,700	1978	621	477	183	1145	2426	7275	67
43,403,000	1979	879	473	219	1082	2653	8158	67
45,350,600	1980	805	524	227	1229	2785	9043	69
44,609,600	1981	968	471	207	1164	2810	8503	67
38,076,200	1982	873	416	126	833	2248	8413	73
34,349,400	1983	684	372	110	960	2125	7608	72
25,901,000	1984	582	332	102	696	1711	5560	69
23,845,650	1985	526	349	142	803	1820	6712	73
32,173,700	1986	1086	482	150	1234	2952	8768	66
26,491,300	1987	790	421	111	1235	2557	8983	72
26,501,200	1988	927	395	120	1149	2591	8808	71

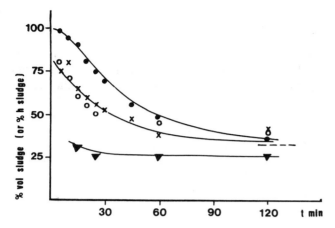

Figure 2 Settling characteristics of sludge after dilution. ■, Without dilution; ×, dilution 5×; ○, dilution 10×; ▼, dilution 20×.

2.2 Direct Recycling

Direct recycling of sludge to the head of the treatment can be favorable for the control and optimization of the flocculation–settling process (see Chapter 9). However, no significant reduction in coagulant has been obtained by this method. As such, it reduces neither the sludge volume to be handled nor the essential properties of the sludge-compacting processes that intervene.

Backwash water has been recycled into the head of the treatment at certain plants. This operation requires appropriate regulation of the raw water intake capacity in view of the fact that the instant flow of wash water can range between 50 and 100% of the raw-water intake (e.g., one filter among 5 to 10 is backwashed at 30 m/h). Furthermore, a sudden increase in suspended matter originating from the backwash may require readjustment of reagents. Although the filter backwash matter is removed by the coagulation–flocculation process, the direct recycling is to be considered objectionable because of the required flexibility in treatment.

2.3 Lagooning

Lagooning remains a popular method for the disposal of water treatment sludges. Lagoons are generally built by enclosure of a land surface with dikes or by excavation. Sludge is added continuously or intermittently until the lagoon is filled. The lagoon should be equipped with flashboards to enable sedimentation and discharge of the supernatant, particularly if the filter wash water is sent to the lagoon. Lagoon effluent characteristics generally enable direct discharge. Lagooning can be economical when land is readily available and inexpensive.

Sludges containing aluminum or iron hydroxide consolidate gradually and rapidly to reach 10 to 15% dry solid content. Sludges from iron removal treatment dewater more quickly than those from aluminum sulfate coagulation. Although this sludge can be removed with a dragline, the material is not suitable for landfill without further dewatering. The operation cost is lowest if the site of the lagoon can be abandoned after filling. After prolonged weathering (e.g., 2 to 3 years), a solid sludge content of 30 to 40% can be attained, in which case the disposal site is dry enough to walk on.

The dewatering occurs primarily by evaporation, freezing, and natural draining. It can be promoted by growing reeds in the lagooning area. Owing to the composition of the sludge, the content of a lagoon is generally stable from the biological point of view. Insect breeding at the surface has been noted. Land requirements for lagooning are high and the technique is attractive only for its low operating costs and its technical simplicity.

For a typical sludge quantity totally ±75 mg/L and a water flow of 10,000 m³/h, the sludge production can attain 18 tons/day or 6600 tons/year of dry material. The assumption of an average concentration of 12 vol% solids settled in the lagoon leads to an annual total sludge volume in the range 50,000 to 60,000 m³. Drying lagoons are generally given a depth of about 2 to 3 m (exceptionally, 5 m) and filled up to 1 to 1.5 m with dried sludge. Hence for a production of 10,000 m³/h of treated water, one reaches an annual lagooning surface of ±50,000 m².

Even if the solids concentration of the raw water intake is lower than the value indicated here as an example, the process is attractive only if a sufficient large area is available. In urban areas lagooning is advisable only for small isolated plants or as a temporary method during a period of maintenance and repair of sludge-drying equipment. The costs for the emptying and transformation of residue are to be added to those of lagooning when the latter may not be abandoned on site. No practical objection exists for its use in agriculture (2,3) and the pelletizing of seeds with fertilizers is one of the possible uses of lagooned sludge.

2.4 Sludge Compacting Techniques

Sludge compacting and mechanical handling is presently part of most existing plants. The principles are variable in nature and mechanical compacting is generally used after the preliminary thickening (see Section 3); however, direct compacting is feasible.

2.4.1 *Compacting by Dehydration Through Heating*

Aluminum hydroxide sludges are not directly combustible and require the supply of an external source of energy such as fuel. Dehydration by heating, including the combustion of organic material, requires heating to 600 to 800°C (see Fig. 1). Consequently, a temperature of at least 600°C is required and is generally impractical from the point of view of energy consumption. The calcinated mass ranges to about 65% of the dry mass (110 to 130°C) of the sludge. This ratio implicates combustion of the activated carbon content as indicated in Section 1 (Fig. 1).

A pulse-jet heating system has been proposed as an attractive and economical method (4). The unit operates in the temperature range 80 to 120°C by injecting heated air with a blower; sonic energy is applied simultaneously. The effluent gas is directed to a cyclone to remove the suspended particles. The pulse-jet system is described as operational in the form of a preliminary treatment to centrifugation or on intensive application to produce a sludge with a high dry solids content. Ninety percent drying has been advanced, but more details on long-term application are not yet known. Energy requirements are described to range 2.3 to 2.8 × 10⁶ J/kg water removed and up to 3.2 × 10⁶ J/kg water removed in the intensive drying technique.

Returning to the previous example (Section 2.3), for a water flow of 10,000 m³/h and a sludge production of 18 tons/day abstracted at a concentration of 5 to 10 g/L after preliminary settling, one can obtain a concentration up to 50 g/L. This is equal to reaching a 25% dry solid with a daily volume of water to be evaporated, producing about 14.4 tons or 60 kg/h, which means an energy consumption of 15,000 kWh.

Thickened, compacted sludge is in some instances suited to self-supporting combustion. When the sludge has been treated with lime in the compacting phase, the composition of the ashes approximates that of cement (20).

2.4.2 Freezing and Thawing

Freezing and thawing may result in marked improvement in lagooned sludges of aluminum or iron hydroxides. In the case of alum sludges, it has been reported that dewatering can reach up to 17.5% dry solid content. The effect involves a disruption of the stability of the gel which is due to freezing. Thus on thawing, interstitial water is released, leaving a granular-like solid, composed of grains 0.1 to 1 mm in size. Its filtration is easy, and when stored on a lagoon or drying bed it does not rehydrate. This natural phenomenon observed in lagooning has been experienced with industrial equipment.

Artificial freezing and thawing of the sludge, obtained by a pressure change in ammonia, leads to an energy consumption of 22 kWh/m^3 sludge. Returning to the previous example, in the treatment of 10,000 m^3/h water flow, with a daily production of 18 tons sludge and a daily average concentration after settling of 10 g/L, the sludge volume reaches 1800 m^3, which means an energy consumption of 40,000 kWh. The freezing process finally delivers a residue with about 20% solids, which can be increased to 30% after subsequent filtration. Except in natural climatical conditions, the freezing–thawing process is not yet economically feasible for sludge compacting and disposal.

2.4.3 Sand Drying Beds

Decantation can be included in the sand drying bed technique, but the operation basically involves the removal of water through drainage and air drying. The beds consist of a sand layer of 15 to 20 cm over a layer of gravel of 30 cm on underdrain tiles with dewatering pipes (Fig. 3).

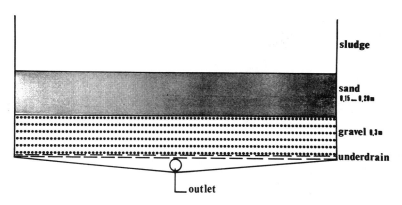

Figure 3 Sand drying bed (schematic).

A typical specification for sand is:

Size range: 0.3 to 1.2 mm
Average diameter: 0.5 mm
Uniformity coefficient: less than 5

The gravel for the underdrain is of 3 to 6 mm grade. A typical solid loading at the surface is 4 kg/m^2. The sludge depth is maintained below 1.2 m, preferably at 0.5 to 0.75 m. Consequently, the method requires large spaces.

Dewatering depends on climatological conditions; long dewatering times and poor performance are noted during winter periods. Up to 97% of the initial solids can be removed, producing a pelletable sludge containing 25% solids after several months of drying. Weak cationic polyelectrolytes of high molecular weight ($\pm 10^7$ units) help the dehydration of mineral sludge. The operation costs, including the removal of the sludge, have been estimated at 60 U.S. dollars per ton of dry solids (1984).

2.4.4 Vacuum Filtration

The sludge must be thickened before vacuum filtration. In vacuum filtration a revolving drum covered with a filter cloth is partially immersed in the sludge (Fig. 4). The ratio of immersion time to drying time is maintained between 0.75 and 0.6. As the drum rotates, a layer of sludge is dehydrated when it emerges from the reservoir. An external vacuum with a differential pressure of up to 0.3 to 0.5 bar can be maintained; however, the values most commonly encountered are 0.05 to 0.1 bar. Before scraping with a blade, pressure is applied to facilitate the release of cake. A sludge layer thickness between 3 and 6 mm is maintained.

Vacuum filters are available in a wide range of sizes: from 1 to 70 m^2 cloth surface. The system can be operated on a continuous basis. The operation involves the action of a vacuum pump, compressor sludge intake facilities, effluent outlet, and a filter cake conveyor.

Alum sludges are known to dewater difficultly on vacuum filtration. The solid loading is usually limited to 2 kg dry solids/m^2 per hour in direct sludge filtration. A maximum concentration of 25% dry solids is obtainable in the cake. Filter clog-

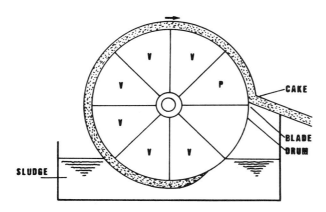

Figure 4   Vacuum filter.

ging is a problem and the effluent quality is not always satisfactory. Therefore, precoating with diatomaceous earth is sometimes necessary. The weight required is approximately 0.5 to 1 kg/m^2 per hour and the cost of the precoating is high. An alternative is the use of a polyelectrolyte polymer, which increases the solids loading. The differential pressure must be maintained at lower values so as to avoid the risk of permanent clogging of the cloths (e.g., 0.03 to 0.05 bar); however, this does lower the dewatering velocity. The filtration cloth, normally synthetic fabric (nylon, polyester), is positioned on perforated plates. As differential pressure is low, compared to pressure–filtration (see Section 2.4.6), the choice of a suitable cloth is difficult given the necessary compromise between permeability and solid retention rate. This requires preliminary investigation with each different sludge.

General specifications for filter cloths are similar to those for pressure filtration (see Section 2.4.6) but monofilament fabrics of higher air permeability are less satisfactory for this application than are multifilament synthetic fabrics (e.g., polyester weaves of air permeability in the range 1 m^3 min per 0.1 m^2). The solids loading for aluminum-based sludge treated with polymers can range 5 kg/m^2 per hour (i.e., low loading) or 30 kg/m^2 per hour (which is high loading). Dosing of the polymers is in the range 0.5 to 2 kg/ton dry solids. The higher the solids loading, the lower the dry solid content in the filtered cake (e.g., 9 wt% at 30 kg/m^2 per hour and 15 wt% at 6 kg/m^2 per hour). This effect illustrates that the solid retention of the cake is far from being excellent. Furthermore, the filtered cake, even if treated with a polyelectrolyte during filtration, is difficult to handle in direct transportation, as it remains adhesive. Therefore, unless significant progress is made, the vacuum filtration is less suitable for aluminum-based sludges.

On a comparative basis, the costs for vacuum filtration, including precoating, can reach up to 200 U.S. dollars per ton of dry solids for alum sludge.

2.4.5 Centrifugation

Similar to vacuum filtration, the sludge must be thickened before concentration by centrifugation. The addition of ±30% quicklime (based on dry material) facilitates dehydration. The final product obtained with drinking water sludges then reaches 25 to 30 wt% in dry product. The lime dosing rate remains set empirically and has not yet been optimized in terms of chemical reactivity as, for example, for pressure filtering (see Section 3). This material can be handled, but it is free of its thixotropy only after drying in open air. The centrifuge is essentially a sedimentation device in which the solid–liquid separation is improved by rotating the liquid. The centrifuges used for sludge treatment are generally cylindrical bowls with conical ends for solid discharge. Most machines use the countercurrent flow of the liquid and the solids (Fig. 5). The sludge is introduced to the revolving bowl through a stationary feed tube. The solids are thrown against the wall of the bowl and removed by a helical screw conveyor which rotates in the same direction as the bowl but at a speed slightly lower or higher (e.g., 0.15 to 0.3 rev/s difference).

Equicurrent centrifuges have also been developed more recently (Fig. 6). The essential advantage of this design, which can have either a central or a tangential outlet, is that the deposited solids are not disturbed by the incoming slurry.

The rating of a centrifuge capacity is a complex problem falling beyond the scope of this book. Many operational parameters, such as rotational speed, shape of the bowl, pitch of the screw conveyor, and differential speed of the bowl and the

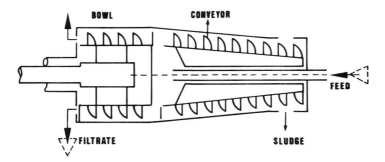

Figure 5   Countercurrent centrifuge system (schematic).

conveyor, influence the rating. Centrifuges for this purpose are known to be in operation at speeds of 15 to 66 rev/s. The force produced may attain 1000 to 4000 times the force of gravity. Optimum results are usually obtained at values of G (times acceleration of gravity) ranging from 2500 to 3200 m/s^2. For a well-preconditioned sludge, the average residence time in the centrifuge can be limited to 1 min. (Slow-velocity centrifuges have a rotational velocity of only 0.005 rev/s.)

Centrifugation implies preconditioning of the sludge by thickening followed by treatment with lime or polyelectrolytes. Pretreatment with lime as optimized for pressure filtering, for instance, makes it possible to obtain a compacted sludge with 25% dry weight content. Polymer treatment of the inflowing sludge is carried out in most cases with nonionic polymers at dosing rates of 1.5 to 2.5 kg per ton of dry matter. This enables to obtain a processed sludge at 15 to 20% dry weight.

The recovery of the solids by centrifugation often depends greatly on dosing of the polyelectrolyte, which means that the recovery yield usually attains only 70 to 80% of the total solids. At the exit the centrifugated sludge remains slightly thixotropic and adhesive. It dries when further exposed to air and can be more easily transported from 2 days to a week after processing and then disposed of subsequently. The stability of the polyelectrolyte on bacterial or chemical oxidation, after long-term disposal, remains an open question.

Overall costs for centrifuging on the base of weight of the dry solids are compa-

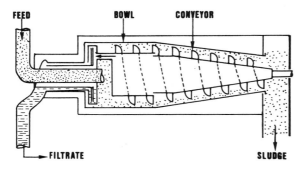

Figure 6   Equicurrent centrifuge system (schematic).

rable to that of the pressure filtering (e.g., 50 to 60 U.S. dollars per ton of dry material). The method is of more easy automation than other, more discontinuous techniques. Although the addition of lime favors the dehydration, this can be substituted by polyelectrolytes, thus reducing the mass to be transported.

2.4.6 Pressure Filtering

In pressure filtration applied forces ranging from 5000 to 20,000 times the gravitational force are used to facilitate filtration. On account of these high pressures, filtration efficiency is improved and the concentration of solids in the final cake increased. The filter press consists of a number of plates aligned in a frame. A filter cloth is mounted on the faces of each plate. The sludge is fed into the press by pumping and passes through the holes into the trays until the filtration chambers are filled with dried solids. All chambers of a given press are mounted in series, and at the end of the pressing cycle, the flow of clarified water drops. The pressure is then released and the press is opened to deliver the cakes. The process is essentially discontinuous; however, an automization of the stopping procedure is possible through the flow control of the effluent. The presses are best mounted at second-floor level. The cakes are then discharged automatically via a hopper into trucks for transportation. With an appropriately preconditioned sludge (see Section 3) and under proper operating conditions, a concentration of 30 to 40% dry weight can be attained. The dehydration is irreversible and the material is suitable for land filling.

The designs of different suppliers vary in the shape and dimensions of the chambers and the system of intake of the slurry. General guidelines can be given as follows:

Frame dimensions: up to 1.5 × 1.5
Construction of frames: preferably corrosion-free armed polyester (eventually cast iron)
Chamber number: less than 100 (preferably ±40 to 50)
Working pressure: 5 to 30 bar (generally, 10 to 16 bars)
Cake thickness: maximum 5 cm (usually, 2 to 3 cm)
Cake volume: 40 to 100 L per chamber of 1.5 × 1.5 m
Pressing cycle: depends on preconditioning (see Section 3)

The typical specifications for the filtering and support cloth are as follows:
Filter Cloth.
Fibers: synthetic (e.g., polypropylene and/or Rilsan warp and weft are often identical but 100% Rilsan can be used for the wrap fiber.
Cloth must be able to resist pH values ranging between 1 and 13.
Construction of fiber warp/weft should be, for example, 478/208 per 0.1 m.
Monofilament fibers perform best for filtering.
Weave: satin
Weight range: 260 to 350 g/m^2
Breaking elongation: 40 to 50% in length and width
Breaking strength: 200 to 300 kg/cm
Air permeability: 1 m^3/dm^2 per minute at 2 mbar
Expansion on wetting: less than 1% in length
Finishing: calendered

Support Cloth.
Fibers: synthetic (most often 100% polypropylene), pH resistant
Fiber construction: multifilament, warp/weft count 64/54 per 0.1 m
Weave: plain
Weight: > 500 g/m^2
Air permeability: e.g., 4 m^3/dm^2 per minute at 2 mbar
Breaking strength: 500 to 800 kg/cm
Breaking elongation: 15 to 20% in width; 30 to 35% in length

Precoating the filter cloth is normally not necessary unless excessive clogging occurs, in which case a precoat of diatomaceous earth can be provided at a rate of 0.3 to 0.5 kg/m^2. Regular cleaning of the filtration cloth is necessary. The latter can be carried out by circulating passivated hydrochloric acid at a concentration of ± 6 wt%, or by a mixture of sodium hexametaphosphate and ethylenediaminetetracarboxylic acid.

Pressure filtering requires higher investment and operating costs than does centrifuging. However, the latter requires more energy. Both techniques appear competitive for sludges from drinking water treatment facilities. The cost is 50 to 60 U.S. dollars per ton of dry material.

3. OPTIMIZATION OF PRELIMINARY SLUDGE THICKENING AND LIME CONDITIONING

3.1 Thickening

The thickening of the sludge in a scraped clarifier must be considered as a pretreatment allowing further compacting by subsequent techniques. In no way can the thickening alone provide a mass containing 20% dry material, which is the very lowest concentration necessary to enable mechanical handling and land filling. The concentration of the thickened sludge after settling is generally maintained between 20 and 30 g/L on the basis of dry solids concentration. On slow settling with prolonged compaction in the sludge layer, concentrations of up to 40 and even 60 g/L can be reached. These concentrations are also more readily attained by the proper use of polyelectrolyte as a coagulation aid. Nonionic high-molecular-weight polymers are more effective as dewatering aids than cationic polymers are for this purpose.

Flow regulation is usually necessary (see Section 1) to operate a primary scraped clarifier which has a classical design. The surface loading is generally selected to be about 0.25 mm/s (always less than 0.4 mm/s). The total depth is at least 3 m, and overflow weirs for clarified effluent are constructed according to the indications outlined in Chapter 9.

Sludge transfer pumps that are set in to regulate the inlet flow will be required to pump a sludge of considerable variability in quantity and concentration. The flexibility can be attained by variable-rate pumping or adjustable time cycles. Over-dimensioned centrifugal pumps have been proven very satisfactory in practice. Obtaining a sludge that is constant or very slightly variable in concentration also implicates very regular abstraction of the sludge from the preliminary tank, that is, hydraulically independent abstraction regardless of the movement and position of the scrapers.

3.2 Lime Conditioning

Preconditioning of the sludge can be improved by a second stage of conditioning. In this operation, the sludge at 20-30 g/L is transferred to a second thickener after addition of quicklime or hydrated lime. According to the indications given by the literature, the lime should be added at 25 to 30% (exceptionally, only 10 to 15%) of the weight of dry solids contained in the prethickened sludge. The second thickener should not be designed as a settler, but rather as a storage reservoir enabling continuous withdrawal of the prethickened sludge. This intermediate storage is often designed for 24 to 72 h and makes it possible to operate subsequent sludge compacting equipment discontinuously (e.g., interruption of the operation during weekend periods). In these general conditions, the concentration expressed on the basis of dry material after thickening attains 60 to 120 g/L on the addition of 25% quicklime. Moreover, filtration becomes easier as the structure of the sludge is modified by the lime treatment.

3.2.1 *Optimization of Lime Treatment*

The filtrability of sludge treated with lime and containing a flocculation residue with aluminum salts can be used to optimize the addition of lime, yet this measuring procedure does not permit automation of the process. Measurement of the specific resistance to filtration is defined by the method of Carman and Ruth (5,6). The specific resistance is defined as the resistance offered by a unit weight of sludge per unit area of a filter at a given pressure difference of filtration:

$$r_s = \frac{2PA^2}{\mu C} \times m$$

The flow rate through a filtration cake can be expressed as the total driving force divided by the sum of resistances:

$$\frac{dV}{dt} = \frac{g \times \Delta P \times A}{\mu(Vr_s C/A + R_m)}$$

where
  V = filtrate volume, m^3
  t = filtration time, s
  g = acceleration of gravity, 9.81 kg/s^2
  ΔP = pressure drop across the cake and the cloth, kg/m^2
  A = filter area, m^2
  μ = dynamic viscosity of the filtrate, $kg\ m^{-1}s^{-1}$
  r_s = specific resistance, m/kg
  R_m = unit filter media resistance, m^{-1}

On integration, at constant pressure, the equation is expressed as

$$\frac{t}{V} = \frac{\mu r_s C}{2g\ \Delta P\ A^2} \times V + \frac{\mu R_m}{g\ \Delta P\ A}$$

In most applications the second term can be eliminated since the resistance of the filter medium is neglectible versus that of the filtrated cake. The equation then becomes

$$\frac{t}{V} = m \times V \quad \text{in which} \quad m = \frac{\mu r_s C}{2g \, \Delta P \, A^2}$$

The factor m can be determined experimentally as the slope of the straight part of the line obtained by plotting t/V as a function of V at filtration in constant standardized conditions.

Remark.   Values for r_s have sometimes been reported in s^2/kg, which means without applying the factor g, that is, the acceleration of gravity ($9.81 \, m/s^2$).

The effect of differential pressure has been quantified in the terms of the compressibility of the cake according to the relationship of Carman (7):

$$\log \frac{r_s}{r_s(P)} = s \log P$$

r_s and $r_s(P)$ being the specific resistance to filtration, respectively, at unit pressure difference and at a pressure difference P, and s is the cake compressibility in m^2/kg. The value of s for alum-based sludges is usually slightly higher than 1 (e.g., between 1.03 and 1.15). The values measured for r_s, even at the same pressure difference, vary widely according to the origin of the sludge. Differences by a factor of 10 have been observed with alum sludges abstracted from different plants (1). As a general rule, the higher the $Al(OH)_3$ content of the sludge, the higher the r_s value. According to the formula for specific resistance, an increase in sludge concentration should lower the specific resistance as a consequence of an inverse function. A typical example with sludge abstracted from the Tailfer plant (Brussels) is given in Fig. 7.

An experimental relationship of the type $r_s = -(DS)^b$ has been observed in the experimental conditions as given. A large dispersion is often observed at low sludge concentrations. At concentrations of 40 kg/m^3 dry solids or more, practical limits to lower the specific resistance are reached by increasing the sludge concentration. Consequently, the yield of preliminary settling in sludge treatment must not be increased to produce a higher sludge concentration if investment costs must be increased. The addition of lime reduces the specific resistance to filtration by increasing the solid content.

Figure 8 indicates on the one hand the calculated value by simple additivity of DS concentration and $Ca(OH)_2$ added, as well as adoption of the curve in Fig. 7, and on the other hand, the measured values. The data indicate that the sludge structure and chemical nature are modified on the addition of lime; moreover, after reaching a certain dose of lime [e.g., 12 kg/m^3 $Ca(OH)_2$ in the example shown in Fig. 8], the specific resistance is no longer improved.

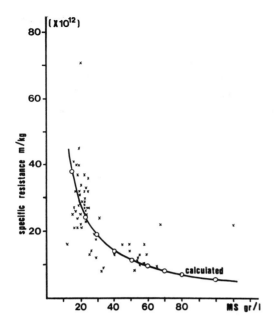

Figure 7   Dependence of r_s on solid content of the sludge (Tailfer plant, $P = 5000$ kg/m^2). (From Ref. 8.)

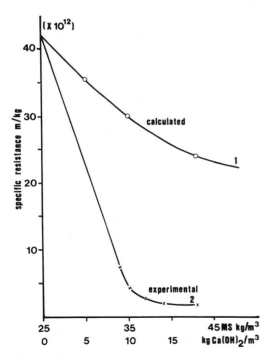

Figure 8   Additivity of DS and Ca(OH)$_2$ concentration on r_S for alum sludge ($P = 5000$ kg/m^2).

485

3.2.2 *Principles of Design and Operation of Lime Thickening*

A systematic investigation of this problem has shown that the determinant parameter for the reaction with lime is the aluminum concentration of the sludge (8). Several compounds have been identified after reaction of $Al(OH)_3$ in sludge and $Ca(OH)_2$, such as

$$Ca_6Al_2(SO_4)_3(OH)_{12} \cdot 25H_2O$$

$$Ca_2Al_2O_5 \cdot 6H_2O$$

$$C_{16}Al_8(OH)_{54}CO_3 \cdot 21H_2O$$

$$Ca_4Al_2O_7 \cdot H_2O$$

$$CaAl_2O_4 \cdot 10H_2O$$

and also, in addition to calcite, portlandite, and crystalline silica.

The overall stoichiometry of the reaction converges to the ratio $2Ca^{2+}/Al^{3+}$. In practice, a slight excess of calcium is necessary to obtain the complete reaction as well as an optimum filtrability. Mixing requires a floc breakage rather than preservation of a structured floc (e.g., at $G \geq 1000 \text{ s}^{-1}$). Breakage is essential for a good reaction. In appropriate mixing conditions, a contact time of 10 to 15 min is sufficient to complete the reaction.

By reacting lime in quantities higher than the stoichiometric amounts, calcium ions appear in the supernatant, the electrical conductivity of which increases. Measurement of the electrical conductivity provides for a suitable continuous method of measurement making it possible to monitor the addition of lime (Fig. 9). In the case of the Tailfer plant, the minimum conductivity necessary for the supernatant is 0.225 S/m. In such conditions the pH is higher or equal to 12.5, but monitoring on this basis is less accurate than using electrical conductivity. The use of a preformed milk of lime is superior to the direct dosing of quicklime. The use of the latter requires an excess that is found partially as hydrated lime in the filtered cake.

Optimation according to these guidelines has made it possible to obtain operational filter pressure cycles at 16 bar of 2 m^3 compacted sludge within 15 to 20 min of pressuring. In adequate lime dosing, too-low dosing particularly, increases the cycle length to more than 90 to 120 min and leads to an insufficiently dry cake. Overdosing increases the costs as well as the amount of transported material and increases the clogging rate of the filter cloths.

3.3 Stability of Lime Conditioning Sludge

At pH > 12.5, reached by lime stabilization and compacting, no significant bacterial activity that could cause a sanitary risk in the environment can create an objection to disposal of the treated sludge. Significant amounts of heavy metals are removed from the raw water and incorporated with the sludge simultaneously with those originating from the chemicals applied in water treatment and sludge condi-

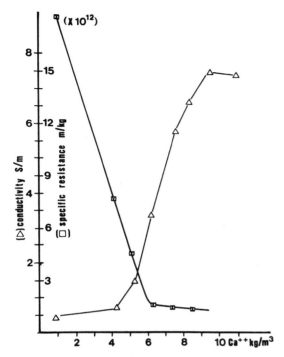

Figure 9   Optimation of the Ca^{2+}/Al^{2+} reaction by monitoring electrical conductivity.

tioning. Therefore, the quality of the conditioned sludge disposed of in landfills or for agricultural use must be evaluated in each case. Typical information for the Tailfer plant (Brussels) is given in Table 2. More complete data have been reported in the literature (2,3). Although the data may be largely variable in function of the raw water source (which has a variable input concentration), the lime-compacted sludge is generally compatible with agricultural use, since the heavy metals content is largely below the limits recommended.

Lixiviation through a column percolation, through a layer 0.2 m thick by distilled water and totally a flow-through of 10 m^3/m^2 (i.e., over 10 years' average rainfall in Belgium) amounts to 1 to 5 % of the metal ion fixed in the compacted sludge. Considering the drinking water standards, copper is the critical element that may require further investigation. In practice, however, lixiviation is not obtained in the same way, as contrary to laboratory experiments, the superficial layer is influenced not only by rainfall but also by evaporation–transpiration. Practical observations kept at Tailfer and carried out when measuring the percolate through a layer of 1-m-thick compacted sludge are as follows.

For an average yearly rainfall of 0.8 m, the yearly volume of water percolated through the sludge layer amounts to 0.012 m^3/m^2, which means 1.5% of the rainfall, with a maximum monthly recovery of 11% from the percolate as has been true in 1981–1984. The pH value of the percolate is 8 ± 0.2, and average concentrations of trace metals measured on the percolated water are, for instance, Fe, <50 $\mu g/L$; Co, <10 $\mu g/L$; Cr, <6 $\mu g/L$; Cu, <70 $\mu g/L$; Ni, <25 $\mu g/L$; and Mn, <15$\mu g/L$.

Table 2   Removal of Heavy Metals from the Water of the River Meuse at Tailfer

	Metal								
	Fe	Mn	Cu	Cr	Cd	Ni	Hg	Co	Pb
Input (mg/m^3)	18.5	34.5	8.1	3.5	5.7	3.9	–	3.5	2.3
Output (mg/m^3)	1.8	1.02	1.2	0.6	0.25	1.95	–	1.7	0.9
Content in compacted sludge (g/kg DS)	4.1	0.8	0.04	0.02	0.001	0.025	8×10^{-6}	0.015	0.035
Limits for maximum agricultural use (g/kg soil) (9)	–	–	0.5	0.5	0.01	0.1	0.005	–	0.5
Laboratory simulation of lixiviation conditions (mg/m^3)	30	12	130	10	0.2	22	0.1	8	1
MAC in drinking water (mg/m^3)	200	50	50–1500	50	5	50	1	–	50

4. WET PROCESSES FOR ALUMINUM-COAGULANT RECYCLING

The solubility of aluminum salts as a function of pH is one of the underlying principles of coagulation and is discussed in Chapter 5. The reaction of alumina-containing sludge with acids or bases such as sulfuric acid and caustic soda makes it possible to dissolve to a great extent the aluminum salt contained in them. Lime enables partial recovery only (1,10) but is in the line of the sludge compaction process.

The wet recovery processes, as reported in the literature, on aluminum-based coagulants (11–19) are mostly founded on the attack of the sludge by sulfuric acid. However, by more careful investigation (10) it is shown that the quality of the coagulant recovered by this process was less pure than that obtained by the hardly known alkaline method (1,11,16). On preliminary investigation, the initial material is sludge as extracted from the settlers and concentrated by preliminary settling at an overflow rate of 0.15 mm/s. The concentration in dry weight ranges from 10 to 45 g/L, and the sludge is reacted with increasing amounts of sulfuric acid, sodium hydroxide, or milk of lime, mixed in a jar test apparatus for 15 min and allowed to settle for at least 1 h. The supernatant is filtered (0.45 μm) and analyzed for aluminum and metal content.

4.1 Acid Recovery Process

According to the reaction stoichiometry of

$$2Al(OH)_3 + 3H_2SO_4 = Al_2(SO_4)_3 + 6H_2O$$

156 weight units of $Al(OH)_3$ can lead to 342 weight units of aluminum sulfate. The aluminum sulfate is recovered as the liquid after reaction and can be recycled in the treatment. On acidification a minimum pH of 3 to 3.4 is necessary, which is obtained in general by the reaction of about 3.2 ion g H^+ per mol Al^{3+} (Fig. 10). Hence the stoichiometric ratio of 3:1 is approached very closely; however, the pH value appears not to be a universal criterion for acid recovery; the ratio of aluminum salt recovered depends on the type and composition of sludge.

Three objections can be formulated against the acid recovery process. Although like heavy metals, most trace elements remain largely within acceptable limits, the recycling of organic compounds characterized globally by the chemical oxygen demand increases so as to reach at a pH of 3.2, about five times the average value measured for the raw river water (Fig. 11). On long-range recycling, the accumulation of organic products may constitute an objection to recycling by the acidification process.

In cases where phosphate is removed from the raw water source with the aluminum floc, the phosphate ion is taken over in the solution on acid recovery of aluminum from the sludge. Hence phosphate ion can accumulate on recycling and determine the limit of the process. The phosphate recovery rate is proportional to the aluminum recovery rate, so there is a stoichiometry relationship Al/P. At pH5, for instance, a formula for the complex has been proposed as

$$Al_{13}(OH)_{30} \cdot 18H_2O \cdot (H_2PO_4)_9 \qquad [\text{ratio Al:P} = 1.44 \ (1)]$$

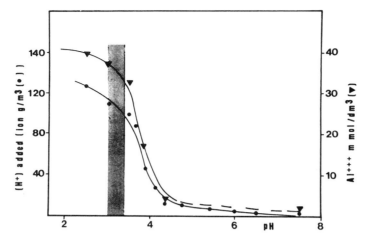

Figure 10   Solubilization of Al^{3+} recovered in the supernatant on acidification of $Al(OH)_3^-$ containing sludge.

The attack with sulfuric acid makes it possible to obtain a pH value of 3. Consequently, special technology and extra protection of the equipment are necessary. After the acid reaction and recovery of aluminum in the supernatant, the solid residue is to be processed with lime up to pH 8 for eventual compaction by pressure filtration, although the direct filtrability of the residue is good ($r_s = \pm 2$ to 3×10^{12} m/kg) (10).

4.2   Recovery with Sodium Hydroxide

The recovery yield as a function of the NaOH/Al ratio is indicated in Fig. 12. A two-stage reaction occurs, forming two different aluminates. Conclusions for the NaOH reaction are the following:

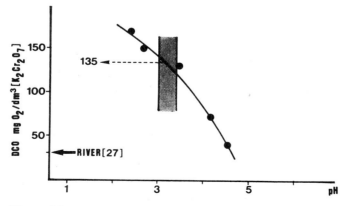

Figure 11   Increase in chemical oxygen demand of the supernatant after sludge acidification.

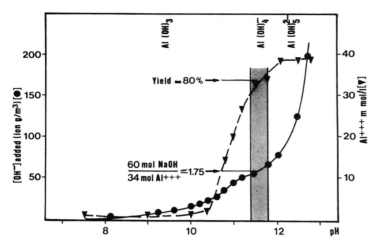

Figure 12   Aluminum recovery by sodium hydroxide.

1. The optimum pH ranges from 11.4 to 11.8 at which a recovery yield of up to 80% is obtained, while, except for iron, the ratio of metals to aluminum is at a minumum value.
2. The molar ratio NaOH reacted to aluminum recovered is 1.75.
3. The liquid to be recycled contains about 40 mmol/L Na.
4. Dissolution of aluminum up to a yield of 98% can be obtained, however, requiring strongly increasing amounts of sodium hydroxide.

In the case of sludge deriving from the treatment of water from the river Meuse, the ratio of trace metals to aluminum in the supernatant after the reaction of sodium hydroxide is given as an example in Fig. 13. No pertinent objection due to heavy

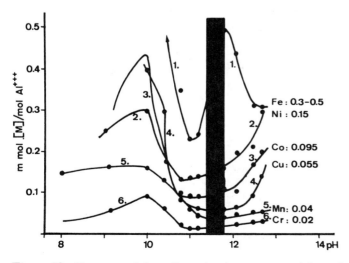

Figure 13   Trace metals in sodium aluminate recovered from the sludge at Tailfer.

metal content exists against recycling the recovered sodium aluminate. If significantly present in the solution recovered as sodium phosphate, phosphates can be precipitated by the addition of calcium chloride (1).

4.3   Recovery with Lime

Reaction of excess of lime with aluminum delivers a poorly soluble product. However, at lower treatment ratios a soluble intermediate is obtained partially as illustrated in Fig. 14. The formation of calcium aluminate as $CaAl_2(OH)_8$ (7) is probable.

1.   The optimum pH ranges from 11.2 to 11.6, at which the recovery rate attained is 25%.
2.   The molar ratio $Ca(OH)_2$ to Al ranges from 7 to 8 or about 75 mmol/L Ca in the liquid to be recycled.
3.   The ratios of metals to aluminum are 10 to 15 times lower than in the case of attack by NaOH. This could make the lime process more adequate in cases where a problem of heavy metal content arises in the raw water (see Fig. 15).
4.   Phosphates remain mainly precipitated as calcium phosphate in the residual solids.

4.4   Optimization Parameters for Alkaline Recovery

A reaction tank equipped with a rapid or flash mixer and a residence time of 15 min enables us to utilize alkaline recovery. The more compact the initial concentration of sludge reacted with alkali, the slower the settling subsequent to the reaction and the lower the liquid volume recovered. The data given in Fig. 16 indicate that for a sludge concentration lower than 20 g/L dry weight, the coagulant recovered in the supernatant amounts to >50 to 60% of the total volume. At higher initial concentration the volume of supernatant recovered by settling is drastically lowered,

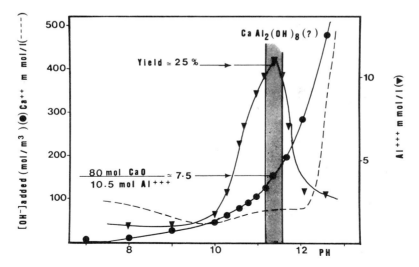

Figure 14   Aluminum recovery with milk of lime.

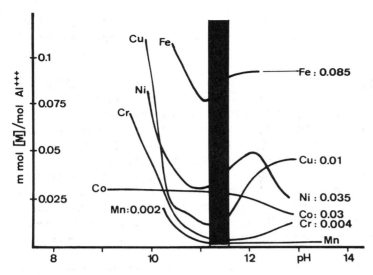

Figure 15   Trace metals in calcium aluminate (?) recovered from sludge at Tailfer.

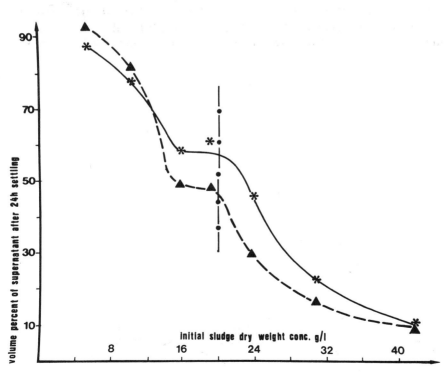

Figure 16   Supernatant volume as a function of initial sludge concentration. Optimum conditions: NaOH, pH 11.6 (dashed curve); CaO, pH 11.4. (solid curve).

which makes the design of the settling tank following the reaction tank difficult. Also, seeing that the dissolved ions are dispersed through the entire liquid phase, the recovery yield in the supernatant decreases when increasing the initial sludge concentration.

The practical limit for the initial sludge dry weight concentration is 20 g/L. If sludge of higher concentration must be handled, alternative methods for the liquid recovery should then be prospected (e.g., centrifugation or pressure filtration). The percent of the total aluminum content found in the supernatant depends on the dry weight concentration of the reacted sludge, as indicated in Fig. 17. The dissolution of trace heavy metals follows a similar trend and is illustrated Fig. 18.

Compared to investigations on sludge with a ±10 g/L dry weight concentration, the concentration of heavy metals in the supernatant has been found to be 10 to 15 times lower in the case of an attack on sludge by lime than when sodium hydroxide is used. Again, at a sludge concentration of less than 20 g/L dry weight, the proportion of the metal recovered increased according to the metal/aluminum ratios given in Figure 13.

The total aluminum content of the sludge can vary from one installation to another. The presettled sludge before addition of alkali, as found in the Tailfer plant, can vary with time according to various parameters, such as type and dosing rate of the coagulant applied, age and concentration of sludge bleed, dilution of wash water of the filters, and so on. Practical data found are illustrated in Fig. 19. The total aluminum content of the sludge before attack with alkali increases with total dry weight concentration. For example, in the range 15 to 20 g/L dry weight solid, the total aluminum content varies between 0.05 and 0.1 mol/L.

Optimum pH ranges of the supernatant have been determined as 11.4 to 11.8

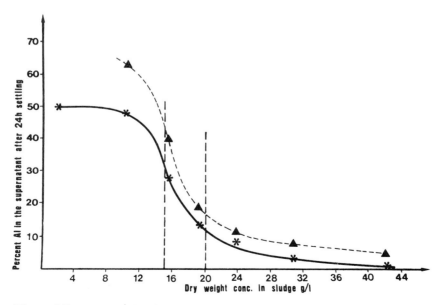

Figure 17   Average aluminum recovery as a function of total solids concentration. Solid curve, CaO (pH 11.4); dashed curve, NaOH (pH 11.6).

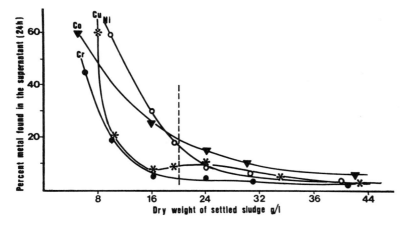

Figure 18 Trace metals recovered in the supernatant after attack of the sludge with NaOH. ●, Cr; *, Cu; ○, Ni; ▼, Ca.

for NaOH and 11.2 to 11.6 for $Ca(OH)_2$. Monitoring by the measurement of the electrical conductivity of the supernatant at values of 15 and 55 $\mu S/m$ for $Ca(OH)_2$ and NaOH, respectively, is a more practical method which enables precise regulation of dosing (Fig. 20).

Conclusions. The optimum sludge dry weight concentration to be treated with alkali in the recovery process must be determined by a preliminary investigation in each case. It appears that the yield drops at concentrations higher than 20 g/L and that a more optimum recovery is obtainable at 10 to 15 g/L or at a total aluminum content in the material of 0.05 mol/L. The reaction basin of a contact time of

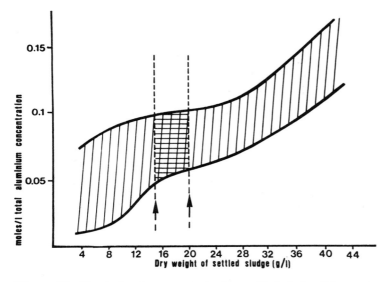

Figure 19 Total aluminum content of presettled sludge (Meuse river water at the Tailfer plant).

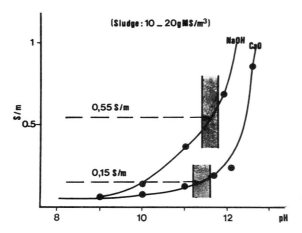

Figure 20   Electrical conductivity of the supernatant on reaction of the sludge with alkali.

about 15 min should be equipped with substantial facilities for dilution with fresh water to enable optimization of the reaction. Control and monitoring by electrical conductivity of the settled supernatant is adequate to the process.

4.5   Performance of the Recovered Coagulant

There exists no sanitary objection to the alkaline recycling process of aluminum-containing coagulants. The use of aluminate instead of aluminum sulfate or polyaluminum chloride of sulfate can determine some difference in efficiency. The performance of partial recycling of coagulant was tested on a reduced scale. A concentrated aluminum solution (e.g., 0.4 g/L) is obtained by attack of sludge from the Tailfer plant with lime in a batch reactor. The coagulation–flocculation is then performed by using the recovered coagulant together with fresh polyaluminum chloride as indicated in Table 3.

The indicative data are reported on a relative basis to compare the performance of virgin and recycled coagulant in the reduced-scale flocculator. They are not in the same order of quality standards reached in full-scale plant operation. However, several conclusions can be drawn from the data on a relative basis:

1. Up to 50% coagulant recycling, no significant decrease in efficiency is observed except during the algal bloom periods.
2. The final pH requires adjusting to reduce the aluminum solubility. This operation causes no difficulty in full-scale treatment but has not been incorporated in this evaluation, which was carried out during the early summer algal bloom period (when the pH of the raw water ranges between 8.5 and 9).
3. Coagulant recycling should remain limited to 25 to 50% of the total dosing rate.

4.6   Benefit Evaluation of Coagulant Recovery

A cost–benefit evaluation of the alkaline recycling processes in the context of the Tailfer plant (10) has demonstrated a potential saving of 20 to 50% of the coagulant

Table 3 Indicative Performance of Recycled Coagulant Versus Fresh Chemical (Experimental Treatment of Water of the River Meuse)

| Run number[a] | Percent recycled | Settled Water | | | |
		Ext. UV, 250 nm (m^{-1})	Suspended solids (mg/L)	Dissolved Al[b] ($\mu g/L$)	Final pH[b]
1	0	3.78	6.1	71	7.41
	25	3.35	3.35	103	7.64
	50	5.0	5.0	88	7.78
2	0	5.3	3.72	141	7.52
	45–50	4.76	2.90	110	8.04
3[c]	0	4.46	3.05	110	7.48
	50	5.25	5.3	247	8.1
	100	7.38	12.1	2280	8.4

[a]The data for each run are average values of about three trials.
[b]Filtered samples.
[c]Experience during algal bloom period (June 1983).

costs. In alkaline recovery, involving sludge compaction and stabilization with lime, the latter process is simplified and the quantity of material to be transported out of the plant to be disposed of is simplified.

SYMBOLS AND UNITS RELATED TO SLUDGE HANDLING

g acceleration of gravity, m/s^2
m experimental slope of the ratio t/V, time (s) to filtrate a volume V (m^3), s/m^6
r_s specific resistance, m/kg
s cake compressibility, m^2/kg
t filtration time, s

A filtration surface, m^2
C sludge concentration, usually expressed as the suspended solids concentration in dry weight units, kg/m^3
DS   dry solids concentration, kg/m^3 or g/L
G times of acceleration of gravity, ng; m/s^2
G velocity gradient, s^{-1}
P differential filtration pressure, kg/ms^2
R_m unit filter media resistance, m^{-1}
S Siemens conductivity unit, S

μ dynamic viscosity of the filtrate, kg/ms
ΔP pressure drop on filtration, kg/m^2

REFERENCES

1. P. H. King, B. H. H. Chen, and R. K. Weeks, *Publ. VPI-WRRC-Bull. 77*, Virginia Water Research Center, Blacksburg, Va.

2. W. J. Masschelein, J. Genot, R. De Vleminck, and C. Goblet, *Trib. Cebedeau, 450,* 245 (1981).
3. B. Regnier, C. Goblet, J. Genot, and W. J. Masschelein, *Water Sci. Technol., 14,* 87 (1982).
4. *Processing Water Treatment Plant Sludge,* AWWA, Denver, Colo., 1974, p. 16.
5. B. F. Ruth, G. H. Montillon, and R. H. Montana, *Ind. Eng. Chem., 25,* 76, 153 (1933).
6. P. C. Carman, *J. Soc. Chem. Ind. (Brit.), 52,* 280 T (1933).
7. P. C. Carman, *J. Soc. Chem. Ind. (Brit.), 53,* 159 T (1934).
8. W. J. Masschelein and J. Genot, *Trib. Cebedeau, 439–440,* 287 (1980).
9. U. Forstner and G. T. W. Wittmann, *Metal Pollution in the Aquatic Environment,* Springer, New York, 1979, p. 354.
10. R. De Vleminck, J. Genot, C. Goblet, and W. J. Masschelein, *Trib. Cebedeau, 470,* 3 (1983).
11. A. Chojnacki, *Bull. Cebedeau, 248,* 351 (1964).
12. J. Salmona, Y. Richard, *T.S.M. Eau, 65,* 3 (1970).
13. G. P. Westerhoff, *Water Works Eng.,* 1973, p. 28.
14. G. P. Fulton, *J. AWWA, 66,* 312 (1974).
15. B. H. H. Chen, P. H. King, and C. W. Randall, *68,* 204 (1976).
16. A. Rosenquist, Thèse, Institutionen för teknisk Abo Akademi, Finland, 1976.
17. G. P. Westerhoff and D. A. Cornwell, *J. AWWA, 70,* 709 (1978).
18. G. P. Fulton, *Water Works Eng.,* June 1979, p. 78.
19. J. Salmona, *Travaux,* 656 (1981).
20. W. J. Masschelein and P. Van Damme, DVGW-Schriftenreihe, Wasser *50,* 159 (1986).

17
Mixing Practice in Water Treatment

1. INTRODUCTION

Mixing is a auxiliary process that takes place in all water treatment designs and operations. This chapter is intended to provide specific information on the numerous mixing technologies applied in the unit processes of a treatment chain.

2. CHARACTERIZATION OF MIXING CONDITIONS IN WATER TREATMENT: VELOCITY GRADIENT

The mixing conditions applied to a fluid are most often characterized by the G value (i.e., the velocity gradient), expressed in s^{-1}. This velocity gradient is defined as the difference of velocity between two points or elementary volumes of the fluid, reported to a direction reference, perpendicular to the displacement. It is related to the friction force between two elementary layers moving at different velocities and is expressed by the following relationship:

$$G \text{ (in s}^{-1}) = \left(\frac{P}{\mu V}\right)^{1/2}$$

in which P is the *net* power input per volume of water, V, and μ is the dynamic viscosity. Typical values are listed in Table 1.

The concept of the velocity gradient applies ideally to "completely mixed reactors," in which steady-state concentrations are maintained (e.g., in rapid mixing basins for coagulation). The expression GT, known as the relation of Camp and Stein, clearly describes the mixing intensity by taking into consideration the deten-

Table 1   Density and Viscosity of Water as a Function of Temperature[a]

Temperature (°C)	Density (kg/m^3)	Dynamic viscosity μ (kg/m^{-1}s^{-1})	Kinematic viscosity ν (m^2/s)
0	999.84	1.787×10^{-3}	1.787×10^{-6}
4	999.97	1.567×10^{-3}	1.567×10^{-6}
8	999.85	1.386×10^{-3}	1.386×10^{-6}
12	999.50	1.235×10^{-3}	1.235×10^{-6}
16	998.94	1.109×10^{-3}	1.110×10^{-6}
20	998.20	1.002×10^{-3}	1.000×10^{-6}
24	997.30	0.911×10^{-3}	0.913×10^{-6}
28	996.23	0.833×10^{-3}	0.836×10^{-6}

[a]At 20°C the total net power input per cubic meter water is given by W (watts) $= 10^{-3}$ (kg m^{-1} s^{-1} $\times$ G^2).

tion time of the fluid in the actual mixing zone (T). Thus the dimensionless product, GT, is known as Camp's number. It is of considerable importance, principally in flocculation (see Chapter 7).

3.   PRINCIPLES OF MIXING IN WATER TREATMENT

The most important functions of mixing as applied in the treatment processes are:

1.   The dispersion of chemicals into the bulk of the water
2.   The homogeneization of the water quality by preventing dead zones by recirculating deposits and by drawdown of floating material
3.   The kinetic control and promotion of reactions involved in the process (e.g., rapid mixing in coagulation and slow mixing in flocculation)
4.   Auxiliary effects such as aeration and gas stripping

3.1   Mixing in Coagulation

The quantity of reagents used in coagulation is very low compared to the water volume. For instance, when one uses 120 kg $Al_2SO_4 \cdot 18H_2O$ (aluminum sulfate per cubic meter as a working solution to treat the water at 60 g/m^3, the dilution ratio is equal to 1 : 2000.

The coagulation process involves the reaction of intermediate polymeric species that are formed within a fraction of a second in fast hydrolysis of the coagulant. The coagulation is irreversible and the reaction times of the entire process last between a tenth of a second and a few seconds. Consequently, inefficient rapid mixing during coagulation results in wasting chemicals. In some cases, up to about 30% extra alum may be required. (The time required for the hydrolysis of coagulants is 10^{-10} s, while that of the formation of polynuclear species is 10^{-10} to 1 s. The adsorption of these species lasts about 10^{-10} s.)

Most coagulants are acids and change the pH of the water. The pH also influences the coagulation process, although in a secondary manner. Local variation of pH may alter the favorable conditions for colloidal neutralization of colloids. Besides avoiding local deficiencies of coagulants, the average pH value is one of the important mixing functions in coagulation. If the pH attains 8.3 and more, as sometimes occurs in surface waters during periods of algal growth, conventional coagulation using aluminum salts implies acidification as an essential function.

According to a survey carried out in the United States, the nominal detention time of coagulation chambers in the most conventional designs has indicated values ranging between 50 and 300 s, with an average value of 100 s. The use of two subsequent compartments for mixing is recommended. In more recent designs, rapid mixing is shortened in time and increased in mixing energy (e.g., 60 s with $G > 200$ s^{-1}). The actual tendency is to increase the velocity gradients and to decrease the detention time in the flash mixing zones (e.g., $G = 1000$ s^{-1} during less than 5 s). However, increasing energy costs may necessitate reconsideration of this option. On the other hand, at optimal flash mixing conditions (e.g., 1000 s^{-1}), according to the literature (4), the longer the residence time in the active mixing zone, the lower the concentration of coagulant required.

Guidelines for Mixing in Coagulation.

Limits: The velocity gradient should never be lower than 100 s^{-1}, and the detention time must be shorter than 60 s.

Average: $G = 200$ to 400 s^{-1} with T 30 to 20 s: Camp's number between 6000 and 8000.

Optional: Flash mixing $G < 1000$ s^{-1} and $T < 5$ s, whence $GT = < 5000$.

General: The flow pattern of a coagulation zone is ideally that of a complete mixing (see Section 4.1.3).

In the nonactive zone of the process, as in transfer channels or pipes, the water velocity should be higher than 0.7 m/s, and a value of 2 m/s can be recommended. In comparative conditions Fe coagulants are faster reacting and may prefer a higher GT value than do Al-containing products. In more charged waters, such as sewage, the impact of mixing conditions is even more decisive.

3.2 Mixing In Flocculation

At the end of the coagulation process, a pinpoint floc is formed. The dimensions of the hydrated particles of this floc would ideally be 100 to 150 μm. Mixing promotes the slow particle increase to form a floc with discrete dimensions capable of settling. In the orthokinetic concept, or flocculation according to Smoluchowski, the collision probability (hence the agglomeration rate) is directly proportional to the velocity gradient (see Chapter 7).

However, a too violent agitation may disrupt the loose hydrated flocs. This condition is major for the maximal admissible velocity gradient. It has been observed, for instance, that when submitting coagulated water to too-high velocity gradients, this can delay subsequent flocculation; for example, visible flocs (pinpoint flocs) only form after 10 min if $G = 4400$ s^{-1} and after 45 min if $G = 12,500$ s^{-1} (5). Flocculation is kinetically a slow process requiring between 900 and 3000 s, during which optionally the velocity gradient can be progressively lowered; this concept is like that of the tapered stirring concept. A sharp decline in velocity gradient from coagulation to flocculation can have a detrimental effect on floc building (6).

Guidelines for Mixing in Direct Flocculation.

Limits: The velocity gradient should always be lower than 100 s^{-1}. The maximum tip speed of paddles is to be fixed between 0.22 and 1.2 m/s so as to avoid floc breakage. A flow velocity between 0.25 and 0.6 m/s in the flocculation tank does not hinder the process. Transferring of flocculated water is allowed with a limit of 0.15 to 0.2 m/s but is best avoided to prevent partial settling. The paddle area should not exceed 20% of the cross-sectional area of the flocculation basin.

Average: $G = 75$ s^{-1}, $T = 2000$ s; $GT = 150,000$.

Optional: Tapered stirring, for example, in six successive compartments each having a residence time of 120 to 150 s and decreasing velocity gradients according to the sequence 100 s^{-1}, 90 s^{-1}, 70 s^{-1}, 50 s^{-1}, 30 s^{-1}, and 10 s^{-1}, which are proved on the basis of practical experience. In practice, it is also important *not to go too abruptly* from rapid mixing to the intensity of stirring in flocculation, and by so doing avoid an unnecessary increase in coagulant doses. (In many plants this point has been overlooked, and it is not seldom that "dead periods" of several minutes occur between coagulation and flocculation.)

General: The ideal flow pattern of a flocculation zone is similar to that of a plug-flow reactor with an intermediate amount of dispersion (see Section 3.2.1). The ability of settling of the floc must not be decreased by introducing air bubbles during mixing.

Remark.   Flocculation with sludge blanket contact must follow specific guidelines (see Chapter 9).

3.3   Mixing in Settling

In the settling process, the specific mixing function involved is contact of the sludge blanket either with flocculated inflowing water or during the flocculation phase. The other aspects are related to floc breakage on excessive mixing. Hence the mixing action intervenes essentially as a backmixing concept in flocculant settling. In typical laminar settling, short-circuiting must be avoided, so mixing by fluid motion is undesirable. Therefore, the question is related to the specific hydraulic conditions of laminar flow in settling rather than to mixing.

Guidelines for Mixing in Settling.   The limit of strict laminar settling is to obtain a Froude's number value of Fr $> 10^{-5}$. This involves a flow pattern with a large amount of axial dispersion. In sludge recirculating systems the tip velocity of the agitators must range between 0.2 and 0.8 m/s with $G < 20$ s^{-1} in the floc recirculation zones. Sludge scrapers in the sludge compaction zone have a velocity (tip velocity in the case of circular constructions) of 35 to 50 mm/s, that is, a negligible velocity gradient. In the particular case of lamellar or tubular settling, the turbulence at the entrance of the settling zone should be sufficiently low so as to correspond to a Reynolds number value such as Re < 200.

3.4   Mixing in Chemical Softening and Water Conditioning Processes

In the processes of softening and alkalinization of water, appropriate mixing is similar to that practiced in flocculation (e.g., $G < 100$ s^{-1}). Reactions are either fast (e.g., with NaOH) or slow (e.g., with quicklime or soda) but always reversible and self-equilibrating. It is recommended to use a contact time of 1500 to 2000 s in a reactor of the completely mixed type.

Guidelines for Mixing in Chemical Softening.   Dosing sodium hydroxide requires $G > 50^{-1}$, with a subsequent action time of 1000 to 2000 s for water quality stabilization. Where lime mixing to water is concerned, it is best to consider a value of G between 100 and 300 s^{-1} for 10 to 60 s, before subsequent agglomeration and settling, as in the case of flocculation–settling. The flow pattern in softening and water conditioning may approach that of a plug-flow reactor. However, backmixing may be favorable for equilibration. The shear strength of the agglomerates being higher than that of the flocs in coagulation–flocculation, the limits for mechanical rupture are less stringent in softening. Water velocities up to 2 m/s are admitted transiently.

3.5 Mixing in Sludge Conditioning

In the presettling of sludge, whose aim it is to obtain a more concentrated sludge than that currently abstracted from sludge accumulation zones of settling basins, the general conditions for mixing are very similar to those of primary settling. Hence velocity gradients are to have a value of about 20 s^{-1}, and the maximum is to be considered 60 s^{-1}. The motor power of the scrapers of the sludge zone is to be overdimensioned, so as to withstand the increasing resistance to movements when the sludge becomes concentrated (e.g., up to 2 to 4 wt% obtained from 0.5 to 1.5 wt% inflowing). In the lime conditioning of sludge, breakage of the sludge structure enabling the reaction with the aluminum (or iron) ion is essential (7). Therefore, the G value of the mixing vessels should attain at least 300 s^{-1}, preferably > 500 s^{-1}, in the active zone of the lime concentrated sludge. The necessary residence time is less than 20 s.

3.6 Mixing in the Filtration Process

By itself, filtration does not require mixing, but there is a need for the homogeneous distribution of the raw water and particles to be filtered. However, in most practices of filtration, the velocity gradient imparted to the water flowing through sand filters is about 120 s^{-1}. Thus good conditions for flocculation are combined. Flocculation–filtration is based on this action promoting interparticle contacts. Coagulation is best performed before filtration according to the principles of mixing discussed earlier.

3.7 Mixing in Adsorption Processes

Activated carbon filters have no specific mixing function apart from that indicated for sand filters. Activated carbon dosing in the form of a slurry in water containing a carbon concentration below 10 wt% necessitates rapid mixing (e.g., G values higher than 200 s^{-1} during a theoretical detention time of approximately 60 s). The flow pattern is preferably that of a complete mixed basin so as to build up the best matrix of adsorbing powder fixed on the forming floc. As the adsorption process itself is under kinetic control, slow mixing is necessary for successful application. The criteria are the same as for flocculation (i.e., $G < 100$ s^{-1}), but the detention time is best extended to at least 3600 s. Hence the optimum GT number can range between 300,000 and 400,000. Criteria to maintain the appropriate matrix structure are the same limits in shear strength applicable in flocculation.

3.8   Mixing in Disinfection

The introduction of highly water soluble disinfectants such as solutions of chlorine, hypochlorite, chlorine dioxide, hydrogen peroxide, or diluted solutions of potassium permanganate do not involve very specific mixing requirements except for homogeneous distribution. To ensure a homogeneous distribution and constant activity as a function of time, the most appropriate pattern is similar to the plug-flow type, which will guarantee the best continuing instant action of the disinfectant.

In the case of post-ammoniation, the guarantee of stabilization of an active residual by mixing is most important, as insufficient distribution of the ammonium ion involves losses due to "local break-point oxidation" instead of the formation of chloramines. In ozonization, the disinfectant is put into contact with the water in the form of bubbles of ozonized air dispersed in the water flow. Different technologies are available (8–10) and discussed specifically in Chapter 3. Similarly, aeration is carried out by different specific techniques without a separate mixing function. Disinfection with UV rays also corresponds to a specific technology discussed in Chapter 4.

3.9   Mixing in Gas Bubbling and Flotation

Mixing in dispersion of a gas into water can be described by the G value computed on the basis of the net power input (turbine, cascade, etc.) per volume of water (8–10). Moreover, owing to the heterogeneity of phase, a specific approach can be applied according to Camp (11), is discussed further in Section 4.4. G values in air flotation on the order of 1000 to 1500 s^{-1} are to be maintained throughout the basin. Thus they do not promote flocculation but maintain fine flocs through breakage.

3.10   Mixing in Biological Treatment Processes

Excessive mixing in biological processes is not suitable to microorganisms that most often live associated in clusters. The G value is usually maintained below 100 s^{-1}. The homogenization of biologically active sludges is most essential. Scrapers are designed according to each specific basin. General rules such as those for settling basins can be adopted.

4.   TECHNICAL EXECUTION OF MIXING

4.1   Static Mixing in Pipes and Conduits

4.1.1   *Homogenization*

In static mixing, homogenization occurs by the flow of water through static systems without mobile parts. The power dissipation can be calculated from the loss of head and the water flow:

$$P = Q\rho gh$$

where P is the power input (in W), Q the water flow (in m^3/s), and g the acceleration of gravity (9.81 m/s^2); or

$$\frac{P}{V} = \frac{\rho gh}{T}$$

where ρ is the mass density (in kg/m^3), h the head loss (in m), and, T the detention time (in s).

Formulas for head loss applicable to laminar flow in straight pipes are:

Extremely smooth pipes: $h = 0.54 \times 10^{-3} \, Lv^{1.75}/D^{1.25}$
Smooth pipes: $h = 0.78 \times 10^{-3} \, Lv^{1.86}/D^{1.25}$
Rough pipes: $h = 1.15 \times 10^{-3} \, Lv^{1.95}/D^{1.25}$
Extremely rough pipes: $h = 1.68 \times 10^{-3} \, Lv^2/D^{1.25}$
Examples. Suppose that $L = 7.2$ m, $D = 1.2$ m, and $Q = 0.5$ m^3/s.

$$v = \frac{0.5 \text{ m}^3/\text{s}}{1.13 \text{ m}^2} = 0.44 \text{ m/s} \quad \text{and} \quad T = \frac{7.2 \text{ m}}{0.44 \text{ m/s}} = 16.3 \text{ s}$$

Rough pipe system:

$$h = 1.15 \times 10^{-3} \times 7.2 \times \frac{(0.44)^{1.95}}{(1.2)^{1.25}} = 1.3 \times 10^{-3} \text{ m}$$

$$G = \left(\frac{\rho gh}{\mu t}\right)^{1/2} = \left(\frac{10^3 \times 9.81 \times 0.0013}{10^{-3} \times 16.3}\right)^{1/2} = 28 \text{ s}^{-1}$$

Highly rough pipe system:

$$h = 1.68 \times 10^{-3} \times 7.2 \times \frac{(0.44)^2}{(1.2)^{1.25}} = 1.86 \times 10^{-3} \text{ m}$$

$$G = \left(\frac{\rho gh}{\mu t}\right)^{1/2} = \left(\frac{10^3 \times 9.81 \times 0.002}{10^{-3} \times 16.3}\right)^{1/2} = 35 \text{ s}^{-1}$$

A handable practical experimental approximation for conduits is

$$G \text{ (s}^{-1}) = 564\left(\frac{f}{D}\right)^{1/2} v^{3/2}$$

in which D is equal to the pipe diameter in meters, v the mean velocity in m/s, and f the Darcy–Weisbach friction based on a rugosity height of 0.26.

$$G \ (s^{-1}) \ = \ 564\left(\frac{0.26}{1.2}\right)^{1/2} (0.44)^{3/2} \ = \ 77 \ s^{-1}$$

for $G < 100$ or for $G > 10$

$100 = 262(v)^{3/2}$ $10 = 262(v)^{3/2}$

$v^{3/2} = \dfrac{100}{262} = 0.382$ $v^{3/2} = \dfrac{10}{262} = 0.0382$

$v < 0.53 \ m/s$ $v > 0.11 \ m/s$

4.1.2 Liquid Injection

The injection of a liquid reagent into water transported in a pipeline necessitates a minimum mixing-length equivalent to at least 10 pipe diameters, and values up to 60 may be required for optimal distribution(2). Mixing by fluid motion, as in pipes, corresponds to axial diffusion in the flowing system and is caused by both molecular (laminar or Fich's) and turbulent (eddy) diffusion and the relative motion of fluid particles. The most important contribution in practice, however, is that of the mean velocity profile. The injection is best located centrally in the axis of the pipe if $l/D = 0.5$, $L/D = 10$. However, in practice, the injection pipe is usually located a few millimeters from the wall of the pipes. Then for $l/d = 0.05$, L/D is equal to 60. The minimum and maximum concentrations taken in Fig. 3 are those measured over the pipe sections at a given distance after the injection point. The C_{min}/C_{max} ratio is indicative for the axial dispersion.

The use of a twofold injection considerably reduces the L/D ratio required for homogeneous distribution. The injection nozzles are best located above the horizontal diameter of the pipe, as indicated in Fig. 1. In this case L/D values of 20 give

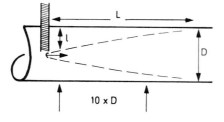

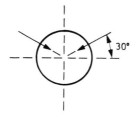

Figure 1 Schemes for injection of chemicals into pipes.

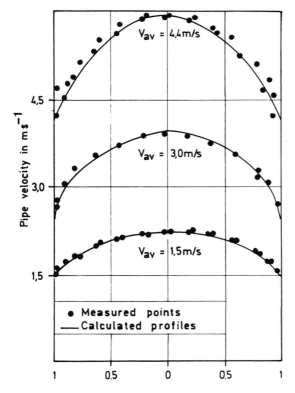

Figure 2   Typical flow-through reactor patterns expressed as effluent response to a step input.

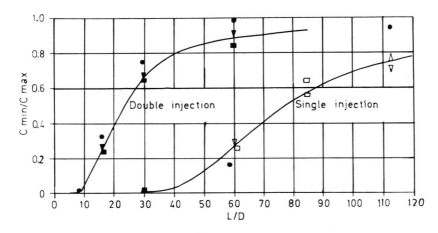

Figure 3   Typical axial mixing profiles.

comparable distribution to that for $L/D = 60$ when a single injection nozzle is used. The use of a multiple-nozzle system with more than two injection pipes does not significantly improve the distribution pattern compared to that obtained by the twofold nozzle system. The rate of mass transfer in the moving fluid depends primarily on the turbulency within the flow. The injecting velocity at the nozzle tip must be higher than 0.75 m/s and attain *at least* half of the large pipe velocity.

4.1.3 Residence Time Distribution

The residence time distribution is an important characteristic in mixing by injection into pipes, conduits, or channels. By injecting a tracer which remains in the water, longitudinal dispersion or "residence time distribution" factors can be determined. Typical flow-through profiles are indicated in Fig. 4. The dispersion coefficient E can be estimated from the retention time dispersion curves by assuming a normal distribution:

$$E = \frac{1}{2} \frac{v \times \sigma^2}{T^2}$$

where v is the linear velocity, x the length of the reactor zone in the pipe, T the mean residence time (10.6 s in the example given in Fig. 4), and σ^2 the variance of the residence time distribution (17.6 s^2) in the example assuming a Gaussian distribution, and for $x = 10$ m and $v = 1$ m/s, E equals 0.72 m^2/s).

A number of indices are derived from the residence time distribution curves:

1. The Morrill dispersion index equals the ratio of the times for 90% and 10% of the tracer to pass. In perfect plug-flow reactors, T_{90}/T_{10} equals unity. The value must always be lower than 2 if a plug-flow system is required (equals 3 in the example given).

2. The ratio of the theoretical residence time to the mean residence time must be

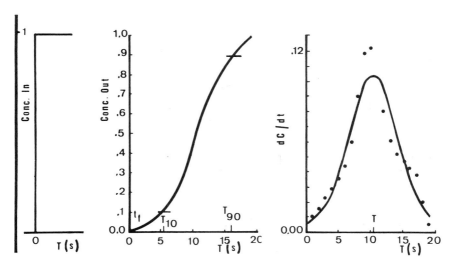

Figure 4   Retention time dispersion curves.

close to unity (t/T). If this ratio is much lower than 1, short-circuiting occurs. If the value is lower than 0.8, an error on the active volume is probable (0.943 in the example given).

3. The ratio of the time after which the first throughput occurs to the theoretical residence time (t_f/t) is a measure of most short-circuiting (1 in perfect plugflow, zero in completely mixed systems, 0.2 in the example given).

4.1.4 Power Dissipation

The volume in which the power input is dissipated is given by $V = (D^2/4)L$, in which L is the mixing length in meters—to be fixed according to the preceding comments (Fig. 3) to between 10 and 60 times the diameter. To increase the efficiency by diminishing the volume of power dissipation and accordingly increase the *net* power input per unit volume, hydraulic accidents are introduced (e.g., pipe bending, expansion and restriction, and static mixing elements). It is accepted as a general guideline to compute the effective mixing volume after injection in a hydraulically disturbed zone on the basis of a mixing length equal to five times the diameter of the mean conduit:

$$V = \frac{D^2}{4} \times 5D - 4D^3$$

Power dissipation through pipe bends is to be computed in each particular case, taking into account the number and angle of each bend and the radius of curvature in relation to the diameter of the pipe (12). The evaluation is expressed in equivalent length of straight pipes for which the head loss is to be computed. The magnitude of the head loss due to sudden pipe expansion is given by

$$h = \frac{v_1^2 - v_2^2}{2g}$$

in which v_1 is the velocity before and v_2 after the expansion.

According to Simpson (13), the head loss due to pipe restriction is given as follows:

$$h = \frac{v_{\text{avg}}^2}{2g} 2.8\left[1 - \left(\frac{d}{D}\right)^2\right]\left[\left(\frac{D}{d}\right)^4 - 1\right]$$

in which D and d are the diameters of the mean pipe and the restricted pipe, respectively. $E/vx = d$ (i.e., the dispersion number characterizing a given pipe system). Rules applicable to closed vessels are:

$d = 0$	No dispersion; perfect plug flow
d lower than 0.01	Low dispersion
d between 0.01 and 0.1	Moderate dispersion
d higher than 0.1	High dispersion
d infinite	Completely mixed

The d value can be used in diagnosing designs that require a plug-flow reactor system (e.g., disinfection with UV light, chlorine dioxide generators, etc.).

The power input by "spray" injection into a pipe can alternatively be estimated by the formula

$$P = \frac{0.75 A v^3 \gamma}{2g}$$

in which A is the open area (in m^2) of the spraying nozzles through which the reagent is injected at a velocity v (m/s) and γ the unit *weight* per cubic meter (e.g., in the case of water 9.81×10^3 kg/m^3). The value of 0.75 is an average assumption for the discharge coefficient.

Remarks.   In practice, the injection of alkaline solutions (e.g., sodium hypochlorite, sodium hydroxide, lime, or ammonia) often causes problems associated with the local precipitation of calcium carbonate at the point of injection. Therefore, the injection of such reagents may require several precautions, which may be contrary to the practice of adequate mixing. Alkaline products are best injected above the water level in an open channel. In the case where direct injection into pipes cannot be avoided in the design as a whole, three possible concepts are indicated in Fig. 5.

In case (a), the principle of injection corresponds to the concepts of direct injection and mixing in conduits. The local deposits of calcium carbonate are inhibited by using plastic tubing for the injection. This material is not susceptible to crystallization by an electrochemical potential. The transit velocity at the injection point seldom attains the critical value of 0.75 m/s for adequate mixing. Therefore, the transit velocity can be increased by an auxiliary dilution water. Furthermore, the use of a supple plastic tube for the injection line enables easy removal for cleaning. Intermediate rinsing is possible by reverse water flow.

In the scheme (b), the dosing is carried out in a gaseous atmosphere obtained in an air-pressurized cupola. The concept is in discordance with the principles of good mixing. This discrepancy can be improved by an appropriate shape of the penetrating tube delimiting the injection chamber.

In the design (c), mixing of the injected chemical is improved by the increased flow obtained with an injector. The dosing is achieved indirectly through an overflow vessel. The reagent is transferred with a high-rate ejector mounted so as to be

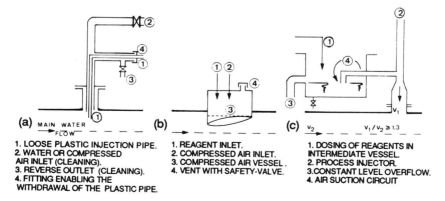

1. LOOSE PLASTIC INJECTION PIPE.
2. WATER OR COMPRESSED
   AIR INLET (CLEANING).
3. REVERSE OUTLET (CLEANING).
4. FITTING ENABLING THE
   WITHDRAWAL OF THE PLASTIC PIPE.

1. REAGENT INLET.
2. COMPRESSED AIR INLET.
3. COMPRESSED AIR VESSEL.
4. VENT WITH SAFETY-VALVE.

1. DOSING OF REAGENTS IN
   INTERMEDIATE VESSEL.
2. PROCESS INJECTOR.
3. CONSTANT LEVEL OVERFLOW.
4. AIR SUCTION CIRCUIT

Figure 5   Systems to inject alkaline reagents into closed pipes or channels.

self-cleaned by aspirated air. The minimum water head loss of the substream is of about 1 m, thus determining an additional impact on the velocity gradient in the mixing zone. This effect depends on the relative proportion of process-liquid flow to water flow and must be computed in any particular case according to the rules for "spray" injection.

To assist cleaning of the injection point of alkaline reagents, it is advisable to incorporate sodium hexametaphosphate and/or polyphosphate in the system.

4.2 Static Mixers

The building-in of elements in pipes so as to promote hydraulic accidents that improve the mixing efficiency is a technique undergoing full development (14–16). In the Kenics mixer (Fig. 6), each element consists of a twisted strip metal or plastic cut at 180° and juxtaposed in a left- and right-handed sequence. The flow inversion, by which a particle travels from the center to the outer wall and back again, promotes mixing. The pressure loss for a Kenics static mixer is approximately

$$h(\text{pipe} + \text{mixer}) = h(\text{pipe}) + 3.24 \times 10^{-3}(1.5 + 0.21\sqrt{\text{Re}})$$

$$\text{Re} = \frac{\rho v d_h}{\mu} \quad (\text{Re} = \text{Reynolds number})$$

$$d_h = \frac{4 \times \text{surface}}{\text{perimeter}} = D \text{ for circular pipes}$$

Q = flow (m³/s) $\quad d_h$ = hydraulic diameter
D = pipe diameter $\quad \mu$ = viscosity (kg/m · s)
L = length of mixing elements (m) $\quad v$ = average viscosity (m/s)
$\quad \rho$ = density (kg/m³)

Figure 6 Kenics static mixer.

With regard to the pipe diameter, the longer the mixing elements, the better the mixing. However, the pressure loss increases. The optimum compromise is obtained for

$$1.5 < L/D < 1.7$$

Kenics mixers have been constructed for pipe diameters between 0.15 mm and 1.2 m. Example.

$$Q = 0.5 \text{ m}^3/\text{s} \qquad t = 10°C$$

$$D = 1 \text{ m} \rightarrow A = 0.75 \text{ m}^2 \qquad v = 0.64 \text{ m/s}$$

$$\text{Head loss} = 1.15 \times 10^{-3} \times 5 \times 1.5 \times \frac{(0.64)^2}{1} + 3.24 \times 10^{-3}(1.5 + 0.21\sqrt{\text{Re}})$$

$$= 4 \times 10^{-3} + 3.24 \times 10^{-3}(1.5 + 0.21\sqrt{488,550})$$

That is, for five elements of $1.5 \times D$ long each,

$$\text{Re} = \frac{\rho v D}{\mu} = \frac{1000 \times 0.64 \times 1}{1.31 \times 10^{-3}} = 488,550$$

$$h_{\text{total}} = 4 \times 10^{-3} + 0.48 = 0.48 \text{ m}$$

$$\text{Detention time} = \frac{L}{v} = \frac{7.5 \text{ m}}{0.64 \text{ m/s}} = 11.7 \text{ s}$$

$$G = \left(\frac{\rho g h}{\mu t}\right)^{1/2} = \left(\frac{1000 \times 9.81 \times 0.48}{1.31 \times 10^{-3} \times 11.7}\right)^{1/2}$$
$$= 550 \text{ s}^{-1}$$

If $D = 1.2$ m,

$$h = h_{\text{pipe}} + 3.24 \times 10^{-3}(1.5 + 0.21\sqrt{\text{Re}})$$

For example, for 0.5 m^3/s, $D = 1.20$-m four elements and $L = 7.2$ m.

$$h_{\text{pipe}} = \frac{128}{\pi} \times \frac{7.2}{(1.2)^4} \times 0.5 \times 1.17 \times 10^{-3} = 0.083 \text{ m}$$

$$\text{Re} = \frac{\rho v d h}{\mu} \qquad \text{section} = \pi R^2 = 3.14(0.6)^2 = 1.13 \text{ m}^2$$

$$\frac{0.5 \text{ m}^3/\text{s}}{1.13 \text{ m}^2} = 0.44 \text{ m/s} = v$$

$$\mathrm{Re} = \frac{10^{+3} \times 0.44 \times 1.2}{1.17 \times 10^{-3}} \qquad (dh = 1.2)$$

$$= 4.51 \times 10^{5}$$

$$h_{\mathrm{total}} = 10^{-3}\ \mathrm{m} + 3.24 \times 10^{-3}(1.5 + 0.21\sqrt{\mathrm{Re}}) = 0.46 + 0.083$$

$$= 0.54\ \mathrm{m}$$

$$\mathrm{Volume} = 1.13\ \mathrm{m}^2 \times 7.2\ \mathrm{m} = 8.14\ \mathrm{m}^3 \rightarrow t = \frac{8.14}{0.5} = 16.3\ \mathrm{s}$$

$$G = \left(\frac{\rho g h}{\mu t}\right)^{1/2} = \left(\frac{10^3 \times 9.81 \times 0.54}{1.17\ 10^{-3} \times 16.3}\right)^{1/2} = 527\ \mathrm{s}^{-1}$$

4.3 Mixing by Flow in Open Channels

4.3.1 Flow Conditions

Mixing in an open channel involves turbulent flow conditions. The Reynolds number must be higher than 500, preferably 2300, Re being obtained as

$$\mathrm{Re} = \frac{v d_h}{\nu} \qquad (\nu \text{ is equal to the kinematic viscosity})$$

The hydraulic diameter in this formula is given by

$$d_h = \frac{W\,(\text{width}) \times H\,(\text{height})}{W + 2H}$$

and v is the average horizontal velocity of the water in the channel. Direct dosing of a chemical into an open plain-flow channel is not an efficient mixing method. Hence hydraulic accidents are created to initiate appropriate dispersion, baffles, hydraulic jumps or cascades, and restrictions causing increased frictional forces. On the other hand, avoiding turbulent mixing may be important in plain-flow settling. The ideal criterion for these flow conditions is to obtain a Froude's number (Fr) higher than 10^{-5}, determining a sufficient stable streaming condition:

$$\mathrm{Fr} = \frac{v^2}{d_h \times g}$$

As a function of the water flow Q, Froude's number expresses for channels of width W and height of water layer H,

$$\mathrm{Fr} = \frac{Q^2}{g}\frac{W + 2H}{W^3 H^3}$$

The general formula for the power dissipation can be applied to baffled chambers or channels:

$$\frac{P}{V} = \frac{\rho g h}{T}$$

in which h is the total head loss. In most practical cases, the head loss in baffled detention tanks ranges between 0.02 and 0.04 m for a detention time of about 60 s. Hence the velocity gradient barely attains 100 s^{-1}. A practical empirical formula is given by (17)

$$G = 2500\left(\frac{h}{T}\right)^{1/2}$$

4.3.2 Coagulation

The use of horizontally baffled mixing basins (Fig. 7a) is exceptional for coagulation. The average acceptable velocity of the water must be between 0.3 and 0.7 m/s, as turbid waters require high velocity to prevent accumulation of sediments. In the absence of specific data, a velocity of 0.5 m/s is suggested. Four to six baffled chambers are generally suitable to avoid short-circuiting in the compartments. The inside baffles can be made adjustable, making it possible to change the velocity if necessary. The alternative "rise and fall" of water obtained in vertically baffled contact basins (Fig. 7b) is generally considered to be more effective in maintaining homogeneous mixing and preventing undesirable deposition of sludge. On the other hand, cleaning the basins is more difficult.

The use of baffled basins with a theoretical detention time of 120 s or less for coagulation alone is a concept that has been superseded in present practice. Equip-

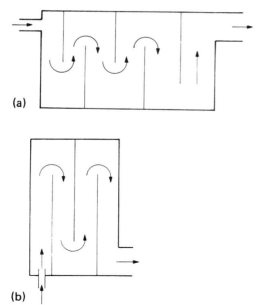

(a)

(b)

Figure 7   (a) Horizontal and (b) vertical baffled chambers (schematic).

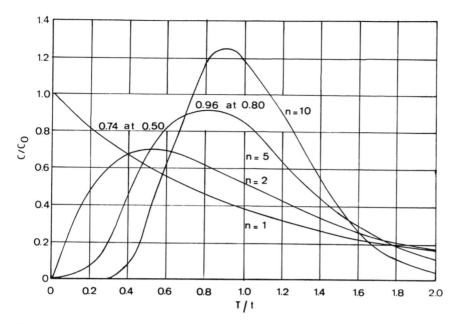

Figure 8 Typical dispersion curves as a function of the number of baffles.

ping the basins with mechanical mixers enables a combination of both mixing actions. In chlorination (e.g., breakpoint chlorination of raw water implicating a longer contact time), the use of baffled contacting basins remains in current practice, although chlorine action is equally obtainable in flocculation basins as in specific chlorine contact chambers. The efficiency of mixing in baffled chambers may be increased by a higher energy dissipation than is usually required (e.g., by cascades).

The equation for the peak concentrations passing as a function of the theoretical detention time is expressed as follows (3):

$$\frac{c}{c_0} = \frac{n^n}{(n-1)!} \; n^{n-1} \left(\frac{T}{t}\right)^{n-1} e^{-n(t/T)}$$

in which n is equal to the number of baffles. By applying Stirling's formula for factorials, one obtains for the maximum concentration and the time after which the maximum concentration occurs,

$$\frac{C_{max}}{C_0} = \frac{n}{\sqrt{2\pi(n-1)}} \; ; \text{and} \; \frac{T_{max}}{t} = \frac{n-1}{n}$$

4.4 Hydraulic Jumps

Mixing through hydraulic jumps is often used for auxiliary injection of chemicals as for additional disinfection, pH adjustment, dosing of active carbon, or alkalinity correction. Control of the mixing conditions, which are flow dependent, make the method sometimes critical for use in coagulation. The mixing intensity is always too high to be admitted in flocculation and results in breakage of the flocs. The system can be represented schematically as in Fig. 9.

The approach depth of the water layer, d_a, and the approach velocity v_a under the gate define the Froude's number in the hydraulic jump:

$$Fr = \frac{v_a^2}{gd_a}$$

The jumps are classified according to their Froude's number:

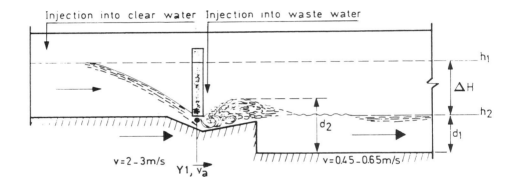

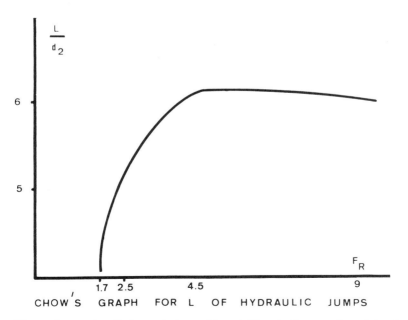

CHOW'S GRAPH FOR L OF HYDRAULIC JUMPS

Figure 9 Hydraulic jump (schematic) and Chow's diagram for mixing length. (From Ref. 18.)

Fr < 1.7   No jump but ondular flow
Fr = 1.7 to 2.5   Weak jump
Fr = 2.5 to 4.5   Wavy jump
Fr = 4.5 to 9   Steady jump
Fr > 9   Strong jump

The downstream water depth d_2 is related to the approach depth by the relation involving the Froude's number:

$$\frac{d_2}{d_a} = \frac{1}{2} (\sqrt{1 + 8\,\text{Fr}^2} - 1)$$

and the mixing length or length of the jump is read from the Chow's graph (18), relating the length to the downstream depth d_2 as a function of the Froude's number. The head loss across the jump is evaluated by the formula

$$h = \frac{(d_2 - d_a)^3}{4 d_2 d_a}$$

The mixing time in the jump is computed by considering an average velocity of the approach velocity through the gate, v_a, and the velocity in the downstream channel, v_2. In practice, the use of hydraulic jumps is often based on a trial-and-error procedure (18). Constructions and velocity gradients designed for use in water treatment often rely on weak jumps.

Example. $Q = 1$ m^3/s; desired velocity gradient = 500 to 1000 s^{-1}; assume an inlet channel width of 0.8 m and $v_1 = 1.2$ m/s; thus $A_1 = 0.83$ m^2 and $d_1 = 1.04$ m. Construct a steady jump (e.g., Fr = 5 and $d_2 = 1.2$ m).

$$d_a = \frac{1.2}{\frac{1}{2}(\sqrt{1 + 8 \times 5^2} - 1)} = \frac{1.2}{6.6} = 0.18 \text{ m}$$

$$\frac{L}{d_2} = 6 \text{ (from graph)} \rightarrow L = 7.2 \text{ m}$$

$$h = \frac{(1.2 - 0.18)^3}{4 \times 1.2 \times 0.18} = \frac{1.06}{0.86} = 1.23 \text{ m}$$

$$\left.\begin{array}{l} v_2 = \dfrac{1 \text{ m}^3/\text{s}}{0.8 \text{ m} \times 1.2 \text{ m}} = 1.04 \text{ m/s} \\[2ex] v_a = \dfrac{1 \text{ m}^3/\text{s}}{0.8 \text{ m} \times 1.18 \text{ m}} = 6.94 \text{ m/s} \end{array}\right\} v_{\text{avg}} = 4$$

$$T = \frac{7.2 \text{ m}}{4 \text{ m/s}} = 1.8 \text{ s}$$

$$G = \left(\frac{\rho g h}{\mu T}\right)^{1/2} = \left(\frac{10^3 \times 9.81 \times 1.23}{10^{-3} \times 1.8}\right)^{1/2} = 2590 \text{ s}^{-1}$$

Conclusion. Steady jumps are often too violent for treatment processes to take place.

If a weak jump is realized (e.g., Fr = 2 and d_2 assumed to be 1 m),

$$\frac{d_2}{d_a} = \frac{1}{2}(\sqrt{1 + 8Fr^2} - 1) \rightarrow d_a = \frac{1}{\frac{1}{2}(\sqrt{1 + 8 \times 4} - 1)} = 0.4 \text{ m}$$

$$\frac{L}{d_2} = 4.5 \text{ (Fig. 8)} \rightarrow L = 4.5 \text{ m}$$

$$h = \frac{(1 - 0.4)^3}{4 \times 1 \times 0.4} = \frac{0.216}{1.6} = 0.135 \text{ m}$$

$$T = \frac{L}{v_{avg}} \qquad v_a = \frac{1 \text{ m}^3/\text{s}}{0.8 \text{ m} \times 0.4 \text{ m}} = 3.1 \text{ m/s}$$

$$v_2 = \frac{1 \text{ m}^3/\text{s}}{0.8 \text{ m} \times 1 \text{ m}} = 1.25 \text{ m/s}$$

$$v_{avg} = 2.2 \text{ m/s}$$

$$T = \frac{4.5 \text{ m}}{2.2 \text{ m/s}} = 2 \text{ s}$$

$$G = \left(\frac{\rho g h}{\mu T}\right)^{1/2} = \left(\frac{10^3 \times 9.81 \times 0.135}{10^{-3} \times 2}\right)^{1/2} = 813 \text{ s}^{-1}$$

Homogeneous distribution of the added chemical into an hydraulic jump zone can sometimes remain questionable. It also occurs that the too strong a jump in limited areas, breakage of suspended matter and/or air binding of colloids occurs, which result in less fast flocculation later. In such cases the jump is to be readjusted empirically.

4.5   Mechanical Mixing

4.5.1   *Types of Mixers*

Any particular type of mechanical mixer (19,20) can be used in water treatment processes, but the most used types at present are the flat-blade paddle agitator, the gate paddle agitator, the disk impeller turbine, and the helical mixer (Fig. 15). For flat-blade paddle agitators and similar systems, Newton's equation determines the force by drag (F_D) and the power (P) impelled to the water:

$$F_D = C_D A \frac{\rho v_R^2}{2}$$

$P = F_D v_R$; hence the power input per volume is $F_D v_R / V$, in which A is equal to the surface of the paddles in m² and v_R the linear relative velocity of the paddles (tips) to the water in m/s. For slowly rotating mixers with less or no turbulency in the water (e.g., up to 0.5 rotation per second), the value of v_R can be assumed to be $0.75 \times v$(absolute) of the paddle tips. Mixers inducing high turbulency in the water are commented on later (Section 4.5.2).

The coefficient of drag depends on the ratio of length to width of the paddle (Fig. 10). In the standard *jar-test experiment*, flat blades or slightly incurved blades are used. The velocity gradient can be computed from the formula of Te Kippe:

$$G(s^{-1}) = \sqrt{\frac{2\pi NT}{V\mu}}$$

where N is the number of rotations per second, T is the torque input in N · m, V the volume in m³, and μ the viscosity in kg/m · s. In practice, the vessels used in the jar test as cylindrical and unbaffled, enable mixing gradients up to 300 s⁻¹ (see Fig. 11), for a rotational speed of 4 to 5 rotations per second. By baffling the vessel the velocity gradient can be increased by a factor of 2.5 in relation to the value of a classical unbaffled jar-test vessel.

Multiple-Paddle Agitators.   In practical full-scale systems, the slow rotating agitators are often composed of several coaxial paddles. The system can possibly be assimilated to the gate paddle agitator mounted on a vertical or a horizontal axis (Fig. 12). The systems hold for immersion of at least 0.3 m under the water level. The power input for systems composed of several flat paddles depends on the distance of the paddles to the axis of rotation. The expression most often applied is

$$P = 1.46 \times 10^{-5} \times C_D \times \rho \times [(1 - v_R)60n]^3 \times L \times \Sigma_a(d_e^4 - d_i^4)$$

where p is measured in kg · m/s and ρ in kg/m³, L is the length of the paddles in m, n the number of rotations per second, and a the number of paddles.

In flocculation, it is suggested to keep the n value between 0.02 and 0.08 and the maximum tip speed between 0.5 and 1.2 m/s.

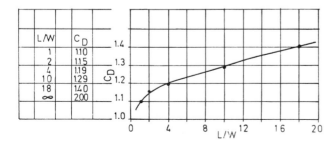

Figure 10   Coefficient of drag as a function of paddle proportions.

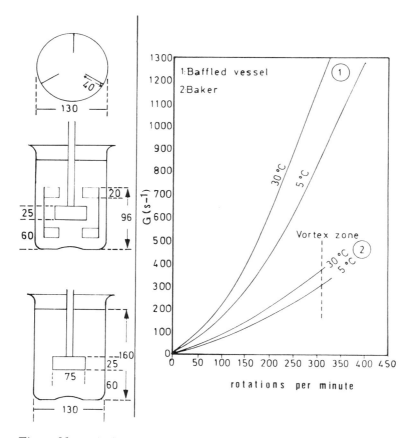

Figure 11   Velocity gradient in the jar test.

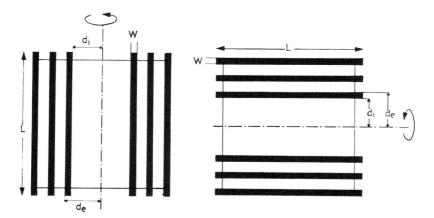

Figure 12   Multiple-paddle agitators.

Example.

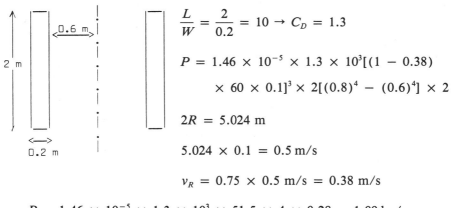

$$\frac{L}{W} = \frac{2}{0.2} = 10 \rightarrow C_D = 1.3$$

$$P = 1.46 \times 10^{-5} \times 1.3 \times 10^{3}[(1 - 0.38)$$
$$\times 60 \times 0.1]^3 \times 2[(0.8)^4 - (0.6)^4] \times 2$$

$$2R = 5.024 \text{ m}$$

$$5.024 \times 0.1 = 0.5 \text{ m/s}$$

$$v_R = 0.75 \times 0.5 \text{ m/s} = 0.38 \text{ m/s}$$

$$P = 1.46 \times 10^{-5} \times 1.3 \times 10^3 \times 51.5 \times 4 \times 0.28 = 1.09 \text{ kg/m} \cdot \text{s}$$

If in basin: $2.0 \times 2.0 \times 2.5 \text{ (depth)} = 10 \text{ m}^3$

$$G = \sqrt{\frac{1.09}{10 \times 10^{-3}}} = 10 \text{ s}^{-1}$$

To-and-Fro Mixers. In a to-and-fro movement, immersed flat blades with a reciprocal displacement dissipate a power in the basin equal to (11)

$$P = 19.74 \times 10^3 L_s^3 n^3 \, \Sigma \, A$$

in which L_s is the average length (m) of the blades' stroke in one reciprocal cycle and n the number of strokes per second. $\Sigma \, A$ is equal to the total cross area of the blades (m^2).

4.5.2 Rapid Mixing

In rapid mixing (e.g., in coagulation), the concept of paddle mixers leads to the design of gate-paddle agitators (Fig. 13, I), disk-impeller mixers (Fig. 13, II), and helical mixers (Fig. 13, III). General guidelines for design of the above-mentioned mechanical mixers are:

	I	II	III
d_2/d_1	0.5	0.33	0.95
h_2/d_2	1.5	0.2	0.36
h_3/d_3	0.2	1.0	0.028
h_5/d_1	1.0	1.0	0.75
s/d_1	(0.1)	(0.1)	—
a/d_1	0.6	—	—
b/d_2	0.1	0.25	0.14

Mixing conditions are turbulent, so that the approximation $v_R \simeq 75v$ does not hold. A yield coefficient C for turbulent conditions is defined and the power input relationship is expressed as follows:

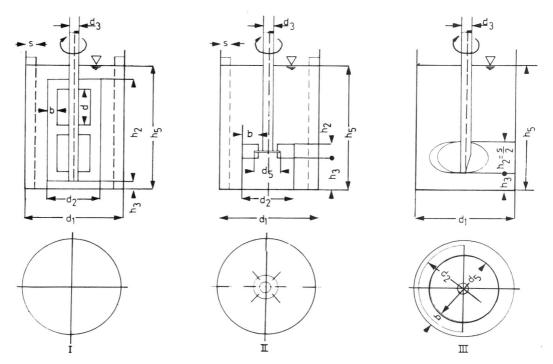

Figure 13   Mechanical mixers (types).

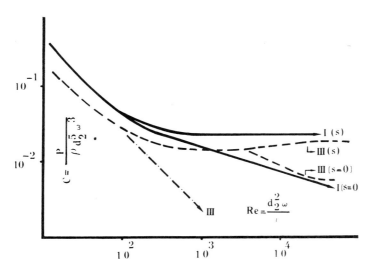

Figure 14   Coefficients of drag versus Re.

$$P = Cpd_2^5\omega^3 \qquad \text{where } \omega = 2\pi n$$

in which C is a yield coefficient, d_2 the overall diameter of the mixer, and n the number of rotations per second. The C value is a function of the Reynolds number and may be construction dependent. Manufacturers dispose of standardization curves. Examples are given in Fig. 14. As a first approximation, a value of C on the order of 0.01 can be accepted.

The yield coefficients are expressed as

$$C = \frac{P}{\rho d_2^5\omega^3} = \frac{1}{\rho d_2^5 \times 248n^3}$$

thus facilitating correlations for both expressions of the rotation velocity of the mixer, either as n (in rotations per second) or as ω (in angular velocity). The corresponding expression for the Reynolds number is

$$\text{Re} = \frac{\omega d_2^2}{\nu} = \frac{2\pi n d_2^2}{\nu}$$

where ν is the kinematic viscosity (m^2/s). (An important aspect of the power input relationship for mechanical mixing with axial turbines is that the power input is related to the diameter of the mixer and the number of rotations given as follows:

$$P \doteqdot d^5 n^3$$

in the expression of the velocity gradient, both parameters are present as

$$G \doteqdot d^{5/2} n^{3/2}$$

By acting on the rotational speed, in particular, the velocity gradient changes according to the power $\frac{3}{2}$.)

Example 1.

$$Q = 0.5 \text{ m}^3/\text{s} \qquad \text{average temp.} = 10°\text{C} \rightarrow \begin{array}{l} \mu = 1.31 \times 10^{-3} \\ \nu = 1.32 \times 10^{-6} \end{array}$$

Assume mixing time = 10 s

→ vol. = 0.5 m^3/s × 10 s = 5 m^3

If desired, the G value = 500 s^{-1}

Required power = $G^2 \times \mu \times V = (500)^2 \times 1.31 \times 10^{-3} \times 5 = 1638$ W

Basin circular diameter (e.g., 2 m) → depth = 1.6 m

$$P = Cpd_2^5\omega^3 = 0.01 \times 10^3 \times d_2^5\omega^3 = 1638 \ W$$

$$\rightarrow d_2^5\omega^3 = \frac{1638}{0.01 \times 1000} = 164$$

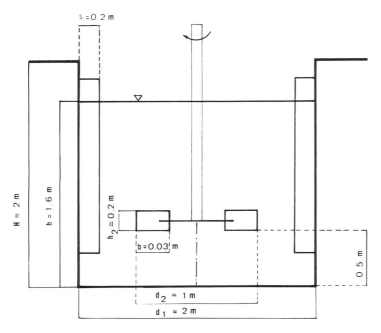

Figure 15   Example of a disk impeller system.

If $d_2 = 1$ m, $\omega^3 = 164$, $\omega = 5.5$

$$\rightarrow n = \frac{5.5}{6.28} = 0.9 \text{ s}^{-1}$$

$$Re = \frac{d_2^2 \omega}{\nu} = \frac{1^2 \times 5.5}{1.32 \times 10^{-6}} = 4.2 \times 10^6$$

The mixing is very turbulent.
Example 2.

$$Q = 0.75 \text{ m}^3/\text{s} \qquad \text{basin } 3 \times 3 \times 5 \text{ m} = 45 \text{ m}^3 \rightarrow t = 60 \text{ s}$$

$$n = \frac{100}{60} = 1.65 \text{ s}^{-1} \qquad \omega = 6.28 \times 1.65 = 10.5$$

$$G = \left[\frac{0.01 \times 10^3 \times 1^5 \times (10.5)^3}{1.31 \times 10^{-3} \times 45} \right]^{1/2} = 443 \text{ s}^{-1}$$

4.5.3 Mixing by Gas Dispersion

In the bubble formation process, the *average velocity gradient* is given by (11)

$$G^2 = 10^5 \frac{Q(\text{gas})h}{\mu V(h/2 + 10.33)}$$

in which h is the height of the liquid column above the diffusion zone, μ the absolute viscosity, and V the volume of the liquid.

The *maximum velocity gradient* is developed in the bubble formation zone and is approximated as

$$G_{\text{max}} = g(\rho_l - \rho_g)\frac{d_B}{6\mu}$$

in which d_B is the average diameter (in meters) of the bubbles, the other symbols having the usual meaning. In flotation, hard-rising bubbles of about 10^{-3} m in diameter are usually formed by expansion of a 10% flow of the water saturated with air at 6 bar introduced under a 2-m water layer. In this case, maximum gradients are more or less equal and on the order of 2000 to 3000 s^{-1} and the average in the range 200 to 300 s^{-1}.

In aeration (ozonization also) soft bubbles are formed with an average diameter of 5 mm while the gas flow is of about 10% of the liquid flow (e.g., 0.1 m^3/m^3). Introduced in these general conditions the velocity gradients due to bubble formation are equal to $G_{\text{avg}} < 200$ s^{-1}, $G_{\text{max}} = 8200$ s^{-1}. Mechanical mixing of the formed bubbles can be further superposed on the action of bubble formation.

Example. Ozonization basin equipped with porous pipe diffusors. $Q_{\text{gas}} = 110$ m^3/h (NPT), $h = 5$ m, basin = 45 m^3 ($3 \times 3 \times 5$), $t = 20°C$.

$$G^2 = \frac{10^5 \times 0.0305 \times 5}{10^{-3} \times 45 \times 12.83} \rightarrow G = 162 \text{ s}^{-1}$$

If $d_B = 5 \times 10^{-3}$ m, the maximum velocity gradient in the bubble formation zone is

$$G_{\text{max}} = \frac{9.81 \times (10^3 - 1.205) \times 5 \times 10^{-3}}{6 \times 10^{-3}} = 8165 \text{ } s^{-1}$$

4.5.4 Wind-Induced Mixing

Wind-induced mixing (Fig. 16) is not directly applied in treatment processes but cannot be overlooked in open-air installations of settling tanks, storage, and so on. The main effects can be disturbance of settling or secondary aeration.

At a wind velocity w, a water current velocity of pw can be induced in which p is proportional to the surface of the water. The mean velocity gradient expresses

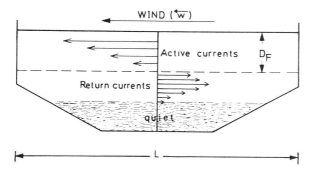

Figure 16   Wind-induced mixing.

$$G^2 = \frac{\rho g (pw)^2}{4g\mu T}$$

in which $T = L/pw$, L the length of the surface exposed in the predominant direction of the wind; hence

$$G^2 = \frac{pg (pw)^3}{4g\mu L}$$

Practical values of p are between 0.01 and 0.05. To evaluate the factor p, one has to rely on the formulation of Ekman (21) and Thorade (22):

$$\frac{v_0}{w} = 0.0127 \sqrt{\sin \phi} \quad \text{and} \quad \frac{D_F}{w} = 3.7 \sqrt{\sin \phi}$$

in which ϕ is the terrestrial latitude in degrees, v_0 the surface velocity (m/s), and D_F the depth of frictional resistance (m). In fact, under practical conditions, G has only a negligible value in the process technology ($G = 0.1$ to 0.01 s^{-1}), but its general environmental effect should not be overlooked (e.g., in aeration and sedimentation).

SYMBOLS AND UNITS RELATED TO MIXING

d	depth, m
d	pipe diameter, m
d	dispersion number, dimensionless
d_B	bubble diameter, m
d_h	hydraulic diameter, m
d_x	distance, m
f	friction factor
g	acceleration of gravity, 9.81 m/s^2
h	height, m
h	head loss, m

l	length, m
n	number of baffles, elements, and so on
n	number of rotations per second, s^{-1}
p	yield factor of air-to-water velocity transfer
t	theoretical residence time, s
t	time, s
v	linear (water) velocity, m/s
v_{av}	average velocity, m/s
v_0	surface velocity, m/s
v_r	relative velocity, ms
w	wind velocity, m/s
x	length of a reactor zone, m
A	surface, m^2
C	yield coefficient
C, C_0	concentrations, moles per volume or weight per volume
C_D	drag coefficient
D	diffusion constant, s/m^2
D	pipe diameter, m
D_F	depth of frictional resistance, m
E	dispersion coefficient, m^2/s
F_D	force by drag, $kg \cdot m/s^2$
Fr	Froude's number
G	velocity gradient, s^{-1}
H	height, m
L	length, m
L_s	stroke length, m
N	number of components
P	power, watt
Q	flow, m^3/s
Re	Reynolds number
T	detention time (average), min (or time units)
T	torque input, $N \cdot m$
V	volume, m^3
W	width, m
γ	specific weight per m^3 (cf. for water 9.81×10^3 kg/m^3)
μ	dynamic viscosity, $kg\ m^{-1}\ s^{-1}$
ν	kinematic viscosity, m^2/s
ρ	mass density, kg/m^3
ω	$2\pi n$, angular velocity, s^{-1}
Δp	pressure loss
ϕ	terrestrial latitude, deg

REFERENCES

1. R. S. Brodkey, *Mixing*, Vol. 1, Academic Press, New York (1966).
2. J. W. Hayden, *J. AWWA, 65,* 593 (1973).
3. H. A. Thomas and J. E. McKee, *Sewage Works J., 16,* 42 (1944).

4. R. D. Letterman, J. E. Quon and R. S. Gemmel, *J. AWWA, 65,* 716 (1973).
5. T. R. Camp, *J. AWWA, 60,* 656 (1968).
6. S. Kawamura, *J. AWWA, 68,* 328 (1976).
7. W. J. Masschelein, J. Genot, C. Goblet, R. De Vleminck, and L. Maes, *Bull. Cebedeau, 440,* 287 (1980).
8. W. J. Masschelein, G. Fransolet, and J. Genot, *Water Sewage Works,* p. 57, Dec. 1975, and p. 35, Jan. 1976.
9. W. J. Masschelein, *Proc. Ozone Symposium in Wasser*, Berlin, AMK, 1977, p. 118.
10. W. J. Masschelein, in *Ozone Technology and Its Applications*, R. G. Rice and A. Netzer, Eds, Ann Arbor Science, Ann Arbor, Mich., 1981.
11. T. R. Camp, *Trans. ASCE, 120,* 1 (1955).
12. P. C. Ziemke, *J. AWWA, 53,* 329 (1961).
13. L. L. Simpson, *Chem. Eng., 17,* 192 (1968).
14. H. Brünemann and G. John, *Chem. Ing. Technol., 43,* 348 (1971).
15. P. Mathys, Seifen, Ole, *Fette Wachse, 99,* 429 (1973).
16. Br. Stevens, *Process Eng.,* p. 76, Apr. 1973.
17. H. E. Hudson, J. P. Wolfner, *J. AWWA, 59,* 1257 (1967).
18. V. T. Chow, *Open Channel Hydraulics*, McGraw-Hill New York, 1959.
19. H. Ulrich, *Aufbereitungstechnik, 1,* 7 (1971).
20. R. J. Te Kippe and R. K. Hamm, *J. AWWA, 63,* 439 (1971).
21. V. W. Ekman, *Ark. Nat. Astr. Fys., 2,* 11 (1905).
22. H. Thorade, *Ann. Hydrogr. (Berlin), 42,* 379 (1914).

18
Sand for Filtration in Water Quality Control: Criteria and Mode of Action

1. INTRODUCTION

Filtration on sand is one of the most widespread processes in drinking-water treatment. For about one century it has been designed as "the perfect imitator of natural filtration," but with higher standards of technological performance. Materials other than sand are used for filtration purposes as well, such as granular activated carbon and conditioned anthracite coal. The filtration on granular activated carbon, which also involves microporous adsorption, is discussed in more detail in Chapter 12. The principles of filtration on sand are applicable as well to the filtration on other materials for which their specific physical properties must be taken into consideration, such as grain size, density, and volumic porosity.

2. FILTRATION

Filtration is a chemical technique of separation and fractioning of substances, particularly the liquid phase from the solid phase. It is one of the important processes in water treatment. There are two alternative methods to be distinguished as a function of the result desired: in the first method the liquid is to be recovered, which is the filtrate. The other method can be used to play an important part in secondary processes; its aim is to recover the solid phase, such as the sludge of the treatment plant. It concerns primarily sludge drainage basins with sand absorption beds. This technique is not directly involved in the water treatment. This chapter is restricted to purification of the liquid phase itself.

In a certain way, sand filtration in the plant is a copy of the natural infiltration of rainwater through sand grounds, a method by which the underground aquifers are renewed. With well-classified sands having a small average diameter of the grains, 0.2 mm, an average infiltration speed or *permeability* of 1 to 2.10^{-4} m/s is obtained. Usually, the main function of the sand in the filtration is that of a screen. However, later we will see that this effect is only partial — not the basic principle of filtration.

In the field of water treatment there are at least three fundamental principles of sand filtration, among which two of them, especially, offer numerous secondary alternatives. First, a distinction should be made for well casings made of an appropriate porous material composed of gravel and sand. In most applications the actual filtration techniques require installations or filters built as superstructures. We distinguish slow sand filters and rapid filters. The former are similar to artificial recharge of aquifers by infiltration on sand beds, while the latter occur mainly in two distinct forms, using gravitational-flow filters or pressure filters. The slow sand filters are characterized by surface loading or by an average velocity of flow in the range 0.10 to 0.30 m/h. On the other hand, rapid filters are characterized by an average velocity of flow of 2.5 to 10 m/h (0.7 to 2.8 mm/s). The ultrarapid pressure filters sometimes reach loadings of 15 to 20 m/h. The above-mentioned velocities are averages estimated with regard to the total surface of the basins in operation. They are thus different from the actual water velocities through interstices of the filtering medium (i.e., from the infiltration velocity).

3. SAND AND ISOLATING GRAVEL OF WELLS

To avoid the carryover of particles of the soil, such as fine sand or chalk, the draining parts are often isolated from the wells themselves. This operation is necessary to make it possible to protect the pumps and avoid the withdrawal of thin material with the pumped water. If this was not avoided, the underground in the neighborhood of the wells could eventually become cavernous and cause unsteadiness in the subsoil. For isolation of the wells, a double casing is set in the ground and the annular space between the two tubes is isolated with a mixture of sand and gravel.

This mixture must be composed of material of at least 50 wt%, of which the size is *four to five times greater* than that of granular dimension of the matter of the geological formation in which the well is bored. If the ratio is higher than 10; in other words, if the isolating sand is larger than the guide value mentioned above, there is a strong probability that the matter of the underground will be carried with the water. On the other hand, if the ratio is lower than 4, there could be the risk of not being able to use the total capacity of the well because of significant clugging of the isolating matter. As a guide, Fig. 1 represents a typical screening curve for a sand of the Brussels subsoil.

4. ESSENTIAL CHARACTERISTICS OF FILTERING SANDS

The hydraulic performances demanded of the sand with slow filters are quite inferior to those for rapid filters. So in the case of slow filters, we can be satisfied with fine sand, since the average filtration velocity that is usually necessary lies in the range 2 to 5 m/day. Exceptionally, the span is between 0.6 and 12 m/day.

In slow filtration, a good deal of the effect is obtained by the formation of a filtration layer, including the substances extracted from the water. At the beginning

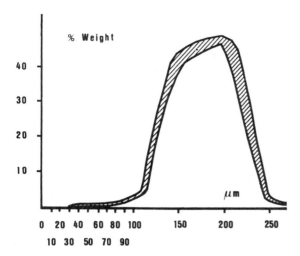

Figure 1　Granulometry of Bruxellian sand.

of the operation, these substances contain microorganisms able to effect, beyond the filtration, biochemical degradation of the organic matter. This effect also depends on the total surface of the grains forming the filter material. Indeed, the probability of contact between the undesirable constituents of the water and the surface of the filter medium increases in proportion to the size of the total surface of the grains. A height of 0.6 m sand with grains 0.15 mm in diameter offers the same surface as does a height of 1.4 m composed of grains 0.35 mm in diameter. Therefore, it is clear that in the case of slow filtration it is not interesting to increase the size of the grains without having seriously weighed the pros and cons.

The actual diameter of the sands used during slow filtration lies between 0.15 and 0.35 mm. It is not necessary to use a ganged sand. It would, nevertheless, be desirable that the uniformity coefficient (see Section 6.2.3) remain below 3, if possible, even under 2. The minimum thickness of the layer necessary for slow filtration is 0.3 to 0.4 m, although the very efficient thickness is only 2 to 3 cm.

The actual requirements for the sand in slow filtration are chemical in nature; in other words, they concern the purity and absence of undesirable matters rather than that connected with grain-size distribution. On the other hand, the performance of rapid filters requires sands with quite a higher precise grain size. In the case of rapid filtration, the need for hydraulic performances is greater than in slow filtration. This means that the grain-size distribution of the medium becomes of first importance (see Section 7).

5.　PURITY CRITERIA OF FILTERING SANDS

It is not advisable to use just any sand whatever, as it contains a series of undesirable impurities. Beyond the fact that its size must be suitable, the sand of a filter will have to be free of clay, dust, and fragments of any kind. The ratio of lime, limestone, and magnesium oxide will have to be lower than 5 wt%. The standard guide value of the quality of fresh sand is to be below 2% soluble matter at 20°C within 24 h in hydrochloric acid of a 20 wt% concentration.

In the treatment plant, the purity of the sand used must be examined regularly. Beyond the indications from the evolution of the head loss of the filter beds, the analysis and aspects of wash water during the operation of washing the filters must be checked. Special attention must also be granted to the formation of concretions or agglomerates. The presence of agglomerates indicates insufficient washing and consequently, the possibility of formation of undesirable microbiological development zones in the filter bed.

The determination of the agglomerates can be made as follows: A representative sample is taken (at least 2 or 3 L) at several spots in the layer and at a depth of 15 cm measured from the free surface of a filter that has just been washed. The collected fractions are assembled on a vacuum screen with mesh openings of 2.5 mm, and the screen is slowly shaken, imposing a vertical movement in a container full of water. After the nonagglomerated sand has come through the screen, the volume of the agglomerates is measured in a test tube and compared with the total screened volume. In this way the Baylis index is obtained, which is the volumetic percentage of the agglomerates in relation to the volume of wet sand taken during the test.

The Baylis index must never be higher than 1; if so, the filter bed will be overdamaged. A good filter mass has a Baylis index below 0.2, and better filtration is obtained if the Baylis index is even lower than 0.1. The degradation of the filter mass, or at least the reduction of its efficiency, can be the result of the agglomeration of its grains. There are several possible causes: insufficient backwashing, the presence of dead zones in the filter, and insufficient pretreatment of the water to be filtered. In particular, a lack of disinfection of the water in the case of the rapid sand filtration is often the real cause of the mass alteration.

6. REGENERATION OF FILTER MATERIALS

In addition to cleaning by washing, for both slow and rapid filters, a degradated mass containing agglomerates or fermentation zones (mud balls) can be regenerated by the appropriate treatment. Among the regeneration techniques we will discuss treatment with sodium chloride, regeneration through application of chlorine, and treatment with potassium permanganate, hydrogen peroxide, or caustic soda. The use of "natural sand" for filtration also involves a preliminary cleaning (1).

6.1   Regeneration with Caustic Soda

The aim of this cleaning method is to eliminate mainly thin clay, hydrocarbons, and gelatinous aggregates that form in filtration basins. After the filter has been carefully washed with air and water or only with water, according to its specific technology, one proceeds as follows. A quantity of 5 to 10 kg caustic soda/m^2 filter surface is spread over a water layer approximately 30 cm thick above the filter bed. The solution is then diffused in the mass by slow infiltration. After a period of action of 6 to 12 h, the filter is washed very carefully.

6.2   Treatment with Sodium Chloride

This method is used especially for rapid filters. In this treatment one spreads approximately 5 to 10 kg salt/m^2 of filter surface in a thin layer of water 10 to 15 cm above the freshly washed sand bed. After 2 or 3 h of stagnation, slow infiltration in the mass is obtained by opening an outlet valve for the filtered water. The brine is then allowed to work for 24 h. The filter is put back into service after a thorough washing. The sodium chloride works mainly on proteinic agglomerates, which are bacterial in origin.

6.3 Use of Potassium Permanganate

This application is particularly convenient for filters clogged with algae. Therefore, one spreads a concentrated solution (3 to 5% $KMnO_4$) containing potassium permanganate at an effective concentration of approximately 5 to 10 mg/L over the surface of the filters to obtain, in a water layer of 10 to 15 cm, a characteristic pink-purple color on the top of the mass. Through infiltration, the solution works for 24 h. After this operation, the filter is carefully washed once again.

6.4 Use of Hydrogen Peroxide

At a concentration of 10 to 100 ppm, hydrogen peroxide is appropriate for the suppression of clogging bacteria of the *Sphaerotilus* type. The cleaning method is similar to that used for permanganate. The addition of phosphates or polyphosphates makes it easier to remove the ferruginous deposits. This method can be used successfully in situ for surging the isolation sands of the wells. Adjunction of a reductor as f.e. sodium sulfite or bisulfite can be useful to create anaerobic conditions for the elimination of nematodes and their eggs when a filter has been infected.

6.5 Use of Hydrochloric Acid

The recurrent cleaning of rapid filters for sand, iron, and manganese removal can be done with a dilute solution of hydrochloric acid. This operation applies the acid at 5 wt% HCl, and has the secondary advantage of causing the formation of chlorine in situ according to the reaction

$$MnO_2 + 4HCl = MnCl_2 + 2H_2O + Cl_2$$

It is, of course, absolutely necessary to proceed so as to avoid all significant attack on the concrete structures of the filters, or to protect the latter by an appropriate coating system.

6.6 Use of Chlorine

For instantaneous cleaning of a filtering sand bed with chlorine, one may spread up to 0.2 kg active chlorine/m^2 filter surface. For this operation it is advisable to use a water layer thickness of approximately 20 to 30 cm as a dispersing medium. Further infiltration of the solution is obtained by percolation into the mass. The action goes on for several hours (at least 24 h), after which the filter is washed. Chlorine is used from concentrated solutions of sodium hypochlorite. As an alternative, chlorine dioxide can be used. This reagent convenes more to stop clogging by algae. One can prevent or eventually arrest the phenomenon of formation of agglomerates of biological origin by permanent treatment of the filter wash water with chlorine. Therefore, the aim is to achieve a residual concentration of approximately 1 ppm in the overflowing backwash water. Because of the formation of organochlorine compounds, the use of chlorine is being called into question more and more. The use of chlorine dioxide may be preferable.

7. GRAIN-SIZE DISTRIBUTION OF FILTERING SANDS OF RAPID FILTERS

The granulometric characteristics of filtering sands are of particular importance for water percolation through filter masses.

7.1 Sieving

The screening of filtering sands is carried out through a series of superimposed screens, the thinner screens being located at the bottom. The equipment is standardized as to intensity and stirring method. A dry sample of filtering sand is distributed over the topmost screen and after a proper agitation period (15 to 30 min) the weight retained by each screen is determined. After conversion for each screen, we obtain the percentage in weight of the mass that has come through screens of a diameter smaller than the value of the sieve-size diameter.

A direct grading curve is obtained by putting the sieve diameter on the X-axis and the percentage in weight of the mass that has come through the screens of the corresponding dimensions of the Y-axis. The scales are linear. Figure 2 shows two examples of the straining curves for filtering sands used in the Tailfer plant of the Brussels Waterboard. An alternative is to draw on a bilogarithmic diagram the percentage in weight of the mass that has come through the various screens, as a function of the size in mm (2,3) (Fig. 3). The result is often close to a straight line. Nevertheless, in most cases we call on the direct grading curve to characterize a filtering sand.

7.2 Immediate Characteristics of Grain-Size Distribution of Filtering Sands

To characterize the filtering sands used in rapid filters, we use the concepts of effective size, diversity, and hydraulic average size as well as the uniformity coefficient.

1. The *effective size* is given by the sieve diameter, which allows 10 wt% of the total mass of the filtering sand to pass during sieving. The corresponding values are indicated as TE in Fig. 2. As a whole, the effective 10% size will not be smaller than 0.3 mm for the slow sand filters and 0.5 mm for the rapid sand filters. The value of this parameter plays a major role in filters that are backwashed with both air and water, in that by progressive assortment, the grains of the smallest sizes are to be found in the upper parts of the filter. Consequently, the tendency for clogging is higher when the proportion of smaller grains increases with regard to the mass of the entire filter.

2. The 60% diversity, in other words, the mesh size that lets 60 wt% of the dry mass pass through, gives an indication of the granulometric homogenicity of the filter material. The homogeneous materials have a diversity 60% smaller than the nonhomogeneous masses for a given effective size. For slow sand filters, the 60% diversity will be approximately 0.5 to 0.6 mm. For rapid filters (5 to 10 m/h global filtration velocity), the 60% diversity is in general fixed to limits below 0.8 to 1 mm.

3. The *uniformity coefficient* is the quotient of the 60% diversity and of the 10% effective size. In any case, for slow filters, no rigorous specification of the

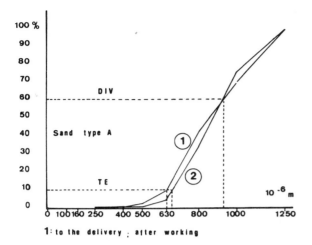

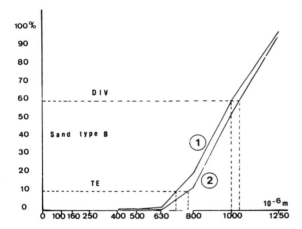

Figure 2   Filter sand types used in the Tailfer plant.

uniformity coefficient is generally imposed since the system has the tendency to form a thin filtration mass (the "Schmutzdecke") at the surface. Nevertheless, in general, the uniformity coefficient will be smaller than 3. For rapid filters the uniformity coefficient will under no circumstances be higher than 3, and generally it must be lower than 2. More accurately, we could consider an upper limit of 1.5 to be a model specification quality.

4. The *average hydraulic diameter* (4) of a filter mass can be globally rated by transformation of the immediate granulometric curve of a filtering sand. We maintain on a linear scale as Y-axis the percentage in cumulative weight corresponding to the passage of a sieve diameter, and on the X-axis the reciprocal of the sieve diameter, or $1/D$, in m^{-1}. Figure 4 gives a graphic determination of the average hydraulic diameter. The average active hydraulic diameter, d_h, is given by the expression

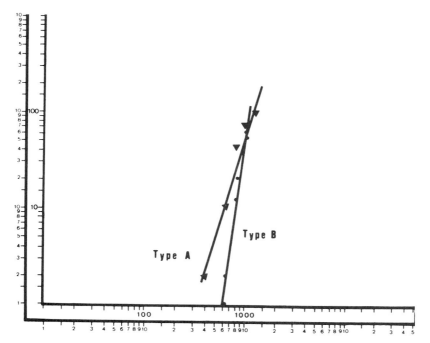

Figure 3   Bilogarithmic graph of grain size (Tailfer plant).

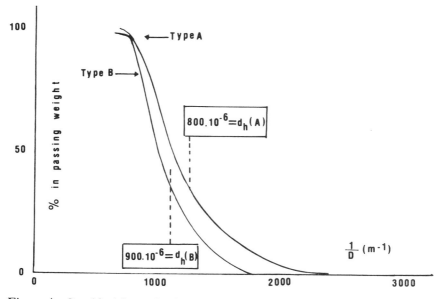

Figure 4   Graphical determination of the average hydraulic diameter.

$$\frac{100}{d_h} = \sum \frac{P_i}{D_i}$$

where D_i is the mesh diameter of the sieve and P_i the percentage of weight passing that sieve diameter (5). On a curve giving the percentage of flow-through as a function of $1/D_i$ we can graphically build the equipotential vertical, limiting equivalently the mass of particles with diameters lower and higher than the average. This vertical squares the X-axis at the reciprocal value of the average hydraulic diameter of the mass. By calculation or by graphical construction, we can deduce the effective average hydraulic diameter of a filter mass.

8. DIMENSIONAL ASPECTS OF FILTER MASSES

A series of parameters of the grains' constitution or shape are used in the specification of filter masses. These parameters may be independent of the aspect of global dimension shown by the average diameter during the sieving test.

8.1 Material Constitution

The silica constituting the filter sands has a density of 2.65. Extraneous material (such as mica, limestone, etc.) in the form of grains can sometimes be assimilated or confused with the sand during the straining test and thus must be analyzed by a density test. This may be done using a test based on Archimedes' law. Chloroform, of density 1.48, allows immersion of the sand when the bromoform, density 2.89, gives rise to its flotation. By mixing, it is possible to obtain fluids of variable density, which permits us to make an immersion test of the filter masses. The density, and eventually the fractionation, of the mass make it possible to characterize its purity.

8.2 Importance of Lamellar Elements

For rapid filters, there must be no lamellar elements such as mica in the filter masses. This point is especially important because when the rapid filters are backwashed with air and water, the lamellar elements are drawn to the surface and cause accelerated clogging during subsequent filtration cycles (see Figs. 5 and 6).

8.3 Shape Factor

The *shape factor* is characterized by a coefficient that we must adjust in terms of the granulometric size resulting from immediate sieving, to take into account the fact that the average grain of the material does not necessarily correspond to identical spheres for all grains. So we can express the diameter corresponding to a weight P of the material as a function of the weight fraction $P_1, P_2, P_3, \ldots, P_n$) corresponding to the diameter $d_1, d_2, d_3, \ldots, d_n$) of mesh spacing in the sieving test.

The specific diameter, which has a restrictive or limiting value for rapid filtration, can be obtained by the following equations:

$$\frac{1}{d} = \frac{P_1/P}{\sqrt{d_1 \times d_2}} + \frac{P_2/P}{\sqrt{d_2 \times d_3}} + \cdots + \frac{P_n/P}{\sqrt{d_n \times d_{n+1}}}$$

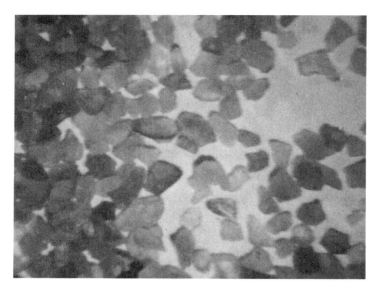

Figure 5   Angulosity of filtration sands.

The specific diameter affects the flow conditions of the sand filters. The correlation among specific diameter, effective size, and uniformity coefficient is expressed by the following empirical formula:

$$d_s = d_{10}(1 + 2 \log \text{UC}) = f \times d_{10}$$

where d_{10} is equal to the effective size diameter and UC is the uniformity coefficient.

The notion of *porosity* of a mass is also involved. This is generally indicated by the symbol ϵ, and corresponds to the volume quotient V_C (or free volume) divided by the volume V_c, the total apparent volume of the shell. This notion corresponds

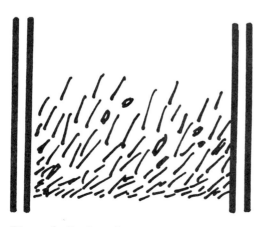

Figure 6   Preferential accumulation of lamellar elements at filter surface.

to the idea of permeability or of passage openings in the sand filter (see Fig. 7). For spherical particles of average diameter D assembled compactly, the diameter of the average cylindrical interstice corresponds to $0.155 \times D$.

Thus it appears that the shape coefficient of the grains of sand making up the filter mass is an important factor. This coefficient completes the estimation obtained by the immediate sieving curve of a given mass. According to the orientation of the grain during straining, this diameter can be minimal or maximal with regard to the screen opening, which has, until now, been the determining factor with regard to the size of the grains of a mass as a whole.

In approximations of the sand filtration hydraulics, we may assume that the water-flow conditions in the bed as a whole remain laminar. That is, the resistance to flow or loss of head of a filter with regard to the filter bed obeys a laminar flow law, in this case Darcy's law. According to this law, the equivalent height of the head loss, Δh, is a linear function of the filtration velocity and the thickness of the bed, h, and an inverse ratio to the permeability of the filter bed. The coefficient k, or permeability coefficient, has the dimensions of velocity and depends on a series of factors, including the water temperature (because of the effect of the viscosity), the porosity of the filtering medium, the specific diameter of the grains of sand making up the filter mass, and finally, the shape factor f, also called sphericity of sands. The shape or sphericity factor is given by a dimensionless figure of value 1 for perfectly spherical grains and less than this value with increased angularity. As a guide we may use the following values:

Shape of the grain	Spherical	Round	Angular
f	1	0.9	0.75

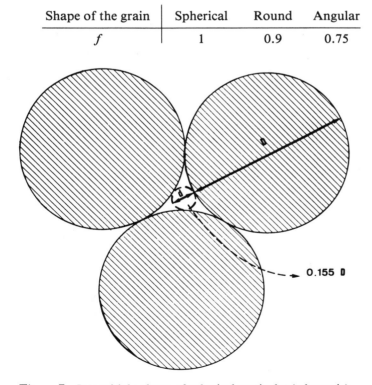

Figure 7 Interstitial volume of spherical particules (schematic).

The corresponding equations of flow are

$$\Delta h = \frac{V}{k} h = 150(0.72 + 0.028t) = \frac{\epsilon^3}{(1 - \epsilon)^2} = f^2 d_s$$

wherein k is in m/h.

Conclusions.   The specifications for particles of a filtering sand must correspond to a series of criteria of composition and purity of the sand and to parameters specifying the dimensions and shape required for the specific use. From our discussion of the "shape-diameter" ratio, it is generally clear that the filtering sands pass through the meshes of a straining interstice 1.05 times the size of the sieve, and in the case of oblong grains, on average, $D = 1.1$ times the sieve opening.

9.   WORKING PRINCIPLES OF SAND FILTERS

The following mechanisms may be listed as preeminent principles for sand filtration (6,7):

Straining
Settling
Centrifugal action
Diffusion
Mass attraction, or the effect of van der Waals forces
Electrostatic attraction
Fixation by magnetic effect (8)
Fixation by specific interaction

9.1   Straining

Straining is the best understood fractioning mechanism for solids and liquids. The action consists of intercepting particles larger than the free interstices left between the filtering sand grains. If we assume spherical grains, an evaluation of the interstitial size is made on the basis of the grains' diameter — that is, the specific diameter, since we have to take into account the degree of nonhomogeneity of the grains.

Porosity constitutes a very important global criterion in a description based on straining. Recall that porosity is generally characterized by the letter ϵ (sometimes by p) and is determined by the formula V_L/V_C, in which V_C is the total or apparent volume limitated by the filter wall and V_L is the free volume between the particles. First, note that the porosity of a filter layer changes as a function of the operation time of the filters. Indeed, the grains become thicker because of the adherence of material removed from the water, whether by straining or by some other fixative mechanism of particles on the filtering sand. At the same time, the interstices between the grains diminish in size. This effect is essential for good functioning of a filtration system, in particular for slow sand filters, where a deposit is formed as a skin or layer of slime that has settled on the bed making up the active filter (the "Schmutzdecke"). Moreover, biochemical transformations occur in this layer which are necessary to make slow filters efficient as filters with biological activity.

For reasons of economy (caused by the obstruction of filters), there has been a tendency during recent years to abandon slow sand filtration in favor of rapid filtration, thus bringing under renewed investigation the biochemical activity of these filters. For reasons that we discuss later, biochemical activity must generally be avoided in a rapid filter. On the other hand, a rapid filter used for the removal of iron and manganese in groundwater requires a running-in period before being effective. Indeed, we often observe that "a new filter mass" does not give satisfaction. It would be wrong to conclude at once that there is a construction or design defect in the filter.

As a general rule, filtration occurs correctly only after buildup of the sand mass. This formation includes a "swelling" of the grains and thus of the total mass volume, with a corresponding reduction in porosity. Indeed, the increases and swellings are due primarily to the formation of deposits clinging to the empty zones between grains. Nevertheless, in this case the porosity effect might be only secondary and the improvement in filtration of a continued mass might be due to electrokinetic phenomena involved in a chemical flocculation process in the filter masses.

However, the porosity of a filter mass is an important factor for its proper operation. This property is best defined by experiment either directly or by a derivative equation. As a guide, for masses with an effective size greater than 0.4 to 0.5 mm and a specific maximum diameter below 1.2 mm, the porosity is generally located between 40 and 55% of the total volume of the filter mass. It must be noted that layers with spherical grains are less porous than those with angular material. According to the experience we have acquired at the Brussels Waterboard (CIBE), in an iron-removal plant for water of a pyrite mine at Vedrin St-Marc (Namur), filters with angular grains have a better iron-removal filtration efficiency than do filters with round grains. From this we deduce that the essential mechanism of the process is not the straining filtration.

9.2 Settling Filtration

The settling velocity S of a particle into a fluid in laminar flow is given by Stokes' law:

$$S = \frac{1}{18} \frac{g}{\nu} \frac{\Delta\rho}{\rho} D^2$$

where

ρ	= volumetic mass density of the water
$\rho + \Delta\rho$	= volumetic mass density of the particles in suspension
D	= diameter of the particles
g	= 9.81 m/s^2
ν	= kinematic viscosity (e.g., 10^{-4} m/s at 20°C)

In the interstices of the filtering sands, which provide a path for the transport of water, there are zones where the water flow velocity is lower, as indicated, for example, in Fig. 8.

In sedimentation zones the flow conditions are laminar. A place is available for

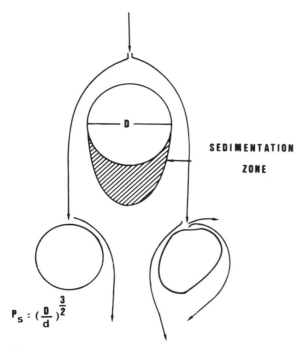

Figure 8   Sedimentation zones in a filter layer.

the settling of sludges contained in the water to be filtered. An estimation of the total free surfaces of deposit can be given as a first approximation by the following equations. If we suppose spherical grains with a specific diameter d_s, the surface/volume ratio is d_s^2, divided by $d_s^3/6$ or $6/d_s$. Therefore, the total surface of the grains is equal to $6/d_s \times$ the total volume of the grains. Now, for a total volume of filter mass $V = 1$ m^3, the volume of the grains is $(1 -$ the porosity) m^3. In other words, the surface of the particles is equal to the specific surface of the filter layer multiplied by $(1 - \epsilon)$. Thus the total surface of a filter mass is $(6/d_s)(1 - \epsilon)V_C$, where V_C is the total volume of the filter mass. An estimation of $d_s = 0.8$ mm and porosity $\epsilon = 40\%$ gives a total surface of approximately 4500 m^2 for 1 m^3 of volume of filtering sand. On average, the filtering sand layers of rapid filters correspond to this surface/volume ratio. On the other hand, for slow filters, having grains with a specific diameter of 0.25 mm and a porosity of 38 to 40%, the total surface of the grains per cubic meter of filter mass can reach 15,000 m^2.

Although the total inner surface that is available for the formation of deposits in a filter sand bed is important, only a part of this is available in the laminar flow zones that promote the formation of deposits. In general, matter with a volumetic mass slightly higher than that of water [i.e., with a small $\Delta\rho/\rho$ value (1%, for instance)] is eliminated by sedimentation during filtration. Such matter could be, for example, organic granules or particles of low density. On the other hand, colloidal material of inorganic origin—sludge or clay, for instance—with a diameter of 1 to 10 μm is only partially eliminated by this process, in which case the settling velocities in regard to the free surface become insufficient for sedimentation.

9.3 Elimination by Centrifugal Force or by Inertial Force of Flow

The trajectory followed by water in a filter mass is not linear. Water is forced to follow the outlines of the grains that delineate interstices. These various changes in direction are also imposed on particles in suspension being transported by the water. This effect, illustrated in Fig. 9, leads to the evacuation of particles in the dead flow zones.

The centrifugal action is obtained by inertial force during flow, so the particles with the highest volumetic mass are rejected preferentially. The rejecting probability by inertia (P_i) is expressed by the equation

$$P_i \frac{\rho_s D^2 v}{\nu d_0}$$

where

ρ_s = volumetic mass of the particle
D = diameter of the particle
v = transfer velocity of the particle
ν = kinetic viscosity
d_0 = average diameter of the drain channel (supposed to be cylindrical)

9.4 Diffusion Filtration

The diffusion of which the microscopic manifestation is Brownian motion obtained by thermal agitation forces is a complementary mechanism in sand filtration. The diffusion increases the contact probability between the particles themselves as well as between the latter and the filter mass. This effect occurs both in water in motion

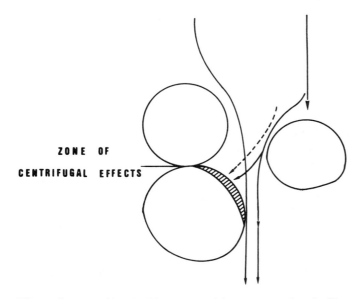

ZONE OF

CENTRIFUGAL EFFECTS

Figure 9 Centrifugal effects on particles on water flow in filters.

and in quiet water. This action is very significant in the mechanisms of agglomeration of particles (e.g., flocculation). Flocculation–filtration is to be included in the mechanisms of filtration.

9.5   Filtration by Mass Attraction

The mass attraction between particles is due to van der Waals forces, universal forces contributing to the transport and fixation mechanism of matter. The greater the inner surface of the filters, the higher is the probability of "attractive action." However, it is true only of forces corresponding to a law in r^{-7}, in which r is the distance between two infinitely small particles, atoms for instance. For larger particles, containing several atoms, the repulsive action takes place at a short distance. These forces are proportional to r^{-13}; in other words, the van der Waals potential (energetic term corresponding to the balance of attractive and repulsive forces) is proportional to r^{-6} and r^{-12}, respectively. From all these considerations we could conclude that van der Waals forces, which imply short molecular distances, must play a minor role in filtration practice. Moreover, they decrease very quickly when the distance between supports and particles increases. Nevertheless, the indirect effects, which are able to provoke an agglomeration of particles and thus a kind of flocculation, are not to be neglected and may become predominant in the case of flocculation–filtration or more generally in the case of filtration by flocculation.

From the preceding example $d_s = 0.8$ mm, $\epsilon = 40\%$, total surface $= 4500$ m^2/m^3; we deduce the average thickness of the film of water to be

$$\frac{0.4\,\text{m}^3}{4500\,\text{m}^2} = 8.8 \times 10^{-5}\,\text{m} < 100\,\mu\text{m}$$

The thickness of the film is thus sufficiently low to allow the intervention of attractive forces.

9.6   Filtration by Electrostatic and Electrokinetic Effects

Filter sand has a negative electrostatic charge. Microsand in suspension presents an electrophoretic mobility that varies between 0 and -4 μm/s per volt and per centimeter (see Fig. 10). The value of the electrophoretic mobility or of the corresponding zeta potential depends on the pH of the surrounding medium. In any case, the sand remains charged negatively. In applications other than real filtration, yet also using sand in the water treatment, we mention the possibility of using microsand (i.e., sands as micrograins, with a dimension of 0.1 to 1 μm) in coagulation–flocculation basins with high surface loading. Moreover, the technique requires a coagulation aid able to condition the surface of the microsand. It therefore has something in common with filtration after chemical conditioning of the filtering material. In filtration without using coagulant aids, other mechanisms may condition the mass more or less successfully. For instance, the formation of deposits of organic matter can modify the electrical properties of the filtering sand surfaces. These modifications promote the fixation of particles by electrokinetic and electrostatic processes, especially coagulation.

In connection with this, the addition of a neutral or indifferent electrolyte tends

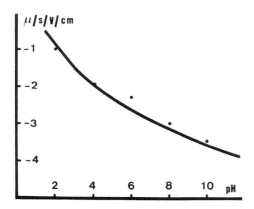

Figure 10 Electrophoretic mobility of microsand.

to reduce the surface potential of the filtering sand by compression of the double electric layer. This concept is based on the principles of electrostatic coagulation. The sand, as the carrier of a negative charge spread over the surface of the filter according to the model of the double layer, will be able to fix the electropositive particles more exhaustively. This has a very favorable effect on the efficiency of the filtration of precipitated carbonates or of flocs of iron or aluminum hydroxide-oxide. Optimal adherence is obtained at the isoelectric point of the filtrated material. On the other hand, organic colloidal particle carriers of a negative charge (e.g., bacteria) are repulsed by the electrostatic mechanism in a filter with a fresh filter mass. In the latter case, the negative charges of the sand itself appear unchanged. But with a filter at a maturate stage or conditioned in advance, there are sufficient positively charged sites to make it possible to obtain an electrochemical fixation of the negative colloids.

This largely explains the tendencies observed in fresh filters where the mass is not conditioned in advance, which give rise to a piercing phenomenon at the beginning of the operating cycle. This effect, which is prejudicial to the quality of the filtered water, disappears during the cycle; it can be eliminated rapidly by proper preconditioning of the filter sand with coagulation–filtration aids such as activated silica. The electrostatic forces are attractive forces between opposite electric charges that vary in inverse ratio to the square of the distance between the elementary charges. They may be added to any of the sand filtration mechanisms.

9.7 Fixation by Magnetic Effect

Most particles found in water are not magnetic but can be "magnetized" by adequate coagulation–flocculation. Therefore, loading with magnetite, Fe_2O_4, to obtain a matrix occupying only 5% of the active filtration surface can be efficient. The magnetic material is added to the water to be filtered at a rate of 200 to 500 ppm Fe_2O_4 (maximum 1000 ppm), together with an adequate coagulant: aluminum sulfate, ferric chloride, or lime. The magnetic complex is generated by coordination. The results are excellent for the removal of living particles such as bacteria, bacteriophages, and viruses. The filter is magnetized to saturation by the application of a field able to reach 100 Oe. Washing is done by interrupting the field. During this operation, the loadstone must be cleaned sufficiently to make recycling possible. The operation includes separation in an agitator or, alternatively, washing at an acid pH (2.8 or less).

9.8 Fixation by Specific Interaction

Filtering sands in the form of hydrated crystals are capable of ionization. The mass then allows the formation of hydrogen bonds. Moreover, the particles capable of being ionized must be filtered, and might form polar bonds, even real ionic bonds, with the mass. This effect is well known in the case of the flocculation–filtration of flocs obtained with iron salts. We must stress here that the degree of adsorption depends on the nature of the ferric ion present, and therefore on the pH. Moreover, the aging or maturation rate of the dispersions that are to be filtered, and of the combined nature of the anion corresponding to the ferric cation, play a predominant role.

Conclusions.   The elimination and fixing mechanisms of particles in suspension in water submitted to filtration on a macroporous sand bed depend on several forces of differing characteristics. In addition to the size of the filter sand itself, other factors are involved, such as the specific weight, the size and electrophoretic mobility of the particles to be removed, the hydraulic transport velocity, the viscosity and the density of the fluid, and finally, the temperature.

If we take the size of the grains and the pores into account, the contact probability of a particle with a filter mass is expressed by Stein's relation,

$$P_c = \frac{D^2}{d_s^3}$$

where P_c is the contact probability, D the diameter of the particles to be filtered, and d_s the specific diameter of the grains of filter sand. With regard to the removal probability of settling, the Hall formula,

$$P_s = \left(\frac{D}{d_s}\right)^{3/2}$$

is closer to reality. Nevertheless, these mechanical models do not take into account a series of surface phenomena that intervene in the fixation on sand, forming the quintescence of its mode of operation in water purification. So the general conclusion is that sand filtration corresponds to a method of fixing water impurities involving a series of principles of biophysicochemistry. These can hardly be split into principles of partial value. From a theoretical point of view, problems involved in the latter are related to the mechanisms of reaction in the heterogeneous phase.

10. FLOCCULATION–FILTRATION

10.1 Place of Rapid Sand Filtration in Treatment Sequences

The sand filtration process, especially rapid filtration, usually comprises a clarification chain including other unit processes. These generally precede filtration in the treatment sequence and can not be conceived of completely independent of the filtration stage. The most classical scheme of treatment is given by the sequence coagulation–flocculation–settling followed by filtration. When the preceding process, in this case flocculation and/or settling, becomes insufficient, subsequent rapid filtration can be a sufficient guarantee for the quality of the effluent treated. However, this action is often obtained at the expense of the evolution of filter head loss. Problems in washing and cleanliness of the mass may arise in this case. It is nevertheless interesting to note that in a significant number of cases, filtration is conceived to act as a coagulant flocculator. This is called flocculation–filtration.

10.2   Conditioning Effects of Filter Sand

10.2.1   Effects of Inadequate Coagulation

The presence of thin, highly electronegative colloids, such as particles of activated carbon, introduced in the form of powder in the settling phase may be a problem for the quality of the settled effluent. The carbon particles, which are smaller than 50 μm, penetrate deeply into the sand filter beds. They may rapidly provoke leakage of classical rapid filters. The same holds for small colloids other than activated carbon.

10.2.2   Effects of Activated Silica

Activated silica, which may have a favorable or an unfavorable effect on filtration, is composed of ionized micella formed by polysilicic acid–sodium polysilicate which become negatively charged colloidal micella. The behavior of activated silicas depends on the conditions of neutralization, the grade of the silicate used in the preparation of the activated silica, and the degree of maturation of the product.

In any case, activated silica is a coagulant aid that contributes to coalescence of the particles and thereby brings about an improvement in the quality of settled or filtrated water, depending on the point at which it is introduced. Even when the preparation of the silica has not been properly checked, we very often observe a maturing effect at the surface of the filter. This gives rise to curdled deposits that cause a rapid increase in head loss. The consequences of this effect are favorable on filtration performance as well as on the quality of the water treated. However, the filtration cycles can be considerably shortened: in a ratio of 1 : 2 to 1 : 3, for example. It is therefore necessary to be aware of the precautions required for the use of activated silica in sand filtration. Activated silica acts as a stimulus for filtration and must be controlled by a properly qualified supervisory staff.

10.2.3   Use of Polyelectrolytes

Preconditioning of the sand surface of filters by adding polyelectrolytes is an alternative use of sand filters as coagulator–flocculator. In the treatment of drinking water the method depends on the limitations of these products in foodstuffs.

10.2.4   Use of Polyphosphates

The addition of polyphosphates (9) to a water being subjected to coagulation usually has a negative effect: the breaking of the agglomeration velocity of the particles during flocculation. In sand filtration, the addition of polyphosphates simultaneously with phosphates can be of value in controlling problems of corrosion. This sometimes makes it possible to avoid serious calcium carbonate precipitation at the surface of filter grains when the water is alkaline. The application concerns very rapidly incrusting water in which one tries to maintain the maximum hardness in solution. It should be noted that the addition of polyphosphates involves deeper penetration of matter into the filter mass. Accordingly, the breaking of flocculation obtained by the action of polyphosphates enables the thinner matters to penetrate the filters more deeply.

Consequently, these products favor the "in-depth effects" of the filter beds. It is obvious that their use necessitates first carefully checking that they are harmless from a hygienic point of view. Moreover, it can be decided only after examination of the costs and benefits of the operation, and by taking into account the staff or equipment required, whether or not to apply these aids.

10.3   Design of Flocculation–Filtration Filters

The penetration in depth of material in coagulation–filtration is somewhat opposite to the concept of the filter as a screen. Precipitation initiated by "germs" plays an important role. There exist a certain number of empirical relations to guide the design of the filters as a function of the penetration in depth of coagulated material. The concentration of those residual matters in filtered water (C_f) depends on several factors: the linear infiltration rate (v_f), the effective size of the filter medium (ES), the porosity of the filter medium (ϵ), the final loss of head of the filter bed (Δh), the depth of penetration of the coagulated matter (l), the concentration of the particles in suspension in the water to be filtered (C_0), and the water height (H). Hence

$$C_f = f\left(v_f \times (ES)^3 \times \epsilon^4 \times \frac{\Delta h}{l} \times C_0 \times H\right)$$

Moreover, it has been shown that the total loss of head of a filter bed is in inverse ratio to the depth of penetration of the matter in suspension.

10.4   Secondary Phenomena of Flocculation–Filtration

The addition of salts causes increased penetration of the colloidal material in a filter bed, favored, in increasing order, by chlorides, sulfates, and phosphates. This sequence also corresponds to the order observed in electrostatic coagulation by the action of ions in solution. Finally, sand mass can be considered as a series of walls with negative electric charges. The electrostatic coagulation is induced by applying a positive load through a source of external potential difference.

Example.   It may perhaps be useful to illustrate these various concepts by an actual case. Among its catchments, the Brussels Waterboard also disposes of water stored in quarries which are open-air storage places. The quarries are filled either naturally, by rainfall, or by artificial feeding. The volume thus stored totals approximately 2 $\times 10^6$ m^3 water. Initially, the water is quite pure. However, after a certain period of exposure to sunlight, it becomes necessary to treat the water, in particular to eliminate algae and undesirable mineral compounds such as iron. Since quarries represent standby or peak catchments that are not used permanently, the means of treatment used must be as simple as possible in operation. Therefore, preoxidation is used to treat the bacterial problem and the algae as well as to promote the coagulation–flocculation–filtration process. So in the treatment plant at Ligny (Fig. 11), we inject chlorine and sodium carbonate for pH adjustment and to obtain nonaggressive water under all circumstances, and finally, ferric chloride. The latter is quickly hydrolyzed.

The water is brought onto a series of rapid sand filters and the impurities are removed very satisfactorily by coagulation–flocculation–filtration. Working at a speed up to 8 m/h with respect to the free surface of sand, the average filtration cycle is of 24 h. The total loss of head admitted is 1 bar. Backwashing is done in the counterflow mode, by air and water. (The respective velocities are approximately 50 and 25 m/h.) A filter used in the Ligny plant, represented in Fig. 12, consists of closed horizontal pressurized filters. The grain-size distribution of the mass has been studied with care and includes the sequence of layers shown.

11. PHENOMENA IN RELATION TO SILTING OF FILTER SAND

11.1 Negative Pressure in Filter Beds

The head loss of a filter bed increases with silting of the filter sand. This evolution is shown schematically in Fig. 13. The head loss of an operational pure sand filter is a few millimeters but can range from 20 to 30 cm for the filter sand and the underdrain system as a whole.

Figure 11   Set of pressurized filters in parallel at Ligny (CIBE).

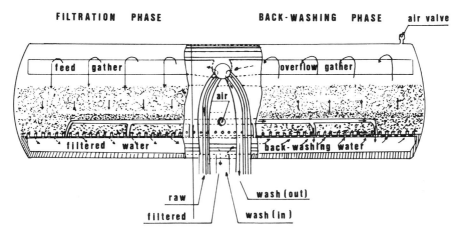

Figure 12   Cross section of a filter at Ligny (CIBE).

The head available decreases with silting. This phenomenon appears most significantly in the upper parts of the sand in a filter with downflowing water. At a given time, in theory, a depression zone will be reached in the upper part of the mass. However, the existence of this zone usually does not appear macroscopically. Indeed, quite a number of compensatory phenomena that prevent the formation of such a depression zone occur in the filter beds. The principal one is cracking of the surface of a filter bed, with or without formation of preferential penetration channels. Also, pathways for oxygen or CO_2 degasification which compensate for the depression can occur. From this effect "air silting" eventually results. The phenomenon is known by this term because it usually provokes a decrease in filtration velocity, thus of the treated water flow. The channels formed by the cracks are a few millimeters in width up to 1 cm in serious cases. They are liable to affect the

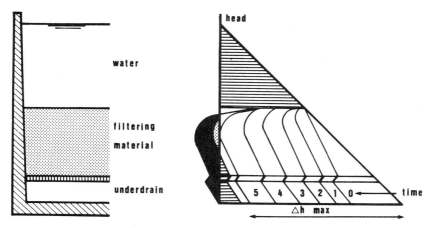

Figure 13   Evolution of head loss in the filter layer.

filter bed to a depth of 10 to 15 cm. In any case, operation of a rapid sand filter cannot result in putting specific zones (e.g., at the surface of the sand) of this filter in depression. Cracking with the formation of channels suitable for deep, direct penetration must certainly be avoided. Therefore, to avoid treatment of this phenomenon, normal use of the filters must leave an available positive load of at least 20 to 30 cm of water column for a filter operated at a v_f value of 5 m/h ($v = 1.5$ mm/s). For example, a filter of total depth 2 m (filter mass + water layer) must not be exploited at a total head loss of more than 1.7 to 1.8 m or there will be a risk to the quality of the filtered water.

11.2 Influence of Filtration Velocity

Figure 13 illustrates rapid filtration *at constant infiltration velocity*. A series of mechanisms, discussed later, make it possible to maintain a constant flow of filtered water during the entire filtration cycle. These methods of exploitation are based on variation of the head loss of the equipment with a change, complementary but opposite to the head loss, which makes possible a well-defined production capacity that can be employed at any time.

Another technique, involving less regulation, is to work at a decreasing filtration velocity with a constant head, without a regulating mechanism (i.e., only by maintaining a constant level of the water on the filter). In this case, a diminution of flow by filter occurs, as a function of both time and the moment of the operation cycle. The filtration velocity can have a significant influence on filter clogging. Empirical equations for the removal of particles in suspension in water through rapid sand filtration have been formulated:

$$l = kd_s^{2.46} \left(\frac{Q}{A}\right)^{1.56}$$

where l is the penetration depth of the filter mass (m) and d_s is the specific diameter of the grains of filter sand, and where $k = 1.6 \times 10^{-3}$ if d_s is expressed in mm and Q/A in m/h. The evolution of the saturation in depth of a sand filter bed as a function of filter operation time and the sites of the relevant layers is schematized in Fig. 14.

Comparative experiences between filtration with constant total flow and filtration with constant velocity between the pores of the sand filter mass have shown that filtration with constant velocity in the pores can present some advantages, the most important being that the length of a cycle could be increased by approximately 20%. In some cases suspended matter could, statistically, be reduced by 20% in comparison with the concentration obtained by filtration at constant flow. In practice, however, filtration at practically constant flow with flow regulation is the usual procedure in most production plants.

The surface loading of filters generally varies from 3 to 10 m/h for rapid gravitational-flow filters, and from 10 to 20 m/h for pressure filters. "In-depth" saturation imposes limits inherent in the risk that the filtered deposits are being drawn upon further on. In practice, the layer thickness of sand filters reaches a maximum of 1.5 m, with a total sand–water layer up to 2.5 m.

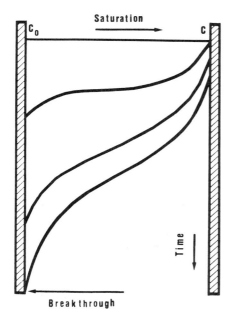

Figure 14 Progressive saturation of the filter layer.

12. SLOW FILTRATION

12.1 Working Parameters of Slow Sand Filters (10)

Slow sand filtration consists of removing material in suspension and/or dissolved in water by percolation at slow speed ($Q/A < 0.3$ m/h). In principle, a slow filter comprises a certain volume of areal surface, with or without construction of artificial containment, in which filtration sand is placed at a sufficient depth to allow free flow of water through the bed. Usually, the water height is 1 to 1.50 m above the sand surface. When the available head loss reaches a limit of approximately 1 m, the filter must be pulled out of service, drained, and cleaned. The thickness of the usual sand layer is approximately of 1 to 1.50 m, but the formation of biochemically active deposits ("Schmutzdecke") and clogging of the filter beds takes place in the few topmost centimeters of the bed. The thickness of the biologically active layer can reach 0.3 to 0.4 m. The really clogged part is far smaller, however, approximately 2 to 3 cm. It is precisely this mass thickness that must be replaced when renewing the filter bed.

12.2 Building Parameters of Slow Filters

Technically, the filter mass is pored on gravels of increasing permeability. Each layer has a thickness of approximately 10 to 25 cm. The lower-permeability layer can reach a total thickness of 50 to 60 cm. So-called gravels 18 to 36 cm in size are used and their dimensions are gradually diminished to sizes of 10 to 12 cm or less for the upper support layer. The total depth of the filter tank can thus be estimated as follows:

Thickness of the water layer on the surface: 1.25 m
Thickness of the filter medium: 1.20 m
Support layer: 0.30 m
Filter bottom: fitted out with bricks, porous stones, or compacted earth, 10 to 15 cm

Above the water layer it is advisable to allow for a security zone of approximately 20 to 30 cm. The total depth of a slow filter thus reaches 3 to 3.50 m. It may be built as a simple or reinforced structure with massive materials such as ferroconcrete, stones, and bricks, according to local conditions. Removal of the filtered water can be by way of a system of drains located under the filter bed or by free percolation in the water table. The latter case necessitates an artificial feeding method through slow filtration.

12.3 Maintenance of Slow Filters

The sand filter is cleaned by the removal of a few centimeters of the clogged layer and by washing this layer in a separate installation. According to local circumstances, the removal of the sand can be done manually or by mechanical means. It should be noted that the sand removed may not be replaced entirely by fresh sand. Placing preconditioned and washed sand is better. This takes into account the biochemical aspects involved in slow filtration. If for operational reasons fresh sand is used, one must allow for a lapse of time before optimum operation of the filter. The sand can be swilled in a hydrocyclone or similar equipment. As an alternative to manual or mechanical removal, cleaning can be done on the spot using a hydraulic system (Fig. 15). In this case the investment cost will be high.

12.4 Performance of Slow Filtration on Sand

If at the beginning, the water does not contain much suspended matter, it is possible to use slow filtration without previous coagulation. However, if the water is loaded (periodically or permanently) with clay particles in suspension, pretreatment by coagulation–flocculation is absolutely necessary. Previous adequate oxidation of the water, in this case preozonization producing biodegradable and metabolizable organic derivatives issuing from dissolved substances, can be favorable because of the biochemical activity in slow filters.

On the whole, control of slow filter operation is easy and requires only a limited qualified staff. The main drawbacks of slow filters are of several different types. They may require a significant surface area and volume, and may therefore involve high investment costs. Their use may lack flexibility—mainly during the winter, when the open surface of the water can freeze. During the summer, if the filters are placed in the open air, algae may develop, leading to rapid clogging during a generally critical period of use. Algae often cause taste and odor problems in the filter effluent. The building costs to cover slow filters are worth taken into consideration in this context. The results of slow filtration are generally not completely satisfactory. They may even be poor, yielding untreated, colored water in which, in the long run, algae are numerous.

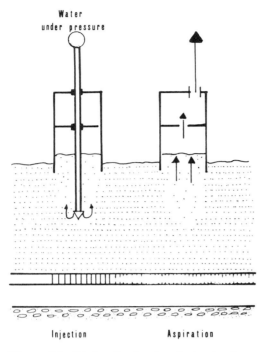

Figure 15   Hydraulic cleaning device for slow sand filters (schematic).

12.5   Slow Sand Filtration as an Infiltration and Artificial Recharging Technique for Groundwater

Slow filtration can be considered as an artificial recharging technique for groundwater when it takes place in basins not equipped with draining substructures but which allow the recovery of filtered water when the basins come in contact with an infiltration zone for groundwater.

The recharging of groundwater is a process in which rainfall reaches porous layers by percolation through the surface ground. In the Brussels Waterboard we may count on an output of approximately 0.3 m/year, or 0.3 m³/m² per year in regions of water infiltrated in the groundwater basins of the Bruxellian sands. Theoretically, the maximum output possible is approximately 0.3 million m³/km² per year. For example, assuming an aquifer 50 m thick with a volumetic porosity of approximately 30% and able to store groundwater, the quantity of water stored can thus reach 15 million m³/km² of surface. We encounter this type of situation in cretaceous deposits. If by artificial recharging techniques, we decrease the reconstitution period of the water layers to 6 months, the annual theoretical storage capacity is 30 million m³/km²: in other words, 100 times the output of rainfall on the same infiltration surface.

Artificial infiltration is thus a first-order technique of sand filtration. Some regions — for example, an important part of the Parisian suburbs — constitute more or less natural infiltration basins. They are made of old sand pits present, but out of service, situated above the aquifer in cretaceous subsoil. The sand layer in the basins can be compared with that of slow filters in terms of granulometry and quality. As the percolation velocities remain weak, slow silting takes place. A superficial annual cleaning of the mass is sufficient.

Infiltration through sand basins requires well-pretreated water, to avoid clogging problems, among others. One must also avoid the possibility of infiltration of impure water. The problems involved in this infiltration technique and other methods, such as forced infiltration, are being studied carefully. The Brussels Waterboard has installed experimental infiltration basins in the Yvoir–Champale plant, where experiments have been carried out to refeed the alluvial layer (Fig. 16).

As to the kinetic aspect of filtration, infiltration basins normally work at lower speeds than do classical slow filters. The average superficial load falls in the range 0.01 to 0.1 m/h, or a maximum of 2.5 m/day. Consequently, the treatment efficiency is high and the activity periods between two washings are long (i.e., longer than 1 year) if the pretreatment of the water is satisfactory.

13. RAPID FILTRATION

The goal of this section is not to give an extended overview of rapid filtration technologies but to relate aspects of interaction of sand and water quality.

13.1 General Filtration Process

Rapid filtration can be carried out either in open gravitational flow filters or in closed pressure filters. The velocities of the surface loadings reach 4 up to 20 m/h. In exceptional conditions they can even reach 50 m/h (11). For filters of the open type, the total charge of the water would normally have to be approximately 2.5 m at maximum. This available water height corresponds to the sum of the height of the filter layer and that of the water layer. Rapid pressure filters have the advantage of being able to be inserted in the pumping system, thus allowing use of a higher effective loading. We can formulate several consequences, such as that, in general, pressure filters are not subject to development of negative pressure in a lower layer of the filter. Moreover, these filters generally support higher speeds, as the available pressure allows a more rapid flow through the porous medium made up by the filter sand. On the other hand, the risks of break-through are greater. Pressure filtration is generally less efficient than the rapid open type with free-flow filtration. Pressure filters have the following drawbacks: The injection of reagents is complicated; one cannot observe the appearance of the water during filtration; and it is

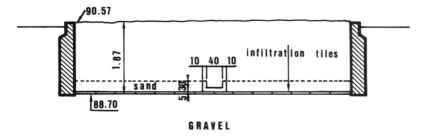

Figure 16   Cross section of an experimental infiltration basin at Yvoir–Champale.

more complicated to check the efficiency of backwashing. Work on the filter mass is difficult considering the assembly and disassembly required. Moreover, the risk of break-through by suction increases. Finally, pressure filters eventually need a longer filtration cycle, as the loss of head available to overcome clogging of the filter bed is generally 10 m (1 bar) of water column and for some sand filters might reach a total height of 20 m of water column (2 bar).

13.2   Open Filters

Open filters are generally built in concrete. Their recommended form is rectangular. A global economic approximation would be to fix the most appropriate dimensions. In this evaluation one must take into account the surface/volume relations and the thickness of wall necessary. Generally, this analysis leads to building filters whose width is approximately one-third to one-sixth of the length of the filter basins.

The filter mass is posed on a filter bottom, provided with its own drainage system, including the bores necessary for the flow of filtered water as well as for countercurrent washing with water or air. Several types of washing bottoms will be described. First, porous plates, composed of corundum or aloxite, directly support the filter sand, generally without a layer of support gravel. Even if the system has the advantage of being of simple construction, it nevertheless suffers from the disadvantages that result from incrustation. This is especially true for softened water or water containing manganese. The porous filters bottoms are also subject to disintegration upon the filtration of aggressive water.

Figure 17   Porous underdrain system of the rapid sand filters at the Vedrin plant (CIBE).

The filter bottom can be made up of pipes provided with perforations of approximately 5 to 10 mm turned toward the underpart of the filter bottom and embedded in gravel. The latter can reach a total thickness of 50 to 60 cm and decrease in size from bottom to top. In general, the lower layers are made up of gravel of approximate diameter 35 to 40 mm, decreasing up to 3 mm. The filter sand layer, located above this gravel layer, serves as a support and equalization layer. Several systems of filter bottoms, including the Leopold and Wheeler bottoms, comprise perforated self-supporting bottoms or false bottoms laid on a supporting basement layer. The former constitutes a series of glazed tiles, including bores of approximately 3 to 7 mm, above which are a series of gravels in successive layers 5 to 10 cm thick. The Wheeler bottom is composed of a layer in concrete perforated with hoppers, opening onto a porcelain drain pipe in which there are porcelain tightening spheres 6 to 10 mm in diameter. On the whole, the repartition gravel specified above is positioned in successive layers 6 to 8 cm thick.

Actually, all these systems are surpassed to some extent by filter bottoms in concrete provided with strainers. In principle the choice of strainers should be the object of a preliminary investigation which takes into account the dimensions of the slits that make it possible to stop the filter sand, which is selected as a function of the filtration goal one decides to reach. In general, the maximum width admitted for the slits is less than 0.35 mm, usually 0.2 mm, if no charging of the bottom with strainers by a repartition gravel is foreseen. Obstruction or clogging occurs only rarely. There is thus no objection to the use of strainers.

Normally, at least 50 strainers/m^2 of filter bottom are required. The strainers may be of the type with an end that continues under the filter bottom. On the other hand, they promote the formation of an air space for backwashing with air. If this air space is not formed, it can be replaced by a system of pipes that provide for an equal distribution of the washing fluids.

13.3 Pressure Filters

Pressure filters are generally set up in the form of steel cylinders placed vertically. An older technology consists of using horizontal filtration groups, such as those at the Ligny plant mentioned previously. The drawback of this technology with horizontal groups is that the surface loading is variable in the different layers of the filter bed; moreover, it increases with greater penetration in the filter bed (the infiltration velocity is lowest at the level of the horizontal diameter of the cylinder).

The filter bottom of pressure filters usually consists of a number of screens or mesh sieves that decrease in size from top to bottom or, as an alternative, perforated plates supporting gravel similar to that used in the filter bottoms of an open filter system. Present technology often makes use of a filter bottom consisting of a supporting plate with filter strainers similar to those of open filters but adapted hydraulically to the conditions of increased filtration velocity in pressure filters.

13.4 Washing Rapid Filters

The hydraulics of rapid filtration does not itself affect the quality of filtered water under normal operational conditions. However, washing of the filter mass can very quickly influence the quality of water filtered. Changes may be consequent to fermentation, agglomeration, or formation of preferential channels liable to occur if backwashing is inadequate.

In backwashing we can distinguish washing with water alone from combined washing with water and air (12). In the general design for backwashing with water, the velocity ranges between an absolute minimum of 10 m/h and a maximum of 55 to 60 m/h. The average conventional value usually lies between 20 and 30 m/h. In fact, the velocity we tend to reach is a function of expansion of the filter bed as a consequence of fluidization of the mass. The critical velocity, for backwashing of the filter sand with water, is equal to the fluidization velocity of the filter mass (13). This depends on a number of factors, among them granulometry and grain shape. The problem of fluidization of filter masses during backwashing can be approached differently, and in particular as a function of the parameters of size and of walls. It is therefore best to consult the literature (14). Hereafter are indicated some practical suggestions that have given satisfaction experimentally. In no case may expansion of the filter bed during backwashing be lower than 15%, or agglomerates will develop. The degree of expansion of a filter sand depends on the grain-size distribution and consequently, on the homogeneity of the sand layer as well as on the temperature of the water, which influences the viscosity.

In Fig. 18 are shown typical data for 20 to 40% expansion of a filter bed made up of sand grains of increasing average diameter and 40% porosity. Various theoretical approaches to the problem are somewhat idealized and thus applicable

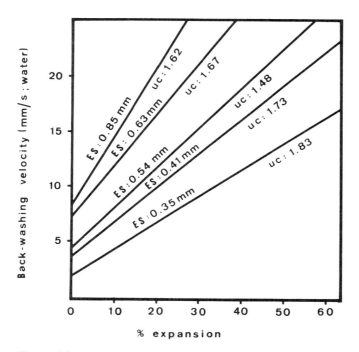

Figure 18   Expansion of the sand layer on backwashing with water.

only under standard conditions. In practice, Leva's semiempirical equation seems to give best results for evaluation of the minimal washing velocity necessary for fluidization of the sand beds:

$$V_{mf} = \frac{kd_s^{1.82}\rho_l(\rho_s - \rho_l)^{0.94}}{\nu^{0.88}}$$

where V_{mf} is the minimal fluidization velocity in $m^3/m^2 \cdot h$, d_s the specific diameter in mm, ρ_l and ρ_s the densities of the fluid and the solid, respectively, in g/L, and ν the kinematic viscosity in stokes (cm^2/s).

From all the preceding considerations it is clear that a sand filter bed with consecutive backwashings with water alone tends to form layers matching the thinner grains at the surface and the thicker ones at the bottom. This effect has both advantages and drawbacks. As to the favorable effects, we note first the higher filtration efficiency and thus better quality of effluent. Indeed, rearrangement of the thinner grains in an upper filtration layer makes it less porous, and thus the filtration effect by screening is more efficient. On the other hand, as a consequence of the aforementioned advantage, clogging is accelerated at the surface of the filters, and important drawbacks can result, such as channeling and mud-ball formation.

13.5 Practical Conditions for Backwashing with Water

For backwashing it is necessary to locate a source able to supply the necessary flow and pressure of wash water. This water can be provided either by a reservoir at a higher location or by a pumping station that pumps treated water. Sometimes an automated system is utilized, with washing by priming of a partial siphon pumping out the treated water stored in the filter itself (systems such as Permutit and Degremont) (Fig. 19). The wash water must have sufficient pressure to assure the necessary flow. It is therefore advisable to take into account the various counterpressures possible.

The following design parameters are involved in establishing the washing devices of filters.

1. The main loss of pressure is the loss of head in the system of drains or in porous plates of the filter bottoms.
2. The loss of head of the filter bed in expansion: This is equal to the apparent weight of the filter mass in water.
3. All the hydraulic head losses in the repartition pipes of the valves, the sudden outlets, and so on, that are specific for each installation and must be calculated on the usual basis.
4. In the alternative of filters provided with a layer of support gravel, the head losses in the gravel, which can be evaluated according to the equations of the porous medium, must also be added. In general, they are less than 20 to 30 cm.

13.6 Washing with Air and Water

At present, the predominant techniques are the washing of the filter sands with air, followed by washing with water and in most cases including a short intermediate

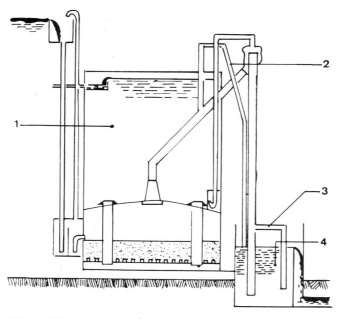

Figure 19 Automatic backwashing filter with a partial siphon system. 1, Filtered water (reserve); 2, partial siphoning; 3, initiation; 4, restitution.

phase of simultaneous washing with air and water. To begin with, it is a specifically European technique. Because of greater homogenization of the filter layer and more efficient washing, it has the great advantage of avoiding the formation of fermentation areas and agglomerates in the filter mass of treatment plants for surface water (mud balls). The eventual formation of a superficial crust on the filter sand is avoided by washing with air. The airflow velocity necessary during the intermediate phase reaches at least 15 m/h and is generally fixed at 40 to 60 m/h.

After washing with air, which normally lasts 2 to 3 min, water flow is gradually superimposed on the air flow. This operational phase ends at the same time that the wash air is terminated, to avoid the filter mass being blown away. The superposition phase is usually shorter than 1 min, after which there is a period of washing with water alone.

In the treatment of surface water previously coagulated, flocculated, and settled, the filtration cycle or duration between two consecutive washings of classical rapid sand filters of the open type with free flow must normally be at least 24 h and under normal working conditions is liable to reach 40 to 60 h or 2 days on average.

13.7 Elimination of Wash Water

The water used in backwashing is eliminated through a series of "gutters." In the case of open filters, the gutters guarantee a free fall of at least 5 to 10 cm with regard to the hydrostatic level of introduction of water to the filter. In the case of closed pressure filters, the gutters are generally located within the water layer and consist of perforated pipes.

The wash water of sand filters contains materials that eventually require treatment in the proper sludge treatment plant. Their concentration in the water varies as a function of the washing cycle. In general, the maximum concentration at the beginning of washing is on the order of 1 g of dry matter per liter of water, and is often close to 0.5 g/L. Very quickly (i.e., after 2 to 3 min, the concentration falls to values close to 0.1 g/L. For a sand filter working normally, the duration of overflow of the wash water will not be longer than 15 min (maximum). If we take into account all the parameters involved—superficial load in filtration, velocity of the wash water, and length of the filtration cycle—we may assume that the water used for washing will not attain 5% of the total production.

For a new installation, the first washing cycles will give rise to the removal of fine sand as well as all the other materials usually undesirable in the filter mass, such as particles of bitumen on the inner surface of the water inlet or other residuals from the crushing or straining devices of filter media. Consequently, it is normal that at the beginning of operation of a filter sand installation, blackish and grayish deposits appear at the surface of the filter mass. In the long term they have no consequence and disappear after a few filtration and wash cycles. If, after several weeks of filtration (e.g., 8 to 10 weeks), these phenomena have not disappeared, it will be necessary to examine the filter sand closely. The elimination of fine sand must stop after 1 or 2 months of activity. If this sand continues to be carried away after the first 30 washings it is necessary to reexamine the hydraulic criteria of the washing conditions, the granulometry of the filter mass, and the filter's resistance to shear and abrasion.

14. CONCLUSIONS

Sand filtration is an economical fractionating technology of first importance in water treatment plants. This operation combines with active treatment phases comprising disinfection, sterilization, coagulation–flocculation, and so on. Sand filtration is generally divided into slow and rapid filtration; however, this distinction is somewhat conventional in terms of principles, but very real as to its modes of application. Slow filtration is a more accurate imitation of the natural processes of infiltration and can also constitute a preparatory step in the artificial recharging of groundwater. On the other hand, rapid sand filtration, usually without auxiliary biological effects, is a more artificial process that as used in treatment plants is generally cheaper.

The goals of rapid filtration are to obtain water containing concentrations of suspended matter lower than 50 μg/L, and certainly, below 100 μg/L, with turbidity below 0.3 to 0.5 ppm SiO_2 (German standard) or approximately 4 drops of mastic (French standard). For water previously coagulated–flocculated the standard value or quality recommendation for iron or aluminum in solution must be below 50 μg/L. In no case may a water that has undergone rapid filtration after the preliminary operations of coagulation–flocculation–settling or even water treated by flocculation–filtration contain more than 100 μg/L of dissolved iron or aluminum. If this value is exceeded, it is necessary to reconsider the filtration and to define more suitable modes of application.

SYMBOLS AND UNITS RELATED TO SAND FILTRATION

d	diameter of interstitial space, m or mm
d_h	hydraulic diameter, m or mm
d_0	diameter of filtering channels, m
d_s	specific diameter, m or mm
d_{10}	effective size diameter, m or mm
f	shape or sphericity factor, dimensionless
g	acceleration of gravity, 9.81 m/s^2
h	filter layer height or thickness, m
k	permeability coefficient, m/s
k	filter penetration constant, s/m^2
l	depth of penetration or of saturation, m
p	volumic porosity, dimensionless ratio
r	interparticle distance, m
v_f	infiltration velocity, m/s
A	filter surface area, m^2
BI	Baylis index: agglomerate volume to sand volume, dimensionless ratio
C_f	final or residual particle concentration, number of particles per unit volume
C_0	initial or inlet particle concentration, number of particles per unit volume
D	particle diameter or sieve diameter, m
DIV	diversity 60%, m or mm
ES	effective size, m or mm
H	height of water in the filter, m
P	permeability, m/s (usual range 10^{-4} m/s)
$P, P_1, \ldots$	weight, (mass) x g
P_c	probability of contact, dimensionless ratio
P_i	probability of removal by inertia, dimensionless ratio
P_i	percentage passing a mesh (i), dimensionless
P_s	probability of sedimentation, dimensionless ratio
Q	water flow, m^3/s or m^3/h
Q/A	surface loading, m/h
S	settling velocity, m/s
SL	surface loading, m/h
UC	uniformity coefficient, DIV/ES, dimensionless ratio
V_C	total equivalent volume of grains supposed to be spherical, m^3
V_L	open volume between filter grains, m^3
V_{mf}	minimal fluidization velocity, m/h (m^3/m$^2 \cdot$ h)
Δh	head loss, m
ϵ	volumic porosity; dimensionless ratio
μ	dynamic viscosity, kg m^{-1} s^{-1}
ν	kinematic viscosity; m^2/s
ρ	density, kg/m^3

REFERENCES

1. P. Frison, *Eau, 36,* 725 (1949).
2. Anon., *Water Sewage Works, 6,* 266 (1968).
3. W. Beyer, *Wasserwirtsch. Wassertechn., 14,* 165 (1964).
4. T. F. Craft, *J. AWWA, 52,* 428 (1966).
5. D. Maeckelburg, G.W.F., *119,* 23 (1978).
6. Ch. R. O'Melia and D. K. Crapps, *J. AWWA, 56,* 1326 (1964).
7. Ch. R. O'Melia and W. Stumm, *J. AWWA, 59,* 1393 (1967).
8. A. J. Drapeau and R. A. Laurence, *Eau Quebec, 10,* 314 (1977).
9. C. V. Smith and S. J. Medlar, *J. AWWA, 60,* 921 (1968).
10. N. P. Burman, *H₂O, 11,* 348, (1978).
11. I. Horvath, *J. AWWA, 67,* 452 (1975).
12. Anon., *J. AWWA, 51,* 1433 (1959).
13. J. L. Cleasby, J. Arboleda, D. E. Burns, P. W. Prendiville, and E. S. Savage, *J. AWWA, 69,* 115 (1977).
14. A. Amirtharajah and J. L. Cleasby, *J. AWWA, 64,* 52 (1972).

19
Reagents for Water Treatment

1. INTRODUCTION

Products added to drinking water or to water during its treatment purification steps to purify, disinfect, or stabilize it are considered to be reagents. Among the reagents some play a role of primary importance; for example, disinfectants are determinants for treatment, such as coagulants and products used for pH correction. Furthermore, some chemicals may have a significant auxiliary importance, such as the use as flocculation aids. Other derivatives or mixtures are injected directly or indirectly into the water for particular purposes (e.g., momentary disinfection of part of the mains). Still other products (e.g., adhesives, lubricants) may be used during the laying of mains.

The reagents used in water treatment are manufactured products seldom produced exclusively for the purpose of water treatment. Consequently, questions related to residual impurities are pertinent and often obviated by general statements. Of course, when beginning with a given water source, in no way may the use of reagents cause the maximal admitted concentrations in drinking water of a given toxic or undesirable substance or element to be exceeded. This principle related to the purity and harmlessness of reagents is the complete responsibility of the water authority, with no obligations of "product responsibility" on the part of the manufacturer. Therefore, these concerns are not in the arena of legal involvement.

2. ORIGIN OF UNDESIRABLE IMPURITIES

Impurities in reagents used for water treatment have several origins:

1. Impurities of extracted or transformed mineral material: for example, impurities of limestone (Sr, Ba) for use as lime, or impurities of brominated or organochlorine compounds in products originating from NaCl for use as chlorine.

565

2. Impurities resulting from manufacturing procedures: for example, the presence of mercury in chlorine or sodium hydroxide, the presence of chromium in quicklime, and so on.

3. Impurities added to processed products for other uses (e.g., stabilizing additives, antioxidants or anticorrosive products; persulfates added to oxidants; etc.).

4. Impurities resulting from the handling of the chemicals due to the lack of cleaning of storage or transportation tanks, use of vessels for a variety of products, or possible confusion occurring in conditioning or transportation.

5. Impurities added to facilitate the handling of chemicals, such as flowing and antibulking additives.

3. USE OF CLASSICAL REAGENTS

Some products have long been in use as reagents for the treatment of drinking water. They can be widely classified into four groups, as listed in Tables 1 to 4. The result of our investigation among the major European water authorities (W. J. Masschelein, Eureau Report, 1977) was to lay down *maximum* doses for treatment that may be considered as a basis for evaluation of the sanitary risks of impurities.

First, one must emphasize the fact that normally, the maximal dosages given here are not applied. In the operation of most treatment plants the dosages used average about 30 to 50% of the maximum dosages listed in Tables 1 to 4. The dosages indicated determine an implicit safety factor of 2 to 3, which constitutes a

Table 1   Reagents for Disinfection and Oxidation

Reagent	Base unit considered	Maximum dose for treatment as a calculation base
Liquid chlorine	kg Cl_2	30 g/m^3
Sodium chlorite	kg $NaClO_2$	5 g/m^3
Sodium hypochlorite	kg Cl_2	30 g/m^3
Calcium hypochlorite	kg Cl_2	30 g/m^3
Magnesium hypochlorite	kg Cl_2	30 g/m^3
Ammonia	kg NH_3	0.5 g/m^3
Sulfureous anhydride	kg SO_2	20 g/m^3
Sodium sulfite	kg Na_2SO_3	7 g/m^3
Sodium bisulfite	kg $NaHSO_3$	4 g/m^3
Sodium metabisulfite	kg $Na_2S_2O_5$	7.5 g/m^3
Copper sulfate	kg $CuSO_4$	5 g/m^3
Potassium permanganate	kg $KMnO_4$	2 g/m^3
Silver (and silver salts)	kg Ag	50 g/m^3
Ozone	kg O_3	12 g/m^3
Ammonium chloride	kg NH_4Cl	1.5 g/m^3
Ammonium sulfate	kg $(NH_4)_2SO_4$	1.8 g/m^3
Oxygen	kg O_2	30 g/m^3
Hydrogen peroxide and other peroxides	kg H_2O_2	10 g/m^3
Pyrethrines	kg of the mixture	0.5 g/m^3

Table 2 Reagents Used for Coagulation–Flocculation

Reagent	Base unit considered	Maximum dose for treatment considered
Aluminum sulfate	kg $Al_2(SO_4)_3 \cdot 18H_2O$	150 g/m³
Ferrous sulfate	kg $FeSO_4 \cdot 7H_2O$	100 g/m³
Sodium aluminate	kg $Na_2Al_2O_4$	30 g/m³
Ferric chloride	kg $FeCl_3 \cdot 6H_2O$	100 g/m³
Ferric sulfate	kg $Fe_2(SO_4)_3 \cdot 9H_2O$	200 g/m³
PAC or WAC	kg $Al_n(OH)_mCl_{3n-m}$	100 g/m³
Flural	kg $AlFSO_4$	10 g/m³
Ferric chlorosulfate	kg $(FeClSO_4)$	70 g/m³

Table 3 Reagents Used for pH Adjustment or Improvement of Mineral Content

Reagent	Base unit considered	Maximum dose for treatment considered
Caustic soda	kg $NaOH$	100 g/m³
Sodium carbonate	kg Na_2CO_3	200 g/m³
Sodium bicarbonate	kg $NaHCO_3$	200 g/m³
Quicklime	kg CaO	200 g/m³
Hydrated lime	kg $Ca(OH)_2$	200 g/m³
Sulfuric acid	kg H_2SO_4	30 g/m³
Chlorhydric acid	kg HCl	25 g/m³
Carbon dioxide	kg CO_2	50 g/m³
Magnesium hydroxide	kg MgO	80 g/m³
Magnesia	kg $CaCO_3MgO$	300 g/m³
Calcium carbonate	kg $CaCO_3$	300 g/m³
Magnesium carbonate	kg $MgCO_3$	175 g/m³
Calcium chloride	kg $CaCl_2$	120 g/m³
Calcium sulphate	kg $CaSO_4$	140 g/m³
Sodium chloride	kg $NaCl$	150 g/m³

Table 4 Reagents as Additives

Reagent	Base unit	Maximum dose for treatment considered
Sodium silicate	SiO_2	10 g/m³
Sodium hexametaphosphate	$(NaPO_3)_6$	20 g/m³
Sodium or potassium salts of mono- or polyphosphoric acids	P_2O_5	5 g/m³
Sodium fluoride	NaF	2 g/m³
Sodium fluorosilicate	Na_2SiF_6	3 g/m³
Hexafluorosilicicacid	H_2SiF_6	2 g/m³

Source: W. J. Masschelein, Eureau Report (1977); W. J. Masschelein, Bull. Cedebeau, *400*, 107 (1977).

guarantee of the safe consequences of momentary overdosing or the need to use high doses under exceptional circumstances. The consequences of the use of these classical reagents can be of several different types.

1. Bacteriological implications appear to be negligible since the possibility of bacterial survival or development in the concentrated chemicals is limited.

2. A treatment process usually requires the use of at least four to six reagents, and always fewer than 10. For chemical applications, the European water authorities have taken parameters of the purest class (I,A_1) of Directive 75/440 relative to the quality of surface waters used to produce drinking water as a first basis to define quality criteria for the chemicals used classically in treatment processes. If no imperative level is given, the basis can be the maximum admissible concentration in drinking water according to Directive 80/778.

3. The quality of a reagent must be defined independent of the very precise scheme of treatment since the effects depend on local circumstances and operating conditions. Consequently, in some cases, use of the maximal applicable dose given in Tables 1 to 4 should be considered. For example, if for quicklime and chlorhydric acid, the maximum treatment rates are 200 and 25 g/m^3, respectively, the maximum permissible content of an undesired element would be eight times lower per kilogram of quicklime than per kilogram of chlorhydric acid.

4. The quality of a reagent must be evaluated independent of the potential elimination of an impurity by the treatment applied, since this effect depends on local circumstances and operating conditions.

5. The proposed quality criterion is that at maximal dosing, each reagent should not introduce to the water more than 10% of the suggested level (or of the maximal admissable concentration if no level is recommended) of each toxic or objectionable impurity.

The practical results of this approach are given in Tables 5 and 6.

Table 5   Toxic Inorganic Factors

Factor	Directive ECC 75/440 I,A1 (g/m^3)	Directive EEC, MAC drinking water (g/m^3)	Maximum permitted level for each reagent (mg/m^3)
Ag	—	0.01	1
As	0.05	0.05	5
Ba	0.1	0.1	10
Cd	0.005	0.005	0.5
CN^-	0.05	0.05	5
Cr	0.05	0.05	5
Hg	0.001	0.001	0.1
Ni	—	0.05	5
NO_2^-	—	0.1	10
Pb	0.05	0.05	5
Sb	—	0.01	1
Se	0.01	0.01	1

Table 6   Undesirable Factors

Factor	MAC (g/m^3)	Maximum level allowed without elimination by treatment, dilution, or mixing of the water (mg/m^3)
Al	0.2	20
Cu	0.1	10
F$^-$	1.5	150
Fe	0.3	30
Mn	0.05	5
NO$_3^-$	50	5.000
PO$_4^-$	—	—
S^{2-}	<0.001	—
Zn	0.1	10
Phenols	0.0005	0.05

In the United States, the *Water Chemicals Codex* (National Academy Press, Washington, D.C., 1982), rather than proposing a maximum dose (MD), emphasizes a safety factor of 10 for the recommended maximum impurity content (RMIC). The latter is defined as a function of the suggested no-adverse-response level (SNARL) or maximum contaminant level (MCL). If a MCL value has been established for drinking water, it should be taken into consideration; if not available, the SNARL value can be taken as a substitute.

$$\text{RMIC} = \frac{\text{MCL (drinking water) (or SNARL)}}{\text{MD} \times 10}$$

No reference dose of reagent is indicated, which means that the RMIC can change as a function of the dose rate applied in the treatment. This makes a unified and harmonized standardization difficult to achieve.

A further fundamental difference is that in the United States, the safety factor is defined by taking the maximum contaminant level in *drinking water* as a basis, while the European water authorities have preferred to refer to the maximum or recommended guide for each relevant contaminant in the raw water to be treated. Moreover, the reference dose value is the maximum expected possible dose in Europe. This facilitates a *uniform certification of quality* applicable independent of the dosing rate in a given particular plant. At present, a safety factor of 10 is also adopted in the Nederlands (guideline: quality of materials and chemicals for drinking water supply, Part B, first version, June 1988, edited by Chief Inspectorate of Public Health, published by KIWA, The Netherlands), where a factor of 100 was applicable before. When available, the value for the admissible concentration of impurities [W. J. Masschelein, *Bull. Cedebeau, 400,* 107 (1977)] and RMIC according to the *Water Chemicals Codex* are indicated in the data files for individual reagents (see Section 13).

4. ADMISSIBILITY OF NEW REAGENTS

If reagents with industrial or economical advantages were to be found out of line as to one or more parameters, it would be wise to:

1. Critically consider the possible advantage of their use in existing treatment schemes
2. Encourage the manufacturing of purer products
3. Increase research to improve products with similar advantages and fewer drawbacks, and make them available.

To enable the advisory committees to take new products into consideration, a minimum of the following information should be made available:

1. The possible use of the product and its highest recommended dose.
2. The chemical constitution of the manufactured products and a full description of the analytical methods to be employed for examination of the bulk reagent and of residual traces in water after treatment, inclusive of the impurities.
3. Complete data on the toxicity of the product(s), its additives and impurities, with reference to testing by official laboratories.
4. The stability of the manufactured reagent, packing and transport of the product, effects of humidity, and effects of bacterial degradation.
5. Data concerning the products that may be formed by hydrolysis or oxidation by reagents normally used in water treatment.
6. In general, a preliminary investigation will be required to determine the rate of elimination of the product by the treatment processes applied and the possibility of maintaining or promoting bacterial growth in the treated water.
7. Manufacturers must certify the constant quality of their production within limits to be indicated by them.

For toxicological examinations of new products, the recommendations for products of "polyelectrolyte" type for potential use as coagulants or flocculation aids should be followed. The description of the procedure is published by the World Health Organization (Report No. 5 of July 1969): *Sanitary Aspects of the Use of Polyelectrolytes for the Treatment of Water Assigned for Human Consumption.* The report recommends several types of research.

1. *Research into acute toxicity* (summarized and simplified)
   a. Determination of oral dose–response curves and LD_{50}, with qualitative description of symptoms
   b. Description of cutaneous toxicity, irritation, and ocular toxicity
   c. Synergy tests; ability to complex heavy metals
2. *Metabolic research:* determination of the rate of absorption and accumulation of the product or its residues in vital organs. The data must report on several doses administered to more than one test animal.
3. *Oral-chronic toxicity* (in case of positive responses on earlier tests): administration to at least two animal species of both sexes is recommended at dose levels near the no-effect level (NEL) of acute toxicity but also at doses ranging from 10 to 100 times those resulting from recommended concentration limits.
4. *Complementary examinations:* carcinogenicity, mutagenicity, and teratogenicity of the newly proposed reagents.

A list of potential accepted new products, particularly flocculant aids of the polyelectrolyte type, is given in Section 14. At present only a few of them are being used in the treatment of drinking water.

5. PACKING AND IDENTIFICATION OF REAGENTS

There is a lack of packaging identification of reagents for drinking water. The following indications, as a minimum, should be written on the packages, containers, or identification documents.

Name, type of active compound, and registered trademark of the product
Label "Grade: quality for drinking water"
Date of manufacture (serial number, etc.) and peremptory date
General recommendations for use and applications
Reservations concerning use of the product
Codes of admittance existing; reference to standards
Full name and address of the firm responsible for the identification

6. EXISTING STANDARDS

Several standards or recommendations of similar value have been formulated for reagents currently used in the treatment of drinking water (Table 7). After more than 10 years' of neglect, the question is again under study. This is due primarily to instigation by the European Economic Community (European Committee of Normalization: CEN).

7. REAGENT DOSING MODES

Reagents may be in solid, liquid, or gaseous in form. Their dosing mode can be very different as a function of their physical state. Gas dosing and dispersion are discussed in Chapter 3. Generally, solutions or suspensions are required or preferred for feeding and dosing into water to be treated. Practice indicates that liquids, solutions, or at least suspensions react more smoothly and predictably in water; cause fewer mixing problems, thus ensuring a more homogeneous contact; and can eventually be conservative as to the consumption of chemicals. Furthermore, "undissolved particles" can exert erosive effects on pumps, valves, and piping systems.

The following circumstances are required, or at least preferred, for the use of solutions or dilute suspensions:

1. If liquid forms of the reagent are commercially available under economically competitive conditions, the design should be based on use of these products. This appears to be the case more and more; for example, in addition to solutions of aluminum sulfate and related coagulants, ferric chloride and "active chlorine" solutions, coagulation acids, and lime are available on the market in the form of solutions or suspensions.
2. If the quantity of reagent to be dosed is very small and its cost is comparatively high, as is the case for potassium permanganate, polyphosphate. This also holds in cases where high performance is required for dosing rates, as, for example, for coagulation aids.
3. If the treatment is of the batch type or intermittent in time, with high performance standards in rapid start–stop procedures.

Table 7 Existing Norms, Standards, or Codes Applicable to Reagents (1989)

Reagent	ECC Code[a]	DIN[b]	APHA AWWA[c]	USA Codex[d]	Cebedeau[e]	NL[f]
Activated carbon		19603	600-66	7440-44-0		
Aluminum (based) coagulants					Yes	code 2.5
Aluminum sulfate	510-527	19600	403-52	10043-01-3		
Ammonia				7664-41-7 1134-21-6		
Ammonium sulfate/chloride	510-527	19602	302-64	7783-20-2	Yes	
Calcium chloride						
Calcium hypochlorite				7778-54-3		
Calcium oxide (hydroxide)						
Carbon dioxide				124-38-9		
Chlorhydric acid		19610			Yes	7647.01-0
Chlorine		19607	301-57	7782-50-5	Yes	
Copper sulfate		19609	602-57			
Ferric chloride		19602		7705-08-0	Yes	7705.08.0
Ferric sulfate			406-61	10028-22-5		
Ferrous chloride						
Ferrous sulfate		19609	402-68	7720-78.7 16961-83-4		7720-78.7
Fluosilic acid (+ salts)		19611				
Lime		19613/14	202-52	1305-62-0 1305-78-8		
Magnesium chloride						
Polyphosphates	E 450 543-544					
Potassium permanganate			603-68	7722-64-7		
Pyrophosphates				65997-17-3		

	Code[a]	DIN[b]	AWWA[c]	CAS[d]	[e]
Silicofluorhydric acid (+ salts)			702-69 / 703-60	16893-85-9	
Sodium aluminate	500		405-60	1302-42-7	
Sodium bicarbonate				144-55-8	497-18-8
Sodium carbonate		19612	201-51	497-19-8	
Sodium chloride		19604	200-69		
Sodium chlorite		19617	303-67	7758-19-2	
Sodium fluoride			701-60	7681-49-4	1310-72-2
Sodium hexametaphosphate			502-67		
Sodium hydroxide	524	19615/16	501-51	1310-72-2	
Sodium hypochlorite		19608	300-53		
Sodium metabisulphite	E 221			(7631-90-5)	
Sodium phosphate	E 339	19620	500-66	7758-29-4	
Sodium polyphosphate				10124-56-8	
Sodium silicate	550		404-55	1344-09-8	
Sulfur dioxide	E 220			7446-09-5	
Sulfuric acid		19618		7664-93-9	
Zinc polyphosphate					

[a]The ECC code concerns a positive list of authorized additives in foodstuffs. Inclusion in the list supposes the existence of a toxicological evaluation.

[b]DIN norms: Deutsche Industrie Norms.

[c]American Public Health Association–America Waterworks Association; standards are published in the *Journal of the American Waterworks Association* and are also available as separate leaflets.

[d]*USA Water Chemicals Codex*, National Academy Press, Washington, D.C., 1982.

[e]Cebedeau: Centre Belge d'Etude des Eaux, Liège.

[f]*Guideline Quality of Materials and Chemicals for Drinking Water Supply 86-01* (first revision), Kiwa, The Netherlands, 1988.

4. If the chemical is sticky or gummy (e.g., ferrous sulfate, solid ferric chloride, calcium hypochlorite, polyelectrolyte, or lime).

5. If microporous materials used require preliminary hydration (e.g., powered activated carbon, diatomaceous earths, or ion-exchange resins).

8. PARTICLE GRADES OF SOLIDS

Lumps: aggregated particles of size varying between 4 and 100 mm.
Granules: group of particles forming a stable entity for handling and dosing. The group includes crystals, pellets, preforms, and grains constituting approximately homogeneous materials varying in size between 0.15 and 0.5 mm.
Powders: particle size of the material is finer than 0.075 mm.

Particle-size grades are often expressed in different units. Without entering exhaustively into many different definitions, the conversions listed in Table 8 can be handled for practical purposes in water treatment. This conversion table is approximate because different definitions of mesh size are cited in the literature.

9. APPARENT OR VOLUMETRIC DENSITIES OF POWDERS

Under dry conditions or with normal hydration, the apparent densities of powders are approximately as listed in Table 9. The data can be utilized when designing dosing systems. When the possible density range is high, gravimetric feeders are preferred to volumetric feeders if direct dosing is applied. As an alternative, batch makeup of solutions or suspensions to be dosed volumetrically as a liquid is often preferable in such cases. (The table is limited to products that can be considered for direct dosing as solids into the water to be treated.)

10. EFFECTS OF HYGROSCOPICITY

A particularly troublesome property of chemicals used as powders in water treatment is the effect of hygroscopicity. This property consists in general of picking up water from the ambient air with or without structural changes in the solid chemical and can be subdivided into several ranges, depending on the principal effect. Definitions are as follows:

Table 8   Particle Sizes

U.S. Standard Sieve	Millimeters	Micrometers	Inches
6	3.36	3360	0.132
10	2.00	2000	0.0787
20	0.84	840	0.0331
40	0.42	420	0.0165
60	0.25	250	0.0098
100	0.149	149	0.0059
200	0.074	74	0.0029
325	0.044	44	0.0017
400	0.037	37	0.0015

Table 9 Apparent or Volumetric Densities of Powders in Loose State
(g/L powder volume)

Activated carbon	250–400
Aluminum sulfate	
Fine powders	610–750
Coarse powders or crystals	950–1100
Ammonium sulfate	650–770
Calcium hypochlorite (70%)	850
Copper sulfate	
Powder	950–1050
Crystal	1200–1440
Diatomaceous earths	140–280
Ferric chloride	900–1000
Ferric sulfate (crystals)	1300
Ferrous sulfate	700–1100
Lime	
Quicklime	880–960
Hydrated lime	400–560
Dolomitic hydrated lime	480–650
Sodium aluminate (anhydrous)	800–960
Sodium bicarbonate	800–1200
Sodium carbonate	370–650
Sodium chlorate	1000–1300
Sodium chloride	900–1200
Sodium fluoride (powdered and crystal product)	800–1400
Sodium fluosilicate	1150–1370
Sodium hexametaphosphate	750–800

Deliquescency: picking up atmospheric moisture such that the product changes to the liquid state. The best dosing form of such products (e.g., ferric chloride and sodium aluminate) is a liquid suspension or solution obtained by prior dissolution of the solid reagent.

Caking-up: buildup of a hard solid mass as a result of local hydration (e.g., aluminum sulfate, ferrous and ferric sulfate). Temperature and relative humidity changes are best avoided in the dosing room.

Sorption: fixation of atmospheric humidity at the surface of grains with modification of flow conditions. A typical example is lime. One must avoid placing the dosing equipment above the dissolution water. Periodic control should be carried out to avoid capillary uptake of water into the bulk of the product.

Efforescency: gradual change in crystallinity accompanied by the development of a powder at the surface of the crystals. The effect occurs with dehydrated products (e.g., lime) and with alkaline reagents (e.g., sodium hydroxide or carbonate). With caustic reagents carbonation occurs simultaneously with efflorescency. To avoid this effect, the product should be stored in hermetically closed containers when not in use. The air let into the storage area during dosing can be kept dry by using drying towers (e.g., $CaCl_2$ or molecular sieves).

Some examples of the effects of hygroscopicity are listed in Table 10.

Table 10   Typical Examples of the Effects of Hygroscopicity

Reagent	Effect of hygroscopicity	Bulk density (kg/m³)			Recommended concentration in makeup solution (less than, kg/m³)
		Loose	Packed	Working	
Aluminum sulfate, $Al_2(SO_4)_3 \cdot (14-18)H_2O$	Caking	800 ↓ 1000	1050 ↓ 1200	800 ↓ 1100	125
Ferric chloride, $FeCl_3 \cdot (2-6)H_2O$	Deliquescency	900	1050	940	500
Ferric sulfate, $Fe(SO_4)_3 \cdot 3H_2O$	Caking	950 1350	1150 1170	1050 1450	200
Ferrous sulfate, $FeSO_4 \cdot 7H_2O$	Caking	700 1000	750 1150	700 1000	110
Ferric chlorosulfate	Deliquescency	≈	≈	1500	400
Quicklime, CaO	Sorption	700 1200	1000 1200		See "Lime" (in Section 13)
Hydrated lime	Sorption	300 700	500 800	400 700	80
Sodium aluminate Anhydrous Hydrated	Deliquescency Caking	900 1000	1050 1250	900 1000	65

11. TYPES OF FEEDERS FOR SOLID REAGENTS

The screw conveyor feeder (Fig. 1) is an endless screw or helix feeder that can deliver a constant stream of solid reagents from a hopper storage tank provided that the flow by gravity to the screw is uninterrupted. To prevent secondary problems due to hygroscopicity, the screw is best mounted upward. Accuracy claimed is 2%.

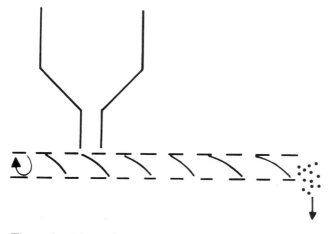

Figure 1   Schematic of screw feeder system. The operational dosing range for screw feeders is on the order of 1 to 20.

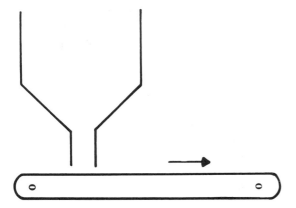

Figure 2   Schematic of belt-feeding system.

In a belt feeder (Fig. 2), the variable gate determines the volumetric dosing for a given belt speed. The dosing range can usually be varied between 1 and 10 for the particular equipment. Oscillating hopper feeders and rotating disk feeders (Fig. 3) can be used as well. Their accuracy in dosing is often somewhat less ($\pm 5\%$ error is acceptable). The dynamic dosing range depends on the chemical (e.g., 1 to 12 for powered activated carbon and 1 to 100 for aluminum sulfate).

Gravimetric feeders are generally of the belt type, sometimes pneumatic for less

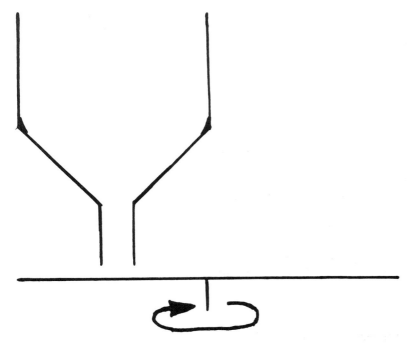

Figure 3   Schematic of oscillating hopper and rotating disk feeders.

hygroscopic materials. The dosing is controlled by weight determination or by a variable counterweight system. The dynamic range is from 1 to 20 up to 1 to 100, depending on the construction. Manufacturers claim an accuracy of 1% of the maximum dosing rate.

Remark. When transferred, powders are capable of heating by friction. Some hazards can result. More comments are given in Section 13 in the discussion on activated carbon.

Further Reading

R. L. Carr, The flow and handling of dry materials used in water and waste treatment, *Water Sewage Works, R95* (1965).

12. SAFETY CHECKLIST FOR RISKS ASSOCIATED WITH REAGENTS

Risks

Explosion
Combustion
Corrosivity
Toxicity
Irritating
Hindering
Highly flammable
Flammable

Identification

Commercial name
Manufacturer
Local dealer
Recommended use

Composition

Chemical formula
Major impurities
Trace impurities

Characteristics

Physical state: at 20°C ⎫
         at 0°C   ⎬ solid, paste, liquid, gas
         at . . . °C ⎭
Aspect: color, odor
Significant temperatures: boiling, fusion, decomposition, explosion, etc.
pH: at delivery, evolution with time
Solubility in water: as a function of temperature
Mixibility in water
Vapor tension: as a function of temperature
Density: liquid, vapor, solid as a function of temperature

Storage and Handling

General recommendations
Cleaning specifications
Incompatible decomposition products
Specific danger of contact with other chemicals

Individual protection equipment required
Special protection measures to be used (cloths, cleaning, venting, rinsing, etc.)
Methods to control leakages (neutralization, venting, destruction, sluicing, etc.)
Risks of fire, incompatibility with other chemicals
Risks of contact with electrical circuits, on heating.

Toxicological Information
Acute toxicity
Chronic toxicity
Specific effects: skin, eyes, and so on
First measures
Necessary medical assistance

Inflammation and Explosion Danger
Flash point
Auto-flash point
Friction risk
Catalytic effects (impurities)
Chock effects
Risks associated with open flames
Firefighting means (pro–contra)
Increasing-risks factors (metals, heat, etc.)

First-Aid Measures
Major risks
Eye risks
Skin contact
Ingestion risks
Chronic risks
Need for medical assistance

Particular Specifications
Norms of discharge
Neutralization products

13.   TECHNICAL DATA ON REAGENTS USED IN WATER TREATMENT

For each major reagent for which data are available, the following order of presentation is followed.

1.   Name
2.   Basic formula; molecular mass (MW)
3.   Use
4.   Commercial forms
5.   Handling and dosing characteristics
6.   Safety
7.   Corrosivity and materials
8.   Purity characteristics and requirements
9.   Other comments
10.   Further literature

For minor reagents only comments are given.

1. Acid (Passivated).   See "Chlorhydric Acid," item 9. Passivated chlorhydric acid is used as a scale-removing agent.

1.   Activated Carbon.   See Chapter 12 for more general information on the use of this product.

6. No significant acute or toxic effects are known due to activated carbon impurities, but it can cause skin irritation. The product is incompatible with concentrated strong oxidants such as chlorites, chlorates, perchlorates, and nitrates.

8. Reference impurities at 20 mg/L dosing, ppm wt, are as follows:

As   25
Cr   25
Pb   25
Hg   0.5
Ag   25

9. When saturated with toxic substances removed from water, activated carbon in its granular form may cause environmental problems in transportation and regeneration that need to be assessed for each particular case. No major problem is known with the use of thermally activated carbons. The use of chemically activated carbon is subject to reservations. Regenerated carbons must be analyzed for heavy metals. On friction, powdered materials in general are capable of self-ignition, particularly when, like powdered activated carbon, they are combustible. Preheating promotes such reactions. Combustion may result from friction of the material during transfer and, also, from fermentation or bacterial growth in general. Naturally, the presence of oxygen is necessary for combustion. In this respect it is worth noting that activated carbon concentrates oxygen upon contact with air.

Respiration (i.e., fermentation of solid materials in the cases to be considered) liberates considerable energy—about 25 to 30 kJ/mol CO_2 formed. When the CO_2 accumulates in a closed vessel to more than 10%, respiration is blocked and the energy produced is less than 1 kJ/mol carbon. The velocity of heat evolved can increase considerably in the presence of humid air. The question has been studied exhaustively in powdered food processing.

The specifications for powdered activated carbon should include data on self-ignition risks and critical velocities of the powder and dust transfer. The risk is higher with chemically activated carbon than with thermally activated carbon. In some cases, inert gases such as nitrogen or carbon dioxide are used for pneumatic transfer instead of air. Manufacturers must give full information on these considerations in their representations.

10.   A. Yehaskel, Activated Carbon, Noyes Data Corporation, Park Ridge, N.J., 1978. G. Giltaire and J. Dangreaux, Les poussières explosives, *Ann. mines* (Jan. 1979); and *Druckentlasterng von Staubexplosionen*, Standard VDI-3673 (June 1979).

1.   Aluminum Chloride (Hydrated Polymer or Polyaluminum Chloride).   These reagents are usually known by their registered trade names, such as in Europe, PAC, WAC, Prodefloc, and Saptoclar.

2. $Al_n(OH)_mCl_{3n-m}$ or $Al_n(OH)_m(SO_4)_kCl_{3n-m-2k}$.

3. Used as a coagulant, also to remove phosphate in wastewater treatment.

4. The reagent exists as a solution in two fundamentally different forms (con-

centration being expressed as Al_2O_3: (1) 18 ± 0.5 wt%: density 1.32 kg/L, $\mu = 22 \times 10^{-3}$ kg/m · s (at 20°C), freezing point $-20°C$, pH 1.5; (2) 10.3 wt%: density 1.2, $\mu = 4.2 \times 10^{-3}$ kg/m · s at 20°C, freezing point $-12°C$, pH 2.7,

$$\left(\frac{n}{3m} \times 100\right): 40 - 60$$

The solutions are stable for at least 1 year. The product of the solutions are gradually lowered with storage. Solutions are delivered in carboys (30 to 60 L), cubitainers of 250 tons, and in bulk. On special order, some manufacturers can deliver the product as a powder.

5. All dosing methods, adequate for weak acids, are applicable. In concentrated solutions the product is stable for about 1 year. Gradual lowering of pH occurs on storage.

6. Risks of weak acids; rinsing spoiled products with water.

7. Corrosive to ferrous metals and alloys. Storing and handling materials are PVC, PE, PTFE, and chlorinated rubber. Storage tanks in steel must be epoxy-lined or coated with PVC.

8. The chemical, obtained by a high-technology process involving successive reactions and precipitations, is generally of a high degree of purity (Table 11). Reference basis: maximum dosing of 15 mg/L Al.

Table 11   Impurities of Aluminum Chloride

Impurity	Maximum permissible (mg/m^3)	RMIC (at 25 ppm Al_2O_3 dosing)	Measured content (mg/kg Al_2O_3)
As	5	200	0.2–3.7
Cd	0.5	20	0.3–5
Cr	5	200	13–50
Hg	0.1	4	0.3–11
Ni	5	200	30–40
Pb	5	200	10–20
Se	1	40	1.5–10
Cu	10	400	3–10
Mn	5	200	3.3–10
Zn	10	400	17–40
Sb	1	40	1.5

9. No major problem is to be expected with these high-technology-manufactured reagents.

10. R. L. Kerckaert, *Trib. Cebedeau, 475–476*, 345 (1983). G. A. Lekimme, *Trib. Cebedeau, 475–476*, 353 (1983).

1. **Aluminum Sulfate.**

2. $Al_2(SO_4)_3$ · $18H_2O$; MW 666.

3. Coagulant and precipitant of heavy metals and phosphates.

4. The basic formula is $Al_2(SO_4)_3$ · $14H_2O$, but the more hydrated form is more

often available. The commercial concentration is expressed as Al_2O_3 wt% content. For the solid product the standard of more than 17 wt% in Al_2O_3 holds. The concentrated solution as marketed is 8.3 wt% Al_2O_3 with a density of 1.305 to 1.35 kg/L.

5. The chemical in aqueous solutions is a weak acid. The pH of the saturated solution ranges from 2.5 to 3.0. The 8.3 wt% Al_2O_3 freezes at −15 to −16°C. Dilute solutions freeze at temperatures above these values. Solubility increases with temperature. The viscosity of the 8.3% Al_2O_3 solution ranges from 18 mPa at 25°C to 55 mPa at 0°C. In the case of dry feeding of the powder, aspiration of the dust is necessary.

6. Provide rinsing facilities to discharge spoiled products. These areas are slithery; covering with calcium sulfate, gypsum, or plaster aids control of the sluicing of spoiled reagent.

7. Aluminum sulfate is corrosive to ferrous metals and must be handled in contact with plastic materials (e.g., PVC, PTFE, or lined materials such as epoxy or PVC lining. Simultaneous contact of acids and aluminum sulfate on the skin must be invoided to prevent oxidation.

8. Purity characteristics are listed in Table 12. Mercury and chromium are the impurities requiring attention.

9. The usual industrial process of manufacturing aluminum-based coagulants is to dissolve bauxite with a caustic soda solution. An alkaline solution of aluminate results, containing about 50 wt% in Al_2O_3. This solution is acidified to generate acid–aluminum reagents such as sulfate (aluminum) or chlorides and polychlorides either as solids or as solutions of various strengths. Major impurities in standard available bauxites are listed in Table 13.

Local production of an acceptable coagulant is sometimes possible starting with kaolin, if available [M. Y. Bakr and H. Mitwally, *Indian Ceramics, 17*, 166 (1973)]. Recently, direct generation of aluminum sulfate from bauxites without passing through an alkaline dissolution step has been introduced on an industrial scale. The chemical is currently available on the European market (PAX-products).

Table 12   Impurities of Aluminum Sulfate

Impurity	Maximum permissible (mg/m^3)	Measured content	RMIC (at 150 mg/L)
As	5	2.2	30
Cd	0.5	0.08	3
Cr	5	30	30
Hg	0.1	2	1
Ni	5	5	
Pb	5	1.5	30
Se	1	0.002	7
Cu	10	1	
Mn	5	12	
Zn	10	0.5	
Ti	?	1	
Ag	1		30

Table 13 Impurities of Bauxite

Secondary component	wt% in bauxite	In alkaline liquor (wt%)	Versus Al_2O_3 (wt%)
SiO_2	2–8	≈0.2	0.01–0.04
Fe_2O_3	15–25	≈0.02	0.01–0.04
TiO_2	2.5–3.5	≈0.005	0.002–0.01
P_2O_5	0.1–0.4	≈1	0.002–0.008
V_2O_5	0.05–0.2	≈0.6	0.001–0.005
Cr_2O_3	0.05–0.2	≈0.7	0.001–0.003
ZnO	0.01–0.05	≈0.15	0.003–0.02

The commercial product is known as **PAX** (registered trade name), marketed as liquid suspensions.

	pH	ρ (kg/L)	μ (kg/ms)	% Al_2O_3	%SiO_2	Freezing
PAX-A	1.4	1.25	0.02	12.2	–	−20°C
PAX-W	1.4	1.30	0.02	11.7	2–4	−20°C
PAX-S	1.5	1.40	0.12	9.2	10–15	−20°C

Interesting is the association of the reagent with active silica and its potential combination with ferric coagulants in various proportions [e.g., up to form $FeAl(Cl,SO_4)_2$]. Maximal levels of known impurities of the yet available reagent are in mg/kg Al_2O_3:

Cd	0.05	Pb	1.1
Cr	32	Cu	1.1
Hg	0.03	Zn	0.6
Ni	1.1		

Further developments are to be expected in this field.

Remark: Aluminum-potassium sulfate is a less commonly used product with coagulation properties similar to those of aluminum sulfate.

1. Amines. Amines can be used in water treatment in the form of chloramines (see "Chloramines"). Amines, inclusively hydrazine (NH_2–NH_2), are used for boiler water treatment as reductor agents to inhibit corrosion. For this purpose they are incorporated in formulations. The water treated with such products is no longer suitable for consumption. Amines can be used as alcaline agents to neutralize aggressive CO_2. The compounds most used are cyclohexylamine, ethanolamine, and morpholine. They are incorporated in boiler water and, while vapor is condensing, they react with the CO_2 to form aminobicarbonates. Dosage is 1 g/m^3 water.

Another category of amines are the linear "fatty amines" containing 4 to 18 carbons, which have film-forming properties on the metal walls of boiler equipment. The dosage is adjusted for the particular case but is often in the range 2 to 20 g/m^3. Application is difficult since the coating is often not entirely homogeneous and corrosion occurs in the less coated areas.

1. Ammonia (Gas).
    2. NH_3; MW 17.

3. Formation of chloramines (bromamines); more stable than free chlorine.

4. Compressed liquefied gas at more than 99.5% NH_3 in cylinders. The equilibrating gas pressures are 6 bar (600 kPa) at 10°C, 12 bar (1200 kPa) at 30°C, and 20 bar (2000 kPa) at 50°C. Liquid density is 680 kg/m^3 at -33°C. The gas is dosed with ejector equipment similar to the chlorinator systems. The dissolution into the water is exothermic by 38 kJ/mol. Solutions must be kept at lower than 10% for technical application. Ammonia solutions are alkaline and can promote scaling by calcium carbonate.

6. The safety of the pressure vessels is set at 2000 kPa and the vessels are tested at 3000 kPa. Ammonia has a pungent odor and attacks mucous membranes and the respiratory system. Larger bulk storage tanks must be placed in a collecting zone for accidental spillage. The storage tanks are to be located at a distance of not less than 7.5 m from open parts of buildings. Electric equipment must be protected since critical ammonia–air mixtures can become explosive on contact with ignition sources.

7. Dry ammonia is not corrosive to metals; moist ammonia slowly attacks copper and copper alloys and zinc galvanized steel. Iron-based steel is not corroded.

8. At a maximum theoretical concentration of 5 mg/L ammonia is not known to contribute contaminants that could affect drinking water. Maximum dosing is limited to 0.5 mg/L. When nitrifying in the distribution system, ammonia can build up nitrite, which needs to be controlled.

9. Ammonia can be important in sludge digestion control.

1. Ammonium Chloride.

2. NH_4Cl (commercial purity 99.7%).

3. Hydrolyzes in water to form ammonia, which forms chloramines with dissolved chlorine:

$$NH_4Cl + H_2O = HCl + NH_4OH$$

$$NH_4OH + HOCl = NH_2Cl + 2H_2O$$

Hence normal dosing is less than 0.5 mg/L as NH_4Cl.

4. Powder or crystals in bags of 50 kg or in bulk. The product must be protected against hydration. Density 1.35; loose volumetic mass 1.0. Thermally stable product up to 335°C (sublimation 337.5°C at 1 atm and thermal decomposition at 350°C).

5. The product is usually dissolved to dose the solutions. Solubility 372 g/L at 20°C. Dosing solutions are made up at 5 to 10 wt%. The pH of a 5% solution ranges from 4.2 to 5 and that of a 10% solution from 3.5 to 4.

6. The chemical must not be brought into contact with chlorites, sulfites, or alkali because of the risk of formation of gaseous compounds: chlorine, sulfur dioxide, and ammonia, respectively. Spoiled product can be rinsed with water, with eventual application of sodium carbonate. Storage and dosing rooms must be vented. In normal conditions of use the product does not pose any specific danger.

7. Contact with electrical connection system relays, programmed systems, and so on, must be avoided.

8. Purity characteristics are listed in Table 14. The data imply that no limiting impurities for normal use of the reagent are known.

10. G. A. Lekimme, *Trib. Cebedeau, 475–476*, 297 (1983).

Table 14 Impurities of Ammonium Chloride

Impurity	Permissible (mg/m^3)	Content in the product [ppm (wt%)]
Cd	0.5	0.1–0.005
Cr	5	5
Hg	0.1	1
Ni	5	0.2–0.1
Pb	5	0.05–0.002
Al	10	0.2–0.1
Cu	10	0.05–0.02
Fe	20	1–0.5
Mn	5	0.01–0.005
Zn	10	0.8–0.4
K	120	20–10

1. **Ammonium Hydroxide.**

2. NH_4OH; MW 35.

3. Formation of chloro- and bromamines. Auxiliary use as pH stabilizer.

4. Colorless solution in water of 29.4 NH_3 wt% of density 0.77 kg/L.

5. Alkaline product.

6. Chemical with pungent odor, releases ammonia gas on standing.

9. Aqueous ammonia and ammonium hydroxide are less used than gaseous liquefied ammonia, which can be dissolved on site.

1. **Ammonium SilicoFluoride.**

2. $(NH_4)_2SiF_6$; MW 178.

3. The product can be used for the auxiliary treatment of fluoridation. The product is less used than other fluorine salts. (See "Sodium Silicohexafluoride" for further comments on this class of products.)

1. **Ammonium Sulfate.**

2. $(NH_4)_2 \cdot SO_4$; MW 132 (commercial purity > 99%).

3. Preparation of chloramines.

4. Crystalline powder, density 1.77 kg/L.

5. To be dissolved in water and dosed as a concentrated solution in the water to be treated. Concentration range of the working solution: up to 20 wt% (solubility 700 g/L).

6. Spoiled product can be rinsed with water. In normal conditions of use, the product does not pose any specific danger.

7. Avoid contact with electrical systems since nonferrous metals can be heavily corroded by the chemical.

8. Purity characteristics are listed in Table 15. Pyridine can be another significant impurity. Concentrations should be less than 50 mg/kg dry reagent.

9. No problems are to be expected with synthetic ammonium sulfate. However, more and more of the technical products available on the market are secondary products recovered from phytopharmaceutical industries. Therefore, attention must be paid to "ether-extractible residues," including detergents, pyridine, and antibulking compounds.

Table 15 Impurities of Ammonium Sulfate

Impurity	Maximum permissible (mg/m^3)	Measured content	RMIC (at 150 mg/L)
As	5	4–8	200
Ba	?	20	
Cd	0.5	2	
Cr	5	10	
Ni	5	10	
Pb	5	0.1–3	200
Al		10	
Fe		0.6–5	
Mn	5	150	
Zn	10	2	
Se			200

1. Barium Carbonate.

2. $BaCO_3$; MW 197.3.

3. Barium carbonate can be associated in boiler water treatment processes.

4. The reagent is usually incorporated as a component of commercially formulated products.

1. Bentonite.

2. Aluminosilicate clay.

3. Bentonite is used to improve the coagulation of waters containing insufficient suspended particles, especially when color removal is the main problem to be solved. It can also improve the removal of heavy metals.

4. Bentonite is available in bags or in bulk.

5. The product is insoluble in water. Handling and dosing of the powder are similar to those of diatomaceous earths.

6. See "Filtration and Diatomaceous Earths (Chapter 14).

8. Bentonite as used in raw waters is not known to introduce objectionable impurities into the treated water.

1. Bromine.

2. Br_2 or HOBr; MW 160 and 97, respectively.

3. Bromine can be used as a disinfectant (minimum residual required is 1 mg/L), and secondarily, for algal control.

4. Bromine is available as a liquid but is less used or not used at all for drinking water treatment in view of its cost versus that of chlorine. In swimming pools a residual concentration of 1 to 2 mg/L is necessary, the pH being slightly alkaline: 7.5 to 8.2.

9. Bromine can be formed on the oxidation of bromide existing in some natural waters (see Chapter 2 on chlorine dioxide and Chapter 3 on ozone).

1. Calcium Fluoride.

2. CaF_2; MW 78.

3. Calcium fluoride can be used in fluoridation.

4. Calcium fluoride is available as a powder or as micronized crystals.

5. Bulk density of the powder ranges from 820 to 1300 kg/m^3 as loose product, and 1100 to 1875 kg/m^3 as packed product. The working density range to be considered is 900 to 1600; calcium fluoride is poorly flowable. Working with primarily madeup slurries is recommended for dosing. Solubility is poor: 16 mg/L at 20°C; hence working solutions are very difficult to make up. The product is dosed as a slurry.

6, 7. Calcium fluoride is a toxic dust. It is alkaline and abrasive. For more information, see "Sodium Fluoride."

8. No major impurity is known in synthetic calcium fluoride that can hinder its use at a maximum dosage of 1.5 mg/L of F$^-$. Additives sometimes used to make the product free-flowing may require attention.

1. Calcium Hydroxide (Hydrated Lime).

2. Ca(OH)$_2$; MW 74.

Comments on this product are included in the section dealing with lime.

1. Calcium Hypochlorite (Chlorinated Lime, Bleaching Powder).

2. Ca(OCl)$_2$; MW 143; technical chlorinated lime could correspond to the formula

$$CaO \cdot 2Ca(OCl)_2 \cdot 3H_2O \qquad (MW\ 293)$$

3. Disinfection, oxidation.

4. Bleaching powder is a solid containing 25 to 37% available chlorine. Pure calcium hypochlorite containing 65 to 70 wt% available chlorine is more stable.

5. Unstable material subject to decomposition by heat or moisture. For intermittent action, dosing is done by dropping the product as tablets, or for continuous treatment, as a solution [e.g., 2% in water for bleaching powder and up to 3% for calcium hypochlorite; solubility 220 g/L (active chlorine)]. Continuous-flow-through systems for dosing operate by dissolving 70 wt% tablets.

6. Storage in a cool, dry room for a limited time not exceeding 6 months (never for more than 1 year). Losses of 3 to 10% active chlorine can occur within 1 year. Can decompose violently when in contact with organic materials and needs to be handled using protective gloves, and when dissolving the reagent in water, eye-protecting accessories such as goggles or facial masks.

7. See "Chlorine" and "Sodium Hypochlorite."

8. The reference impurity to be expected is mercury RMIC (dosage basis 20 mg/L), of 10 ppm (wt).

9. Technical reagents may contain insoluble impurities which are to be subject to limitations (e.g., less than 0.15 wt%). No dirt or other foreign material should be present in the product.

10. *AWWA Standard B-300-80* (1980).

1. Calcium Oxide (Quicklime). This reagent is also used in the form of calcium hydroxide (hydrated lime). See "Lime."

1. Carbon (Activated). See Chapter 11.

1. Carbon Dioxide (Gas).

2. CO$_2$; MW 44.

3. Carbon dioxide is used for the adjustment of pH and mineralization of waters (recarbonation). The chemistry of carbon dioxide is the basis of carbonic acid equilibrium in water:

$$CO_2 + H_2O \overset{K}{\rightleftharpoons} H_2CO_3 \overset{K_1}{\rightleftharpoons} H^+ + HCO_3^- \overset{K_2}{\rightleftharpoons} 2H^+ + CO_3^{2-}$$

$$K = 4 \times 10^{-2}, \quad K_1 = 4.07 \times 10^{-7}, \quad K_2 = 4.17 \times 10^{-11} \quad \text{(at 20°C)}$$

4. Carbon dioxide is available in liquefied form resulting from bacterial metabolism or combustion of gas. Burners using propane or natural gas can produce CO_2 locally as well as combustion of fuels and calcination of limestones.

5. Colorless, odorless, inflammable gas of specific weight 1.98 g/L at 0°C and 1 atm. Relative density versus air is 1.529. Temperature of liquefication at 1 atm is −78°C. Solubility in water (acid solution) is about 1 volume per volume. Dosing is by blowers or by expanding liquid carbon dioxide to gas. Both techniques suppose gas-to-liquid contacting systems.

6. No particular safety measures are necessary compared to those required for inert gases. Instructions for handling compressed gases are available in the manufacturers' instructions. Danger of asphysia can exist in closed rooms or areas of use.

7. Acid compound on hydration. Corrosion by weak acids in solution or in the gas phase containing water vapor must be considered.

8. Radon must be taken into consideration when carbon dioxide resulting from combustion or burning procedures is used. Residual hydrocarbons may be a problem in the case of manufacturing by combustion of heavy fuels. Maximum dosing requirements are set at less than 100 mg/L as CO_2. Purification by adsorption on activated carbon is recommended to obtain a foodstuff grade.

10. Major manufacturers provide full technical information.

1. Chloramines. Chloramines for drinking-water treatment usually consist of a mixture of monochloramine (predominantly) and dichloramine: NH_2Cl (MW 51.5) and $NHCl_2$ (MW 86), respectively. Usually, the chloramines are generated "on site" (i.e., in the water to be treated). The most efficient process is to dose chlorine to obtain bactericidal action for about 2 min and to dose the complementary amount of ammonium [either as NH_3 or as $(NH_4)_2SO_4$ or NH_4Cl, for example]. The rate of formation of monochloramine is second order.

$$v = k_2 |NH_3| \cdot |HOCl| \quad ; \quad k_2 = 5.1 \times 10^6 \, M^{-1} \, s^{-1} \text{ (at pH 8.5)}$$

Different investigations converge toward the equilibrium

$$\frac{|NH_4^+| \cdot |NHCl_2|}{|H^+| \cdot |NH_2Cl|} = 6.7 \times 10^5$$

From this relation it appears that at pH 7 to 8, monochloramine is the dominant species, while at pH 4 nearly only monochloramine is present. At still lower pH values, disproportionations occur with the formation of noxious trichloramine. The rate of formation of dichloramine is about 100 times slower than that of monochloramine. The weight ratios used in practice are for 1 Cl_2, 0.25 to 0.33 NH_3. Chloramines are weak oxidants only (e.g., NO_2^- and M_n^{2+} are not oxidized by NH_2Cl).

Chloramines are weak disinfectants compared to free chlorine or chlorine dioxide. However, they have interesting bacteriostatic effects. Dosages can range from 0.2 to 3 mg/L, the highest levels being used in warm waters distributed through a long main system (e.g., in Australia). In that case there is a notable taste in the water.

Some concern may exist as to the toxicity of chloramines, as dissolved gases, to fish. The "ammonia nitrogen" of chloramines has been set to a maximum limit of 1.5 mg/L. To what degree the nitrogen of monochloramine can act as a source for nitrification bacteria and, consequently, support bacterial aftergrowth remains a question. Bactericidal properties of chloramines can be enhanced by association with other oxidants, such as hydrogen peroxide.

Controlled reactions to form chloraminated compounds with glycine (H_2N-CH_2-COOH), methionine [$CH_3-S-CH_2-CH(NH_2)-COOH$], and gelatin result in disinfectants with 40 to 50% of the disinfecting power of monochloramine. However, some compounds are stable and capable of being pelletized and thus are suitable for use on travel. A well-known component is chloramine T:

$$p - CH_3-C_6H_4SO_2NClNa \cdot 3H_2O$$

Taste is associated with the use of these reagents.

On hydrolysis, monochloramine can release hypochlorous acid. However, equilibrium favors chloramine:

$$NH_2Cl + H_2O = HOCl + NH_3 \qquad K_H = 2.8 \times 10^{-10}$$

Similarly, for chloramine T, $K_H = 4.9 \times 10^{-8}$. To increase this possibility, a new generation of chloramines has been developed: chloroisocyanuric acids.

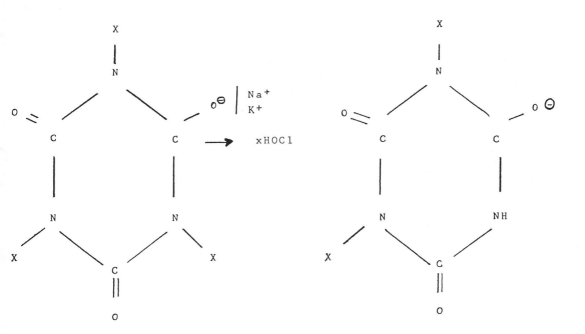

X = H or Cl

monochloro compound $K_H = 6.71 \times 10^{-4}$

dichloro compound $K_H = 3 \times 10^{-4}$

trichloro compound $K_H = 1.9 \times 10^{-4}$

Chloroisocyanuric acids are not acceptable for use in drinking water but can be appropriate for disinfection of swimming pools and general hygienic use.

1. Chlorhydric Acid or Hydrochloric Acid.

2. HCl; MW 36.5.

3. Acidification, correction of pH, and auxiliary uses such as generation of chlorine dioxide, activation of silica, and regeneration of ionic exchangers.

4. In aqueous solutions of 27 to 38 wt%, the freezing point is between $-70°$ and $-40°C$, respectively. Specific weight of the concentrated solution at 20°C is 1110 kg/m^3 and 1190 kg/m^3 for the 27 and 38% solution, respectively.

5. Strong acid. Acceptable materials for handling are glass, PTFE, PVC, HD-PE, polybutene, and alloys (i.e., AISI 318, etc.). The product is usually dosed directly as delivered, eventually into process dilution water on a continuous basis.

6. Identification code (ECC): 1789, danger code: 88 (corrosive material). Venting of the premises is essential. Action on metals is capable of promoting the formation of explosive hydrogen; contact with peroxides must be avoided. Precautions are as for the handling of strong acids; rinsing with water is a first-aid measure. Vapors are irritating, but spontaneous recovery occurs after limited exposure. The limit of exposure on a long-term basis is about 10 ppm (vol); concentrations between 50 and 100 ppm (vol) are limiting within 1 h. Dental erosion occurs on prolonged exposure. Severe burns occur on ingestion, which requires medical assistance.

7. Highly corrosive product (see item 5).

8. Purity characteristics are listed in Table 16.

9. Passivated chlorhydric acid. Passivated chlorhydric acid is used as a scaling remover and cleaning agent of objects clogged by calcium carbonate or similar precipitates. Filter cloths, water heaters, and so on, can be cleaned by this formula: 65 mL of concentrated chlorhydric acid with 60 g of commercial formaldehyde

Table 16   Impurities of Chlorhydric Acid

Impurity	Permissible (mg/m^3)	Content in the product [ppm (wt)]
Fe	20	10
As	5	1
Pb	5	1
Hg	0.1	0.1
Cd	0.5	0.2
Cr	5	1
Zn	10	1

solution containing at least 35 wt% formaldehyde mixed and diluted to 1 L to obtain a "passivated chlorhydric acid" solution. Formaldehyde is added as a corrosion inhibitor.

10. G. Lekimme, J. Tielemans, P. Smeets, and R. Husson, *Trib. Cebedeau*, *475-476*, 261 (1983).

1. Chlorinated Copperas (or Ferric Chlorosulfate).

2. $FeSO_4Cl$ (aq). Product resulting from the oxidation of ferrous sulfate with chlorine (can also be prepared by reaction of acids on ferric oxides).

3. Used as a coagulant–flocculant, particularly in the treatment of drinking water; can be associated with lime softening.

4. The chemical is currently available as a solution containing either 180 g/L Fe^{3+} or 200 g/L Fe^{3+}. The density at 20°C is 1.48 and 1.54, respectively.

5. Freezing of the liquid starts at -8°C for the product of 200 g/L Fe and at -22°C at 180 g/L Fe. The viscosities at -5°C are 160 mPa · s and 86 mPa · s for the concentrated and dilute solution, respectively.

6. Acid solution, rinsing with water.

7. The reagent is corrosive to ferrous metals; it can be stored and handled in glassware, plastics (e.g., PE, PVC, polypropylene), and titanium-containing stainless steels. Hastelloy C is convenient for some components.

8. Purity characteristics are listed in Table 17.

10. R. L. Kerkaert, *Trib. Cebedeau, 475-476*, 313 (1983).

1. Chlorine (Liquefied Gas).

2. Cl_2 is rapidly hydrolyzed in water ($k_1 = 12.8$ s^{-1} hypochlorous acid and chlorydric acid):

$$Cl_2 + H_2O = HCl + HOCl \ (K_H) \text{ in water at the usual pH}$$

Table 17 Impurities of Chlorinated Copperas

Impurity	Permissible (mg/m^3)	Content in the product[a] [ppm (wt)]
As	5	0.1–0.3
Cd	0.5	0.15–3
Cr	5	1–30
Ni	5	5–30
Pb	5	2–7
Al	10	5
Cu	10	0.5–5
Mn	5	700–1500
Zn	10	0.3–14
Sb	1	1–7
Sn		7–15
Ti		1700–2200
V		3–15
Mo		1.5

[a]The concentration of impurities depends essentially on that of the ferrous sulphate used to generate the ferric chlorosulphate. Test impurities are Cr, Ni and Mn.

Dissolved chlorine is a mixture of hypochlorous acid and hypochlorite ion according to the dissociation constant of hypochlorous acid (K_a).

Temperature (°C)	0	5	10	15	20	25
$K_H \times 10^4$	1.5	1.8	2.4	3.0	3.7	4.5
$K_a \times 10^8$	2.0	2.3	2.6	3.0	3.8	3.7

3. Chlorine is used primarily as a disinfectant. The maximum admissible concentration remaining in treated water is usually set in the range 0.25 mg/L. Chlorine is also used in the generation of chlorine dioxide and chloramines for the purpose of disinfection (see also "Chloramines" and "Chlorine Dioxide"). A secondary use is in the oxidation of mineral compounds (e.g., ferrous ion).

4. Liquefied chlorine is available in pressurized vessels (e.g., 40 kg net) tanks of 500 to 1000 kg net, and in bulk railway transport of 25 metric tons and more.

5. Chlorine can be abstracted directly from the gas phase of the pressurized vessels at an equilibrium pressure depending on the temperature of the stored liquid (Fig. 4). Abstraction from the liquid phase with subsequent heating of the liquid makes it possible to increase the possible dosage from a given unit storage. It is generally carried out at a consumption level of 3 kg/h and higher. The chlorine gas must have a temperature of at least 10°C above the equilibrium temperature to prevent misting in the gas phase.

6. Chlorine is highly toxic on inhalation (Fig. 5).

7. Chlorine can be handled in SM steel, Hasteloy, and Monel alloys and with appropriate stainless steel (AISI 318 to 322). Special care needs to be taken to keep the storage, transfer, and dosing premises free of humidity. This can be accomplished by flushing all standby equipment with dry nitrogen.

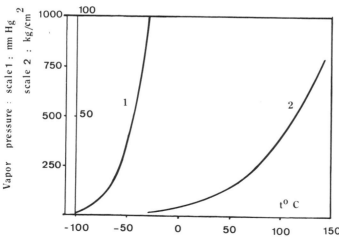

Figure 4   Vapor tension of liquid chlorine as a function of temperature.

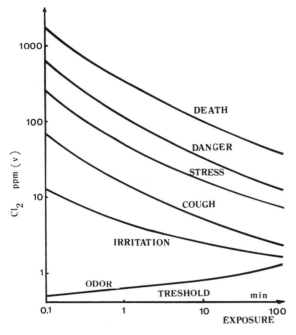

Figure 5 Effects of exposure to chlorine.

8. Trace substances in liquid chlorine are, in ppm (vol) (gas):

CO_2 500 to 2000
H_2 50
N_2 100
O_2 50

Hydrated ferric chloride in well-conditioned liquefied chlorine remains less than 20 mg/L liquid chlorine at 20°C, depending on the electrode material and the salt involved in electrochemical generation of the chlorine.

10. M. Francois, *Trib. Cebedeau, 475–476*, 283 (1983). J. Clifford White, *Handbook on Chlorination*, Van Nostrand Reinhold, New York, 1986. J. Ph. Buffle, *Tech. Sci. Munic., 60*, 373 (1965).

1. Chlorine Dioxide.   Chlorine dioxide is an oxidation–disinfection reagent generated on the site of its use. It is a toxic gas which, in water treatment, is handled in aqueous solution. The most commonly used generation processes are based on the reactions

$$2NaClO_2 + Cl_2 = 2ClO_2 + 2NaCl$$

and

$$5NaClO_2 + 4HCl = 4ClO_2 + 5NaCl$$

Secondarily, to avoid the risks of storage of chlorine on site, a tertiary process can be used:

$$8NaClO_2 + 2NaClO + H_2SO_4 = 8ClO_2 + 4NaCl + Na_2SO_4 + H_2O$$

However, this process requires very accurate mixing conditions and appropriate proportions of the chemicals reacted.

The industrial process by which chlorine dioxide is generated, by acidification of chlorate in the presence of a reductor, is not generally used in water treatment because the operation is complex and more appropriate for larger-scale applications. In direct processes, the chlorine dioxide generated is not free of chlorine. For use in water treatment, the chlorine dioxide generated must be pure. More details on chlorine dioxide are provided in Chapter 2.

1. Copper Sulfate.

2. $CuSO_4 \cdot 5H_2O$; MW 249.54.

3. Copper sulfate is used as an algicidal treatment either in settlers constructed and operated in open air or in stored (impounded) water sources. The treatment is best applied before explosive algal growth occurs. The dosing rate is to obtain a copper concentration in the range of 0.1 mg/L, but further study is needed on "combined copper" and enhanced oxidation processes used in advanced water treatment schemes.

4. Copper sulfate is available as a hydrated solid either as a powder, superfine powder, or crystals. Loose density ranges from 960 to 1380 kg/m^3, packed density from 1150 to 1440 kg/m^3, and working density from 1041 to 1400 kg/m^3. Flowability is good.

5. Copper sulfate is either dispersed as a solid at a water storage surface or dosed as a made-up solution.

6. Contact with skin and eyes must be avoided. Rinsing with water is necessary if contact with skin occurs. If copper sulfate has come into contact with an open wound, medical assistance is necessary.

7. Dissolved copper initiates electrochemical corrosion of ferrous metals. Copper sulfate must be handled without contact with ferrous materials (e.g., in plastics or glassware). Copper sulfate dust is harmful.

8. At a maximal dosing of 5 mg/L $CuSO_4 \cdot 5H_2O$ no significant impurities appear to be introduced in the water. A more usual dosing rate is 3 mg/L or less. Studies are needed on the secondary products contained in technical grades of copper sulfate.

1. Diatomaceous Earth.   Fossil natural product as a marine or freshwater sediment formation containing the silica skeletons of microscopic marine organisms. For further details, see Chapter 13.

5. Working densities of the powders range from 140 to 280 kg/m^3. Special care must be taken to avoid breathing the powders, which can cause silicosis. Wetting the powders is particularly important.

10. J. E. Bouvier, *Trib. Cebedeau, 475–476*, 333, 349 (1983).

1. Disodium Phosphate.

2. The product exists as anhydrous Na_2HPO_4 (MW 142) or as the monohydrated $Na_2HPO_4 \cdot 12H_2O$ (MW 358) solid.

3. Disodium phosphate is used on a small scale for pH adjustment and for

control of algal growth. It can be associated in formulations for corrosion control and for boiler water treatment.

4. The solid products have a working density of 835 to 1370 kg/m^3 according to the level of hydration and degree of crystallinity.

5. Rotor feeding is desirable, although the product is not highly floodable. The chemical is best made up as a working solution and dosed by metering pumps.

6. Avoid contact with electrical circuits. The solid product must be stored in a dark, cool, dry room to avoid hydration with coking. To avoid lumping, bags must not be piled too high. The reagent cakes upon hydration.

7. Disodium phosphate has a slight passivating effect against the corrosion of ferrous metals.

8. Each blend needs a specific analysis since the origin of the phosphate rock determines the type and nature of impurities. Known "guide impurities" are arsenic, fluoride, and to some extent, mercury and secondarily, zinc.

1. Dolomitic Hydrated Lime; Dolomitic Lime; Dolomitic Limestone. See "Lime."

1. EDTA (Ethylenediaminetetracetic Acid; Sodium or Potassium Salts). This product is used as an antiscaling agent (e.g., in membrane filtration and similar structures. A solution is made up of 120 kg trisodium phosphate, 6 kg EDTA, and 3 L Triton 100. The injection rate is 10 mg/L in phosphate. The chemical itself is not suitable for drinking water treatment.

1. Exchange Resins. Information is given in Chapter 15.

1. Ferric Chloride.

2. $FeCl_3$, usually hexahydrated, obtained by dissolution of iron in chlorhydric acid and subsequent oxidation with chlorine.

3. The coagulant is also used as a phosphate precipitating agent or as a sludge conditioner.

4. Exists as a crystalline product, $FeCl_3 \cdot 6H_2O$ (MW 270) (i.e., a 60 wt% product with a bulk density range of 900 to 1540 kg/m^3, or as an aqueous solution of 40 wt%), of density 1.425 kg/L at 20°C, hence a concentration of 570 g $FeCl_3$/L (the pure form is Fe_2Cl_6). Ferric chloride is marketed in bulk, and the solid form, also in bags or drums.

5. The 40% solution starts to freeze at −11°C and can be handled at temperatures above −8°C. The solid $FeCl_3 \cdot 6H_2O$ has a density of 1.632. The solid product is subject to caking and hydration. It is less suitable for dry feeding and is best dissolved in process water before dosing. Dissolution of the solid product in water is exothermic.

6. Ferric chloride in aqueous solution behaves as a strong acid. It stains porous surfaces and is damaging to electric contacts. Sluice with water.

7. In addition to its acid properties, ferric chloride is a strong oxidant, reacting with most metals. With aluminum and magnesium, hydrogen gas is liberated. The use of traditional plastic materials is the best method of coping with corrosion.

8. Ferric chloride should not contain more than 1% $FeCl_2$. Other purity characteristics are listed in Table 18.

9. Organic compounds known to be present as impurities are amines, aliphatic and aromatic, at concentrations of 3 and 0.5 ppm (wt), respectively.

10. R. Husson and A. Couttenier, *Trib. Cebedeau, 475–476*, 301 (1983). B. Balikdjian, *Trib. Cebedeau, 340*, 121 (1972).

Table 18   Impurities of Ferric Chloride

Impurity	Permissible (mg/m³)	Content in the product[a] [ppm (wt)]
As	5	1–33
Ba	10	5
Cd	0.5	0.5–3.25
Cr	5	30
Hg	0.1	0.1–0.75
Ni	5	33–50
Pb	5	5–33
Cu	10	50
Mn	5	700
Zn	10	20
Se	1	5
Sb	1	1
CN		0.2
Co		15
Aromatic amines		0.7–5

[a]mg per kg $FeCl_3$.

1. Ferric Chlorosulfate. See "Chlorinated Copperas." A new generation of manufactured products, combining aluminum and iron salts as coagulants, will probably be introduced on the European market. Some preliminary comments in manufacturers' leaflets on "polyaluminum coagulants."

1. Ferric Sulfate.

2. $Fe_2(SO_4)_3 \cdot nH_2O$ of net formula molar weight 400. Most hydration numbers (*n*) are 2 or 3, but up to nine "hydration waters" exist in some commercial products. The products are known under such commercial (registered) names as Ferriclear and Ferrifloc.

3. Ferric sulfate is a coagulant. A standard exists: AWWA-B.406-64.

4. The reagent is available as a solid with a working density between 960 and 1440 kg/m³ product.

5. The product is usually made up as a solution of 30 to 50 wt%.

6. The product and its solutions are weak to strong acids and must be handled accordingly. The product is staining.

7. Like other ferric salts, ferric sulfate is corrosive through oxidation reactions (see "Ferric Chloride").

8. Maximum accepted dosage as a coagulant is 200 g/m³ as $Fe_2(SO_4)_3 \cdot 9H_2O$. Accordingly, "steering impurities" have following RMIC values expressed in mg/kg reagent.

As	25	Hg	1
Cd	5	Se	5
Cr	25	Ag	25
Pb	25		

9. Ferric sulfate is often a secondary product of metallurgy which requires constant evaluation of quality.

1. **Ferrous Sulfate.**

2. $FeSO_4 \cdot 7H_2O$; MW 278.

3. Coagulation–flocculation after oxidation with chlorine or with dissolved oxygen in alkaline solution. Also used, together with lime, in dephosphatation and sludge conditioning.

4. The end product of the reagent of sulfuric acid on iron, it is often a waste solid from the steel-plating industry. Water content up to 6%. Available as bulk or in polyethylene bags of 30 to 1000 kg each.

5. On hydration the product has a tendency to cake and to leave a solid residue in oxidation tanks. A free-flowing alternative is marketed. About 1% of the commercial product is not water soluble. Solutions must be kept at a temperature of 10°C or higher. A working solution can be made up at 25%, and up to 1% acid is often added to limit oxidation.

6. Solutions of the product are slightly acid pH 2 at saturation and leave dirty, highly colored residues when spoiled; rinsing facilities must be provided in the working area.

7. Corrosive to ferrous metals, handleable in acid-resistant concrete.

8. Purity characteristics are listed in Table 19.

9. The product itself cannot be used as a coagulant but must first be oxidized by chlorine to form chlorinated copperas (see this reagent) or by oxygen in an alkaline medium (see item 10).

10. J. C. Schippers, H_2O, *14*, 320 (1971).

1. **Flural.**

2. $AlFSO_4 \cdot xH_2O$. Versions proposed contain 11 to 18 wt% fluor and 21 to 26 wt% available coagulant expressed as Al_2O_3.

3. The product is intended as a compromise between coagulation–flocculation and fluoridation. To date, extensive applications have not been developed.

9. The product does not seem to be used sufficiently to provide more information at this time.

10. W. E. White, J. C. Gillespie, and O. M. Smith, *J. AWWA, 44*, 71 (1952).

1. **Hydrazine.** See "Amines."

1. **Hydrochloric Acid.** See "Chlorhydric Acid."

1. **Hydrofluosilicic Acid.**

2. H_2SiF_6; MW 88.

Table 19 Impurities of Ferrous Sulfate

Impurity	Permissible (mg/m^3)	Content in the product[a] [ppm (wt)]
As	5	2
Cu	10	2
Zn	10	40
Pb	5	10
Mn	5	600
Ti		2000

[a]High variations can exist seeing it concern a recovered waste product.

3. Used as a fluoridating agent.

4. Normally supplied in small quantities or in a 20 to 30% aqueous solution.

5. The solution can be dosed directly into the water.

6. The product has a pungent odor, is irritating to the skin, and is toxic on ingestion. Chronic symptoms can appear on very long term exposure: necrosis of eyes and tissues, pulmonary edema, spasmosis, and paralysis. Appropriate venting of the premises is required as well as individual protective equipment (cloths, masks, spectacles, gloves, etc.). On contact, rinsing is recommended. If ingested, abundant water consumption is recommended, but vomiting is to be avoided. Medical assistance is recommended.

1. Hydrogen Peroxide.

2. H_2O_2; MW 34.

3. Hydrogen peroxide is used as an oxidant (e.g., of sulfide ions), as an indirect oxygenation agent, and as a bacteriostatic agent. Recent development in the chemistry of hydroxyl radicals in water treatment places hydrogen peroxide in a new perspective. (See Chapter 3.)

4. The usual form for commercial delivery is that of a 50 wt% solution, but strengths of 35 and 70 wt% are also available. The product is available in carboys (e.g., of 65 kg) and in bulk in quantities of 10 or 20 metric tons. The density at 20°C of a 50 wt% solution is 1.195 kg/m^3.

5. Hydrogen peroxide is highly corrosive and can be stored and handled in PE, PVC, series 316 stainless steel, and aluminum of at least 99.5% purity. Uncontrolled contact with organic compounds, as well as with heavy metals such as Cu, Ni, Co, and Fe, must be avoided. Venting and decompression of gases capable of being formed on decomposition are to be provided. At 50 wt%, concentrated H_2O_2 remains liquid at temperatures as low as $-50°C$. It is recommended that the plant handle 5% solutions made up by dilution (DVGW, *Information 22-4/90*), and dose after further dilution to 1.5%.

6. The product is an irritant to the skin. Clean with water. Handling with individual rubberized protective devices is recommended: gloves, spectacles, boots, and so on. Mixing concentrated hydrogen peroxide with flammable products is prohibited.

7. See item 5.

8. Hydrogen peroxide is a very pure chemical reagent. In Belgium its use is allowed up to 10 mg/L. In Germany, concentrations are set on a conservative basis at 0.1 mg/L. No problem of trace impurities is known to occur through the use of hydrogen peroxide in water treatment.

10. Y. Denutte, *Trib. Cebedeau, 475–476*, 337 (1983). W. Masschelein, M. Denis, and R. Ledent, *Water Sewage Works* p. 69 (Aug. 1977). P. A. Giguere, *Peroxide d'hydrogène*, Masson, Paris, 1975.

1. Iodine.

2. I_2; MW 254.

Iodine can be used as a disinfectant, particularly in swimming pools. It has not yet been used extensively in drinking water treatment. Secondary formation of iodine can occur on oxidation of iodide-containing waters. Generally, there is formation of iodamines, and even of iodoform, which is associated with disagreeable taste.

Lime

Lime is a chemical used since ancient times, for example, for the preparation of "medical water." As such, it is a chemical reagent that is of assistance in environmental conservation. But handling and dosing of limes involve considerable difficulty, although these can be controlled by the appropriate technologies.

1. Lime. Either quicklime, CaO (MW 56), or hydrated lime, $Ca(OH)_2$ (MW 74): calcium oxide and calcium hydroxide, respectively. The product is sometimes called calcinated limestone, but should not be confused with limestone, which is "natural" calcium carbonate.

2. Quicklime, CaO, is also called "burnt lime," "lump lime," "unslaked lime," and "caustic lime." Calcium hydroxide, $Ca(OH)_2$, is known as hydrated lime and sometimes as "slaked lime," which is a questionable terminology.

Remarks: Lime is obtained by heating limestone to calcination at 1200 to 1400°C. By burning dolomitic limestone (i.e., a double carbonate of calcium and magnesium) a *dolomite lime* results that contains 35 to 40% magnesium oxide. If the magnesium oxide in the calculated product exceeds that of calcium, the starting product was "dolomite magnesite." *Carbide lime* is a waste product of the manufacture of acetylene through the calcium carbide process:

$$CaC_2 + 2H_2O = Ca(OH)_2 + C_2H_2$$

Both products are less well suited for drinking water treatment.

3. Lime has several uses in water treatment, among which are the following:

a. Water softening by precipitation using lime alone or in conjunction with other alkaline reagents, such as sodium carbonate and sodium hydroxide.

b. As coagulant–flocculant or as an alkaline complement in coagulation–flocculation processes using aluminum sulfate or iron salts. Coagulation of colored low-mineralized water is often assisted by the addition of lime, especially in the case of water with low alkalinity.

c. As an alkaline chemical to control the pH of the treated water, to neutralize the acidity and carbonate aggressivity.

d. In conjunction with aeration, to remove iron compounds such as ferric hydroxide coprecipitated with calcium carbonate.

e. Used as coagulant, lime significantly reduces the heavy metal content of the water and eliminates or reduces to some extend inorganic compounds such as fluorides and phosphates.

f. Lime as a sludge conditioner, enabling compaction of alum sludges from water treatment plants.

g. By the treatment called "excess lime treatment," consisting of the addition of enough lime to attain a pH value of 10.5 to 10.6, producing an hydroxide alkalinity of 10 mg/L, a disinfecting action is obtained, at least for *E. coli*, coliforms, and other enterobacteria. The use of lime may be considered as an emergency disinfecting technique.

4. Quicklime is available in all grain sizes, from 60 mm to a few micrometers. Typical specific terminologies are:

Pebble lime: average diameter lower than 60 mm
Lump lime: average diameter up to 20 to 25 mm
Ground lime: average diameter under 6 mm
Pulverized lime: diameter under 2.8 to 3 mm
Powdered quicklime: diameter under 90 to 100 μm
Super-quicklime: diameter under 10 μm (maximum 20 μm)

Actually, the major manufacturers are able to supply lime in any size specification except for the super-quicklime of small grain size. Hydrated lime is produced by reacting stoichiometric amounts of water [$CaO + H_2O = Ca(OH_2)$] with quicklime. This "hydration" process is often erroneously called "slaking." Slaking involves treatment of quicklime with an excess of water in comparison with stoichiometric hydration.

Remark: On hydration, quicklime expands to about twice its volume (apparent density of quicklime 880 to 1000 kg/m^3 and of hydrated lime, 400 to 560 kg/m^3). The absolute densities of both materials range from 3200 to 3400 kg/m^3 and from 2300 to 2400 kg/m^3 for quicklime and hydrated lime, respectively. Moreover, there is a structural difference between quicklime (cubic lattice) and hydrated lime (hexagonal lattice). Hence hydration is a slow process that implicates a crystalline rearrangement. The specific surface of quicklime ranges from 4 to 7 m^2/g product, while that of hydrated lime is about 12 to 16 m^2/g. Lime, both quicklime and hydrated lime, can be purveyed in 40- to 50-kg multiwalled paper sacks or moisture-proof bags, in 50-kg barrels, or in bulk, usually in hoppered trucks of capacity 20 to 23 tons. A ready-made solution (Aquacal Reg.) of slaked lime of selected granulometry under 20 μm is available on the European market. Concentration is 30 wt%, density at 15°C is 1.2 ± 0.02 kg/L. The solution is stable when kept free of freezing and can be dosed with liquid metering pumps.

5. Solid lime must be stored dry. Quicklime in bags must not be stored for more than 60 days, as, on gradual hydration with moisture in the air, the bags will rupture. The bags should be placed on pallets so as to prevent moisture being absorbed from the bottom structures. By means of dry storage, hydrated lime can be kept for periods up to a year without deterioration. Lime is generally stored on site, in vertical cylinders with an inverted conical bottom. The free-flowing slope should be at least 55°, preferably 60 to 70°, as the repose angle for lime ranges from 30 to 40°. The height/diameter ratio of the storage tank is to be kept in the range 2.5 to 4. Although the lime remains free-flowing, the conical outlet zone is best equipped with vibrators to prevent possible hanging-up. Storage facilities are to be airtight, and a supplementary atmospheric dust elimination facility (e.g., cyclone) is advisable. To prevent clugging of the stored lime, vibration is recommended. The best standard capacity of storage bins depends on the local conditions, but a minimum content of 30 to 35 tons is practical for pneumatic filling from a bulk transport carrier of 20 to 25 tons. For safe operation, the equipment for storage and dosage is best duplicated. When delivered in bulk the lime is transferred by means of elevators, screw conveyors, or pneumatic transfer equipment. The latter causes the least dust during transfer. It is at present the most used technology in water treatment facilities. The size of screw conveyors ranges between 15 and 45 cm, thus enabling, at a rotation speed of 50 rpm, a transfer range of 3 to 50 tons/h. High air moisture is objectionable for pneumatic transfer, as it can cause air slaking.

Air for transfer is on the order of 1 to 1.2 bar positive pressure to enable a 20- to 23-ton truck to be unloaded in less than 1 h (airflow is approximately 1.5 to 2 m³/ min).

The storage silo must be equipped with an air release valve, and a dust removal cyclone system is optional. Liquid slaked lime is subject to slow carbonation with the CO_2 in the air. Continuous mixing in the storage tank is advised and on-site storage is best limited to less than 1 month. Continuous slaking and dosing is the most widespread technology recommended to avoid the effects of carbonatation.

Quicklime can be dosed directly by dry feeders into the water (e.g., to correct pH). However, a highly reactive lime is required (granulometry size lower than 20 μm) and reaction time must be available since hydration followed by ionization is, overall, a slow process. Therefore, with direct dosing, hydrated lime is used more often.

Batch Slaking

In batch slaking quicklime is mixed with water in proportions of 1 : 3 for highly reactive limes to 1 : 2 for slowly reactive limes. The suspension that results is a paste of 30 to 35 wt% lime solids. The suspension can be pumped as is or, better, diluted before dosing. Temperature control is essential in batch slaking. The maximum limit recommended is 90°C. Degritting of slaked lime can be necessary when low-grade material is used to manufacture quicklime. In such cases screening at a sieve aperture of about 5 mm is suitable to remove coarse material. Batch slaking is a neglected technology but could still be an innovative method for the application of lime.

Continuous slaking is subject to high-technological requirements. Feeders for lime dosing or continuous slaking can be either gravimetric or volumetric. Gravimetric feeders include belt feeders and hopper feeders, with dosing ranges from 250 g/h to 2000 kg/h. Volumetric feeders are of various types: screw conveyor, belt, rotating paddle, oscillating hopper, vibrating feeders, and so on. An acceptable accuracy for continuous feed of quicklime is on the order of 5%, but some guidelines must be maintained relating to the reactivity of lime with water and the physical condition of the bulk storage.

Hydration of lime is an exothermal reaction with $\Delta H = 3.7$ kJ/mol, where the hydration causes the water to heat. If water vapor is formed, the vapor can climb up into the solid lime in the dosing mechanism and block the system. Hence provision must be made for the evaporation of water and venting of the slaking vessel.

Milk of lime, or *creamy lime*, is an oversaturated suspension of lime in water. It is obtained by slaking. Suspensions up to 420 g/L expressed as $Ca(OH)_2$ (or 320 g/L expressed as CaO) can be obtained. The usual water/quicklime ratios used in slaking are (in weight proportions), 3 : 1 for highly reactive lime and 2 : 1 for slowly reactive lime. Two extreme conditions should be avoided:

1. If too much water is used, a "drowning" effect occurs by which the particles hydrate quickly at the surface but the hydrated layer formed hinders further penetration of water into the mass of the particles. Thus rupture of the particle into more reactive microparticles is delayed.
2. If insufficient water is added, a "burning" effect occurs due to high temperature, hydration water is evaporated, and unhydrated particles remain in the

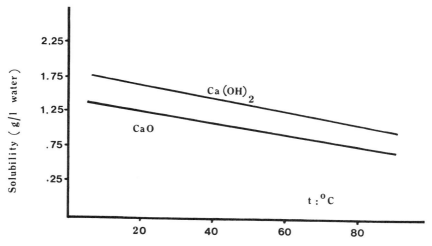

Figure 6   Solubility diagram of Ca(OH)$_2$ as a function of water temperature.

suspension. Under normal conditions the entire slaking process last 10 to 20 min under steady mixing ($G = 300\,s^{-1}$ and more).

Water of lime or *lime-water* is a solution containing dissolved and ionized calcium hydroxide. A saturated solution (25°C) has a pH of about 12.4. The slaking process is less good with cold slaking water. Heating the process water to 25 to 40°C is recommended to improve the result. A vapor vent system must be mounted at the top of the slaking vessel.

Once the slaking process is complete, the milk of lime can be diluted to the

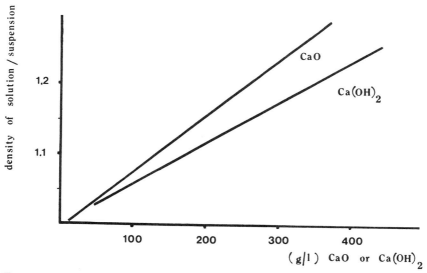

Figure 7   Density at 15°C of milk of lime suspensions.

appropriate strength (e.g., 5 to 10 wt%) for dosing in a separate mixing vessel. However, carbonation and scale formation can limit the feasibility of dilution. By transferring the milk of lime into the dilution vessel through an overflow system, the latter also makes it possible to separate the undissolved coarse material. For dosing, diaphragm metering pumps are the preferred technology at present, although piston pumps are also suitable. Dipper-wheel feeders are an alternative technology that is applicable. Centrifugal pumps are less appropriate, as they are subject to clogging when used for the transfer of slaked lime. All pumps must be mounted with the necessary circuits for flushing with water during maintenance and standby periods.

The closer the point of dosing into water, the better. Dosing with open troughs instead of pipes to convey the lime suspensions simplifies scale removal. The best pipes to use smooth plastic (hard PVC) or asbestos–cement pipes, both without vertical allows. Inspection and rinsing facilities should be provided (manholes or at least hand holes should be installed at a minimum of every 15 m).

Sedimentation in pipelines must be avoided. Recommended transfer velocity is more than 0.75 m/s at all dosing regimes. The use of sodium hexametaphosphate (1 to 10 ppm) is recommended to "deflocculate" the calcium carbonate precipitates. Alternative methods are acid (HCl) treatment followed by mechanical cleaning.

Temperature rise on hydration of quicklime can be used as a quality criterion for its reactivity. A test of reactivity is to pour 50 g of quicklime into 200 g of water maintained at 25°C in a stoppered vacuum bottle and to record the temperature while shaking the bottle continuously. According to the AWWA standard, the temperature should rise 40°C within 3 min. Standard apparatus for this slaking test is described by AWWA-B202-77. The reactivity depends on the manufacturing process and the purity of the limestone. It can be shown by such tests that lime obtained from other sources, such as oyster shells, is less reactive in the hydration test and therefore less well suited for water treatment.

When milk of lime is dosed into water the electrical conductivity increases due to dissolution of the lime slurry. The more reactive the lime, the faster the conductivity increase. A tentative standard is in development in Belgium using a 5 wt% $Ca(OH)_2$ lime suspension dosed at ratio of 1.6 vol% into distilled water. The conductivity increases to over 400 to 450 mS/cm. The speed of increase is a characteristic of the reactivity of the lime.

6. Quicklime is a caustic product, causing skin burns, particularly burns to eyes. Hydrated lime is irritative to a lesser degree. For heavy-duty operations or interventions, protective PVC clothing, including head and neck covering and eye protection, is necessary. Protective creams or oils can be applied to prevent burns to potentially exposed areas.

In accidental situations, a respirator mask should be worn during delayed interventions so as to avoid respiratory irritations resulting from the inhalation of lime dust. Eyes should be protected by tight-fitting safety glasses with side shields. After removing most of the dust by a soft brushing, Vaseline or other ointments applied to skin surfaces protect temporarily against the consequences of accidents during handling of quicklime. Safety goggles and long PVC gloves are recommended as a minimum protection for inspecting or cleaning slakers. After handling lime, operators should shower. A 6% solution of acetic acid or vinegar is able to neutralize the caustic effects of lime on the skin. Its application shall be followed by copious washing with soap and abundant rinsing with water. In the case of accidents to the eyes, a saturated solution of boric acid is recommended to counteract the caustic effect, but medical assistance must be called for as soon as possible. Abundant, continuous rinsing with water is recommended. (Remember: Heat develops on hydration of quicklime.) Do not rub eyes irritated by quicklime. Fire can result on mixing solid ferrous sulfate and quicklime. When mixed with solid aluminum sulfate quicklime hydrates by reaction with the water of crystallization of aluminum sulfate. The whole expands in volume, and as a consequence, in closed containers, explosion may result. (The explosions occur at a temperature of 590°C, at which hydrogen forms.)

A preventive cleaning system is recommended for those parts of the plant where lime is handled. Thermostatic valves *must be provided*, mounted on the slakers to shut off dosing in case of overheating, or as an alternative measure, to flush the system with a large flow of cold water. Dosing equipment of powdered lime, either quicklime or hydrated lime, must be specific and not used occasionally for dosing other reagents.

7. Quicklime and hydrated lime are not significantly corrosive to iron and steel. Because of their caustic properties the chemicals are not to be brought into contact with aluminum or zinc metals or coatings. Ordinary black-carbon steel (Siemens-Martin grade) is suitable as long as the equipment is kept dry. In addition to iron, suitable materials for handling lime are asphalt, concrete, rubber, and higher-grade steel. Cast aluminum is not recommended but is acceptable for small parts of the equipment, such as stops, flanges, and valves. When wet or slaked, lime is more corrosive; choice materials for transfer are carbon steel, concrete and PVC. However, the temperature limit for the latter is 80°C.

8. Lime is obtained from limestone, a natural product. Hence it can contain impurities derived from the limestone source. Also, trace elements issued from materials used for grinding, calcination, and storage can secondarily contaminate the reagent.

Traditionally, the absolute rejection basis for CaO content is 75%, and for $Ca(OH)_2$, 62%. A bonus formula can be proposed for each increment of calcium oxide content. The amount of insoluble material in water must never be more than 5%, and the reagent should be free of core, ash, dirt, and insoluble silicious material. The fluoride content should be below 0.5 g F/kg CaO, or at least not increase the fluoride content of the treated water by more than 0.1 ppm. Major impurities to be considered in applications are MgO, SiO_2, and Fe_2O_3 and $CaCO_3$ (Na_2O, K_2O), $CaSO_4$, and Al_2O_3. Good-quality lime must have a CaO content higher than 90 wt% or more than 95 wt% as $Ca(OH)_2$. Typical available compositions are listed in Table 20. Trace impurities of basis $Ca(OH)_2$ to be considered are listed in Table 21. The elements of limestone available in Belgium are given for comparison.

Metals are widely removed on precipitation with lime. The steering impurity is chloronium. For mineralization purposes the dosages are considerably lower. A significant increase in the content of some heavy metals can result from the manufacturing procedure. Chromium, barium, and zinc are the impurities that require attention if lime is used at high dosages.

9. Dry feeding and continuous slaking are difficult operations that require maintenance. The manufacturers or specialized designers are best consulted while

Table 20   Compositions of Quicklime and Hydrated Lime

Basic formula	Quicklime, CaO	Hydrated lime, Ca(OH)₂
% CaO	91.3	—
% Ca(OH)₂	2.3	96.2
% CaCO₃	1.8	0.7
% CaSO₄	0.08	0.08
% MgO	1	0.5
% Fe₂O₃ + Al₂O₃	0.8	0.25
% SiO₂	1.25	0.25
% H₂O		1.2

planning a facility to be constructed. Although less used, batch slaking, for example in a daily operation tank, is an attractive alternative to the more common continuous slaking and dosing.

10. *Lime: Handling Application and Storage Bulletin 213*, National Lime Association, Washington, D.C., 1976. A. A. Hirsch, *J. AWWA., 54*, 1531 (1962). Ph. Dumont and E. Bechoux, *Tech. Eau, 376*, 39 (1978). Technical data sheets: *AWWA-standard B202-83*; Belgian standards, *Cebedeau 475–476*, 275–277 (1983) (ISSN-0007-8115).

1. **Magnesite (Dolomite) or Dolomite Lime.**

2. Magnesian quicklime contains 5 to 35% magnesium oxide, and dolomitic quicklime contains 35 to 40% magnesium oxide.

3. Dolomitic limestone is used as an inorganic coagulant, for pH adjustment, and in softening. It can be employed in conjunction with phosphates for fluoride removal. To some extent, it makes possible the removal of silica. Solid dolomitic limestone can be a suitable filtration material for deferrization and mineralization of soft waters. Partial removal of silica can be achieved using dolomite.

Table 21   Impurities of Lime

Impurity	Permissible (mg/m³)	CaCO₃ (mg/kg)	CaCO₃ × 0.74 [equivalent to Ca(OH)₂] (mg/kg)	Ca(OH)₂ measured (mg/kg)	RMIC Ca(OH)₂ 200 mg/L dosing (mg/kg)
Ag	1	4	3	5	5
As	5	26	19	12	25
—	10	600	444	450	
Cr	5	160	118	200	25
Ni	5	3	2	4	25
Pb	5	5	4	52	25
Al	20	1100	810	2640	100
Cu	10	3	2	3	50
Fe	30	900	666	2000	150
Mn	5	100	74	700	25
Zn	10	50	37	200	50

4. Commercial forms of dolomitic lime are the same as for lime.

5–7. Handling and dosing characteristics are similar to those applicable to lime.

8. Dolomite is usually less pure than limestone. The steering impurities are the same (see "Lime").

1. Octylamine.

2. $C_8H_{17}NH_2$; MW 129.

3. Octylamine is used in formulations of commercial corrosion inhibitors and neutralization agents of CO_2 in closed-loop boiler water.

See "Amines" for further comments.

1. Ozone, O_3. Ozone is a strong oxidant–disinfectant that must be generated on the site of its use. It is a toxic gas that is contacted in a mixture with air or oxygen into the water. In practical systems, ozone is usually generated from a dry process gas containing oxygen that is submitted to a laminar electrical discharge. The latter generates activated radical oxygen either 3P or 1D, which on combination with molecular oxygen forms ozone:

$$O (1D) + O_2 = O_3$$

$$O (3P) + O_2 + M = O_3$$

In the case of the less energetic radical (3P), a scavenger (M) is generally regarded to be necessary. The oxygen radicals can also be generated by irradiation of oxygen containing gases. γ and X irradiation have not yet found practical use in the field. Vacuum-UV light is used in small-scale applications. The photochemical generation of ozone is based on absorption of the 185-nm resonance line. More details on ozone are given in Chapter 3.

1. Phosphoric Acid.

2. H_3PO_4; MW 98.

3. Orthophosphoric acid is an adduct that can be used in formulations of products used as corrosion inhibitors and water-conditioning products. By itself it has no direct application in water treatment, but its salts are of importance.

5. Passivating solutions are made up with phosphates so as to contain between 25 and 50% orthophosphate and 50 to 75% hexametaphosphate. For initial treatment a dosage of up to 2 g of phosphate mixture per liter can be applied. The formation of a compact protection layer can be improved by adding 10 mg/L of slightly alkaline metasilicate ($Na_2O_3SiO_2$). When disinfection is also required (e.g., when iron bacteria are present), hydrogen peroxide, 100 mg/L, can be added. A typical formula for passivating stainless steel surfaces is composed of 400 g oxalic acid, 7 kg H_3PO_4 (80 to 90%), and 26 L ethyl alcohol, made up with water to 40 L.

6. Contact of phosphoric acid with skin and eyes can cause irritation and burning. Respiration of vapors must be avoided, as it can cause damage to the upper respiratory system. The maximum allowable concentration in a working area is 1 mg/m^3. On mixing with concentrated alkali, explosions can result. Rinsing and cleansing with water are recommended. Lung intoxication requires medical assistance.

1. Polyelectrolytes. There exist a very wide variety of polyelectrolytes, each with its own dosing characteristics. An exhaustive list of possible reagents, to be used as a reference basis for further information, is given in Sections 14 and 15.

1. **Potassium Permanganate.**

   2. $KMnO_4$; MW 158.

   3. Potassium permanganate is used for iron removal, disinfection, algal treatment, and taste and odor control. The deferrization reactions correspond globally to the following reactions:

$$KM_nO_4 + 3Fe(HCO_3)_2 + 2H_2O = MnO_2 + 3Fe(OH)_3 + KHCO_3 + 5CO_2$$

and

$$2KMnO_4 + 3Mn(HCO_3)_2 = 5MnO_2 + 2KHCO_3 + 2H_2O + 4CO_2$$

Under practical conditions, the MnO_2 formed is a mixture that varies between hydrated $MnO_{1.3}$ and $MnO_{1.9}$, a mixture of MnO and MnO_2. A realistic representation is

$$O{=}Mn \left\{ \begin{array}{c} \diagup O^{2-} \\ \\ \diagdown O \end{array} \right. Mn^{2+}$$

In this structure Mn^{2+} can be exchanged with other cations such as Fe^{2+}, Fe^{3+}, and even H^+. In addition to the exchange capacity of MnO_2, the adsorption capacity must be considered since the BET surface of the product is 300 m^2/g. Overdosing must be avoided to prevent the release of an excess of soluble manganese salts. Filtration is essential to remove precipitated MnO_2. Subsequent filtration on activated carbon (see Chapter 11) can prevent the development of excess $KMnO_4$, which is reduced by carbon according to the reaction

$$4KMnO_4 + 3C + H_2O = 4MnO_2 + 2KHCO_3 + K_2CO_3$$

Excess carbon is necessary.

   4. Potassium permanganate is generally purchased as a solid in iron drums and under those conditions may be kept indefinitely. It can be obtained in prismatic or conical crystals. The crystals are not hydrated and have very little hygroscopicity. At a high moisture content of air, caking up occurs. A free-flowing product containing silicates is also available. Loose densities of commercial products range from 1100 to 1450 kg/m^3, and packed densities, from 1300 to 1700 kg/m^3. The solid product is to be dissolved and dosed as a solution. Its solubility depends strongly on temperature; expressed in percent weight it equals 3.5 at 5°C, 4.4 at 10°C, 5.1 at 15°C, 6.1 at 20°C, and 7.2 at 25°C. Practical working concentrations are kept between 0.5 and 3 wt%. Working solutions are best made up batchwise and dosed continuously subsequently.

   6. Potassium permanganate is toxic and is irritating to the skin and mucous membranes. As a solid it can initiate fires by oxidation. Rinsing with water is the major safety measure to be taken. It can damage mucous tissues, and on strong ingestion, a drop in blood pressure and anuria and chemical jaundice can result. Immediate medical assistance is required in such cases.

7. Potassium permanganate is alkaline. It presents no particular corrosion risk. The use of rubber, hemp, and other textiles is best avoided.

8. Normal use of potassium permanganate remains limited to 2 mg/L, but doses up to 10 mg/L have been considered to meet RMIC values. The product should be analyzed for heavy metals. Steering impurities are cadmium, chromium, and mercury.

10. Carus Co., technical documentation.

1. Sodium Aluminate.

2. NaAlO$_2$(dim) (MW 82) or Na$_2$OAl$_2$O$_3$(anhyd) MW 164.

3. Coagulation–flocculation can be associated in alkaline softening processes (lime and soda softening). Loose density, 960 to 1010 kg/m^3; packed density, 1150 to 1250 kg/m^3.

4. Bulk solid, 200-kg drums, or 50-kg bags.

5. Made-up solutions at concentrations higher than 10% are stable for only 2 to 4 h. At concentrations lower than 2.5% the stability is about 4 days at ordinary temperatures. When exposed to air, the solid product gradually hydrates and cakes.

6. Alkaline product, irritating to the eyes and mucous membranes. The reagent contains usually 9 to 10% excess Na$_2$O. In the case of dry feeding, aspiration of dust should be anticipated. Contact with skin can cause burning. Water must be available for cleaning.

7. Corrosive for ferrous metals, can contain abrasive constituents. Sodium aluminate can be used at a dosing rate of 30 mg/L (exceptionally, 40 mg/L).

8. Purity characteristics are listed in Table 22.

9. Can be used in conjunction with aluminum sulfate to adjust the pH in coagulation–flocculation processes.

10. F. Briffeuil, *Trib. Cebedeau, 72,* 466 (1956). J. T. Burke, *Water Works Wastes Eng.,* p. 64 (Feb. 1965).

1. Sodium Bicarbonate.

2. NaHCO$_3$; MW 84.

3. Sodium bicarbonate is used for pH control and water stabilization. Secondary use for silicate activation is described (see Chapter 8).

4. The product is available in bulk or as a solid in bags of 25 to 50 kg or in containers of 1 to 1.5 tons. The loose density is 800 to 1200 kg/m^3, and the packed density ranges from 1000 to 1400 kg/m^3.

Table 22   Impurities of Sodium Aluminate

Impurity	Maximum permissible (mg/m^3)	Content in the product (wt ppm)	RMIC (at 30 mg/L) dosing
As	5	0.2–4	165
Cd	0.5	0.3–5	16
Cr	5	10–50	165
Pb	5	10–20	165
Hg	0.1	2–10	3
Se	1	2–10	30
Ag	1		30

5. The product is best used as a made-up solution of less than 50 g/L (solubility at 20°C is 96 g/L; dissolution is endothermic). Caking can occur by absorption of ambient humidity as a consequence of pressure (e.g., by excess piling). The product is slightly alkaline (pH 8.5 for saturated solutions).

6. Sodium bicarbonate is a mild chemical that is also used in health and beauty products. It is best stored isolated from all sources of heat. It has the ability to concentrate environmental odors.

7. No major corrosion risk is associated with bicarbonate.

8. Possible impurities are Na_2CO_3 (0.5%), NaCl (0.01%), Fe_2O_3 (0.001%), and water (0.1%). The maximum admitted dosing is 200 mg/L (150 mg/L in the United States). No trace impurity is known to affect the quality of the drinking water at those dosages. The steering impurity is mercury.

1. Sodium Bisulfite.

2. $NaHSO_3$(anh); MW 104.

3. Sodium bisulfite can be used as a reducer.

4. The solid product has a loose density of 880 to 1090 kg/m^3 and a packed density of 1100 to 1340 kg/m^3.

5. The product is free-flowing unless packed excessively. It cakes by hygroscopicity.

6. Sodium bisulfite is toxic by ingestion.

7. The product is mildly acid and slightly corrosive to ferrous metals. Contact with electric circuits or electronic components must be avoided.

8. Maximum admissible dosing is set at 4 mg/L. Known trace elements are listed in Table 23.

1. Sodium Carbonate. Also called calcinated soda, light soda or soda ash, or Solvay salt).

2. Na_2CO_3; MW 106.

3. Used for pH adjustment and softening (lime-soda process).

4. Commercial crystalline powder of at least 99% Na_2CO_3. Density at 20°C is 2.533. There exist a wide range of apparent density products: loose 500 to 1000 kg/m^3 and packed 720 to 1250 kg/m^3.

5. The product has the tendency to cake on compression or hydration. Dosing in the form of working solutions in the concentration range of less than 50 g/L is recommended (solubility of sodium carbonate is complex and associated with several hydrated forms of the chemical).

Table 23 Impurities of Sodium Bisulfite

Impurity	Permissible (mg/m^3)	Content in the product [ppm (wt)]
As	5	1
Pb	5	2
Se	1	5
Cu	10	1
Mn	5	1
Zn	10	2

6. Moderately alkaline product—rinse with water. Prolonged contact with the skin (or eyes) can cause burning. Dermatitis can develop chronically.

7. Store and handle in plastic or plastified metallic basins. Polyethylene, PVC, ABS, and PTFE are convenient.

8. As leading impurities one has the following maximal concentration of impurity (in mg/kg reagent) on the basis of a maximum reference dosis of 200 mg/L of the reagent:

As   25   Hg   0.5
Cd   2.5   Pb   25
Cr   25   Ni   25

Steering impurities are chromium and lead.

9. The maximum reference dosage in the treatment is probably an overestimation since maximum dosages in the range 60 to 100 mg/L are more likely to occur in practice.

1. Sodium Chlorate.

2. $NaClO_3$; MW 106.5.

3. Sodium chlorate itself is not used directly in water treatment; however, the chemical can be significant as part of an inorganic herbicide, secondary reaction product of chlorine dioxide and can be used in alternative methods for generation of chlorine dioxide: for example, according to the reaction

$$2NaClO_3 + SO_2 = 2ClO_2 + Na_2SO_4$$

However, high concentrations are necessary, making this reaction less well suited for drinking water treatment (see Chapter 2).

4. Sodium chlorate is available commercially as a crystalline solid of melting point about 250°C, containing more than 99% $NaClO_3$. "Herbicide" variants containing 60 to 66 wt% $NaClO_3$ are also available. The product is available in unit packages up to 25 to 100 kg or in solid bulk.

5. The specific density of the product is 2.49 kg/L; the apparent density is about 1.3 kg/L. Sodium chlorate is soluble in water (endothermic dissolution). At 0°C up to 45 wt% solution of $NaClO_3$ is obtained. Neutral or alkaline solutions are stable. The average grain size is 0.25 to 0.35 mm in diameter. For small-scale uses the product is best stored as a solution.

6. Cloth and leather are readily oxidized by sodium chlorate. Avoid contact with flames or open fires. Do not smoke while working with chlorate. Contaminated areas should be rinsed immediately with water. Firefighting equipment with water must be installed. Chlorate is of low toxicity (see Chapter 2) but ingestion of 10 to 15 g can cause nausea. Vomiting must be promoted in such cases, and medical assistance is required. Contact with acids must be avoided since chlorine dioxide can develop in such cases. In case of a cloth fire, an extinguishing shower is required; blankets are not efficient. Thermal decomposition of chlorate releases oxygen at 300°C.

7. Solid sodium chlorate is stable, but contact with products that can induce catalytic decomposition must be avoided [e.g., organic greases, points, sugar, solvents, and mineral reducers (e.g., sulfur, cyanide, metal powders, etc.]. The chemi-

cal is best handled in plastic materials (e.g., PE, PVC). Plastic barrels with chlorate are best equipped with a polyethylene lining. Cisterns are of series 316 stainless steel. Polyester or epoxy-coated normal steel can be used for static vertical tanks. Storage in concrete tanks is a possible option.

8. A typical composition of the crystalline product is

$NaClO_3$	>99%
NaCl	<0.6%
H_2O	<0.2%
Insoluble	<0.01%

The product may contain phosphate-based antibulking agents. No specific data are available on trace impurities that may be relevant for drinking water since the chemical as such is still of less relevance for drinking water treatment. The chlorate herbicide is a less pure product: standard concentrations in chlorate are 87 to 92 wt%, but products containing only 57 wt% of sodium chlorate are also present on the market. Important secondary components are sodium carbonate and sodium chloride in the proportion 10:1. Herbicidal chlorate is generally not suitable for drinking water treatment. A known critical impurity is arsenic, but other organic substances or adducts can be present.

10. $NaClO_3$ brochure, Solvay & Cie, Belgium. *Le Chlorate de Soude*, Atochem, France.

1. Sodium Chloride.

2. NaCl, MW 58.5.

3. Sodium chloride is used to regenerate cationic resins and secondarily as a mineralization agent of soft waters and cooling brines. It can have an indirect impact from its use as a snow removal agent on roads.

4. The chemical exists as fine powders of less than 50 μm or in a 170-μm grade. To regenerate ionic exchangers, 10/15-grade crystalline salts (2 to 3 mm size) are recommended or a briquetted powder of centimeter size. The density of NaCl at 20°C is 2.163 kg/L; the apparent density or liter weight of the powders ranges from 0.9 to 1.2 kg/L.

5. Most variants of the product are available in bags from 1 to 50 kg and in bulk up to truck capacity. The salt is best dosed as a preliminary make-up solution or for small-scale use introduced in a static container and flushed by a water stream into the ionic exchanger column. NaCl solubility is (in g/kg water) at 0°C, 356; at 20°C, 358.5; and at 100°C, 392. Making-up concentrated brines is a specific technology if operated on a continuous basis.

6. Foodstuff and pharmacopea grades are nontoxic. Massive ingestion must be avoided. Some technical salts contain adducts such as eosin or corrosion inhibitors. Their use is not recommended for drinking water treatment.

7. Sodium chloride is extremely corrosive for ferrous metals, including stainless steel. It must be handled in plastic, ebonited, epoxy-coated, or glass material. Concrete is best coated with butyl rubber or epoxy coatings when in contact with solid salt or concentrated brines.

8. Natural salt can contain from 85 to 98.5 wt% NaCl. Purified salt usually contains more than 98% NaCl. Nutritional-quality salt contains up to 99.9% NaCl. A typical composition for salts used to regenerate water-softening resins is:

NaCl	98.48%
H_2O	0.104%
Mg	0.032%
Ca	0.302%
K	0.113%
SO_4^{2-}	0.95%
Insoluble	0.016%

Trace elements are Co, Fe, Cu, Pb, and Ba, the latter two being steering impurities. When of appropriate grade (e.g., nutritional quality), sodium chloride is not known to be objectionable for drinking water treatment.

9. The chloride content of discharge effluents must be limited (see Chapter 15).

10. NaCl brochure, Solvay et Cie, France, Br. 1071b-B-2.274.

1. Sodium Chlorite.

2. $NaClO_2$; MW 90.5.

3. Generation of chlorine dioxide; disinfection of ion-exchange resins.

4. Solid containing 80% $NaClO_2$
Solution containing 300 to 310 g/L $NaClO_2$, or 25 wt%; density of the solution, 1.21; solidification at -8 to $-10°C$ or a 31% solution of density 1.27 containing 310 g/L, which solidifies at 5 to 6°C. The density of the solution varies linearly as a function of the concentration of $NaClO_2$. Both forms can be stabilized (e.g., with $Na_2S_2O_4$).

5. Handle only solutions in treatment plants, by dissolving the solid product upon arrival.

6. The solid product is spontaneously explosive on heating at 80°C and higher, or in the presence of any ignition source. Rinse spoiled product immediately. Chlorine oxides and chlorine result upon explosion or fire. These gases are toxic when inhaled. For a discussion of safety, see "Chlorine."

7. Sodium chlorite is corrosive to ferrous alloys, including ordinary stainless and to a lesser extent, copper and its alloys. Recommended materials are glass, PE, PVC, PTFE, nickel, and tantalum alloys (polyester materials may have a tendency to scale off).

8. Normally, secondary components are expressed relative to 100% $NaClO_2$: NaCl, 1 to 3%; NaOH, 0.5 to 1.5%; Na_2CO_3, 1.5 to 2.5%; Na_2SO_4, Na_3PO_4 traces + $Na_2S_2O_4$ (eventually). Trace impurities are listed in Table 24.

10. J. L. Colas and Y. Denutte, *Trib. Cebedeau, 475-476,* 291 (1983).

1. Sodium Chromate (Dichromate).

2. Na_2CrO_4; MW 162.

5. Exists as anhydrous or hydrated salt. Flowability is poor; the chemical becomes delinquiscent on contact with water.

6. Sodium chromate is toxic and thus is not used for drinking water treatment. It is incorporated in formulations for corrosion control and stainless steel conditioning.

1. Sodium Fluoride.

2. NaF; MW 42.

3. Sodium fluoride is a fluoridating agent.

4. Available as a white to nile blue powder of loose density 880 to 1600 kg/m^3 and packed density 1280 to 1600 kg/m^3.

Table 24 Impurities of Sodium Chlorite

Impurity	Permissible (mg/m³)	Content in the product [ppm (wt)]
Ag	1	0.03
Al	20	3.6
As	5	0.01
Ba		0.03
Cd	0.5	1
Co		0.03
Cr	5	0.1
Cu	10	0.9
F	150	0.13
Fe	30	1.6
Hg	0.1	0.03
Mn	5	0.03
Mo		0.2
Ni	5	0.07
Pb	5	0.7
Sb	1	0.03
Se	1	0.2
Sn		0.03
Te		0.03
Ti		1.5
V		0.03
Zn	10	0.2

5. The product is hygroscopic and has a tendency to lump or to mass and arch. Dosing is best obtained through a preliminary made-up solution (saturation at 20°C is about 48 g/L). The solubility is not strongly dependent on temperature.

6. Dust of sodium fluoride is highly toxic on repeated exposure. It is skin irritating. When ingested it causes vomiting and diarrhea with dehydration; it may cause muscular weakness and pseudoepileptic convulsions. Heavy-duty protective clothing is required to handle the product. Venting of the premises is essential; eye-rinsing facilities are necessary. If ingested accidentally, drinking milk is the first-aid treatment before medical assistance arrives.

7. Solutions are corrosive to metals, and contact of sodium fluoride with electrical hardware must be avoided.

8. As for other fluorides, steering impurities are arsenic and lead.

1. Sodium Fluorosilicate.

2. Na_2SiF_6; MW 188.

3. Slight acidifying agent with insecticide and fungicidal action and with potential auxiliary effects such as fluoridation and flocculation aid.

4. By-product of the phosphate industry. It is a crystalline hygroscopic salt to specific weight 2.679. Its tendency to cake is minimal. Crystalline solid of loose density 990 to 1375 kg/m³ and packed density 1450 to 1700 kg/m³. Solubility in water at 25°C ranges from about 7.6 to 5.4 g/L at 10°C. The chemical hydrolyzes in water to give fluorhydric acid and silica (SiO_2) of acid pH (e.g., 3.5 to 4.0 in concentrated solutions).

6. The product is toxic and irritating, although not directly severely poisonous to humans. (Toxicity limits are between 120 and 270 mg/kg body weight). Dust must be avoided; the product is best handled as a solution in closed vessels.

7. Solutions of sodium fluorosilicate are acid corrosive to most metals and aggressive to glassware. The product is incompatible with much metals and can promote the development of hydrogen, with the corresponding danger of fire and explosion.

8. Steering impurities are arsenic and lead, but the levels are acceptable.

9. Product with interesting possibilities not yet widely applied.

10. Cl. A. Hampel, *Chem. Eng. News, 27,* 2420 (1949).

1.   Sodium Hexametaphosphate.

2. $(NaPO_3)_6$; MW 612).

3. Sodium hexametaphosphate is incorporated in formulations for corrosion inhibition on the cathodic side, and as a dispersing agent in boiler feed water. To avoid precipitation, the product can be incorporated into alkaline reagents which are diluted by process water before injection into the main stream. Also, injection of a small amount (less than 1 mg/L) generally prevents scaling at the point of injection of alkaline reagents.

5–7. No special problems are encountered in dosing the solutions. The product is not toxic, although massive ingestion is not recommended. Solutions are not corrosive to metals, but handling in plastic vessels is generally recommended. Plumbosolvency is a risk to be considered.

8. Trace elements are as follows (in mg/kg):

As	1	Pb	3
Cr	1	F	3
Ni	1	Fe	10

No major problem occurs up to the maximal permissible dosage of 20 mg/L.

1.   Sodium Hydroxide.   Also called caustic soda.

2. NaOH; MW 40.

3. pH correction; softening; water conditioning.

4. Solid anhydrous or hydrated (six forms containing 1, 2, 3, 4, 5, or 7 water molecules per molecule). Solutions 50% wt% (dissolution in water is exothermic). The density of the 50 wt% solution 1.5 at 20°C. The 73 wt% solution is less used in Europe because it crystallizes at 63°C.

5. Solutions must be stored and handled without risk of freezing, preferably at a temperature higher than 12°C. On bulk delivery the solution can be hot. Specification should be to limit the temperature to less than 45°C. Contact with air must be minimized to avoid hydration and carbonation. Dissolution must be carried out at reduced speed and under cooling because considerable heat is evolved during this operation.

6. Causes burning of skin. Wear eye and skin protection equipment. Wash with water at low pressure. Consult a doctor immediately in case of ingestion or contact with eyes. Emergency showers and eye wash units must be installed in the working area.

7. The solution can be stored at temperatures below 50°C in ordinary SM steel and at less than 30°C in PVC. At higher temperatures AISI 316 or 316 stainless

steel is required. Zinc and aluminum or their alloys are not resistent. Skin burns can ulcerate and require medical assistance.

8. The technical reagent should contain not less than 96% NaOH and not more than 2% Na_2CO_3 reported on the basis of the solid reagent. Purity characteristics are listed in Table 25. Reference maximal dosing 100 mg/L as NaOH.

9. The RMIC concept could allow higher contents of impurities than recommended here, which are easily achievable by current manufacturing processes of the reagent.

10. P. Smeets and A. Francois, *Trib. Cebedeau, 111,* 317 (1983).

1. Sodium Hypochlorite. Also called Javelle water; chlorine bleach.

2. NaClO; MW 74.5.

3. Disinfection, oxidation.

4. Equimolecular solution of NaClO and NaCl of variable strength. The solution most commonly used for water treatment contains 150 g/L active chlorine; 1 mol NaClO corresponds to 2 equivalents active chlorine. (The solid crystalline product is insufficiently stable to be of practical use.) Concentrated solutions are available in bulk (20 tons and more), in drums (100 to 120 L), and in carboys (30 to 60 L).

5. Density of the solution is 1.2 to 1.22; freezing at $-6°C$; must be stored in the dark to avoid decomposition. The chemical is subject to carbonation on storage in open air.

6. Reacts violently with acids, with the formation of gaseous chlorine. The product, which contains excess caustic soda (5 g/L), has a pH of about 11. Under accidental decomposition radical oxygen is liberated, which can cause explosion danger. The reaction can be accelerated by light, heat, organic matter, and heavy metals (e.g., copper, nickel, cobalt) in their cationic form.

7. All metals are corroded by sodium hypochlorite. Storage is best done in plastic tanks (PVC or even PE). Corrosive for electrical and electronic material. Water vapor is to be avoided.

Table 25 Impurities of Sodium Hydroxide

Impurity	Permissible (mg/m^3)	Content in the product [ppm (wt)]
Ag	1	0.2
As	5	1
Cd	0.5	0.1
Cr	5	1
Hg	0.1	0.02
Ni	5	1
Pb	5	1
Sb	1	0.2
Se	1	0.2
Fe	30	6
Mn	5	1
Cu	10	2
Zn	10	2
Al	20	4

8. Major secondary components are as follows:

Chlorate	8–9 g/L
Chloride	10 g/L
NaOH	8 g/L
Na_2CO_3	6 g/L

Purity characteristics are listed in Table 26.

9. Avoid contact with sulfur compounds and mixing with organic liquid compounds.

10. P. Smeets and A. Francois, *Trib. Cebedeau, 475,* 325 (1983). *AWWA Standard for Hypochlorites,* ANSI-AWWA-B.300-80. DVGW, Wasser Information 23 (4/90), Germany.

1. Sodium Nitrate.

2. $NaNO_3$; MW 85.

3. Sodium nitrate is a very secondary reagent that is used in formulated mixtures for boiler water treatment. The product is also used as conditioning agent for stainless steel surfaces.

1. Sodium Pentachlorophenate.

3. Sodium pentachlorophenate, C_6Cl_5ONa (MW 288.5) has been used in algal control and as an inhibitor of slime formation. Use of the product should be prohibited where drinking water resources are involved.

1. Sodium Silicate (Meta).

2. Na_2SiO_3 with excess alkali, $nSiO_2 \cdot Na_2O$.

3. Basis of activated silica and is also used as a corrosion conditioner.

4. Exists in vitreous solid or as a solution. Most used in water treatment are of the type $3.3SiO_2 \cdot Na_2O$ (molar ratio SiO_2Na_2O, 3.3), of density 1.35 to 1.39 containing 27 to 29 wt% SiO_2 and 8 to 8.95 wt% Na_2O with a viscosity of 140 to 180 mPa.

Table 26   Impurities of Sodium Hypochlorite

Impurity	Permissible (mg/m^3)	Content in the product [ppm (wt)]
Ag	1	0.6
As	5	6
Cd	0.5	0.3
Cr	5	3
Hg	0.1	3
Ni	5	6
Pb	5	15
Sb	1	6
Se	1	6
Fe	30	30
Mn	5	3
Cu	10	3
Zn	10	9
Al	20	6

Table 27 Impurities of Sodium Silicate

Impurity	Maximum permissible (mg/m³)	Content in the product [ppm (wt)]	RMIC [ppm (wt)] (at 10 mg/L dosing)
As	5	3	500
Cd	0.5	0.1	50
Cr	5	1	500
Hg	0.1	0.2	10
Ni	5	2	500
Pb	5	10	500
Sb	1	3	100
Cu	10	25	1000
F	(1.5)	30	
Fe	20	150	2000
Mn	5	5	500
Zn	10	25	1000
Ti	?	40	

5. Temperature must be maintained above 10°C; heating can cause a risk of scaling. Alkaline liquid (pH > 12); can be stored and handled in steel or plastic vessels.

6. Risks as for alkali; spoiled product is slithery. Facilities for rinsing and sluicing are to be provided.

7. Not corrosive; risks of scaling exist.

8. Purity characteristics are listed in Table 27. At concentrations used in water treatment, activated silica (sodium metasilicate) is not known to affect the potability of drinking water.

1. Sodium Silicohexafluoride.   See "Sodium Fluorosilicate."

1. Sodium Sulfite.

2. Na_2SO_3(anh); MW 126.

3. Sodium sulfite is used for dechlorination and is also incorporated in formulas for corrosion control and boiler water treatment.

4. Sodium sulfite is available as a solid of loose density 865 to 1340 kg/m³ and packed density 1280 to 1710 kg/m³.

5. Flowability of the product is poor; it has a strong tendency to arch. The reagent (when anhydrous) is highly hygroscopic, with a tendency to mass, with possible formation of hard lumps. The product is best used in the form of a preliminary made-up solution in a batch system. The chemical is unstable, as oxidation can occur on storage.

6. Dusts of sodium sulfite are toxic on inhalation; however, dustiness of the product is very limited. Spoiled product can be rinsed with water.

7. The product is a weak base with a slight corrosion tendency when dissolved. Contact with zinc, copper, and their alloys and must be avoided, as must electrical contact.

8. Purity characteristics are listed in Table 28. The reference basis for *maximum* dosing is 7 mg/L. In conclusion: Properly manufactured sodium sulfite poses no

Table 28   Impurities of Sodium Sulfite

Impurity	Maximum permissible (mg/m³)	Content in the product [ppm (wt)]	RMIC [ppm (wt)] (at 7 mg/L dosing)
As	5	1	700
Pb	5	2	700
Se	1	5	140
Cu	10	2	1400
Fe	30	10	4200
Mn	5	1	700
Zn	10	2	1400

problems when used in the treatment of water. The possible availability of waste products requires attention.

1. Sodium Thiosulfate.

2. $Na_2S_2O_3$(anh); MW 158. The salt exists more often as a crystalline pentahydrate: $Na_2S_2O_3 \cdot 5H_2O$; MW 248.

3. Sodium thiosulfate is used to eliminate residual oxidants, particularly residual chlorine. Its use is frequent in laboratory investigations up to pilot scale. The product is normally not applied full-scale on drinking water that is distributed.

4. Often, the pentahydrated form is readily available.

5. Loose density of the product is about 850 kg/m³; packed density is 950 kg/m³. The product has excellent flowability with little or no dustiness. Some tendency to efflorescency exits. The product is unstable through deterioration by oxidation. Formation of colloidal sulfur can occur by side reactions.

9. The product is used very rarely for large-scale treatment processes.

1. Sulfamic Acid.

2. $HSO_4 \cdot NH_2$; MW 113.

3. Sulfamic acid is used in water conditioners for boiler water feed control and secondarily for acid neutralization and pH control.

4. The product is available in the form of granules of loose density 990 to 1140 kg/m³ and packed density 1300 to 1440 kg/m³. It is often incorporated in made-up solutions ready for use.

5. Flowability of the powder is very variable and depends strongly on the packing of the chemical. The reagent is best handled as a solution, since on hydration, the solid becomes deliquiscent and cakes.

6. The product is harmful when inhaled in the form of dust. It can cause burns on skin.

7. Sulfamic acid is a corrosive acid when used in a concentrated solution. Appropriate dosing is important in its application.

1. Sulfur Dioxide (Gas).

2. So_2; MW 64.

3. Sulfur dioxide is used almost exclusively for "deoxidizing," principally as a dechlorination agent. A secondary use that has been considered is as an activating agent of silicates used to obtain activated silica. Dechlorination is used more often in the United States than in Europe since high-level chlorination as an intermediate step in chlorination is less common in Europe.

4. Sulfur dioxide is available in cylinders and bulk pressurized vessels of 1 metric ton and more.

5. Vapor density is 2.264 kg/m^3 (NTP), which is approximately that of chlorine (2.482). Hence dosing equipment for chlorine is appropriate for gaseous sulfur dioxide dosing. Sulfur dioxide is soluble in water up to 120 g/L at 21°C. The equilibrating vapor pressure of sulfur dioxide at 21°C is about 2.4 bar (241 kPa), and that of chlorine is 6.2 bar (650 kPa). Therefore, the withdrawal capacity at equal cylinder capacity for sulfur dioxide is about 30% of that for chlorine. The reliquifaction risk is of the same relative order. When "chlorinators" are used for dosing of sulfur dioxide, they must be thoroughly cleaned beforehand. Low concentrations of gaseous SO_2 cause a sensation of suffocation, tearing of the eyes, and the symptoms of a heavy chest cold. Rapid recovery occurs after a few minutes in the open air. Liquid sulfur dioxide causes severe freezing of the skin and eyes due to the rapid evaporation (the heat of evaporation at 0°C is about 376 kJ/kg). The lowest "irritating" concentration in air ranges between 10 and 20 ppm (vol). The maximum allowable concentration for prolonged exposure is in the range of 10 ppm. Short-term exposure limits for 30 to 60 min range from 50 to 100 ppm. Dangerous short-term exposure is 400 to 500 ppm. First-aid treatment is similar to that applicable for chlorine.

7. All materials used for liquid chlorine dosing equipment are suitable for handling and dosing of liquid–gaseous sulfur dioxide except stainless steel grades AISI 316 to 318.

8. Sulfur dioxide is not known to contaminate drinking water when used at conventional dosing rates. Steering impurities are arsenic and selenium.

9. Reduction of residual oxidants may require pH values lower than usual for "finished" drinking water.

1. Sulfuric Acid.

2. H_2SO_4; MW 98.

3. pH adjustment, water conditioning, preparation of activated silica, and regeneration of ion-exchange resins.

4. Concentrated solution (e.g., 98 to 99% H_2SO_4, density 1.9, and eventually any dilution of this strength).

5. Can solidify between +8 and +10°C. Kinematic viscosity at 15°C is 1.7 × 10^{-5} m^2/s and at 25°C, 1.3 × 10^{-5} m^2/s.

6. Strong acid. Do not mix water into sulfuric acid as violent heat will evolve. Avoid contact with skin and ingestion. In case of contact, wash with a dilute solution of sodium carbonate or bicarbonate. Reacts violently with organic products. Chronic effects are dermatitis and dental erosion. Maximum permissible concentration in the gas phase is about 1 mg/m^3.

7. Strong sulfuric acid (more than 80 wt% concentration) can be stored in ordinary SM steel vessels. Lead lining is convenient for all concentrations lower than 80%. Plastics (i.e., PE) are convenient for diluted solutions. When dosing concentrated sulfuric acid the inlet air must be dried (e.g., on $CaCl_2$).

8. Sulfuric acid can be obtained by catalytic oxidation of sulfur or by burning iron on zinc sulfide. The former is of higher purity and is preferred for the treatment of drinking water. Purity characteristics are listed in Table 29.

10. G. Crocq, G. Lekimme, P. Smeets, and R. Husson, *Trib. Cebedeau, 475,* 267 (1983).

Table 29   Impurities of Sulfuric Acid

Impurity	Content in the product [ppm (wt)]		Permissible (mg/m^3)
	Standard	Purified	
Ag	0.05		1
As	0.2	0.1	5
Ba	?	0.05	10
Cd	0.05	0.005	0.5
Cr		0.04	5
Hg	0.1	0.005	0.1
Ni	5	0.5	5
Pb	5	0.1	5
Sb	?	2	1
Se	5	15	1
Cu	0.1	2	10
F	?	0.5	150
Fe	50		30
Mn	0.1		5
Zn	0.1	40	10
Sn	2		
Co	1		

1. **Tetrasodium Pyrophosphate.**

2. $Na_4P_2O_7$; MW 226.

3. Used essentially for boiler water treatment and corrosion control.

4. Basically, available as a solid but can also be incorporated in formulated solutions for limited uses.

5. Loose density ranges from 640 to 900 kg/m^3 and packed density from 720 to 960 kg/m^3. The solid is of low dustiness and remains easily flowable unless packed excessively. At high relative air humidity, the product can cake.

6. Dust is irritating to eyes and respiratory airways; at high dosages the product can be toxic.

7. Concentrated solutions are mildly corrosive for ferrous metals.

8. Steering impurities are cadmium and lead from the point of view of traces in the reagent, but the levels are below relevant significance.

14.   ADDITIVES APPROVED BY THE EPA FOR TREATMENT OF DRINKING WATER

In the United States, EPA agreement to use a given additive or auxiliary product to be used in drinking water treatment can be given on the basis of information provided by the manufacturer. Several aspects are taken into consideration, such as safeguarding the public health while protecting the global environmental, and whether the product carries a guaranty of efficiency in the water treatment process applied. Additional information regarding biodegradability and toxicity to fish can be obtained from the manufacturers.

Manufacturer or dealer	Product	Maximal accepted concentration (mg/L)
Allied Colloids, Inc.	Percol LT-20	1
One Robinson Lane	Percol LT-21	1
Ridgewood, NJ 07410	Percol LT-22	1
	Percol LT-24	5
	Percol LT-25	5
	Percol LT-26	1
	Percol LT-28	1
	Percol LT-28	1
	Percol LT-29	1
	Percol LT-30	1
Allstate Chemical Co.	Allstate No. 2	3.0
Box 3040	Allstate No. 6	1
Euclid, OH 44117		
Allyn Chemical Co.	Claron	1.5
2224 Fairhill Road	Claron 207	2
Cleveland, OH 44106		
American Cyanamid Co.	Superfloc 127	1
Berdan Avenue	Magnifloc 513-C	65
Wayne, NJ 07470	Magnifloc 515-C	50
	Magnifloc 517-C	40
	Magnifloc 521-C	10
	Magnifloc 570-C	10
	Magnifloc 571-C	10
	Magnifloc 573-C	10
	Magnifloc 575-C	10
	Magnifloc 577-C	10
	Magnifloc 579-C	10
	Magnifloc 581-C	10
	Magnifloc 843-A	1
	Magnifloc 845-A	1
	Magnifloc 846-A	1
	Magnifloc 847-A	1
	Magnifloc 848-A	1
	Magnifloc 860-A	1
	Magnifloc 971-N	1
	Magnifloc 972-N	1
	Magnifloc 985-N	1
	Magnifloc 990-N	1
	Magnifloc 1848-A with Activator 478	4 : 0; 10 : 1
	Magnifloc 1985-N with Activator 478	4 : 0; 10 : 1
Atlas Chemical Div.	Sorbo	20
ICI America Inc.		
Wilmington, DE 19899		
Berdell Industries	Berdell N-489 Floccul.	1
28-01 Thomson Avenue	Berdell N-821 Floccul.	1
Long Island City, NY 11101	Berdell N-902 Floccul.	1

Manufacturer or dealer	Product	Maximal accepted concentration (mg/L)
Betz Labs, Inc.	Betz Polymer 1100P	1
Somerton Road	Betz Polymer 1110P	1
Trevose, PA 19047	Betz Polymer 1120P	1
	Betz Polymer 1130P	1
	Betz Polymer 1140P	1
	Betz Polymer 1150P	1
	Betz Polymer 1160P	1
	Betz Polymer 1190	10
	Betz Polymer 1200P	1
	Betz Polymer 1205P	1
	Betz Polymer 1210P	1
	Betz Polymer 1220P	1
	Betz Polymer 1230P	1
	Betz Polymer 1240P	1
	Betz Polymer 1250P	1
	Betz Polymer 1260P	1
	Betz Polymer 1290	10
	Betz Entec 610	10
	Poly-Floc 4D	25
Bond Chemical, Inc.	Bondfloc No. 1-101	5
1500 Brookpark Road		
Cleveland, OH 44109		
Brennan Chemical Co.	Brenco 879	100
704 N. First Street	Brenco 880	100
St. Louis, MO 63102		
The Burtonite Co.	Burtonite 78	5
Nutley, NJ 07110		
Calgon Corp.	Coagulant Aid 2	1
Box 1346	Coagulant Aid 18	15
Pittsburgh, PA 15222	Coagulant Aid 233	1
	Coagulant Aid 243	1
	Coagulant Aid 253	1
	Coagulant Aid 961	5
	Cat-Floc	7
	Cat-Floc	7
	Cat-Floc B	10
	Cat-Floc T	5
	Polymer M-502	5
Commercial Chemical	Coagulant Aid	
11 Patterson Avenue	Speedifloc 1	10
Midland Park, NJ 07432	Speedifloc 2	5
Dearborn Chemical Div.	Aquafloc 408 (liquid)	50
W. R. Grace & Co.	Aquafloc 409	1
Merchandise Mart Plaza	Aquafloc 411	2
Chicago, IL 60654	Aquafloc 422	1
Dow Chemical U.S.A.	Dowell M-143	5
Barstow Bldg	PE1-600	5
2020 Dow Center	PE1-1090	5
Midland, MI 48640	Purifloc A-22	1

Manufacturer or dealer	Product	Maximal accepted concentration (mg/L)
	Purifloc A-23 (PWG)	1
	Purifloc C-31	5
	Purifloc N-17	1
	Separan AP-30	1
	Separan AP-273 Premium Water Grade	1
	Purifloc N-20	1
	XD-7817	1
Drew Chemical Corp.	Amerfloc 2	10
701 Jefferson Road	Amerfloc 265	1
Parsippany, NJ 07054	Amerfloc 275	1
	Amerfloc 307	1
	Amerfloc 420	10
	Drewfloc 1	1 : 8 Alum 0.5 : 10 Lime
	Drewfloc 3	3
	Drewfloc 4	5
	Drewfloc 21	5
	Drewfloc 922	10
	Himofloc SS-100	1
	Himofloc SS-120	1
	Himofloc SS-500	1
DuBois Chemicals	Flocculite 550	2
Div. of W.R.Grace & Co.	GOP-16A-LT	4
3630 East Kemper Road	Split	19
Sharonville, OH 45241		
E. I. DuPont de Nemours and Co.	Carboxymethyl cellulose	1
Eastern Lab		
Gibbstown, NJ 08027		
Environmental Pollution	DynaFloc 631	5.0
Investigation and	DynaFloc 632	1.0
Control Inc.	DynaFloc 633	1.0
9221 Bond Street	DynaFloc 634	1.0
Overland Park, KS 66214	DynaFloc 661	1.0
	DynaFloc 662	5.0
	DynaFloc 664	5.0
	DynaFloc 691	1.0
	DynaFloc 692	1.0
	DynaFloc 693	5.0
Fabcon Intl.	Zuclar 110 PW	0.5
1275 Columbus Avenue	Fabcon	0.5
San Francisco, CA 94133		
Henry W. Fink & Co.	No. 102	1
6900 Silverton Avenue	No. 109 Kleer-Floc	1
Cincinnati, OH 45236	No. 116 Kleer-Floc	1
	No. 119 Kleer-Floc	1
	No. 730 Kleer-Floc	1
	No. 735 Kleer-Floc	1

Manufacturer or dealer	Product	Maximal accepted concentration (mg/L)
Fuel Economy Engineering Co. 3094 Rice Street St. Paul, MN 55113	Feecolite 201	1
Gamlen Sybron Corp. 321 Victory Avenue San Francisco, CA 94080	Gamlose W	5
	Gamlen Wisprofloc 20	5
	Gamafloc N1-702	4
Garrett-Callahan 111 Rollins Road Millbrae, CA 94031	Coagulant Aid 72A	50
	Coagulant Aid 74B	30
	Coagulant Aid 76	40
	Coagulant Aid 76A	50
	Coagulant Aid 78B	50
	Formula 70A	50
	Formula 73	50
	Formula 74E	20
General Mills Chemicals 4620 North 77th Street Minneapolis, MN 55435	Supercol Guar Gum	10
	Guartec F	10
	Guartec SJ	10
W. R. Grace & Co. Research Div. 7379 Route 32 Columbus, MD 21044	PO-107	1
	PO-115	1
	GR-962	1
	GR-963	1
	Copolymer GR-989	1
	Copolymer GR-996	1
	Copolymer GR-997	1
	Homopolymer GR-999	1
Hercules, Inc. 910 Market Street Wilmington, DE 19899	Carboximethylcellulose	1
	Hercofloc 818 (Potable Water Grade)	1
	Hercofloc 821 (Potable Water Grade)	1
Frank Herzl Corp. 299 Madison Avenue New York, NY 10017	Perfectamyl A5114/2	10
ICI America, Inc. Wilmington, DE 19899	Atlasep-PWG-11	1
	Atlasep-PWG-44	1
	Atlasep-PWG-77	1
	Atlasep-PWG-255	1
	Atlasep-PWG-1010	1
Illinois Water Treatment Co. 840 Cedar Street Rockford, IL 61102	Illco IFA 313	10
Kelco Co. 8225 Aero Drive San Diego, CA 92123	Kelgin W	2
	Kelcosol	2
Key Chemicals 4346 Tacony Philadelphia, PA 19124	Key-Floc-W	25
Metalene Chemical Co. Bedford, OH 44014	Metalene-Coagulant P-6	5

Manufacturer or dealer	Product	Maximal accepted concentration (mg/L)
The Mogul Corp.	Mogul-CO-940	10
Chagrin Falls, OH 44042	Mogul-CO-941	10
	Mogul-CO-980	2
	Mogul-CO-982	1.5
	Mogul-CO-983	50
	Mogul-CO-984	50
	Mogul-CO-985	3.5
	Mogul-CO-986	5
	Mogul-CO-9003	10
	Mogul-CO-9007	1.5
	Mogul 9013 (revised)	125
	Mogul 9016	1
	Mogul 9020P	1
	Mogul 9021P	1
Nalco Chemical Co.	Nalcolyte 110A	5
6216 West 66th Place	Nalcolyte 607	40
Chicago, IL 60638	Nalcolyte 671	1
	Nalcolyte 7870	1
	Nalcolyte 8101	10
	Nalcolyte 8113	10
	Nalcolyte 8114	10
	Nalcolyte 8170	1
	Nalcolyte 8171	1
	Nalcolyte 8172	1
	Nalcolyte 8173	1
	Nalcolyte 8174	1
	Nalcolyte 8175	1
	Nalcolyte 8180	1
	Nalcolyte 8182	1
	Nalcolyte 8184	1
	12CO6	10
	BX-50	22
Narvon Mining & Chemical	Sink-Floc Z3 & AZ3	10
Co., Affiliate of Irl	Sink-Floc Z4 & AZ 4	10
Dafflin Assoc.	Sink-Floc Z5	10
Keller Avenue and Fruitville	Zeta-Floc C	20
Pike	Zeta-Floc K	20
Lancaster, PA 17604	Zeta-Floc O	20
	Zeta-Floc WA	20
National Starch and Chemical	O'B Floc	10
Corp.		
1700 West Front Street		
Plainfield, NJ 07039		
Olin Water Service	Olin-4500	1
120 Long Ridge Road	Olin-4502	1
Stanford, CT 06904		
Oxford Chemical Div.	Oxford-Hydro-Floc	10
Consolidated Foods Corp.		
Box 80202		
Atlanta, GA 30341		

Manufacturer or dealer	Product	Maximal accepted concentration (mg/L)
W A Scholten's Chemische Fabrieken, NV Foxhol, Postbus 1 The Netherlands	Wisprofloc P	5
Scroil Chemical Corp. 1375 Linden Avenue E Linden, NJ 07036	Flocgel	10
Standard Brands Chemical Industries, Inc. Div. of Standard Brands Drawer K Dover, DE 19901	Tychem 8035	1
A. E. Staley Mfg. Co. Box 151 Decatur, IL 60525	Hamaco 196	5
Stein, Hall & Co., Inc. 605 Third Avenue New York, NY 10016	Hallmark 81	1
	Hallmark 82	1
	Jaguar	0.5
	MRL-14	1
	MRL-22	2
	Polyhall M-295 P.W.	1
Tretolite Div., Petrolite Corp. 369 Marshall Avenue St. Louis, MO 63119	TFL 324	5
	Tolfloc 333	7
	Tolfloc 334	10
	Tolfloc 350	1
	Tolfloc 351	1
	Tolfloc 355	1
	Tolfloc 356	1
	Tolfloc 357	1
James Varley & Sons, Inc. 1200 Swtizer Avenue St. Louis, MO 63147	Varco-Floc	150
Warren Cook Chemical Co Oak and Astor Ast. Monee, IL 60449	Crownlite 1	1
W. E. Zimmie, Inc. 810 Sharon Drive Westlake, OH 44145	Zimmite ZM-100	1
	Zimmite ZC-301	130
	Zimmite ZT-600	1
	Zimmite ZT-601	1
	Zimmite ZT-603	1

15. ADDITIVES FOR USE IN THE UNITED KINGDOM

In the early 1970s a list was published in the United Kingdom [*J. IWE, 3,* 146 (1973)] [revised in 1982 by the Department of Environment (Water Quality Division)] of additives, chemicals, and materials of construction for use in public water supplies and swimming pools.

1.   Polyacrylamides.   Maximal acrylamide monomer content of 0.05%; maximal treatment doses: 0.5 ppm; exceptionally: 1 ppm.

Manufacturer or Dealer	Product
Cyanamid of GB	Superfloc A100 (PWG)
Industrial Products Division	Superfloc A110 (PWG)
Fareham Road	Superfloc A130 (PWG)
Gosport	Superfloc A150 (PWG)
P.O. Box 7	Superfloc C100 (PWG)
Hants PO13, OAS	Superfloc C110 (PWG)
	Superfloc N100 (PWG)
	Superfloc 992
Fospur Ltd.	Decapol A11P
Alfreton Industrial Estate	Decapol A30P
Nottingham Road	Decapol A33P
Somercotes, Derbyshire	Decapol A39P
	Decapol A45P
	Decapol C300P
	Decapol C330P
	Decapol N11P
	Decapol N100P
	Decapol N50P
	Decapol N10P
	Decapol C11P
	Decapol C10P
Allied Colloids Manufacturing Ltd.	Magnafloc LT20
Low Moor	Magnafloc LT22
Bradford, Yorkshire BD12 OJZ	Magnafloc LT22S
	Magnafloc LT24
	Magnafloc LT25
	Magnafloc LT26
	Magnafloc LT28
	Magnafloc LT29
Nalfloc Ltd.	Nalfloc A373
P.O. Box 11	Nalfloc A375
Mond House	Nalfloc A378
Winnington, Northwich, Cheshire CW8 4DX	Nalfloc N8170
Hercules Powder Co. Ltd.	Hercofloc 812
1 Great Cumberland Place	Hercofloc 815
London, W1H 8AL	Hercofloc 839
Float Ore Ltd.	Flocbel F2
Apex Works	Flocbel F3
Willowbank, Uxbridge	
Chemical Dept.	Sanpoly 305 (PWG)
Mitsubishi Corporation	Sanpoly A510 (PWG)
Bow Bells House	Sanpoly A520 (PWG)
Broad Street	Sanpoly N500 (PWG)
Cheapside	
London EC4M 9BQ	
Ash Spinning Co.	Purifloc N127
Chamber Mill	Purifloc N17
Heron Street	Purifloc A22
Oldham Lancashire	Purifloc A23
	Purifloc A24

Manufacturer or Dealer	Product
Chemviron	Polymer 233
Division of Baltimore	Polymer 243
Aircoil—Chemviron S.A.	Polymer 253
Brusselesteenweg 359	
1900 Overijse	
T.R. International (Chemicals) Ltd.	TRW 90AP
Cheadle Heath	TRW 91AP
Stockport, Cheshire, SK3 ORY	TRW 93AP
	TRW 95AP
	TRW 90CP
	TRW 91CP
	TRW 80NP

2. Polyacrylic Acids. Monomer must be lower than 0.5%; maximal permitted dosage: 10 ppm.

Manufacturer or Dealer	Product
Buckman Laboratories S.A.	Bufloc 30
Wondelgemkaai 157	TAPA
B-9000 Gent, Belgium	
Oakite Ltd.	Oakite Clarifier SF3
West Carr Road Industrial Estate	
Retford, Notts DN22 7SN	

3. Products Containing Starches. Maximum permitted dosage: 3 mg/L for products of category A; 5 mg/L for products of category B.

Manufacturer or Dealer	Product	Category
W A Scholten's Chemische Fabrieken N.V.	Wisprofloc P	A
Foxhol, Postbus 1, Holland	Wisprofloc 20	B
Starch Products Ltd.	Stadex WTA	A
Stadex Works	Stadex WTB	A
Middle Green Lane	Stadex WTC	A
Langley, Slough, Bucks		
Fospur Ltd.	Fostarch 4P	A
Alfreton Industrial Estate	Fostarch 5P	A
Somercotes, Derby DE5 4LR	Fostarch 6P	A
Tunnel Avebe Starches Ltd.	Perfectamyl	A
Avebe House	Perfectamyl A5114/2	B
Otterham Quay	Perfectamyl A5114/2	B
Rainham, Gillingham, Kent		
Gamlen Chemical Co. (UK) Ltd.	Gamlose-W	B
Wallingford Road		
Uxbridge, Middlesex UB8 2TD		

Manufacturer or Dealer	Product	
Royal Scholten Honig (Trading) Ltd. Moss Lane Trading Estate Moss Lane Whitefield, Manchester M25 6FM	Flocgel E30 Stadex WS 612	B B

4.   Seaweed-Based Products (Alginates).

Manufacturer or Dealer	Product
Alginate Industries Ltd. 22 Henriette Street London WC2E 8NB	Welgum S

Index